Green Energy and Technology

Climate change, environmental impact and the limited natural resources urge scientific research and novel technical solutions. The monograph series Green Energy and Technology serves as a publishing platform for scientific and technological approaches to "green"—i.e. environmentally friendly and sustainable—technologies. While a focus lies on energy and power supply, it also covers "green" solutions in industrial engineering and engineering design. Green Energy and Technology addresses researchers, advanced students, technical consultants as well as decision makers in industries and politics. Hence, the level of presentation spans from instructional to highly technical.

Indexed in Scopus.

More information about this series at http://www.springer.com/series/8059

Durmuş Kaya · Fatma Çanka Kılıç ·
Hasan Hüseyin Öztürk

Energy Management and Energy Efficiency in Industry

Practical Examples

Springer

Durmuş Kaya
Head of the Thermal Energy Systems Division, Department of Energy Systems Engineering, Technology Faculty
Kocaeli University
Umuttepe, Kocaeli, Turkey

Fatma Çanka Kılıç
Head of the Renewable Energy Division, Department of Energy Systems Engineering, Technology Faculty
Kocaeli University
Umuttepe, Kocaeli, Turkey

Hasan Hüseyin Öztürk
Department of Agricultural Machinery and Technologies Engineering
Faculty of Agriculture
University of Cukurova
Balcalı, Adana, Turkey

ISSN 1865-3529 ISSN 1865-3537 (electronic)
Green Energy and Technology
ISBN 978-3-030-25997-6 ISBN 978-3-030-25995-2 (eBook)
https://doi.org/10.1007/978-3-030-25995-2

This Springer imprint is published by the registered company Springer Nature Switzerland AG
The registered company address is: Gewerbestrasse 11, 6330 Cham, Switzerland

We have dedicated this book to the memory of our valuable colleague, Prof. Dr. Durmuş Kaya, who always believed in our ability to be successful in the academic arena. He was a beloved father, an excellent scientist, and a wonderful friend. We will never forget you. Rest in peace.

Prof. Dr. Fatma Çanka Kılıç and Prof. Dr. Hasan Hüseyin Öztürk

Preface

Countries that are targeting sustainable development are evaluated according to their energy development levels, measured based on energy consumption per capita and energy intensity values. The high-energy consumption per capita means both the vitality of the economic activities in the country and the high level of prosperity. Most of the countries in the world, especially which are still in the fast development phase, industrialization activities, efforts to reach new technologies, increase of living standards, and fast increasing population cause more energy consumption with each passing year.

The studies to assure the sustainability in energy, reduce the external energy dependence on foreign countries, and the struggle with climate change are some of the reasons to increase the importance of efficient use of energy and energy resources all over the world. These studies have given rise to an important development of consciousness about applying environmentally friendly policies and methods in energy production, resource diversification, and consumption. To increase the energy efficiency by preventing the wastage has been a very important policy for all the countries in the world. Ensuring the energy supply security and being sustainable in energy policies are not easy topics, which need to be handled carefully. In addition, the dependency on the outside energy resources makes the situation even more difficult. For these reasons, energy efficiency has gained a special importance in today's energy world.

While energy savings is an important concept, the energy efficiency, which is our main topic, is different and wider. Energy efficiency is the use of energy with a high efficiency and saving the energy without sacrificing the quality of our lives, needs, and productions. Today, it is accepted all over the world that energy saving is the result of the using energy in an efficient manner, which can also be accepted as the fastest, cheapest, and cleanest energy source. As for the developing countries, the energy that can be saved by increasing the energy efficiency is a domestic and clean energy source, which is cheaper than the others and needs to be applied in first row of the energy studies. The increase in energy demands and costs has made energy-saving compulsory.

It is obvious that the efficient use of energy is so important not only for all the countries in the world, but also especially for the countries, which have to import a significant part of their energy needs and have the power plants that produce electricity by using fossil fuels.

Energy efficiency is an answer to some vital problems such as gradually diminishing fossil fuels, insecurity against nuclear power installations, meeting energy demands insufficiently by the alternative energy sources, increasing pollution of the environment, and climate changes. In addition, energy efficiency is the most important component of sustainable development and competitiveness. In developed countries, energy efficiency means the least costly energy, resulting in efficient use.

This book, prepared by considering various sources, aims to support and guide everyone who is interested in energy management and energy efficiency.

Kocaeli, Turkey Prof. Dr. Durmuş Kaya
Kocaeli, Turkey Prof. Dr. Fatma Çanka Kılıç
Adana, Turkey Prof. Dr. Hasan Hüseyin Öztürk

Acknowledgements

We would like to thank everyone who helped us with our scientific book beginning with all family members, advisers, teachers, co-workers, assistants, students, inspirations, editors, and people who worked on the book production,

We are eternally grateful to you all.

Kindest Regards,
Prof. Dr. Durmuş Kaya
Prof. Dr. Fatma Çanka Kılıç
Prof. Dr. Hasan Hüseyin Öztürk

Contents

List of Figures

List of Tables

Chapter 1
Energy

1.1 Energy

Energy is one of the basic sizes of the physical science. Energy is a value that is not directly measured. To change the state of a physical system, it can be determined by *the calculations* that are required to be performed by *the work* or *the energy type.* In more technical terms, energy is the name given to the ability of materials to do work. Albert Einstein (1879–1955) defined the energy as proportional to mass with the following famous equation:

$$E = m \times c^2 \tag{1.1}$$

In the equation;

$E = $ energy ($kg \times m^2/s^2 = N \times m = J$),
$m = $ mass (kg), and
$c = $ the speed of light (it is approximately 3×10^8 m/s or 299,792,458 m/s).

Energy is a state function that determines how much work a physical system can do or how much heat it can transfer. The unit of energy in the International System of Units (SI) system is the Joule (J). This unit was named after James Prescott Joule, a British scientist who lived between 1818 and 1889 and it is a derived unit. Transformations to other energy units and Joule units are given in Table 1.1 [1–3].

D. Kaya et al., *Energy Management and Energy Efficiency in Industry*,
Green Energy and Technology,
https://doi.org/10.1007/978-3-030-25995-2_1

Table 1.1 Energy units and Joule unit transformations

Energy unit	Definition	Equivalent
kilowatt hour (kWh)	It is the amount of energy consumed in 1 h	1 kWh = 3.6 × 10^6 J
British Thermal Unit (BTU)	It is the amount of energy required to heat a pound (~453.6 g) of water by one-degree Fahrenheit at one atm.	1 BTU = 252 cal 1 BTU = 1055.07 J 1 BTU = 0.000293 kWh
Therm	It is a unit used in the gas production industry	1 Therm = 10 × 10^4 BTU 1 Therm = 1.055 × 10^8 J = 105.51 MJ
Tons of oil equivalent petroleum (TOE)	It is a unit used in petroleum industry	1 TOE = 4.5 × 10^{10} J
Calorie (cal)	It is the amount of energy required to increase the temperature of 1 g of water by 1 °C at a pressure of one atmosphere	1 cal = 4.184 J
Barrel (Barrel)	It is a unit used in petroleum industry	1 TOE = 7.5 barrel 1 barrel = 6 × 10^9 J

1.2 Types of Energy

1.2.1 Mechanical Energy

Mechanical energy is motion energy that can do useful work. This energy is associated with the motion and position of an object. When motion energy (kinetic energy) does a work, it turns into mechanical energy. For example, cutting the cable with pliers, tightening a screw with a screwdriver, water that strikes turbine in electric power plants, all these motions turn into the mechanical energy. With the obtained mechanical energy, any work can be done or electricity can be produced. Mechanical energy is the sum of kinetic and potential energy.

1.2.1.1 Potential Energy

Potential energy is defined as the energy that a system or a body possesses because of its elevation in a gravitational area. This energy exists in an object that is connected to other objects due to the its position. In addition, it is called as stored energy because of their physical condition. Potential energy can also be defined as energy stored in systems that can do work at any time.

One of the common potential energy types is the gravitational potential energy of an object that depends on its mass and its distance from the center of mass of another object. The gravitational potential energy is also defined as the energy that the object has due to its position.

$$E_p = m \times g(h_2 - h_1) \quad h_2 > h_1 \tag{1.2}$$

$$E_p = m \times g \times h$$

In the equation;

- E_p = potential energy of the object (kg m^2/s^2),
- m = mass of the object (kg),
- g = the acceleration due to the gravity (m/s^2),
- h = the altitude of the object (m).

Another type of potential energy is elastic potential energy, which flexible objects contain when they are in a stretched or tightened position. In the case of compression of a spring as x distance, the energy stored in the spring is determined as follows:

$$W = \frac{1}{2}k \times x^2 \tag{1.3}$$

In the equation; k = spring constant, x = displacement.

1.2.1.2 Kinetic Energy

Kinetic energy is the energy caused by the motion of the object. That is to say, kinetic energy is the energy form, which that moving objects, possess. Energy is defined as the ability to do work. The kinetic energy is obtained as follows by taking into account the work equation:

If the value of force,

$F = m \times a$ is substituted for the force in the relation of work,
$W = F \times s$, which specifies the quantity of work, the work relation is obtained as,
$W = m \times a \times s$. If the acceleration relation is replaced by,
$a = (v_2 - v_1)/t$ and the distance at a constant acceleration,
$s = (v_1 + v_2) \times t/2$, then the kinetic energy of an object which its mass is m, and its velocity is v, is determined as follows:

$$E_k = \frac{1}{2}m \times v^2 \tag{1.4}$$

If an object is both rotating and moving (if it is moving linearly), the kinetic energy of this object is determined by the following equation:

$$E_k = \frac{1}{2}m \times v^2 + \frac{1}{2}I \times \omega^2 \tag{1.5}$$

In the equation;

I = Moment of inertia (inaction) of the object and ω = Angular velocity.

1.2.2 Heat Energy

Heat is the average kinetic energy of the sum of the atoms or molecules in the substance. Heat is a type of energy. Heating is the process of increasing collisions between the particles in the matter. According to the first law of thermodynamics, heat is a form of energy, which can be derived from either the mechanical work or transformed into the mechanical work. Heat energy is released by the combusting of fuels such as coal, oil, lignite, natural gas.

1.2.3 Chemical Energy

Chemical energy is a type of energy that is formed due to the chemical bonds between the atoms, which is also stored by the chemical bonds. Energy, which is resulting from the chemical reaction, is called chemical energy. Batteries and accumulators are the devices that convert chemical energy into electricity. Chemical energy can be converted into mechanical, heat, and light energy.

1.3 Energy Resources

The energy that is defined in the form of the ability of a material or a system of materials to do work, which is produced from a variety of sources. Energy sources are the sources that enable energy to be produced by any means. Depending on their formations, the energy resources in the world can be divided into two groups as fossil and renewable resources (Fig. 1.1).

1.3.1 Fossil Energy Resources

Fossil energy sources are the energy resources that are consumed as they come out of the source, such as hydrocarbon-bearing coal, oil, and natural gas. These sources occur when dead living organisms dissolve in an oxygen-free environment for millions of years. Fossil energy sources are widely used in today's industrial world. Detailed information on fossil fuels is given in Chap. 8, Fuels and Combustion.

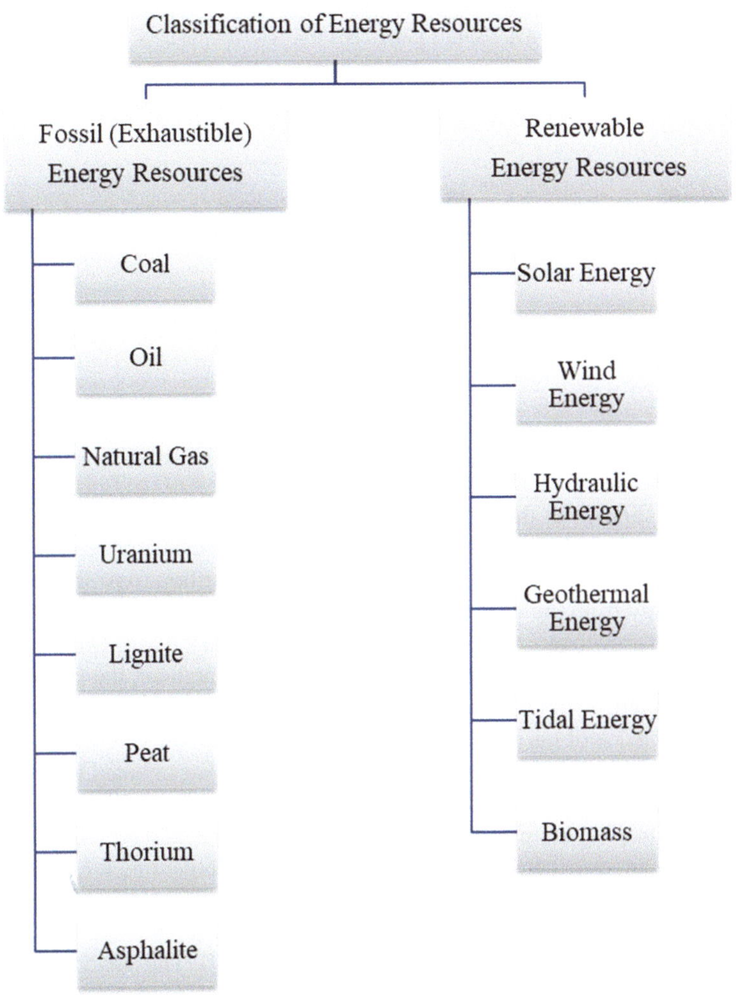

Fig. 1.1 Classification of energy resources

1.3.1.1 Status of Fossil Energy Resources

The studies that have been realized by the International Energy Agency (IEA) have showed that the global energy claim will increase approximately 60% by the year 2030 with an annual increase of 1.7% and fossil fuel reserves can respond to the demand during this period. In the future, fossil resources will continue to be important in the world's energy future, as they are today. In this period, oil will be the most consumed energy source. According to forecasts, production will increase by 60% globally and in 2030, the production will reach 120 million barrels per day [4].

Table 1.2 The World fossil fuel reserves

Region	Oil (Billion tons)	Natural Gas (trillion m^3)	Coal (Billion tons)	
			Hard coal	Lignite
North America	8.3	7.6	120.2	137.6
Central and South America	13.7	7.2	7.8	14
Europe	2.6	4.9	47.5	77.9
Former USSR countries	9.1	56.1	97.1	132.6
The Middle East	93.3	56.9	1.7	
Asia and Oceania	5.9	12.3	189.3	103.1
Africa	10	11.2	55.2	0.2
Total world	142.9	155.1	519.1	465.4

It is expected that the biggest demand increase in fossil resources will be natural gas usage. It is calculated that the production amount of 2080 MTOE in 2000 will be doubled in 2030 with the aim of meeting the high demand of Western Europe, especially. The world has entered a new period of dependence on natural gas. On the other hand, with an estimated global reserve of 1 trillion tons, the amount of produced energy from the coal is expected to reach 3600 MTOE by 2030, with an average annual growth rate of 1.4%. It is also estimated that the world primary energy supply will reach 16.5 billion TOE by 2030, and in this amount, the shares will be 35% of petroleum, 25% of natural gas, 21.8% of coal, 11.3% of renewable energy resources, and 2.2% of nuclear resources.

When the regional distribution of fossil fuel reserves is examined (Table 1.2), it has been seen that Middle Eastern, Central Asian countries have the largest reserves in natural gas, and the Middle East has the biggest petroleum reserves. Coal and hard coal resources are distributed more evenly in the world. Regarding the interregional oil and natural gas trade, it is seen that oil is exported from Russia, Africa, and the Middle East, where oil is still produced, to Europe, North America, India, China, and Japan, which are the main consumption regions [5–7]. The same trend is valid for natural gas trade. It is estimated that the oil trade will be doubled, and the gas trade will be tripled between these regions by the year 2030.

The world fossil fuel reserves and availability of the world fossil fuel reserves can be seen in Table 1.2 and Table 1.3, respectively.

1.3.1.2 Environmental Impacts of Fossil Energy Resources

The natural environment is affected by all human activities. One of the most effective ones of these activities is the energy field. The use of fossil fuels, which is the main cause of energy-environmental problems, continues to grow from the beginning of the industrial revolution to today and reaches extreme dimensions, continuing at the expense of its extinction. The environmental impacts that are caused

Table 1.3 Availability of the World fossil fuel reserves

Region	Oil (year)	Natural Gas (year)	Coal (year)
North America	12	10	231
Central and South America	41	52	269
Europe and Eurasia	22	60	241
The Middle East	81	100	399
Asia and Oceania	14	41	92
Africa	32	88	270
Total world	41	65	155

by energy productions can be listed as acid pollutants, global warming, human health and safety problems, particles, heavy metals, hazard-disaster probability, waste problem, environmental deterioration, noise, light pollution, radiation pollution, and land requirements.

Greenhouse Effect and Global Warming: The intensive use of fossil fuels has caused the increase of greenhouse gases into the atmosphere, along with the increase of carbon dioxide (CO_2), and thus has caused the warming of our planet, all is defined as global warming due to the greenhouse effect. Among the gases that caused the greenhouse formation, we can count CO_2, methane, CO, hydrocarbons, and chlorofluorocarbons (CFCs). According to the Kyoto Protocol, six main gases were identified as causing of greenhouse effect, which can be seen in Table 1.4.

Greenhouse gases generally let the rays, which are in the form of short wavelength coming from the sun pass through the earth. However, some of the reflected infrared radiation from the earth is absorbed, some of them are diffused back by greenhouse gases molecules. This is called greenhouse effect (Fig. 1.2). The contribution of greenhouse gases to greenhouse effect is different. If the carbon dioxide effect is taken as a unit, the relative effect of each element in the atmosphere over a period of 100 years is calculated as Global Warming Potential (GWP). The GWP values of greenhouse gases and their atmospheric durations are given in Table 1.4.

Although the greenhouse effect is an extremely important mechanism for balancing the world temperature, the proportion of these gases has increased due

Table 1.4 Properties of greenhouse gases

Green houses	Global warming potential (GWP)	Atmospheric duration (Year)
Carbon dioxide	1	5–200
Methane	21	12
Dinitrogen Monoxide	310	114
Perfluorocarbons	140–12,000	2–>50,000
Hydrofluorocarbons	140–12,000	2–>50,000
Sulfur Hexafluoride	23,900	3200

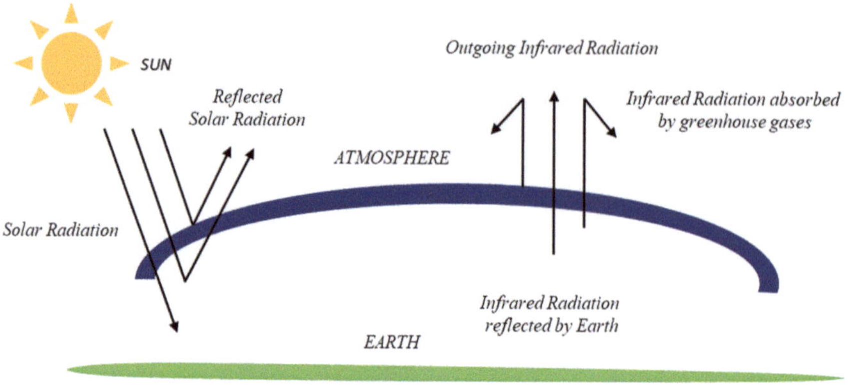

Fig. 1.2 Greenhouse effect

to industrialization and excessive use of fossil fuels and because of misapplications that made in recent years. Thus, the greenhouse effect is negatively associated with climate change. If there was no greenhouse effect in the atmosphere, the world average temperature would be 255 K or −18 °C, perhaps life on earth would not be possible. Due to this mechanism in the atmosphere, some of the long wave reflection is captured and the world average temperature is +15 °C.

The changes in the economic and social lifestyles of human beings caused the increase of various greenhouse gases in the atmosphere. In general, it is estimated that 50% of CO_2 formation, which caused the greenhouse effect is due to the activities of human beings. The contribution of various human activities to global warming consists of 49% of energy use, 24% of industrial applications, 14% of deforestation, and 13% of agricultural production operations. Agricultural operations account for about 20% of the world is growing greenhouse gases. In agricultural production processes applications, like energy consumption, animal breeding, fertilization, drug use, etc., cause especially the increase of CO_2, CH_4, and N_2O gases that are the reasons for the increase of greenhouse effect.

Due to the increasing CO_2 concentration in the atmosphere over the last century, it is determined that there has been an increase in the average temperature of our world as a result of the greenhouse effect. This increase will create the warming effect in the lower parts of the earth and the cooling effect in the upper parts of the air globe, altogether will affect the high-pressure systems, related to these changes, extreme climate conditions are expected to be seen. As a result of the effects of global warming and climate change; the ecological systems, agricultural and socioeconomic sectors will be affected directly or indirectly, and all of these are given in Table 1.5.

Energy resources and activities related to the production and use of energy are influential in the formation of CO_2 and other potential greenhouse gases. By making some changes in the production and use of the energy, it is possible to reduce the amount of these gases, which cause to occur the greenhouse effect. Administrative and individual measures that need to be taken against the global warming are given in Table 1.6 [8].

Table 1.5 The effects of global climate change

General impacts	Agricultural impacts
• Shift of climate zones, • The rise of sea level, • More frequent occurrence of floods and overflows and strengthening of their effects, • Drought, • Erosion, • Desertification, • Epidemics, • Food shortage and hunger, • Degradation of natural balance, • The deterioration of human health.	• Drought, • Lack of water, • Soil erosion, • Decrease in agricultural lands, • Increase of swamp areas, • Migration of agricultural population, • Decrease of soil quality, • Agricultural processing difficulties, • Increase in diseases and difficulty of their control, • Increase in the use of water and medicine, • Decrease in product yields and quality.

1.3.2 Renewable Energy Resources

Renewable energy sources (Table 1.7) are natural energy sources such as solar, hydraulic, wind, geothermal, biomass, and marine energies that are less harmful to humans and the environment than fossil energy sources. Renewable energy is the energy obtained from natural resources, continuously or repeatedly. Renewable energy is also defined as sustainable energy that obtained from natural sources. In contrast to fossil energy sources, these sources are inexhaustible over the time and they are an alternative to non-renewable energy sources such as coal, gasoline, and natural gas.

Due to the negative impacts of fossil fuels on the environment and its depletion, the concept of sustainable energy has come to the agenda, which is related to the energy productions with high efficiency. Clean energy production technologies to evaluate fossil fuels environmentally friendly methods, replacing these resources with the renewable energies, using the resulting energy as an input in another cycle can be seen as solutions in today's energy world.

The advantages and disadvantages of renewable energy sources are given in Table 1.8. The major renewable energy types in the world are classified as wind, hydraulic, solar, hydrogen, biomass, geothermal, and wave. The use of renewable energy in the world electricity generation has an important position in terms of not harming nature and producing inexhaustible energy by its own natural methods [9, 10].

Today, the need for the development of new technologies for the use of renewable energy sources is increasing.

Table 1.6 The precautions against global warming

Precautions that the state can take	Precautions that local governments can take
• Determining the real values of greenhouse gas values, • To make more utilization of hydraulic energy, • Promoting the use of renewable energy sources, • Planting vacant lands and reforesting as much as possible, • Controlling forest fires, • Developing good combustion methods and using high-quality fuel in thermal power plants, • Controlling heating fuels, • Raising public awareness, • Making a realistic agricultural policy, • Adopting recycling as a lifestyle for consumers, • Making tourism planning.	• Organizing educational programs on climate change in schools, • Applying energy and water savings precautions, • Finding new water resources and renewable energy sources, • Increasing measures and controls to reduce greenhouse gases, • Considering the effects of climate change on infrastructure and settlement planning, • To attract immigrants from big cities to small cities, to encourage remigration, • To lead the people to the way of sustainable living, • Collecting waste by degrading, • To apply internal and external lighting correctly.
Precautions that individuals can take	
• Applying correct lighting, • Using low energy bulbs, • Using ventilator instead of air conditioner, • Insulating houses against heat losses, • Changing air-conditioning filters when necessary, • Buying low energy consumption appliances, • Avoiding splurge in water use, • Take precautions against water spills, • To reduce the hobby gardens, • In homes and businesses, when not in use, unplug electronic devices such as TV, radio, computer, • Planting trees as much as possible, • Walking short distances, • Traveling by public transport, • Reducing consumption, • Using tools at low speed, • Reducing the amount of laundry water consumption, • Repair leaking installations.	• Use lime and bacteria-free filters for water, • Do not wash and irrigate with hose, • Preferring unleaded gasoline vehicles, • Do not exceed the carrying capacity of the vehicle, • Switching off the vehicle during long pauses, • Using long-term products, • Avoid as much as possible from purchasing products that are sold in non-recyclable packages, • Avoiding plastic packaging, especially PVC, • Use glass products such as bottles, • Reuse products such as plastic bags and food containers, • Do not use plastic bags in shopping, • Throwing away garbage by degrading (separating), • Do not leave computers on standby mode.

Table 1.7 Evaluation of renewable energy resources

Evaluation criteria	Solar Heat	PV	Hydraulic	Wind	Tide
Investment cost	Much	Much	Very much	Medium	Very much
Operating cost	Medium	Medium	Negligible	Low	Negligible
Efficiency	15%	5–10%	80%	42%	25%
Renewability	Yes	Yes	Yes	Variable	Yes
Storage	Not Necessary	It is not clear	–	Important	It is not clear
Pollution	No	Waste heat	No	Visible	No
Unit Cost (*As an example for a country)	25 cent/kWh	16 cent/kWh	4 cent/kWh	4.5 cent/kWh	Unknown
Environmental Effects	Medium	Much	Very much	Little	–
Small scale	No	Rough	Available	Available	No
Large scale	Very expensive	Very expensive	Good	Very likely	In separate regions
Unit capacity	1000 MW	–	2000–6000 MW	Very variable	250 MW

Table 1.8 Advantages and disadvantages of renewable energy sources

Source of energy	Advantages	Disadvantages
Hydraulic energy	It does not create environmental pollution, Step-in very fast in need of excessive energy, Can be quickly deactivated in emergency situations, Natural source, not dependent on outsources, The investment that is to be made can be used not only for energy production, but also for irrigation and flood control.	Investment costs are high, Total construction time is long, It depends on the amount of rainfall, Dams change the ecology of their surrounding areas, Settlement zones and ancient sites are likely to be inundated.
Geothermal energy	Environmentally friendly, Fossil energies are not needed for water heating and evaporation during the electricity generation, It is a natural resource and not dependent on the foreign resources, Efficiency is very high, Cost is low because it can be obtained directly.	Due to some harmful chemical substances in their constitutions, re-injection is necessary, It is not possible for the consumed part to be reproduced in the same proportion and in a short time, Fluid coming out from geothermal sources, usually contains corrosive and polluting minerals, Cost of preliminary research and preparation is high, Sufficient free space is required for drilling and machinery, Energy transmission may not be efficient, The system should be close to the settlement areas.
Wind energy	Stable, reliable, and continuous resource, It is not dependent on foreign resources, The unit cost of the produced energy with the developing technology is decreasing, It is a renewable energy source that creates less pollution and little harm to the environment, Wind energy can be obtained on 95% of the earth. Other activities can also be continued in these areas at the same time.	Large areas are required for turbines, Turbines are visually and aesthetically unfavorable, They need to be established away from the settlement centers and sensitive rural living areas, Desired efficiency cannot be obtained in urban centers and valleys, Safety problems may occur related to overturning and disintegration of turbines.

(continued)

Table 1.8 (continued)

Source of energy	Advantages	Disadvantages
Solar energy	It is a renewable energy source, Natural materials are used, It is economic, It is not dependent on foreign resources, Environmentally clean energy source.	Efficiency is low, Seasonal and daily discontinuity may occur, The initial cost is high, Cost is high for consumers, Storage may be required, Capacity drops in the areas that in the shade.
Biomass energy	Production/conversion technologies are well known, Low light intensity is enough, Storable, Requires a temperature range between the degrees of 5–35 °C for the process, It has socioeconomic importance, There are less harmful effects on the environment.	Cycle efficiency is low, Creates competition for agricultural areas, The water content is excessive, Only applicable for large settlement areas.
Tidal energy	Energy source is infinite and plentiful, Reduce dependence on fossil fuels, It does not pollute the environment, It creates job opportunities, Provides electricity to non-network areas, Allows the use of technology in other work to be done in the marine environment, Salty water is converted into fresh water and pumped to the required areas, The natural sources of the bottom of the sea are brought to the surface, Provides protection of coasts.	The systems should be close to the settlement areas, Systems can be damaged through bad weather conditions, As being a new technology, it is difficult to find sufficient technical staff.

References

1. https://www.birimler.info/birimler.php
2. Republic of Turkey Ministry of National Education, Industrial Automation Technologies Force, Weight and Pressure Measurement, 523EO0362 Ankara/Turkey, 2012 (In Turkish: T.C. Millî Eğitim Bakanlığı, Endüstriyel Otomasyon Teknolojileri Kuvvet, Ağırlık ve Basınç Ölçümü, 523EO0362 Ankara/Türkiye, 2012)
3. McGraw-Hill Dictionary of Scientific & Technical Terms, 6E, Copyright © 2003 by the McGraw-Hill Companies, Inc. All rights reserved
4. IEA (2020) Global Energy Review 2020, IEA, Paris https://www.iea.org/reports/global-energy-review-2020

5. Energy Situation in Turkey and the World, GAZBİR, (In Turkish: Dünyada ve Türkiye'de Enerji Durumu, GAZBİR).https://www.gazbir.org.tr/uploads/page/Dunya-ve-Turkiye-Enerji-Gorunumu.pdf Accessed on 20th December 2019
6. https://enerji.gov.tr/enerji-en
7. Republic of Turkey Ministry of Energy and Natural Resources, Directorate General of Energy Affairs, Primary and Final Energy Intensity, (2000–2017) (In Turkish: T.C. Enerji ve Tabii Kaynaklar Bakanlığı, Enerji İşleri Genel Müdürlüğü, Birincil ve Nihai Enerji Yoğunluğu, (2000-2017))
8. Kaya D, Ozturk HH (2012) Air Quality Management, Umuttepe Publications, 2012. (In Turkish: Kaya D, Ozturk HH (2012) Hava Kalitesi Yönetimi, Umuttepe Yayınları)
9. Çengel YA, Boles MA (1998) Thermodynamics an engineering approach, 3rd ed. McGraw-Hill
10. Karakoç TH, Karakoç N, Erbay B, Aras H (2012) Energy analysis, Republic of Turkey Anadolu University, Publication No: 2486, Distance Education Faculty, Publication No: 1457 (In Turkish: Karakoç TH, Karakoç N, Erbay B, Aras H (2012) Enerji Analizi TC Anadolu Üniversitesi, Yayın No: 2486, Açık Öğretim Fakültesi, Yayın No: 1457)

Chapter 2
Energy Management

Today, the efficient use of energy has gained importance in terms of controlling energy prices, reducing external dependence, ensuring energy sustainability, and protecting environment and natural resources. Moreover, the importance of the topic is increasing day by day in parallel with the industrialization activities and the rise of the energy demand depends on the population increase in the world. Energy management is a type of work that consists of education, inspection, measurement, monitoring, planning, and implementation in order to ensure efficient use of energy resources and energy. Some sub-goals of energy management programs can be listed as

- Reducing energy consumption by using energy efficiently,
- Provide good communication among energy issues,
- Develop and implement effective monitoring, reporting, and management strategies for the methods of energy use,
- Investigate new and better ways to increase recycling from energy investments with the R&D studies,
- Ensuring all users are interested in the energy management programs and making them be part of it,
- Reducing restrictive effects or interruptions in energy supply.

The following activities are carried out within the energy management [1]:

- Establishing an energy policy that defines targets and priorities in energy management; defining the place, duty, authority, and responsibilities of the energy management unit and the energy manager in the hierarchical structure. All content should be written and published in the form of rules and regulations, all employees and persons involved in energy management activities should be informed about it,
- Determination of measures and methods for the improvement of consumption habits, prevention of unnecessary and unconscious use, introduction of training and training programs to increase the level of the knowledge of all employees,

D. Kaya et al., *Energy Management and Energy Efficiency in Industry*,
Green Energy and Technology,
https://doi.org/10.1007/978-3-030-25995-2_2

- Identification and implementation of the modifications that may be applied to energy-consuming systems, processes, or equipment,
- Preparation of the energy audits, preparation and implementation of the projects,
- Monitoring the efficiency of energy-consuming equipment, timely maintenance, and calibration,
- Preparing plans for energy needs and productivity-enhancing applications, budget needs, benefits, and cost analysis to be presented to management,
- Monitoring, evaluation, and preparation of periodic reports of energy consumption and costs,
- Supply and make the installation of metering equipment and measuring devices that are needed to monitor energy consumption, and make their timely calibration,
- The monitoring of specific energy consumption, the relationship between product or service output and energy consumption, energy costs, the energy intensity, and the preparation of remedial proposals for them,
- Changing the energy composition and exploring possibilities for alternative fuel use, to reduce harmful emissions to the environment and to prepare and implement measures for not exceeding the limit emission values,
- Make preparation of alternative plans to be applied in case of energy supply interruption to reduce the use of oil and natural gas,
- Presenting annual information on energy usage and energy management work to the management,
- Determining the amount of carbon dioxide emissions per total and unit product or benefit and the amount of emissions that can be reduced by applying energy efficiency measures.

In order to meet the requirements of the production in the most appropriate way, the first rule in energy management is to keep the energy producing systems in good state during the operations. In the studies carried out without a specific program, up to 10% of energy savings can be achieved by taking simple operational measures in some industrial establishments. In addition, with the implementation of comprehensive energy management programs, energy-saving efforts can be sustained and the savings rate can reach 25%.

2.1 Energy Manager

The energy manager is the person who is responsible for managing the energy management activities in industrial enterprises and buildings on behalf of management and has the energy manager certificate [2]. It is targeted that the persons who will receive the energy manager certificate have the following competencies:

- Having knowledge about primary energy sources, secondary energy types, and supply-demand developments in the world and in their own country,
- To be able to distinguish between energy saving and energy efficiency,
- To know what energy-saving potential is and how it can be predicted,

- To know the concepts of energy density and specific energy consumption, calculation methods and trends in industrial sectors and industrial enterprises countrywide,
- To know how the activities related to energy management will be carried out and how to be reported,
- To know the technical specifications, operation, and maintenance procedures of energy-consuming equipment and systems, knowing how these energy losses and inefficiencies can be formed, how to prevent them, how to measure them, and how to interpret the measurements,
- To know the losses, measurement methods, and insulation measures in the systems where heat is produced, stored/transported,
- To know the energy savings opportunities with simple measures,
- To recognize efficient production processes and efficient products using energy in the market with their technical and economical features,
- To be able to prepare pre-feasibilities that can save energy or increase energy efficiency which require significant expenditure for the precautions,
- Having knowledge about the methods of survey and project preparation.

2.2 Energy Management Policy

Energy management policies and strategies will guide the company and ensure the continuity increase in business productivity [3]. The energy management policy consists of two parts:

(1) The policy statement, which explains the company's comprehensive energy management goals,
(2) A strategy that organizes how the company will fulfill its energy management program.

Monitoring an energy management policy provides the following main benefits:

- It enables the company to adopt its goals in one single thought,
- It guides for a systematic energy management program,
- It demonstrates the company's commitment to energy management,
- It acts as a catalyst for changes in the behavior of people,
- It guarantees to allocate sufficient resources to energy management,
- It forms the concept of energy efficiency that continues in the company's structure.

2.2.1 Goals of the Energy Management Policy

Each operation has its own energy management policy. This policy explains the purpose and objectives of the energy management in the business. This should be open and able to be seen by everyone in the annual reports and other company's literature.

Without going into too much detail, this policy should include certain measurable objectives. An ambiguous and generalized goal is not personal and inspired by a small active action. The energy management policy should include

- Disclosure of liability,
- Comprehensive goals (What is desired to be achieved?),
- Basic logic (Why is energy managed?),
- Organization (How to organize responsibilities and reports?),
- Milestones (Where are the primary objectives aimed at?),
- Environment (What are the environmental impacts related to energy?).

The responsibility in energy management policy can be explained as *"as a part of our company's environmental strategy, we commit our commitment to energy management and our responsibility to apply in our building, in our factories and in our equipment by considering the cost efficiency"*.

The main measures taken into account in policies to control energy consumption may be

- Avoiding unnecessary expenses,
- Improving financial efficiency, productivity, and working conditions,
- Protecting the environment,
- Extending the period of the use of fossil fuels.

The long-term goals of energy policies may be

- To buy fuel at the most economical prices,
- Practically using fuels in the most efficient way,
- Reducing the amount of pollution, especially reducing CO_2 emissions caused by energy consumptions,
- Reducing dependence on fossil fuels as much as possible and using renewable energy sources and their by-products instead.

The immediate targets of energy policies are

- To control energy consumption by monitoring, developing what have been purchased, operating, and educational practices,
- To create the energy manager report for the top management,
- To invest in energy saving,
- To establish a reporting system that will provide timely and useful information to decision-makers and other personnel.

2.2.2 Characteristics of Energy Management Policy

An energy management policy is a proof of the commitment to energy efficiency. Targets taken into account in energy management policies should be authentic, measurable, achievable, realistic, and timely. In order for the energy management system to mature, the policy that is to be followed must be revised periodically, including changes in business objectives and business plans. The energy management policy, as an integrated strategy with the production program in the enterprise cannot contain information for everyone because it can contain commercial information. Factors, which are considered in energy management policies, are given in Table 2.1.

In the energy management policy, inputs are provided from different personnel at various levels. In order to be effective in politics, everyone who is affected must contribute and feel involved. The politics must be compared with policies, which are created for other jobs. For other objectives, adaptation should be aimed at strategies, policies, and jobs.

Some companies prefer to incorporate environmental policies into their energy policies. In an integrated policy, energy, and environmental factors must be taken into account. The management of the business officially does adoption of the approved policy. The top management's support will ensure that the relationship between the energy management staff and other staff is well organized. The process steps to improve energy management policy are

- **Developing an Energy Forecast:** An energy forecast is developed by taking into account current structure of the operation,
- **Determining Strategic Elements:** The impacts on energy and more important factors than energy predictions are determined,

Table 2.1 Factors, considered in energy management policies

Feature	Process
Strategy	• Short-term strategic articles and how to be addressed, • Links with the production plan, • The sustainability of the availability of energy resources in the future.
Application	• Budgetary sources and objectives, • Investment criteria and life cycle costs, • Budget resources.
Management	• Energy managers and functions, • Energy committee.
Structure	• Energy management integration with other management processes, • Division of responsibilities, • Responsibility, • Reporting and communication.
Resources	• Personnel needs and responsibilities, • External consultants and energy suppliers.
Reporting	• Within the unit study groups, • Business, detailed, board reports, external reports.
Information	• Monitoring energy usage, • To separate energy cost centers, • Review, • Interaction with the reporting system.

- **Setting Detailed Goals:** These objectives define strategic items that the company can take,
- **Outline of the Process:** Support the policy and develop the objectives,
- **Thinking about How to Integrate Actions:** The characteristics of current culture and organization structure are taken into consideration.

2.3 Energy Management Program

The energy management policy depends on the implementation of the energy management program. The energy management program will continue throughout the targeted policies, depending on the company goals and plans. The following factors are taken into consideration when creating energy management programs:

(1) **Implementation**: The energy manager will develop an annual business plan for energy management. Working in partnership with the financial manager, the energy manager will create the annual budget for the program and review it. They will classify and organize investment criteria and financial targets according to the conditions.

(2) **Responsibilities**: End users are responsible for controlling energy consumption. The closest budget owner to the point of use is responsible for energy expenditures. The energy manager is responsible for coordinating energy management activities and presenting them to the energy management committee. The energy management committee is responsible for formulating and implementing the energy policy for each report on energy use and it is accountable to the management board. This committee meets four times a year and is managed by the general manager.

(3) **Structure**: Energy manager should form energy management program with company management structure in the same line. Unit managers must work with the energy manager while setting targets for energy use. This program should include training in energy management, energy efficiency, and motivation for energy to be conducted by staff energy management. The energy manager will keep an eye on every update or new capital development work.

(4) **Reporting**: Energy manager must track monthly energy management operations related to consumption and provide reports to the energy committee four times a year. This committee will regularly submit annual reports to the governing board.

(5) **Communication**: The official communications shall be provided through the energy manager among the end users, costumers, co-managers, senior management, and communication on energy issues related to the energy management committee.

(6) **Action Plan**: Annual study cost schedule will be prepared. The designated personnel will make the detailed schedule together with the specific events.

(7) **Resources**: The number of energy management personnel for the next year will be a full day/person for business. The annual budget will be reduced by considering a minimum amount of 5% of each unit's energy expenditure.
(8) **Control**: Energy management policy, program, and activities will be checked periodically. The energy management committee prepares the annual account of the activities and the relevant parts are submitted to the board of directors with the opinions of the general manager, budget owners, and end users.

2.4 Energy Management Method

There are some important steps in energy management method data collection, evaluation and planning, implementation, reporting and evaluation, and ensuring continuity. In addition, several basic features are effective during all these stages. These features are listed below:

(1) **Function of the Top Management**: The top management should continuously monitor the energy consumption target values and determine short and long-term planning. Establishment of networking and monitoring of the creation of savings activities and the evaluation of the results are the tasks to be undertaken by the top management.
(2) **Realistic Targets**: Targets should be realistic, achievable, and measurable. The energy units of the targets will be more effective and measurable than the currency in which kWh is accepted as the energy unit.
(3) **Designated Responsibilities**: Top management is responsible for the organization of the energy management and the sharing of responsibilities. Energy management is a major loss reduction program, not a small additional business or ordinary practice. Although top managers watch these types of operations clockwise, this process can be carried out by taking the persons designated by the authorization transfer into the organization. However, the persons to whom the authority is transferred must be defined directly.
(4) **Coverage of All Employees**: Each employee is assigned to this project should be included in the studies. The energy manager must produce some tools to collect employee opinions and suggestions. All employees should be aware of the energy manager's energy savings plans and role. Here, the following tools can be used to increase participation:

- Lead with examples,
- The framework of the approach taken is drawn carefully,
- Team leaders are selected,
- The suggestion system is made attractive,
- The competition environment is created,
- Short-term and long-term goals are identified,
- Training activities are carried out,
- Incentives may be applied to move applications to their homes,

- Energy use is questioned,
- Priorities are determined.

(5) **Data Collection, Evaluation, and Planning**: Data collection is the first step in creating goals and references for any project to be initiated. For this, it is necessary to create a database with healthy measurements and documents. In the database for energy management, the following values should be combined:

- Distribution of energy consumption in process or operation,
- The monetary value of the consumed energy in the company,
- Energy cost and production cost,
- Monitoring energy consumption and production costs and accessing the information,
- The size of the company, the types of products, the number of types of energy-consuming equipment,
- Equipment for monitoring energy consumption,
- Comparisons of energy savings and current corporate earnings.

2.5 Energy Management System

The energy management system is a disciplined work structured and organized in line with the more efficient use of energy, without sacrificing product quality, safety, and all environmental conditions and without reducing production [4, 5]. There are four main objectives for a successful energy management system:

(1) Increasing the productivity of the producer,
(2) Reducing consumer use,
(3) Keeping high power consumption points under constant control,
(4) Using energy in the most economical way.

An appropriate and well-designed management approach is required to establish an energy management system. The following factors guide a well-organized energy management system:

- Top management's contribution and support,
- Ensuring the participation of all employees and training,
- Energy accounting, monitoring, and goal setting,
- Conducting energy-saving surveys and feasibility study,
- Assessment of measurement results,
- Monitoring and goal setting,
- Improvement of reports,
- Implementation of engineering and design changes,
- Supply of necessary equipment.

2.5.1 ISO 50001 Energy Management System

ISO 50001 is the new international energy management system standard. It is the first global energy management standard published after regional and national standards such as EN16001. Energy management systems in industrial enterprises, organized industrial zones, and buildings that are obliged to appoint an energy manager or set up an energy management unit shall be established in accordance with ISO 50001 Energy Management System-Operator's Guide and Terms Standard.

The ISO 50001 standard promotes good overall energy management practices and strengthens good energy management behavior for institutions while setting up a framework for industrial facilities, commercial, institutional, and public facilities. This standard provides guidance in reporting, recording, measuring, benchmarking of energy development, and on the subject of the discount of greenhouse gas (GHG). ISO 50001 also enables the implementation, prioritization, and evaluation of new efficient energy technologies in the facilities.

ISO 50001 is designed to provide the following benefits:

- To increase energy efficiency,
- To reduce costs,
- To improve environmental performance,
- To reduce risks,
- To increase security.

The time required to establish and implement the ISO 50001 Energy Management System can vary depending on the type of operations to be managed, the types of energy resources that is to be used and whether the organization has any management system or not (e.g., ISO 9001: 2008). Developing and putting into practice of ISO 50001 Energy Management System takes at least 5–6 months. The establishment of the system requires an energy management and the identification of a team that will form this system. This process will take about 15% of the time that is required to install the energy management system. After the system is installed, the time will vary depending on the strategies and the projects that the company will implement.

References

1. Aksoy S, Çalıkoğlu E, Aras H, Karakoç N (2013) Energy management and policies, Republic of Turkey Anadolu University, Publication No: 2787, Distance Education Faculty, Publication No: 1745 (In Turkish: Aksoy S, Çalıkoğlu E, Aras H, Karakoç N, Enerji Yönetimi ve Politikaları TC (2013) Anadolu Üniversitesi, Yayın No: 2787, Açık Öğretim Fakültesi, Yayın No: 1745)
2. Kaya D, Öztürk HH (2014) Sanayide Enerji Yönetimi ve Enerji Verimliliği, Umuttepe Yayınları, Yayın No: 114, Mühendislik Dizisi: 12, Kocaeli/Türkiye
3. REN21 (2019) A comprehensive annual overview of the state of renewable energy. Renewable Energy Policy Network REN21 Secretariat for the 21st Century c/o UN Environment 1 Rue Miollis, Building VII 75015 Paris/France

4. Küçükçalı R (2005) Energy economics, Isısan Studies No:351 (In Turkish: Küçükçalı R (2005) Enerji Ekonomisi, Isısan Çalışmaları No: 351)
5. Republic of Turkey Ministry of Energy and Natural Resources, World and Turkey's Energy and Natural Resources View, Directorate of Strategy Development, No:15, Ankara/Turkey, As of 1st January 2017. (In Turkish: T.C. Enerji ve Tabii Kaynaklar Bakanlığı, Dünya ve Türkiye Enerji ve Tabii Kaynaklar Görünümü, Strateji Geliştirme Başkanlığı. No: 15. Ankara/Türkiye. 1 Ocak 2017)

Chapter 3
Energy Audit

An energy audit is a supervision, surveillance, and analysis of energy flowage for energy preservation in a system, process, or building to decrease energy input amount to the system, without losing or harming any output or product amount. It is a series of study performed to increase the energy efficiency, which consists of the steps like gathering necessary information, measurements, evaluations, and reporting. That is to say, an energy audit is performed to increase energy efficiency and reduce the energy consumption in a building or facility by using a variety of measurement techniques with appropriate equipment. The energy audit determines where and how energy should be used throughout the plant and where it can be saved. The energy audit is a need to be realized in the facilities that use excessive amount of energy. It gives the opportunity to lower the energy cost and protect the environment.

The energy audit is realized in accordance with the procedures and principles published in the related communication. This contains processes like collecting information, measuring, evaluating, and reporting; energy-saving potentials to increase energy efficiency, and measures for the recovery of these potentials by measurements, calculations, and market surveys [1–3].

3.1 The Aims of Energy Audit

As a result of the energy audit, energy consumption of the plant is assessed, energy-saving opportunities are identified, investment costs are calculated, and project proposals including the repayment period are presented in the report. It is aimed to identify energy wastes, greenhouse gas emissions, and energy-saving potentials so that to provide technical and economic measures to realize successful implementations. In a successful energy audit, information is provided for the performance of the building and financial analysis is made for each measurement in a comprehensive and efficient manner.

D. Kaya et al., *Energy Management and Energy Efficiency in Industry*,
Green Energy and Technology,
https://doi.org/10.1007/978-3-030-25995-2_3

Fig. 3.1 Energy audit studies

An energy audit is conducted to assess the energy efficiency of a plant, determine energy-saving opportunities, and implement projects. Energy audits help to create energy efficiency awareness and energy management systems and identify energy savings and investment potentials.

Some of the examples about energy audit studies can be seen in Fig. 3.1.

3.2 Businesses that Need to Conduct Energy Audits

It is necessary to provide successful energy management, organization, and a suitable energy audit to reduce the cost of rising energy prices and at the same time to use the existing energy efficiently in industrial establishments, buildings, or processes. Businesses need to perform energy scans to identify and recover energy losses. The businesses can learn how much energy they use and how efficiently their energy being used by applying energy audit studies.

3.3 Energy Audit Levels

During the preliminary energy audit, all the units of the operation are navigated to determine the energy sources that are being wasted, the bad insulation, all kinds of leaks like steam, water, and fuel leaks, ineffective and inefficient equipment. Instant measurements are made if necessary. As a result of the preliminary energy audit report, energy-saving possibilities are revealed in detail. In the establishment, it is determined that whether a detailed energy audit is needed or not.

Periodic measurements are made in the areas that determined in the preliminary energy audit. The areas where energy losses are present and areas where energy is not used efficiently are identified carefully. As a result of these detailed measurements, the energy consumption of the systems can be precisely determined.

After the energy audit is completed, the required improvement studies for energy saving are applied according to the demand at the establishment. Along with the measures that need to be taken and the applications that need to be done for energy savings, the productivity-enhancing, and energy efficiency increasing project is prepared in line with demand in the industrial facility. While maintaining an energy management program in an establishment accounts for 1–2% of the organization's energy costs, the energy consumption can be reduced by 10–40% at this facility.

The information about daily or monthly energy use of a building or facility is determined by realizing a good energy audit. Especially for the industry, the energy audit can be carried out in three phases:

The First-Level (Primary) Energy Audit: It is a superficial audit, which is prepared to report the use of unnecessary energy in the system.

The Second-Level Energy Audit: This audit is made by using measuring devices. It requires more comprehensive and more technical knowledge than the first-level energy audit. In this audit, the costs of the proposed energy audit studies are determined, and the most suitable optimal solution is selected from among the proposals by determining the depreciation period.

The Third-Level Energy Audit: The energy audit conducted in the second level is supported by modeling on a computer. Related programs are used for this purpose. The calculated costs are also found by adding the life cycle cost to the account.

3.4 Energy Audit Profiles

An energy audit is designed to identify energy-saving potentials, energy wastes, and greenhouse gas emissions, as well as technical and economic aspects of their associated rehabilitative or preventive measures. Within the scope of energy audits, the following survey profiles are evaluated annually.

Input Profile: This profile consists of the types of energy that is incoming in the facility or building (natural gas, fuel oil, coal, electricity, steam/hot water, etc.), the unit energy quantities (quantities of energy input carried by unit masses or volume, in kWh) amount of usage—time charts.

Waste Profile: This profile consists of the types of waste (flue gas, hot gas/water, steam, etc.) from the heating/cooling systems, energy conversion systems, or production process, the reasons for their formation, their quantity-time graphs.

Loss-leakage Profile: This profile identifies and covers the amount of avoidable energy losses due to insufficient heat (thermal) insulation in buildings and the steam/gas/water/fuel leakages in the equipment, and incompatibilities in electrical systems.

Inefficiency Profile: This profile identifies the amount of avoidable energy losses and energy waste due to insufficient and inefficient equipment or process applications.

Waste Profile: This profile identifies the amount of wasted energy due to excessive use or in the equipment that is pending or being on standby position in heating, cooling, lighting processes, in office needs or in the similar areas.

Emission Profile: This profile identifies the amount of greenhouse gas quantities based on energy types in the input profile.

Energy Management Profile: This profile identifies the ability of the energy manager/management unit in the operation, the applied procedures, the level of employees' awareness, and the senior management view of energy management.

3.5 Steps of Energy Audit

Energy audits consist of the phases like determination of energy-saving foci, preliminary interview, data collection, making necessary measurements and evaluation of measurement results, preparation of economic analyzes related to investments for energy savings points, and reporting.

The steps of energy audit can be seen in Fig. 3.2.

3.5.1 The Preliminary Audit

Survey profiles are analyzed with the help of documents, interviews, observations, and necessary point measurements in the establishment or building. Prevention and/or recovery potentials are estimated carefully, the measures that can be applied for these potentials are determined together with their approximate costs and recovery times. Some important recommendations are developed to overcome the shortcomings that are observed in the energy management profile. The studies and study programs are determined to be included in the detailed survey. Preliminary audit is completed within ten-business days maximum for the report.

The necessary data are collected according to the priority measures in the preliminary energy audit. Depending on the condition of the facility, the team consisting of

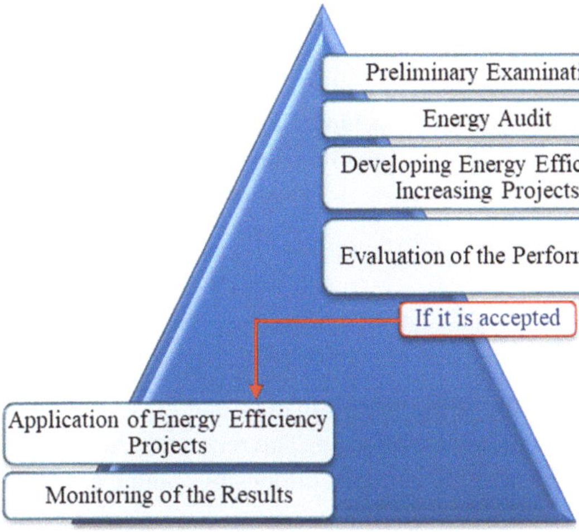

Fig. 3.2 The steps of energy audit

technical staff collects the data and various measurements are made within a period ranging from 1–3 weeks in the framework of working with preliminary audit.

3.5.2 *The Preliminary Audit Briefing*

Preliminary audit briefing is given to the top manager and the managers in the business and the employees who are determined by the top management, for a day. In this briefing, the information about energy efficiency benefits and costs as well as general introduction, examples of applications in the world and in the related country, preliminary audit results, and precautions to be taken are given in detail. In addition, useful documents are distributed to the participants, questions of them are answered, and their opinions are received for the study. Then it is expressed that the detailed audit is needed for this inspected establishment and the productivity-enhancing project can be prepared according to this audit. In this direction, the decision of the top management is requested.

3.5.3 *The Detailed Audit*

In a detailed energy audit, the measurements are made with the devices specified in the regulation. Measurements are realized in intensive energy-consuming devices, the places where energy conversion is carried out and, in the areas, where this type of departments and processes are present. According to the results of the preliminary audit, the prevention and/or recovery potentials are estimated with a maximum error rate of $\pm 10\%$ by measuring and calculating in the relevant operating conditions. By using preliminary audit and detailed audit results, the precaution options that can be applied are analyzed with their technical and economic characteristics. In this context, the information of guidance on efficiency-enhancing projects, which can be prepared later, is provided by selecting the most appropriate measures.

3.5.4 Reporting

The reports are prepared by using the evaluation and conclusion analysis that has been made according to the measurements. The report explains the current state of the business and energy-saving potentials, dimensions, and measures.

3.6 Energy Audit Measurements

Energy-consuming equipment and systems are examined in energy audits. Generally, these equipment and systems can be listed as engines, furnaces, fans, steam installment, boilers and their equipment, lighting, electrical energy system, harmonics, compensation and automation, air-conditioning system, compressed air system, pumps, insulation systems [4–6].

There are different measuring instruments that are used to make energy audit measurements in the energy audits. Detailed information on measuring instruments used in energy audits is given in Chap. 7, Measuring Techniques and Measuring Instruments.

Flue Gas Analysis: Flue gas analysis is performed to determine the combustion efficiency of systems such as boilers or furnaces at the building and to report on the application feasibility of efficiency increasing equipment. It is also determined whether the combustion system is in the correct settings.

Thermal Imaging Camera Measurement: A thermal imaging camera is used to predict facility maintenance. It makes thermal leaks visible in the building exterior, mechanical installations, and other equipments in the facility. In addition, overheating points are determined by using thermal imaging cameras of the equipment and connections in the electric panels to prevent the danger of overheating. With this measurement, heat leakages are determined locally and the risks that may occur in the electric panels are avoided effectively.

Flow rate Measurement Using Ultrasonic Flowmeter: Velocity and flow rate of the fluid passing through the fluid lines are measured by using ultrasonic flowmeter. When the measurement is conducted with ultrasonic devices, no destructive action is required in the fluid line. It is determined whether the current flow rate values in the mechanical systems match the project values or not.

Leakage Detection in Pressurized Air Lines: The location, quantity, and cost of pressurized air leaks in the compressed air line are determined by this application.

Measuring the Electrical Values by Using an Energy Analyzer: Electricity consumption values are recorded, and energy quality is monitored and reported by using an energy analyzer from a part of the equipment to the end of the plant.

Measuring of the Heat Transmission Coefficient in Building Elements [(U-value: Heat transmission coefficient)]: The heat transmission coefficient (W/m^2K) of the walls is designated according to the measurements taken from the wall surface of the building. With this measurement, it is determined whether the outer walls of the building comply with the necessary standards. Thus, the minimum insulation thickness that can be calculated to achieve the required standards.

Luminance Level Measurement: Light intensity is measured by the lux meter. This measurement determines the adequacy of the lighting system for operating conditions and quantifies the state of the changes made in the lighting system.

Sound Level/Noise Level Measurement: The noise level, which is generated by the equipment in the facility, is measured carefully. This measurement process determines the noise level to which plant employees are exposed during the processes in the facility.

Relative Humidity and Temperature Measurement: Temperature and humidity are measured in the desired parts of the plant and the measurements are recorded for the desired period. Humidity measurement and recording can be performed for the desired environment. Temperature measurement and recording can be performed in the environment, on the surface, or within the fluid.

Air Velocity/Air Flow Rate Measurement in Air Ventilation Ducts/Diffusers: Air velocities and air flow rates are measured in the air ventilation ducts and diffusers in the facility.

Air Velocity/Flow Rate Measurements in Air Ventilation Grilles/Diffusers: Air velocity and air flow rate measurements are performed in the air ventilation grilles and air ventilation diffusers in the facility.

Illumination Working/Occupancy Measurement: These measurements are performed to determine unnecessary working time of lighting systems in the facility. A light meter provides a luminance reading based on either lux or foot-candles. Lux (lux) is a unit of illumination of one square meter, which is one meter away from a uniform light source. This is a European standard of measurement. Foot-Candles (FC) is also a unit of illumination, which measures one square foot that is one foot away from a uniform light source, which is a U.S. measurement standard.

Rotational Speed Measurement in Mechanical Systems: The rotational speed is measured in fans and motors and the rate of feed is measured in revolving equipment such as conveyor/tape. With these measurements, it will be possible to check whether the equipment is working on the label value or not.

Efficiency Analysis in the Cooling Groups: The efficiencies of the cooling group are determined by performing this analysis. With this analysis, it is checked whether the cooling groups are working in the current operating conditions or not.

3.7 Preparation of an Energy Audit Report

3.7.1 Purpose of the Energy Audit

The energy audit can be briefly stated as the process of determining the energy used in the facility. The main purpose of the energy audit is how to reduce the energy and its usage cost. For this purpose, the following operations are performed:

- Energy usage mapping is prepared,
- Energy analysis is performed,
- Energy-saving potential is determined.

The following four questions are asked in the energy audit:

- According to the energy types, how much energy is used?
- What is the cost of each energy type?
- How much energy is used?
- What are the possibilities to reduce energy use and costs?

3.7.2 The Energy Audit

- Cost-cutting and energy-saving opportunities are identified,
- If the use of energy resources and energy usage is determined as an audit strategy, the energy audit that is to be performed depends on the targets and conditions,
- It is important to determine energy audit as a priority for energy savings in the facility,
- It is very important to have staffs who are experts in energy audit,
- All necessary information must be clearly stated and listed in the energy audit,
- Correct values increase the efficiency of an energy audit,
- It is very important to correctly identify some deficiencies in the energy audit and transfer it to the decision-makers,
- It is very important to consider an energy audit only at the required levels to determine energy-saving measures and make decision, accordingly.

3.7.3 Energy Audit Steps

(1) The investigation about the determination of the present condition of the plant,
(2) Mapping and analysis of the current situation of the facility,
(3) The intensive capital change is realized by conducting a detailed study for the projects determined in the second phase (Table 3.1).

Table 3.1 An example of the third phase of an energy audit

Project name	Available energy consumption (TOE)	The project cost	Post-change energy consumption (TOE)	Yearly energy saving (TOE)	Payback period (month, year)
–	–	–	–	–	–
–	–	–	–	–	–

3.7.4 *The Method of the Energy Audit*

There are two different ways to perform an energy audit:

(1) The system-based approach,
(2) The solution-based approach.

Each approach has its own advantages and disadvantages. The most appropriate method depends on the purpose of the audit.

3.7.4.1 The System-Based Approach

In the system-based approach, there is no limitation on the entry into the energy system and its evaluation. Thus, it is possible to evaluate each component in terms of efficiency in the energy system. The efficiency of the energy system is compared to a standard reference point. For example, energy audit performed with system-based approach in a house:

- Lighting level,
- Heating and cooling efficiency,
- Kitchen equipment,
- Washing and drying equipment,
- Hot water systems and other electrical equipment are evaluated according to the study.

To evaluate these, comments are made according to the set values in the regulations. As a result of these evaluations, the energy manager brings the energy consumption to the specified levels that have been determined in the standards for homes.

On the other hand, a system-based approach to industrial plants is carried out with the following measurement and analysis in the energy audit:

- Lighting measurement,
- Tariff analysis,
- The measurements with energy analyzer approximately for 5-7 days (maximum and minimum power drawn, harmonics, power shots according to peak hours, etc.),

- Transformer measurements (determination of transformer capacity),
- Heat, steam, compressed air analysis (boiler, compressor, fan, etc.),
- Fuel analysis (coal, natural gas, electricity, etc.),
- Flue gas measurements and analysis,
- Waste analysis,
- Electric motor efficiency analysis,
- Set Point Value and Energy Manager Calculations,
- Analysis of pump systems.

3.7.4.2 The Solution-Based Approach

The most important advantage of a solution-based approach is that it is performed very easily. For example, There is no need to redo all the calculations and cost analysis for a coating project if previously applied for the insulation purposes. The solution-based approach to energy audit can be illustrated by the following examples:

- The replacement of incandescent light bulbs with fluorescent, mercury vapor, metal halogen, or high-pressure sodium vapor lamps,
- Elimination of the cracks due to seasonal conditions and openings due to building designs,
- Double glazing of windows and insulation of the outside of the building,
- Preventive maintenance according to manufacturer's recommendation.

3.7.5 Preparation of the Energy Audit

The retrospective energy consumption and cost values of at least 3-5 years are determined and prepared as a report as the first stage. The studies are organized according to importance. As a result of these studies, the savings/investment ratio, which is included in the energy-saving projects, is calculated. This calculation is also important in terms of bringing together the information and the plan to be made in the final energy audit report.

3.7.6 The Energy Audit Team

After the content of the audit is determined, the energy audit group and their tasks are defined as a first step. In addition to the competencies of the energy managers, people receiving audit-project certification should have knowledge about; energy audit project preparation methods, energy efficiency legislation in their countries; experiences, technologies, and other similar applications in other countries, the methods, standards, devices, industrial processes of the measurements, and evaluations.

The energy audit team should consist of engineers and technicians and they should decide the working time whether will be full time or not. The leader of the team who is responsible for the audit determines this process. According to the capacity of the audit that is to be performed, the experience of the members of the audit team is very important. The duration of the study depends on field and the office work. If necessary, supportive services can be obtained from related companies. The training and explanation of the contents of the study should be given to the team members before the audit.

3.7.7 Energy Audit Instruments

The most important tool in the energy audit is to collect and evaluate the information. The members of the energy team make use of some measuring devices and equipment to gather information. In this context, the measuring instruments can be simple or expensive used. The members of the energy audit team are obliged to comply with the Occupational Health and Safety Rules during the audit. Energy audit tools can be grouped as follows:

(1) **Energy Analysis Equipment:** Electrical analyzer, thermal imaging camera, temperature measurement device, flue gas analyzer, luminous intensity measuring device (lux meter, luminous intensity meter), etc.
(2) **Energy Analysis Software:** Accounting statement analysis for level 1 and 2, and detailed energy simulation for level 3 is prepared by using this software. Economic analysis is carried out based on level 2 and 3 analyses.

3.7.8 The Energy Audit Report

The content of an energy audit report is determined by the following factors:

- Energy consumption of the plant (TOE),
- Production quantity,
- Energy structure of the plant.

The content of an energy audit report should cover all examined topics. The following topics should be included in a general content of an energy audit report [7–9]:

- Executive summary,
- Current situation analysis,
- Energy amounts and costs used in the facility,
- Comparisons with regional and national values,
- Energy consumption target with the Efficiency Increasing Projects to be performed,

- Introduction of the facility,
- Business conditions and transaction processes,
- Energy consumption equipment,
- Suggestions,
- Attachments,
- Past studies,
- Program accounts,
- Detailed calculation tables,
- Production information,
- Estimated cost detail,
- Equipment list.

In the energy audit report, the results should be written clearly, and the following points should be taken into consideration when writing the report:

- Equipment that has been used in the examinations and related measurements made with this equipment,
- Energy consumption model,
- Potential energy-saving areas.

This information is very important because it can be used when real energy-saving calculations are being made in the future. A comparison of the energy consumption of the plant with the past values is also the most important data of energy management. Making broad and clear comments on the projects makes it easier to understand the report. The report should be written in a clear way by using graphics and pictures in its content [10].

3.8 Energy Audit Examples

3.8.1 Steam System

Steam system consists of three sub-systems: production, distribution, and consumption. The process flow in a typical steam system is given in Fig. 3.3.

Current Situation Analysis-1: The necessary information for studies conducted in a medium-scale establishment is

- Fuel information: Fuel type, fuel consumption, the outlet temperature, and the pressure of the fuel pump,
- Combustion: The flue gas temperature and composition, combustion efficiency,
- Steam: Steam pressure, steam temperature, and its quantity,
- Feeding water: Water temperature, quality, the outlet temperature of the pump,
- Additional water: Mildness, quantity, quality, and temperature of the water,
- The amount of steam and flash steam that thrown into the outside with blowdown: The steam amount that thrown into the outside manually and continuously,
- Boiler water: Water quality.

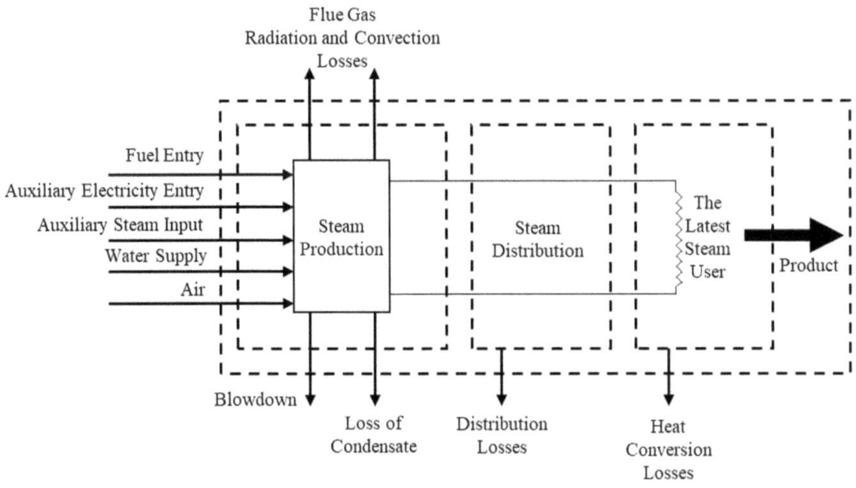

Fig. 3.3 Process flow in a typical steam system

Current Situation Analysis-2: The values that should be checked during studies conducted in a medium-scale establishment are

- Boiler body temperatures,
- Burner maintenance chart,
- Chemical analysis of fuel,
- Control of blowdown systems,
- Measurements of blowdown amount and its temperature,
- Control of feeding water system and the chemical analysis of feeding water,
- The detection of steam consumption points and consumption quantities,
- Control of steam distribution lines,
- Feeding water and steam costs,
- Knowledge and experience of employees.

The information about the establishment can be seen in Table 3.2 (as an example).

3.8.2 The Mass Balance Calculation

Feeding water Feeding water = Steam + Blowdown
= 10,000 + 800 kg/h
= 10,800 kg/h
Returning condensate Returning condensate = Feeding water—additional water
= 10,800−3,000 kg/h
= 7,800 kg/h
Condensate losses Condensate Losses = Additional Water—Blowdown

Table 3.2 The information about the establishment

Variable	Value	Unit
Boiler outlet flow rate	10,000	kg/h
Working pressure	1500	kPa
Working time	6000	h/year
Feeding water temperature	105	°C
Ambient temperature	20	°C
Flue gas temperature	280	°C
Fuel oil Higher Heat Value (HHV)	38.68	MJ/L
Fuel cost	0.50	$/L
Measured combustion efficiency	78	%
Blowdown ratio	8	%
Radiation losses	3	%
Measured additional water flow rate	3000	L/h
Additional water temperature	15	°C
The cost of water	2.00	$/m^3
The cost of water treatment plant	1.00	$/m^3
The cost of electricity	0.10	$/kWh

= 3,000−800 kg/h
= 2,200 kg/h

The energy balance equation can be written as:

$$E_{in} = [(\text{Steam} + \text{Blowdown})]/\left[(\text{Burning Efficiency} - \text{Radiation Loss})(\%)\right] \tag{3.1}$$

$$\text{Steam} = \text{m} \times (\text{h}_1 - \text{h}_2) \tag{3.2}$$

$$\text{Steam} = 10,000\,(2,793.7 - 440.4)$$

$$\text{Steam} = 23.533\,\text{GJ/h}$$

where

m = Steam amount (10,000 kg/h),
h_1 = Vapor enthalpy (2793.7 kJ/kg), and
h_2 = Feeding water enthalpy (440.4 kJ/kg)

$$\text{Blowdown} = \text{m} \times (\text{h}_3 - \text{h}_4) \tag{3.3}$$

$$\text{Blowdown} = 800 \times (856.3 - 62.8)$$

$$\text{Blowdown} \; = \approx 0.635 \quad \text{GJ/h}$$

where

m = Amount of blowdown (10,000 kg/h × 0.08 = 800 kg/h),
h_3 = Boiler water enthalpy (856.3 kJ/kg) and
h_4 = Additional water enthalpy (62.8 kJ/kg)
E_{in} = (23.533 + 0.635)/(0.78–0.03) = 32.224 GJ/h
Thermal efficiency = 23.533/32.224 = 73%

Additional energies are also included to the boiler efficiency of the plant as the lighting of the plant, combustion air, water pump, etc. The total electric charge is estimated as follows (approximately):
20 kW × 3.6 MJ/kWh = 72 MJ/h or 0.072 GJ/h
In this case, the total energy input is calculated as follows:

32.224 GJ/h + 0.072 GJ/h = 32.296 GJ/h
Boiler efficiency of the plant = Steam production/Total energy input
= (23.533 GJ/h)/(32.296 GJ/h)
= 72.9%

3.8.3 Steam Quality

The enthalpy value is not directly derived from the steam charts. The reason is that there is moisture in the vapor. The steam quality is determined by the following equation:
Steam quality = Condensed steam mass/Condensed steam mass and water mixture
When the steam quality is 98%, it indicates that there is 2% water in vapor. The vapor heat can be calculated for the steam that has the pressure of 1000 kPa and a quality of 0.98 by using the steam table as

- Sensible heat = 763.0 kJ/kg
- Latent heat (2015.1 × 0.98) = 1974.8 kJ/kg
- Total heat (h_g) = 2737.8 kJ/kg
- The required heat amount for the removal of the humidity;
- Dry saturated steam which has the amount of 1000 kPa (10 bar)
- Sensible heat (h_f) = 763.0 kJ/kg
- Latent heat of vaporization (h_{fg}) = 2015.1 kJ/kg
- Total heat (h_g) = 2015.1 + 763.0 = 2778.1 kJ/kg = 2778.1-2737.8 = 40.3 kJ/kg

3.8.4 *Calculation of the Blowdown Amount in Boilers*

Blowdowns are conduct based on the concentration ratio. Concentration rate is the number of concentrations of the solid particles in a given volume of water. For example, if water with a dissolved solids content of 100 ppm is evaporated to half its initial volume, the dissolved solids concentration reaches the value of 200 ppm. If half of this water is evaporated again, the dissolved solid concentration reaches the value of 400 ppm. In this case, for the water that is given as an example:

$$\text{Concentration rate} = 400/100 = 4$$

Despite the decrease in the amount of the water, solid matter amount remains constant, and therefore, the concentration increases.

$$\text{Blowdown percent} = 100/\text{Concentration Rate} \tag{3.4}$$

The concentration is allowed to increase 10-fold in a boiler that has a concentration rate of 10. In this case, the blowdown percentage is determined as follows:

$$\text{Percent blowdown } (\%) = 100/10 = 10\%$$

3.8.5 *Feeding Water and Properties*

The cost of the feeding water is directly influenced by the quality and quantity of the returning condensate in the power plant. Feeding water preparation and feeding system must be continuous. The characteristics of the feeding water for a low and medium-pressured boiler are given below:

- Feeding water:

- Total hardness of $CaCO_3$ = 10 ppm
- pH value = from 8.5 to 9.5
- Soluble oxygen amount = 0.1 ppm
- Silica from SiO_2 = 0.0 ppm
- Total dissolved solids = from 100 up to 500 mg/L

- Boiler water:

- Total alkali content = 700 ppm
- Caustic alkalis = 350 ppm
- pH value = from 11 to 12

- Phosphates = from 30 to 50 ppm
- Total dissolved solids = from 1000 up to 2000 mg/L
- Silica = 40 ppm (max)

There is also a condensate tank and a flash tank.

3.8.6 Calculation of the Steam Cost

Natural gas, electricity, water, additives, workmanship, consumables, etc. that are used in steam production are considered for the determination of the steam cost. The cost of steam is calculated as follows:

Steam cost (For example USD or Euro) = Total cost/Produced steam amount

References

1. Çengel YA, Çerçi Y (2000) Opportunities to Save Energy in Industry, 12. Turkish National Conference on Thermal Sciences and Technologies with International Participation, Conference Proceeding, Sakarya/Turkey 2:392–399
2. Hadra M, Hergül AS, Kaya D, Eyidoğan M, Çanka Kiliç F, Özdemir NC (2019) An energy audit and optimization in bar Mill Annealing Furnace. International Journal of Ecosystems and Ecology Science (IJEES) 9(1): 33–42
3. Ertem ME, Şen S, Akar G, Pamukçu C, Gürgen S (2010) Energy balance analysis and energy saving opportunities for Erdemir Slab Furnace #3. Energy Sources Part A 32:979–994
4. Gençoğlu MT, Özbay E (2007) Energy efficiency methods in lighting, XII. Electrical, Electronics, Computer, Biomedical Engineering National Congress, Eskişehir/Turkey (In Turkish: Gençoğlu MT, Özbay E (2007) Aydınlatmada Enerji Verimliliği Yöntemleri, XII. Elektrik, Elektronik, Bilgisayar, Biyomedikal Mühendisliği Ulusal Kongresi, Eskişehir/Türkiye)
5. Güleç M (1999) Saving measures and methods in pneumatic systems, I. National Hydraulic Pneumatic Congress and Exhibition (In Turkish: Güleç M (1999) Pnömatik Sistemlerde Tasarruf Önlemleri ve Yöntemleri, I. Ulusal Hidrolik Pnömatik Kongresi ve Sergisi)
6. Talbott EM (1993) Compressed air systems: a guidebook on energy and cost savings, 2nd ed. Fairmont Press, Inc., Liburn
7. Özdabak A Preparation of energy efficiency report. (In Turkish: Özdabak, A., Enerji Verimliliği Raporunun Hazırlanması)
8. Özdabak A (2008) Energy savings in metal melting process (In Turkish: Özdabak, A., Metal Eritme İşleminde Enerji Tasarrufu)
9. Özdabak A (2008) Efficient use of energy in industry, VII. National Clean Energy Symposium, UTES'2008 17–19 December 2008, İstanbul/Turkey (In Turkish: Özdabak A (2008) Sanayide Enerjinin Verimli Kullanılması, VII. Ulusal Temiz Enerji Sempozyumu, UTES'2008 17–19 Aralık 2008, İstanbul/Türkiye)
10. West T (2002) From mechanical to electronic control in industrial burners. Technical Bulletin, Energy Technology and Control Ltd

Chapter 4
Energy Efficiency

Energy efficiency is the reduction of energy consumption per unit of service or product quantity, without causing a decline in living standards and service quality in buildings, production, and quality industrial enterprises.

The level of development based on energy can be measured with energy consumption and energy intensity values per capita in the countries that target sustainable development. The high-energy consumption per capita demonstrates both the vitality of the economic activities and the high level of prosperity in the country. Industrialization activities, new technologies, the increase of living standards, and the fast-increasing population cause more energy use in many countries, which are especially in the fast development stage, each year.

The importance of the efficient use of energy and energy resources has gained great importance to assure the energy sustainability, reduction of external dependence, and struggle with climate change in all over the world. With the recent studies, there has been an important development of consciousness to apply environmentally friendly energy policies and its usage methods including diversification and consumption, to increase efficiency by avoiding the wastage. Ensuring energy supply security and having sustainable energy policies are not as easy as it is supposed to be thought considering the resource dependence on the outside, which makes it even more difficult. For this reason, efficient use of energy has gained a special importance and the related energy policies are needed to be addressed carefully.

Although energy saving is an important concept, the energy efficiency, which is our main topic, has also a great importance and wider contents. Energy efficiency can be described as using energy without compromising life standards, needs, and quality of the products by considering high efficiency and saving. Energy saving is a result of efficient use of energy, the fastest, cheapest and cleanest energy source. In terms of protecting the future of the planet, the energy, that can be saved by increasing the energy efficiency.

The importance of energy efficiency is increasing every day due to the decrease of fossil fuels, which constitute a significant part of energy demand, the insecurity

D. Kaya et al., *Energy Management and Energy Efficiency in Industry*,
Green Energy and Technology,
https://doi.org/10.1007/978-3-030-25995-2_4

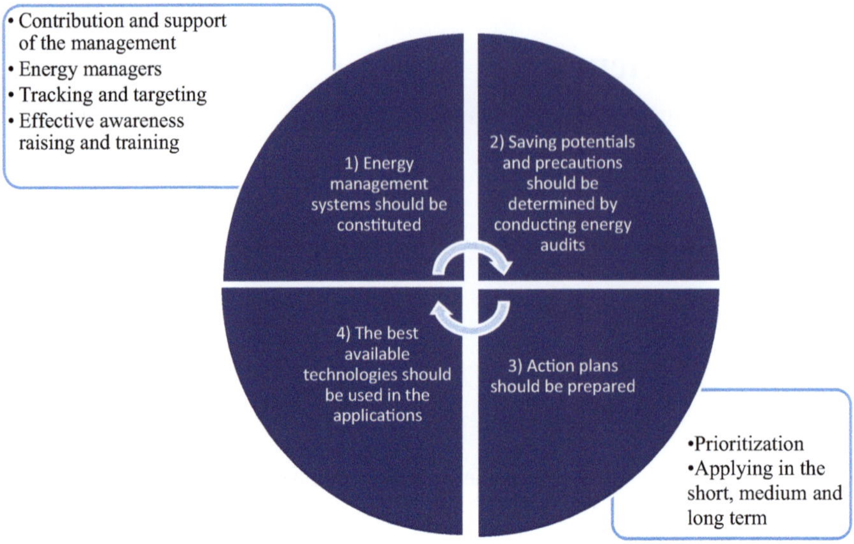

Fig. 4.1 Stages of the energy efficiency applications

towards nuclear power plants, the insufficient supply of alternative energies, which their demands are not being met at the desired level, pollution of the environment, and climate change. Energy efficiency is the most important component of sustainable development and competitiveness. In developed countries, energy efficiency means the least costly energy, resulting in efficient use.

The stages of the energy efficiency application can be seen in Fig. 4.1.

4.1 The Measures for Energy Efficiency

The following measures are primarily taken into account in the operation of existing plants, in the establishment of new plants, in capacity increasing and modernization studies, in energy efficiency studies and projects [1–4]:

- Burning control and optimization in combustion systems and efficient burning of fuels,
- To obtain the highest efficiency in heating, cooling, air conditioning, and heat transfer,
- To make the heat insulation in accordance with the standards on hot and cold surfaces, to isolate all the units that generate, distribute, and use heat, to minimize unwanted heat losses or gains,
- Waste heat recovery,
- Increase the efficiency of the conversion of heat to work,
- Preventing losses in electricity consumption,

- Increasing efficiency in the conversion of electricity to mechanical/heat energy,
- Reduce the human factor by using automatic control,
- To pay attention to the selection of inputs to provide uninterrupted energy,
- To choose machines with high-energy efficiency, paying attention to the requirements of standardization and quality assurance system,
- Ensure that the projecting and application are carried out according to the project in such a way that the undesired heat losses or heat gains are at the lowest level,
- Provide and install measuring instruments related to energy efficiency during the design and assembly phase,
- To analyze renewable energy, heat pump, and cogeneration applications,
- To use high-efficient luminaires and lamps, electronic ballasts, lighting control systems in lighting and to make more use of daylight,
- To provide minimum efficiency criteria for energy-consuming or converting equipment,
- To use low diffusion heat-controlled coating double glazing systems.

In industrial plants, energy efficiency can be improved by taking the measures outlined in Table 4.1 [5, 6].

Table 4.1 Measures to increase energy efficiency in industrial plants

Energy efficiency enhancement measures in existing facilities	Energy efficiency enhancement measures in newly established facilities
• Combustion of fuels in high-efficient combustion systems • Increase the efficiency in heating, cooling, air-conditioning, and heat transfer operations • To make proper heat insulation, to reduce heat loss to the minimum level in all the units that generate, distribute, and use heat • Applying waste heat recovery • Increase the efficiency in heat-work conversion • Preventing losses in electricity consumption • To increase the efficiency in the work and heat conversion from the electric power, to change to the combined heat-power generation • Reduce the human factor by using automatic control applications • Make efforts to reduce air pollutant emissions to the minimum levels and to keep the consumed energy wastes in a way that will not harm the environment as much as possible	• New machines to choose from among the most energy efficient technologies, paying attention to the requirements of standardization and quality assurance systems. Also, to pay attention to the selection of inputs that will provide uninterrupted energy supply • Projecting the plant in a most efficient way in terms of thermal insulation • Supplying and installing all energy efficiency-related measurement devices in the establishment phase of the installation • To give importance to the combined heat-power generation • Make the necessary arrangements to minimize air pollutant emissions and the consumption of energy wastes that will pollute the environment

4.2 Definitions for Energy Efficiency

Total Primary Energy Supply (TPES): Total primary energy supply (TPES) is defined as the total amount of primary energy that a country has in possession. This includes imported energy, exported energy (subtracted off), and energy extracted from natural resources (energy production). Principally, TPES is thought of as being the sum of all primary energy sources, but some end-use energy must be paid attention to. This comes from the fact that TPES contains both imports and exports. The net amount of electricity traded (imports—exports) and the net secondary fuels (for example, amount of gasoline imported—amount of gasoline exported) become part of the TPES [7].

In contrast to the total primary energy supply, countries have a total final consumption of energy, which pays attention to the end-use energy.

Gross Domestic Product (GDP): Gross domestic product (GDP) is the value of a nation's finished domestic goods and services for a specific period. The gross national product (GNP) is the value of all finished goods and services owned by a country's residents over a timeframe. Both GDP and GNP are two of the most generally used measures of a country's economy, both of which symbolize the total market value of all assets and services produced over a determined period. Each one defines the scope of the economy differently. While GDP limits its explanation of the economy to the geographical borders of the country, GNP extends it to comprise the net overseas economic activities realized by its nationals.

Total Final Energy Consumption (TFEC): Sum of energy consumption in different end-use sectors, except non-energy uses of fuels. TFEC is separated energy demand in the following sectors: industry, transport, residential, services, agriculture, and others. It exempts international marine and aviation bunkers, apart from at the world level where international bunkers are inclusive of the transport sector.

Value Added: Value added is the net output of a sector after adding up all outputs and subtracting intermediate inputs. It can be calculated without making subtractions for the depreciation of fabricated goods or exhaustion and degradation of natural resources.

Fossil Fuel Electricity Generation Efficiency (Percentage): Ratio of the electricity output from fossil fuel power production (coal, oil, and natural gas) and the fossil fuel input to power generation [7].

$$\text{Generation Efficiency} = \frac{\text{Electricity Output (from oil, coal, and natural gas)}}{\text{Fuel (coal, oil, and natural gas)Input}} \tag{4.1}$$

4.3 Energy Intensity

One of the important indicators of energy efficiency is energy density. Energy density is an indicator used worldwide, representing the amount of primary energy consumed per GDP (Gross Domestic Product). The amount of TOE (tons of oil equivalent) consumed generally for the output of $1000 is preferred as an indicator of energy density in international publications. Here, TOE is a definition used to express the different units such as kg, m^3, ton, kWh in the same platform used to define the amounts of various energy sources. 1 TOE corresponds to the energy to be obtained by burning 1 ton of oil, which corresponds to about 10^7 kcal (kilocalories), 4.184×10^{10} J and 1.162722×10^4 kWh.

In this case, the lower the energy density of a country, the lower the energy used to produce unit output, which means that energy is used efficiently.

Within the energy intensity indicator, economic output, increase, or decrease in energy efficiency, changes in fuel substitution are expressed together and it is not possible to distinguish the changes individually within this indicator. However, energy density is a widely used tool for monitoring and comparing energy efficiency in the world.

Generally, energy intensity approaches are based on GDP, which shows the positive or negative monetary exchange of a country. The use of this indicator can lead to some problems, especially considering the importance of its components rather than the level of GDP in terms of energy; yet it is still possible to say that the most important benchmark on a universal scale is still energy density.

When global data on energy intensity are analyzed, it is seen that there are serious differences between countries and these differences are due to their development levels.

4.3.1 Primary and Final Energy Intensity

Primary energy supply can be described as energy generation plus energy imports, minus energy exports, minus international bunkers, then plus or minus stock changes. The International Energy Agency (IEA) energy balance methodology is based on the calorific content of the energy commodities and a common unit of account: ton of oil equivalent (TOE). TOE is defined as 10^7 kcal (41.86 gigajoules), as it can be described above. This quantity of energy is equal to the net heat content of one ton of crude oil. The difference between the "net" and the "gross" calorific value for each fuel is the latent heat of vaporization of the water produced during combustion of the fuel. For coal and oil, net calorific value is about 5% less than gross, for most forms of natural and manufactured gas the difference is 9–10%, while for electricity the concept of calorific has no meaning. The IEA calculates balances using the physical energy content method to find the primary energy equivalent. This indicator is measured in million TOE and in TOE per 1000 USD.

$$\text{PEI} = \text{TPES} \ / \, \text{GDP} \tag{4.2}$$

The density calculated by the ratio of primary energy consumption to GDP is called the primary energy density, and the density calculated by the ratio of final energy consumption to GDP is called final energy density. Primary and final energy density units. Primary energy density is an energy efficiency indicator that measures how much energy is needed to create a unit of GDP on a regional and country basis. The level of this indicator shows the economic structure, energy consumption structure, climatic conditions, and technical energy efficiency of countries or regions [7, 8].

4.3.2 Average Yearly Rate of Improvement in Primary Energy Intensity (As in %)

Average yearly rate of improvement in primary energy intensity is calculated by using compound annual growth rate (CAGR).

$$\text{CAGR} = \left(\frac{\text{PEI}_{t_2}}{\text{PEI}_{t_1}}\right)^{\frac{1}{(t_2 - t_1)}} - 1 \tag{4.3}$$

where PEI_{t_1} is primary energy intensity in year t_1 and PEI_{t_2} is primary energy intensity in year t_2 negative values stand for declines (or improvements) in energy intensity (less energy is used to produce a unit of economic output or per unit of activity), while positive numbers demonstrate augmentations in energy intensity (more energy is used to produce one unit of economic output or per unit of action).

4.3.3 Industry Energy Intensity

Industry energy intensity (IEI) is the ratio between industry TFEC and industry value added measured in MJ per USD PPP. Industry corresponds to ISIC divisions (International Standard Industrial Classification) 10–45 and includes production (ISIC divisions 15–37), non-fuel mining, and construction.

$$\text{IEI} = \frac{\text{Industrial TPES}}{\text{Industrial Value Added}} \tag{4.4}$$

4.3.4 Services Energy Intensity

Services energy intensity (SEI) is the ratio between services TFEC and services value added measured in MJ per USD PPP. Services correspond to ISIC divisions 50–99. They include wholesale and retail trade (including restaurants and hotels), transport, government, financial, professional, and personal services such as education, health care, and real estate services.

$$\text{SEI} = \frac{\text{Services TPES}}{\text{Services Value Added}} \tag{4.5}$$

4.3.5 Agriculture Energy Intensity

Agriculture energy intensity (AEI) is the ratio between agriculture TFEC and agriculture value added measured in MJ per USD PPP. Agriculture corresponds to ISIC divisions 1–5 and includes forestry, hunting, and fishing, as well as cultivation of crops and livestock production [9, 10].

$$\text{AEI} = \frac{\text{Agricultural TPES}}{\text{Agricultural Value Added}} \tag{4.6}$$

4.3.6 Passenger Transport Energy Intensity

Passenger transport energy intensity (PTEI) is the ratio between passenger transport TFEC and passenger transport activity measured as in MJ per passenger-kilometers.

$$\text{PTEI} = \frac{\text{Passenger Transport TPES}}{\text{Passenger} - \text{Kilometers}} \tag{4.7}$$

4.3.7 Freight Transport Energy Intensity

Freight transport energy intensity (FTEI) is the ratio between freight transport TFEC and activity measured as in MJ per ton-kilometers.

$$\text{FTEI} = \frac{\text{Freight Transport TPES}}{\text{Ton} - \text{Kilometers}} \tag{4.8}$$

4.3.8 *Residential Energy Intensity*

Residential energy intensity (REI) is the ratio between residential TFEC and square meters of residential building floor area measured in MJ per m^2 [7, 8].

$$\mathrm{REI} = \frac{\text{Residential TPES}}{\text{Residential Floor Area}} \tag{4.9}$$

4.3.9 *Energy Intensity of the Countries*

The amount of primary energy consumption per unit of production or per gross domestic product (GDP) is called energy intensity. For a given calendar year, energy intensity of the country is the ratio between the gross inland consumption of energy and the gross domestic product (GDP). This measures the energy consumption of the country's economy and its overall energy efficiency.

The gross inland consumption of energy is determined as the sum of the gross inland consumption of five energy types: coal, electricity, oil, natural gas, and renewable energy sources. The GDP figures are taken at chain linked volumes with reference year. The energy intensity ratio is calculated by dividing the gross inland consumption by the GDP.

Energy intensity is a tool that represents the amount of primary energy consumption per gross domestic product, it is widely used in the follow-up, and comparison of energy efficiency values across the globe. The energy intensity of the countries is found by dividing the total energy consumption by the domestic, gross national income. The energy intensity is the amount of energy that consumed to produce a unit of the economic value. The low energy density indicates that more value addition to the economy is produced with the same amount of energy. In this case, the indicator of energy development in a country is high-energy consumption per capita and in response to this low energy intensity.

4.4 Energy Use and Energy Efficiency in the World Countries

4.4.1 *Overall Assessment*

Energy consumption increases with the increasing population in the world, urban development, and industrialization. By 2040, the world's population is expected to reach 9 billion, with an increase of 1.6 billion. This increase requires providing more energy to more people. The amount of this increase in the world economy is expected to be about 3% in the next two decades.

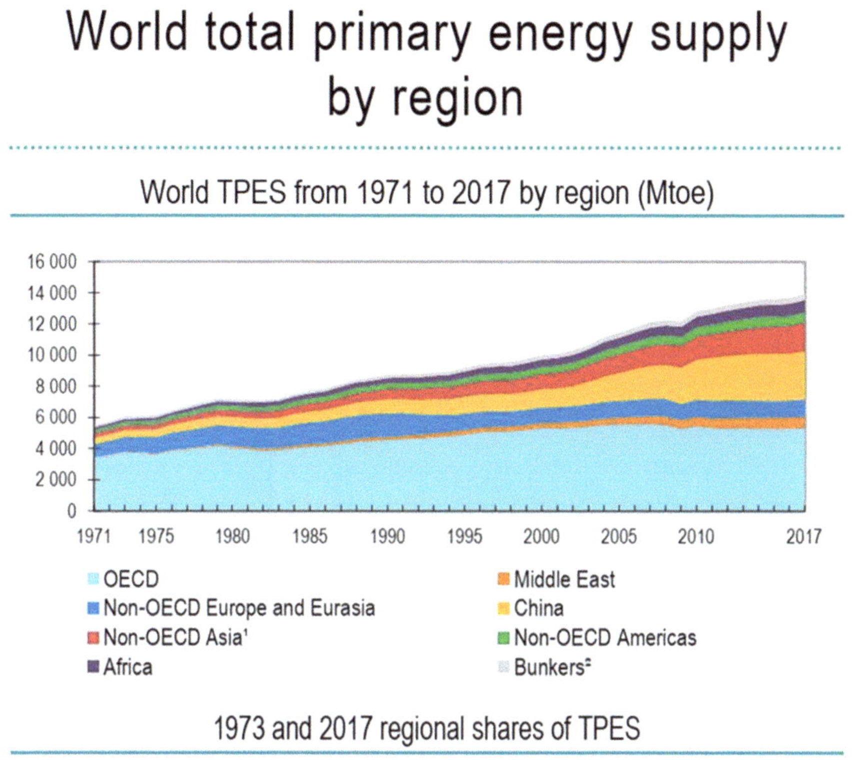

Fig. 4.2 The world total primary energy supply by regions. *Source:* IEA (2019) Key World Energy Statistics. All rights reserved [11]

Figure 4.2 shows the world total primary energy supply by regions [11].

According to the studies of the International Energy Agency, the world primary energy demand, which is currently 14 billion tons of equivalent oil (TOE), (if current energy consumption rate and policies continue), will be increased by 45% to 20.3 billion TOE levels by the next 20 years [11]. China ranks as the first country among the countries that consume the most of energy. Table 4.2 shows the primary energy consumption values and rankings of some countries (As in million TOE) [12].

Although the demand for energy is different in every country, it is continuously increasing on a global scale. In order to meet this increasing demand, energy investments must also increase in parallel.

Investment plan rates expected to be made in the field of energy in the world within 20–25 years can be seen in Fig. 4.3 [13].

Fossil fuel reserves are rapidly decreasing in response to the ever-increasing global energy demand. In terms of natural gas reserves, conventional gas reserves are around 178 trillion m^3. Approximately 40% of these reserves are located in the Middle East (Iran, Iraq, Kuwait, Oman, Qatar, Saudi Arabia, and the United Arab Emirates). Unconventional (Shale gas, coal bed methane, and compressed gases) reserves and

Table 4.2 The primary energy consumption values and rankings of some countries (As in million TOE)* [12]

Country	2013	2014	2015	World total share (%)	Rank
China	2,903.9	2,970.3	3,014.0	22.9	1
USA	2,271.7	2,300.5	2,280.6	17.3	2
India	626.0	666.2	700.5	5.3	3
Russia	688.0	689.8	666.8	5.1	4
Japan	465.8	453.9	448.5	3.4	5
Canada	335.0	335.5	329.9	2.5	6
Germany	325.8	311.9	320.6	2.4	7
Brazil	290.0	297.6	292.8	2.2	8
South Korea	270.9	273.1	276.9	2.1	9
Iran	247.6	260.8	267.2	2.0	10
Saudi Arabia	237.4	252.4	264.0	2.0	11
France	247.4	237.5	239.0	1.8	12
Indonesia	175.0	188.3	195.6	1.5	13
United Kingdom	201.4	188.9	191.2	1.5	14
Mexico	188.9	190.0	185.0	1.4	15
Italy	155.7	146.8	151.7	1.2	16
Spain	134.2	132.1	134.4	1.0	17
Australia	130.7	129.9	131.4	1.0	18
Turkey	120.3	123.9	129.3	1.0	19
Thailand	120.3	123.4	124.9	0.9	20
South Africa	124.6	128.0	124.2	0.9	21
Taiwan	109.9	111.4	110.7	0.8	22
UAE	97.2	99.0	103.9	0.8	23
Poland	96.0	92.4	95.0	0.7	24
Ukraine	114.7	101.0	85.1	0.6	25
And other countries	2,194.7	2,216.0	2,284.1	17.7	–
TOTAL	**12,873.1**	**13,020.6**	**13,147.3**	**100.0**	–

**Source:* (As of January 1, 2017, this is the most current data from MENR: Republic of Turkey Ministry of Energy and Natural Resources).

unexplored reserves, which have been shown as the new face of the gas sector in recent years, are around 354 trillion m^3 [12, 14].

Natural gas consumption increased by 78 billion m^3 (bcm), or 2%, below the exceptional growth experienced in 2018 (5.3%). Nonetheless, the share of gas in primary energy rose to a record high of 24.2% [15]. Natural gas consumption per capita in 2019 can be seen in Fig. 4.4.

The demand grew by 78 billion m^3 (bcm), led by the U.S. (27 bcm) and China (24 bcm) in volume terms. The growth in U.S and Chinese gas consumption was much

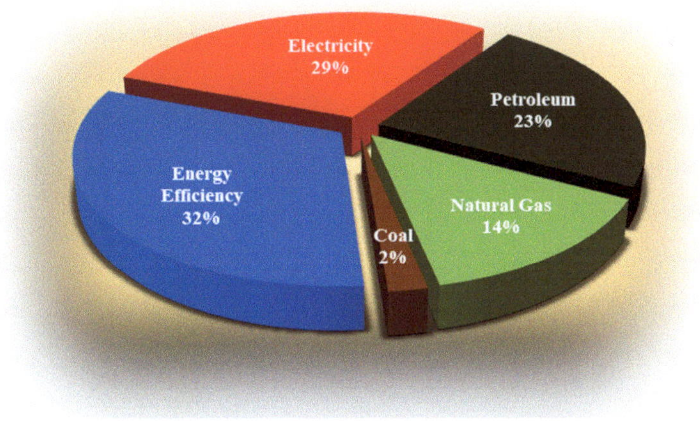

Fig. 4.3 Investment plan rates expected to be made in the field of energy in the world within 20–25 years. *Source:* https://www.gazbir.org.tr/uploads/page/Dunya-ve-Turkiye-Enerji-Gorunumu.pdf [13]

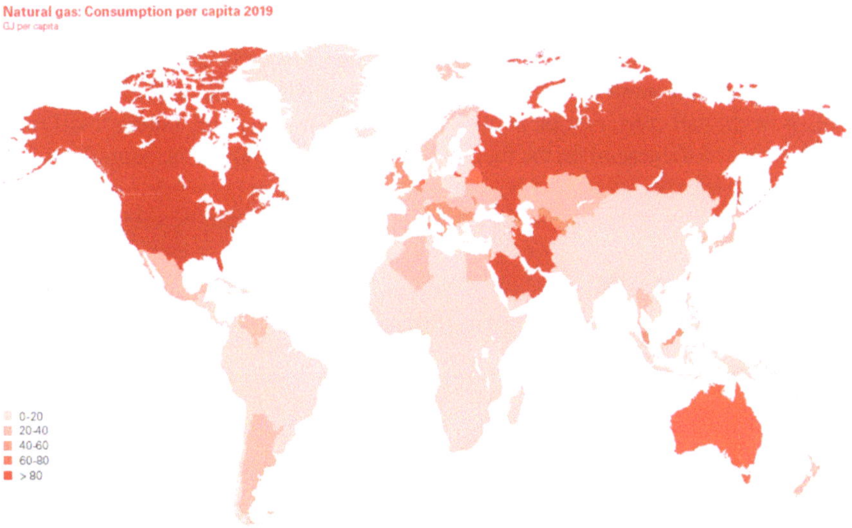

Fig. 4.4 Natural gas consumption per capita (2019) [15]. *Source:* BP Statistical Review of World Energy 2020

slower than in 2018. A decrease in the number of extraordinarily hot and cold days also caused a decline in gas consumption of Russia (10 bcm), the largest decline of any country last year. Gas production grew by 132 bcm (3.4%) surpassing growth in consumption. The U.S. accounted for virtually two-thirds of net global growth. Supply was also augmented by strong growth in Australia (23 bcm) and China (16 bcm) [15].

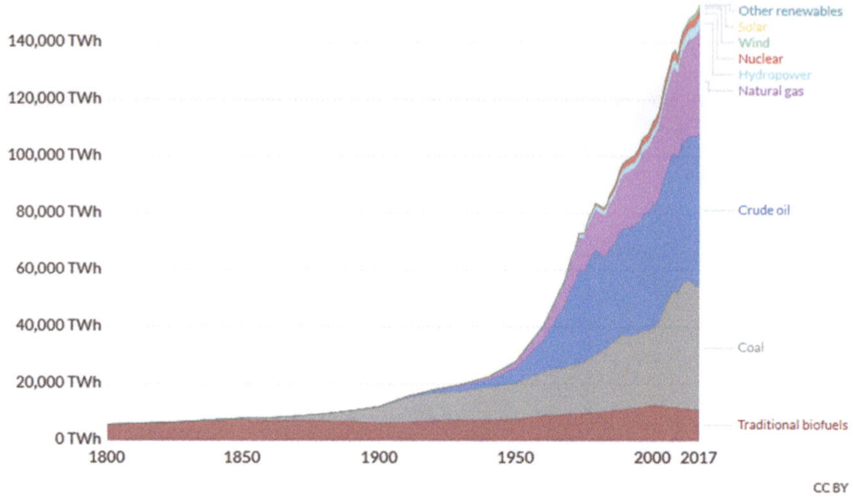

Fig. 4.5 Primary energy consumption by fuel type in the world [16, 17]. *Source:* Vaclav Smil (2017) and BP Statistical Review of World Energy 2017

As of 2016, 81% of the world's primary energy resources are fossil fuels. It is estimated that this number will be around 79% in 2040. Natural gas has a share of 21% in world fuel consumption at the end of 2016. The only fuel among fossil fuels that is expected to increase in the fuel mixture in the year 2040 is the natural gas (the increase rate is expected to be 4%).

Primary energy consumption by fuel type in the world can be seen in Fig. 4.5 [16, 17].

4.4.2 Energy Efficiency

4.4.2.1 Energy Efficiency Overview

Energy has been one of the most important factors that countries have benefited from in providing competitive advantage since the beginning of the last century. In the new age we entered into, technological innovations in the world, increasing the permeability of international borders, the abolition of the borders for capital movements and the enormous developments in the field of communication both increased the amount and speed of energy use in the world and made energy one of the most important problems to be addressed.

Some of the current energy problems are thinning in the ozone layer, reaching the threatening levels of greenhouse gas emissions and rapidly depleting natural energy resources. These problems create more research areas on clean energy productions.

Considering environmental standards, producing energy in the context of safe energy policies and minimizing risks for productions are the main objectives of energy issues.

Continuous change in the energy supply system, the development of new technologies, the dramatic changes in the prices of energy materials in short periods, especially the lack of price stability of the petroleum that displaced the strategic balances in the world from time to time. Therefore, energy and electrical energy systems require significant savings practices and efficiency approaches.

According to the World Energy Outlook (prepared by the International Energy Agency), the necessary energy investments in the field of energy in the world by 2030 are examined and some of the prominent findings that have been found can be listed as

- Between 2001 and 2030, the total investment required for the energy supply infrastructure in the world is US$16 trillion. This amount is necessary to increase the energy supply capacity and to replace the existing or future supply facilities that will be exhausted or become unusable in the coming period.
- Approximately half of the compulsory investments in the energy sector will need to be made in developing countries, where production and demand are increasing rapidly.
- Significant portions of the envisaged investments are necessary to maintain the current supply level. It is estimated that 51% of the investments to be made by 2030 will be spent for the preservation of current or future capacity.
- The financial resources required for energy resources should be obtained from private and foreign sources more than the rate of in the past. Providing the financing needed for the investments that developing countries need is the biggest challenge.
- In developing countries, the private sector needs to be more involved in the markets. On the other hand, given the fact that private investments have been declining since 1997, the extent to which these countries will successfully attract private capital is one of the greatest unknowns of future electricity investments.

Although the use of energy, investments to be made in the field of energy and the amounts to be spent for investments in this field are of interest to all countries in the world, these data are of greater importance for developing countries that have difficulty in financing their own investments.

Figure 4.6 shows the per capita energy use in the world and Fig. 4.7 shows the energy densities of the countries [18].

Improving energy efficiency, preventing unconscious use and waste, and reducing energy intensity on a sectoral and macro-level throughout the entire process from energy production to final consumption are important components of world energy policies.

4.4.2.2 Energy Efficiency in Electricity Generation

Internal energy consumption of a power plant is the energy consumed by the plant itself to produce energy [19]. Ash/coal systems of thermal power plants, water cooling

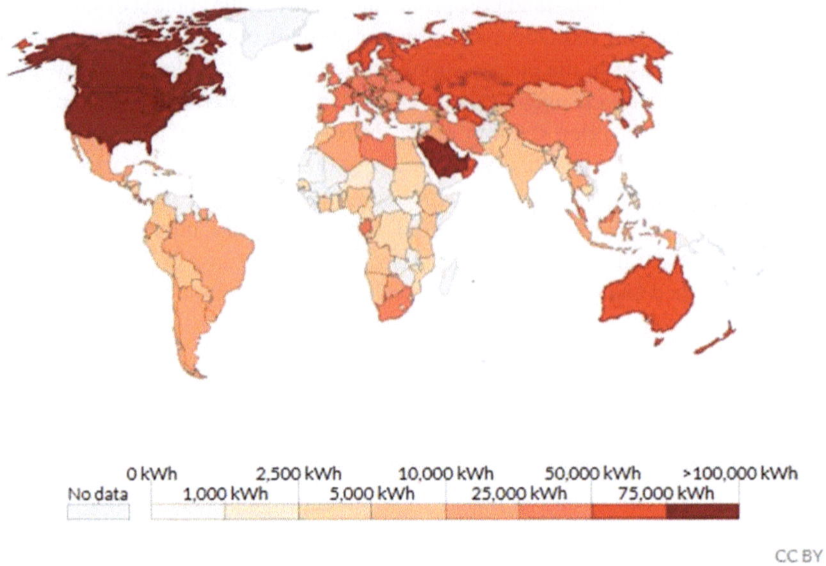

Fig. 4.6 Energy use per capita per year in the world (Energy use per capita, 2015. Annual average per capita energy consumption is measured in kilowatt-hours per person per year) [18]. *Source:* International Energy Agency (IEA) via The World Bank

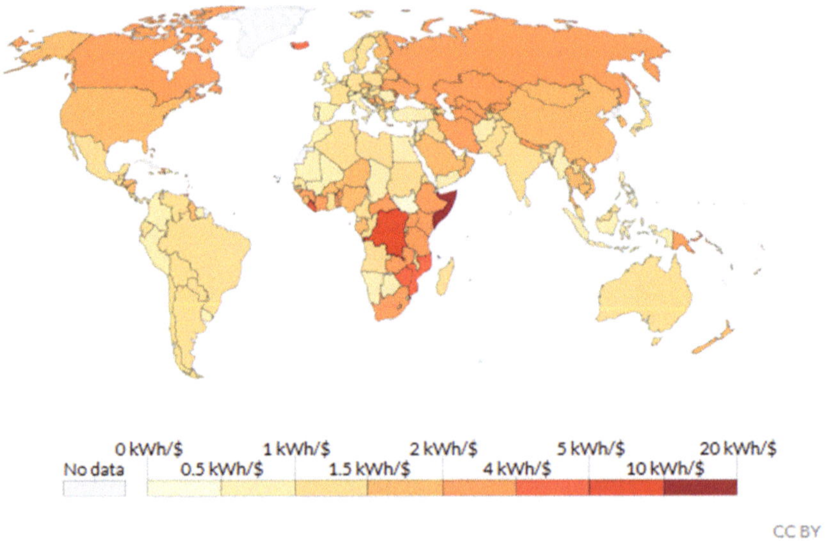

Fig. 4.7 Energy densities of countries [18]. *Source:* World Bank, Sustainable Energy for All (SE4ALL), 2018

and feeding systems, waste and raw water treatment plants, flue gas treatment plants are the systems that increase the internal consumption the most. The biggest share of these is belong to the flue gas treatment plants, which consume 6–7% of the total electricity produced by a power plant. In hydraulic power plants, internal consumption is quite small.

A number of rehabilitations and modernizations are required to ensure energy efficiency in power generation plants. Technologies to increase boiler efficiency in thermal power plants, supply of fuel in accordance with boiler design, frequency control and coordination of pumps and fans and recovery of waste heat are the main saving examples in this field.

Huge developments are observed in efficient power plant technologies in the world. For example, cogeneration is now widely accepted as an option that provides heat and electricity successively. Cogeneration applications can improve the overall efficiency by up to 90% by using emerging technologies and any type of fuel. Given the orientation towards smaller power generation units, cogeneration in the developing world is particularly suitable for new power plants.

In conventional power plants, a number of efficient coal combustion technologies have entered the market and other combustion technologies seem to be close to commercial viability. It can be expected that natural gas combined technologies and natural gas combined cycle technologies, which are currently being used widely all over the world, will enable greater efficiency improvements. Combined cycle gas turbines can increase plant efficiency by up to 55%; future improvements in design and material use may lead to extra productivity gains and allow for higher operating temperatures.

In recent years, one of the most important technologies rapidly spreading all over the world has been combined cycle power plants. The operating principles of these plants are based on two main principles. The first one consists of completely gas turbines. At this stage, compressed air is compressed in a compressor and natural gas (which may be fuel oil, other petroleum derivatives or biomass fuels) is burned in the combustion chambers with the air at high pressure and temperature, resulting in combustion and very high temperatures (usually > 1000 °C). A gas turbine is provided to rotate the turbine, which produces electricity in the generator located on the same shaft as the turbine.

In the second stage, combustion waste gases are passed through heat exchangers to produce steam, thermal energy is converted to the rotational movement by the expansion of the superheated steam as it passes through the steam turbines, which in turn produces electricity in a similar manner.

The following figure shows the energy cycle in a natural gas combined cycle plant (Fig. 4.8).

This approach in natural gas, diesel, and fuel oil combined cycle plants can be used in coal plants with a slightly different approach, which is called Integrated Gasification Combined Cycle-IGCC. The reason that coal cannot be directly brought into the combined cycle is that the ashes formed by combustion stick to the gas turbine blades and make the blades unusable in a short time. As a first step, the primary fuel, which is coal, is gasified in integrated gasification cycle technology.

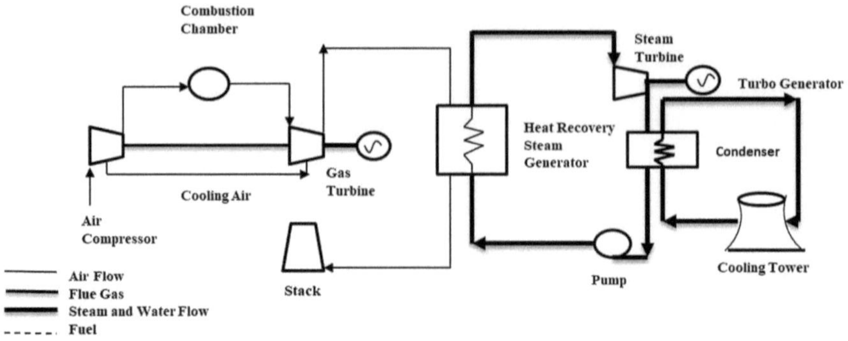

Fig. 4.8 Simple flow chart of natural gas combined cycle power plant

Coal gasification can be carried out in four different ways (fixed bed, fluidized bed, drag bed, and molten bed). In fixed bed gasification, it is possible to obtain high efficiency in terms of heat and to produce gas at the high pressure. Drift bed and fluidized bed gasification approaches allow the use of all types of coal, while gasification in the molten bed is advantageous in that it does not limit the coal size.

The synthesis gas (CO-H_2) obtained after the coal gasification step is cleaned. The cleaned gas is burned in a similar way to the combined cycle and the superheated steam coming from the waste heat boiler is evaluated in the steam turbine by reducing the pressure. In this way, a combined cycle is carried out. In such plants, there are gas cleaning units where coal gas is adapted to environmental conditions in order to have less NO_x and SO_x components. By passing coal gas through these units, these plants reduce the damage to the environment. As a result of passing the coal gas through the gas cleaning units, it is possible to clean the sulfur by 95% by various desulphurization methods. In addition, the formation of nitrogen oxide is minimized by diluting the coal gas before entering the gas turbine. Thus, conversion of small amounts of nitrous oxide compounds to elementary nitrogen is achieved successfully.

When combined coal gas fired combined cycle power plants are compared with conventional power plants, a 12% saving is achieved in the power plant area, 20% in water use and 7% in coal consumption due to increase in efficiency. With the measures taken, it is possible to significantly reduce flue gas emissions. Investment costs are lower than coal-fired power plants. In addition, the unit energy cost in combined cycle power plants with coal gas fuel is lower than those with flue gas treatment plants.

The investment cost of the integrated gasification technology is higher than that of the natural gas combined cycle technology because an additional coal gasification unit is needed in this technology; however, in countries where the cost of coal as fuel is significantly lower than natural gas, it is possible to close the investment cost gap in a short time.

Another advanced technology related to coal is fluidized bed combustion technology.

This technology is called a fluidized bed, since the bed, in which all particles remain suspended in the air, exhibits fluid properties. The reason for the fluidization in combustion is that oxygen is supplied sufficiently, and combustion is carried out with high efficiency.

The amount of fuel in the fluidized bed is very low, and the ratio between the fuel and the bed material (usually the fuel itself) is mostly 5%. Because of this feature, even very low-quality fuels can be burned with high combustion efficiency in the fluidized bed. Peat, lignite, hard coal, coal mining waste, urban waste, heavy oils, biomass solid fuels can also be used.

Other advantages of combustion in fluidized bed are the high combustion efficiency obtained by the homogeneous distribution of coal particles into the boiler. Another advantage of this technique is that the SO_2 produced as a result of combustion in the 800–900 °C range is kept with limestone fed to the combustion chamber without the need for an additional flue gas treatment plant.

In fluidized bed technology, the high heat transfer coefficient in the bed (3–5 times higher than conventional boilers) makes the boiler dimensions smaller than conventional boilers. The relatively low combustion temperature prevents ash from sticking to the heat transfer surfaces and the availability of the boiler reaches up to 98%.

The diagram of a fluidized bed boiler can be seen in Fig. 4.9.

Pre-combustion coal improvement technologies are another important issue to be considered in terms of plant efficiency. Reducing the amount of mineral, moisture, sulfur, and ash by making coal useful (enrichment) increases the efficiency of thermal consumption. Treatment of low-quality (low grade) coals prior to combustion means the removal of contaminants such as sulfur and sometimes ash from coal, which is usually carried out by physical methods.

4.4.2.3 Energy Efficiency in Buildings

The energy consumed in buildings worldwide has always played an important role in total energy consumption. Approximately one-third of the total final energy consumption of IEA member countries occurs in residential and commercial buildings. Residential and commercial buildings account for only 11% of oil demand and consume less oil than the transport and industrial sectors. However, their share in electricity demand is around 60% across the IEA and at least 40% across the EU. In many IEA member countries, space heating has the highest share in final energy use in residential buildings; similarly, water heating is one of the important items in energy consumption.

Large amounts of energy are saved by various technical measures to be applied in buildings. During the project phase, designs should be carried out for the stack and installation pipes in accordance with the physical conditions. Including the correct application of the external insulation, preventing of cooling flue gases and freezing of the pipes. In addition, importance is given to the design of the engine rooms for energy saving, the correct selection of boiler capacities, and the correct use of heat

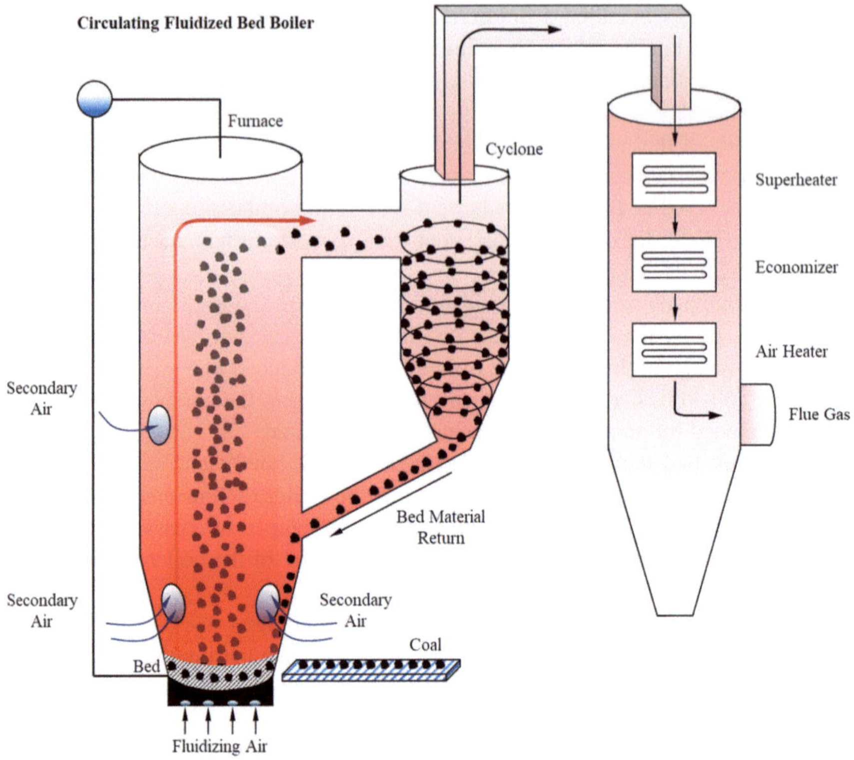

Fig. 4.9 The diagram of a fluidized bed boiler

recovery units (plate, drum, or serpentine) are the main savings areas that can be addressed in this area.

Improving the energy efficiency of the building's outer shell (walls, roof, floor, and frames), which is one of the most important pillars of energy efficiency in buildings, is a matter of increasing the thermal resistance by decreasing the heat transfer coefficients of the building elements.

In many countries, the highest heat transfer coefficients of walls and roofs have been standardized to limit the heat lost by housing from the building envelope. One of the most important studies carried out in this field is the foam insulation approach.

The aim of thermal insulation is to reduce the heating energy consumption by reducing the escape of building heat in winter and to maintain a balanced temperature environment throughout the interior. One of the tools used for this purpose in buildings is double glazing units.

Thermal insulation can be defined as a resistance, which prevents heat loss in hot environments and heat gain in cold environments. In the world literature, materials, which are below 0.065 W/mK, are accepted as thermal insulation material and materials whose thermal conductivity coefficient (λ or k) is above this value are accepted as building material.

It is important that the heating devices of the buildings are equipped with advanced systems and that the architectural designs be made in a way to minimize losses in the heat transfer devices. Applications that are more efficient are possible in the processes of generating and transmitting this heat as well as preserving the heat in the building.

90/70 °C hot water systems, which were preferred in the twentieth century, have turned into low temperature operating systems in buildings including medium capacity district heating, and zoning, pressurization, and flow controlled pumping systems have gained importance in heat transfer. The use of thermostatic valves has become widespread. Especially in heat loss and gain calculations, more advanced and sensitive methods based on computer programs have begun to be used. It is seen that the installed power can reduce the energy consumption by 30–35% and the operating costs can be reduced by 45–50%.

General measures related to the electrical installation for energy efficiency in buildings can be summarized as follows:

- Compensation installation to electrical energy input,
- Use of materials complying with standards,
- Fluorescent bulbs should be preferred instead of incandescent filament bulbs,
- Use of motion sensors in suitable buildings,
- Using photocells that are adjusted to daylight in the field lighting,
- Establishment of computer-controlled automation systems for energy saving in buildings with very special features.

With the comfort provided by increasing prosperity and developing technology, it is an expected development that energy-consuming devices will be used more in houses and other buildings. With the rising level of income, it is highly likely that the trend towards luxury will increase, leading to an increase in energy consumption; because it can be observed from the society behaviors that energy saving is not a priority compared to concepts such as comfort, health and/or effective use of time. One of the reasonable ways to minimize this disadvantage is to raise efficiency standards in household electrical appliances with the help of information and communication technologies.

Information and communication technologies have a wide range of applications. Minimizing surplus consumption in heating, ventilation, and lighting systems by means of sensors that detect the presence of people anywhere is one of the most common examples of these applications for now. Designs in which the same systems can be applied even to the rooms inside the house, whereby heating or ventilation is realized according to the perception of the sensor, has been in the interest of modern architecture as options that do not compromise on comfort but also provide energy efficiency.

Another important pillar of energy efficiency studies in buildings is energy labeling. The labeling method, which can be applied to both existing and new buildings, is of great importance in terms of demonstrating the energy performance of the buildings and encouraging the property owners to work towards energy efficiency.

It is possible to summarize the basic reasons for establishing labeling systems as follows:

1. The energy label provides information to the consumer (the person who will buy or rent the building) about the performance of a product (building).
2. The customers of the product (occupants, buyers, or tenants) make their decisions rationally; they use this information in the decision-making process along with operating costs and product quality.
3. As a result, each actor reassesses his preferences. Residents make their investments considering the comfort conditions. Future buyers or tenants prefer energy efficient buildings, considering their expenditure for energy. Sellers invest to make buildings energy efficient and add a certain amount to the rent/sale price to amortize this investment cost over a period. Investments made at the end of this cycle naturally increase the value of the building.

4.4.2.4 Energy Efficiency in Industry

It is a generally accepted assumption that the most important factor triggering energy efficiency policies is the energy crisis in the 70s and the resulting price increases. However, as the harsh competition in the market forced the production costs to decrease, the firms have started efforts to provide energy savings and energy efficiency related to handling the matter in a very effective way.

Despite this fact, it is not possible to say that the improvements in energy efficiency are solely due to the desire of firms for profit and that they are self-generated; behind these developments are the policies, institutions, organizations, and detailed technical studies supported by the states in the world.

In a world dominated by fierce international competition, the main energy-intensive industries (iron-steel, aluminum, chemicals, cement, paper, etc.) can only survive when they adapt their energy efficient production processes. These industries have sufficient incentives to engage in energy efficiency programs without state aid. However, especially in OECD countries, government aid has been frequently introduced to accelerate the modernization processes and increase the competitiveness of national firms.

Since the late 1970s, energy-saving activities have been concentrated particularly in the industrial sector in many industrialized Western countries. The activities focus primarily on these sectors:

- Even if the energy-saving potential is less than in other sectors, for example, private housing, these savings can be realized at relatively low cost (one-third of the average housing sector) and the investment pays for itself in less than three years on average,
- Energy-saving measures can often play a significant role in modernizing the industry and improving competitiveness, energy accounting, and understanding of energy costs is a better-known phenomenon in industry than in other areas.

Four types of measures can be taken to achieve productivity gains in the industry:

1. Better management, operation, and maintenance practices, often in connection with the industrial organization's efforts to improve energy accounting and metering,
2. Simple investments, such as control systems and insulation, often requiring low spending and repayment can be realized in less than a year,
3. Equipment renewal, addition of new technologies, energy substitutes, and similar investments that require large expenditures,
4. New industrial processes.

Nevertheless, these measures alone are not sufficient, because various factors such as changing raw material properties, product types, product characteristics, climatic conditions, environmental effects, and capacity utilization also affect energy efficiency.

The prominent measures in energy efficiency applications are

1. The use of highly efficient engines in industrial facilities,
2. Prevention of leaks in the compressed air systems,
3. Taking compressor suction air from outside,
4. Heating of combustion air,
5. Heat recovery from used fluid,
6. Recovery of condensate by sending it to the boiler,
7. Insulation of hot and cold surfaces,
8. Giving importance to internal and external insulation,
9. Reducing idle time,
10. Speed control with frequency control of drive motors, fans, and pumps,
11. Reducing heat losses from the boiler surface,
12. Improving steam systems,
13. Insulation of steam pipe systems,
14. Improvements of the furnaces,
15. Correction of the electric power factor in the facilities,
16. Control of excess air,
17. Reducing the amount of bluff in boilers,
18. Heat recovery from blowdown.

One of the most widely applicable of these measures is to increase engine efficiency. As it is well known, electric motors, like all motors, cannot convert all the energy they use into mechanical energy.

The ratio of the mechanical power output of the motor to the electrical power drawn is called motor efficiency and ranges from 70 to 96% depending on the type and size of the motor. In addition, the efficiency of the motors operating at partial load is low. These efficiencies also vary from engine to engine. For example, one motor can operate at full load efficiency at 90%, half load at 87%, and 1/4 load at 80%, another motor with the same characteristics at full load at 91% and at 1/4 load at 75% efficiency.

The efficiency of the advanced motors at various loads is higher than that of the old-type technology and therefore more energy efficient. The use of more efficient motors in the industry provides great energy savings.

Compressed air systems are another area where significant energy savings can be achieved through improvement in work. Because energy loss due to insufficient installation and controlled maintenance in compressed air systems can reach 50% of the energy consumed by the compressor and half of this can be prevented with simple operating measures. Considering that almost all industrial plants today have compressors and their failure can halt or slow down production, the magnitude of potential energy savings in compressed air systems can be seen clearly.

In particular, studies on boiler components and new boiler technologies have led to significant progress towards more efficient operation of boilers. For example, since it is seen that the temperature of the flue gas thrown from the flue in conventional boilers adversely affects the efficiency of the boiler. Air preheaters have been developed with the heat exchangers called recuperator. Lost heat is recovered by means of these systems and thus, the amount of fuel consumed is improved by reducing the lost energy.

Fluidized bed combustion technology, which provides high efficiency in power generation plants, is another important innovation in the industry. Fluidized bed technology is known to give successful results especially in hot water, steam, and hot gas production processes for drying.

4.4.2.5 Energy Efficiency in Transportation

Efforts to achieve economic growth, environmental development, and energy security together in the transport sector are adopted by many governments as the main objectives. Many organizations in the world are carrying out various activities in order to provide sustainable transportation and to realize these basic purposes. Governments and public institutions are developing transport sector policies in parallel with these studies.

It is possible to mention three main factors that affect energy efficiency in transportation. The first is the quality of the need. There are many side factors depending on whether the goods to be transported or passengers; like settlement plans in cities, the characteristics of the roads between work areas and living areas, the necessity of timely transportation of loads, distribution management (shortest road applications, etc.). The second main factor is the mode of transportation. The means by which the passenger or load is transported is a decisive factor in energy efficiency.

The third main factor is the characteristics of the vehicle and the way the driver uses it. The main determinants of the vehicle's characteristics are the model of the car, energy consumption and saving capacity, engine power, engine performance, and maintenance status. All these elements need to be taken into account in the process of developing energy efficiency strategies in transportation.

Today's forecasts show that energy consumption in transportation will increase in the coming years. Therefore, one of the most important policies for energy efficiency is to reduce the energy intensity against the expected energy consumption in the sector.

The most important development that necessitates energy efficiency in the transportation sector is the rapid depletion of oil reserves in the world. Although the oil reserves have increased according to today's values comparing to the forecasts in the 1970s and 1980s, the increase in the oil consumption rate indicates the rapid depletion of the proven reserves to be put into use.

In spite of the rapid depletion of the oil supply, no significant reduction in oil dependence in the transport sector is expected. The fact that alternative sources have not yet become commercially economic has a major role in this situation. For example, in OECD countries where the transport sector has a significant share in total oil consumption, the share of the transport sector in oil use is expected to grow steadily.

Passenger cars, trucks, and vans used in domestic transportation are responsible for half of the transportation oil demand in OECD countries, while the demand for airway fuel is the fastest growing item in this matter.

The share of transportation in oil use will continue to increase for some time, and this is a common point of almost all estimates. However, it is predicted that there will be a pause or even decline in the oil demand of the transportation sector with the developing new technologies over time.

In other respects, it is claimed that the existing measures and improvement in work may reduce the energy intensity in the sector in the long term. The energy information administration (EIA) estimates of transport energy density across different regions can be seen graphically in Fig. 4.10(a) Delivered transportation energy consumption by OECD and non-OECD countries in between the years 2012–2040 and (b) World transportation sector delivered energy consumption by considering energy source, in between the years 2010–2040 [20].

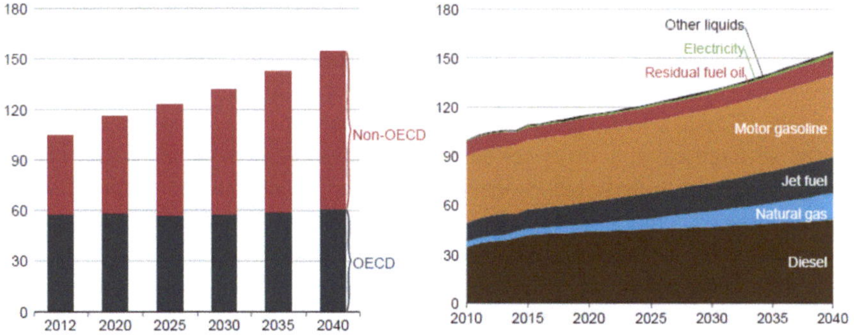

Fig. 4.10 **a** Delivered transportation energy consumption by OECD and non-OECD countries 2012–2040 (Quadrillion Btu). **b** World transportation sector delivered energy consumption by considering energy source, 2010–2040 (Quadrillion Btu). *Other liquid fuels include natural gas plant liquids, biofuels, gas-to-liquids, and coal-to-liquids. *Source:* U.S. Energy Information Administration | International Energy Outlook 2016 [20]

One of the assumptions on which the transport sector predicts a decrease in energy intensity is the prediction that new fuel and vehicle technologies will enter the market. Both the improvements in the efficiency of the existing technologies, especially the combustion efficiency of the engines, and the improvements in the new fuel technologies strongly support this prediction.

There are still many studies in many countries around the world for energy efficient use of vehicles. The most important points of these studies are to obtain the more efficient operation of the engines and to realize the increase of the combustion efficiency of the fuel.

The studies on vehicle technologies are not limited to the researches on the engine. Car body technologies for using low-weight bodywork and reducing body air resistance, wheel surfaces with low rolling resistance, friction reducing materials, and designs are also being studied.

Another field of study is electric vehicles. Electric vehicles, which have been studied for many years, are becoming increasingly widespread in both civil and defense fields. Electricity remains a minor fuel for the world's transportation energy use, nevertheless its importance in passenger rail transportation remains high: in 2040, electricity will account for 40% of total passenger rail energy consumption. The electricity share of total light-duty vehicle energy consumption increases to 1% in 2040 in the mentioned case, since increasing sales of new plug-in electric vehicles step in the total light-duty stock [20].

Work is underway for the use of electric vehicles in land, air, and sea transport. The most prominent developments in these studies are experienced in electric road vehicles and especially in electric railway vehicles. Maglev (it is derived from magnetic levitation, (magnetic lift)) trains hanging in the air with electromagnetic repulsion force exceeded the speed of 500 km/h. Superconducting technology is used in these trains prepared for Berlin-Hamburg line in Germany and Tokyo-Osaka line in Japan. In electric land vehicles, however, there is a need for high-capacity batteries and advanced power electronics circuits that allow the batteries to be charged at regular intervals. Studies in this area are expected to take longer, but the developments in fuel cells, in particular, have reached to a remarkable level.

It is estimated that hybrid and fuel cell vehicles will have better road performance than today's results in the future. Following the spread of these instruments in the market, differentiation of oil prices applied for efficient and inefficient vehicles may direct consumers to these instruments. Government incentives for financial incentives, such as receiving less tax from firms producing such energy efficient vehicles, will also encourage manufacturers to manufacture such vehicles.

In today's conditions, it is stated that hybrid and fuel cell vehicles need to be generalized for applications that are more commercial. One of the most important obstacles to the spread of fuel cell vehicles is the infrastructure problem for many countries. The gasoline fuel cell infrastructure is now available, but it must also be developed for methanol or hydrogen. As the high-tech vehicles have not yet become widespread today, the most important measure that governments can take in this regard is to try to reduce the use of large and high-fuel vehicles with a deterrent approach, such as high taxes. Another important step towards improving energy efficiency in transportation is the expansion of the use of public transport systems. In this area, especially urban transport systems gain significant importance.

One of the prominent alternatives in transportation is cycling in the downtowns in some countries, especially in populated and/or low-income countries. In this context, in cities where there are roads and routes reserved for bicycles, drivers avoid driving over short distances and they provide significant savings in energy consumption (Fig. 4.10).

4.4.2.6 Policy Recommendations

The global primary energy intensity observed in 2010–2016 was related to greater energy efficiency in large energy-using countries. China, India, Japan, Northern America, and Europe all expedited their related processes or maintained their policy goals regarding energy efficiency. There are three wide scale types of energy efficiency policy used by governments to direct progress:

(1) Regulation-mandatory necessaries to advance energy efficiency or to meet specified goals or standards, which comprise minimum energy performance standards for appliances and equipment, vehicle fuel efficiency standards, building codes, and mandatory energy efficiency advancement objectives for industrial firms or sectors;
(2) Incentives-financial or fiscal incentives to energy consumers to enhance efficiency;
(3) Information-labels, training, websites, and capacity building regarding the performance of products or ways to progress energy efficiency.

The extent of energy efficiency regulations at a global and national level is reflected by the percentage of final energy use that is covered by mandatory efficiency laws and standards. This demonstrates the energy use of equipment, appliances, and vehicles that were required to conform to minimum energy performance standards before being sold. Figure 4.11 gives the detailed information about incremental growth in energy use covered by mandatory efficiency policies globally, 2010–2016 (on the left), and coverage in the 10 countries with the highest total primary energy supply (on the right) (EPR 2019) [21].

Financial and fiscal incentives to enhance energy efficiency are policy tools being used by governments to supplement direct regulation and encourage greater levels of efficiency. In 2017, incentives for energy efficiency in some of the world's major economies totaled $27 billion. These incentives included subsidies, grants, loans, tax relief, and rebates, with the transport sector being the largest single receiver, due to $8 billion in incentives for the adoption of electric vehicles.

National government incentives for energy efficiency by sectors (on the left) and types (on the right), 2017 (EPR 2019) can be seen in Fig. 4.12 [21].

Grants, subsidies, and tax relief represent almost 80% of the energy efficiency incentives in the countries examined. Grants and subsidies are used to reduce the

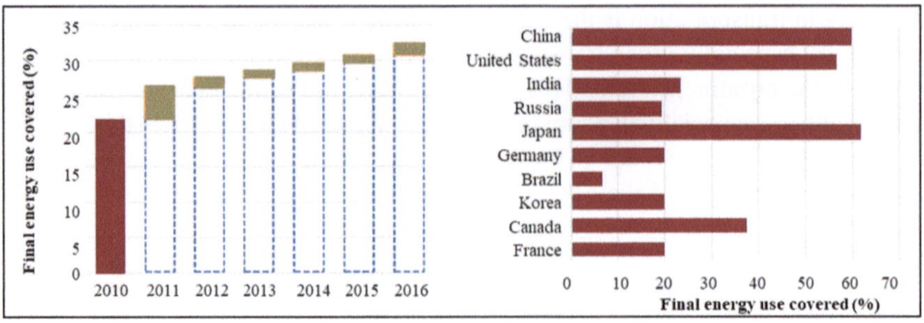

Fig. 4.11 Incremental growth in energy use covered by mandatory efficiency policies globally, 2010–2016 (on the left), and coverage in the 10 countries with the highest total primary energy supply (on the right) (EPR 2019) [21]

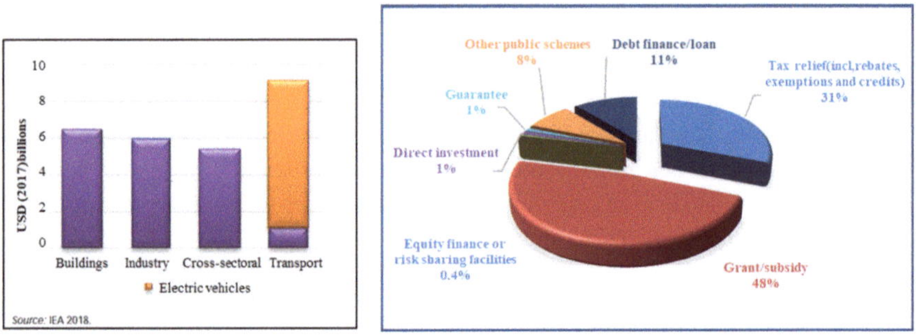

Fig. 4.12 National government incentives for energy efficiency by sectors (on the left) and types (on the right), 2017 (EPR 2019) [21]

capital cost of energy efficient appliances, therefore, make their purchase more attractive for the consumers. Fiscal incentives in the form of tax relief are intended to appeal to consumers, particularly businesses, by lowering their tax bills. Although other forms of incentives based on debt or loan finance are less featured, there are a growing number of incentives to reduce the risk related to energy efficiency projects.

Financial incentives for energy efficiency are often supplied through market-based instruments, in which a government uses regulation to determine a desired outcome, typically energy savings, and then establishes a framework for market actors to deliver the outcome.

4.4.2.7 Energy Efficiency Investment

The increasing amount invested in more efficient buildings, appliances, vehicles, and industrial equipment totaled $231 billion, the majority of which was in the buildings sector in 2016 (Fig. 4.13) [21]. The presence of low-cost and iterable energy efficiency measures, such as lighting upgrades and improvements to heating, ventilation, and

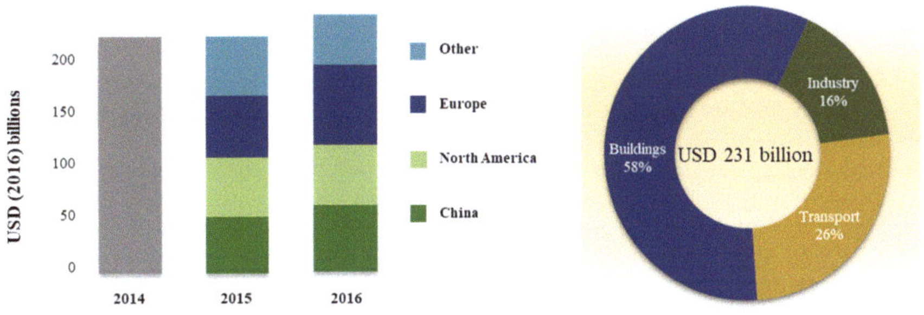

Fig. 4.13 Energy efficiency investment by regions and sectors, 2016 (EPR 2019) [21]

air-conditioning system performance, contributes to the buildings sector receiving the most augmenting investment.

4.4.2.8 Conclusions for Energy Efficiency

In the future, governments' policies will continue to be central to global efforts to actualize the profits of enhanced energy efficiency. Supportive policy decisions are implemented in developed and developing countries. Important actions include

- Applying and fortifying mandatory energy efficiency policies that push equipment, appliances, and vehicles toward to best possible technologies.
- Supplying projected and convenient fiscal or financial incentives to encourage energy users to maintain greater grades of efficiency.
- Leveraging the power of the market, through implementation of market-based mechanisms, to deliver energy efficiency advancements minimum cost.
- Providing aimed and high-quality information and capacity-building measures, to maximize market availableness to deliver higher levels of energy efficiency.

Government policy will also need to create an environment that is suitable for the development of new finance and business models, needed to raise levels of energy efficiency investments [11, 22]. One factor that will have an incremental influence on energy efficiency in all sectors is the growth and application of digital technologies. Digitalization summarizes an increase in the amount and accuracy of energy use data, a developed ability to conduct data analysis, and improvements in connectivity, which improve the interaction between devices and consumers, providing greater control, and flexibility of use.

Digitalization is creating new business models for the delivery of energy efficiency that capture benefits both for individual consumers and the broader energy market.

References

1. Kaya D, Güngör C (2002) Energy saving potential in industry-I, engineer and machine, 514, 20–30. (In Turkish: Kaya D, Güngör C, Sanayide Enerji Tasarruf Potansiyeli-I, Mühendis ve Makina, 514, 20–30)
2. Kaya D, Güngör C (2002) Energy saving potential in industry-II, engineer and machine, 515, 36–44. (In Turkish: Kaya D, Güngör C, Sanayide Enerji Tasarruf Potansiyeli-II, Mühendis ve Makina, 515, 36–44)
3. Kilic MK, Abut N, Canka Kilic F, Coban V, Kaya D (2018) Capacity increase and energy efficiency improvement studies in a biogas power generation plant. Int J Ecosyst Ecol Sci (IJEES) 8(1):45–56, ISSN: 2224-4980, https://doi.org/10.31407/ijees
4. Kaya D, Ozturk HH, Taylan O, Çanka Kiliç F, Alidrisi H (2017) An evaluation of energy and exergy efficiencies for a biogas cogeneration plant. Int J Ecosyst Ecol Sci (IJEES) 7(1):27–34
5. Kavak K (2005) Energy efficiency in the world and Turkey and investigation of energy efficiency in Turkish industry, Publication No: DPT: 2689, General directorate of economic sectors and coordination, September 2005. (In Turkish: Kavak K, Dünyada ve Türkiye'de Enerji Verimliliği ve Türk Sanayiinde Enerji Verimliliğinin İncelenmesi Yayın No: DPT: 2689, İktisadi Sektörler ve Koordinasyon Genel Müdürlüğü, Eylül 2005)
6. Kilic MK, Abut N, Çanka Kiliç F, Eyidoğan M, Kaya D (2018) An examination of measuring and control systems in an industrial biogas plant in Turkey. International conference of ecosystems (ICE2018), International conference of ecosystems (ICE2018), paper No: 008, 64–69, ISBN: 978-9928- 4443-1-8
7. IEA, IRENA, UNSD, World Bank, WHO (2020) Tracking SDG 7: The Energy Progress Report. World Bank, Washington DC. © World Bank. License: Creative Commons Attribution-NonCommercial 3.0 IGO (CC BY-NC 3.0 IGO)
8. https://trackingsdg7.esmap.org/data/files/download-documents/chapter_4_energy_efficiency.pdf
9. Öztürk HH (2011) Energy management in herbal production, Hasad Publications, İstanbul/Turkey, ISBN: 975-8377-78-7. In Turkish: Öztürk HH, Bitkisel Üretimde Enerji Yönetimi, Hasad Yayıncılık, Istanbul, ISBN: 975-8377-78-7
10. Öztürk HH (2013) Knowledge of Climate and climatic measurement technique, Birsen Publisher Code No: Y.0028, ISBN: 978-975-511-590-0, 2013. (In Turkish: Öztürk HH, İklim Bilgisi ve İklimsel Ölçme Tekniği, Birsen Yayınevi Kod No: Y.0028, ISBN: 978-975-511-590-0)
11. IEA Key World Energy Statistics (2019) All rights reserved
12. Rebuplic of Turkey Ministry of Energy and Natural Resources, World and Turkey's Energy and Natural Resources View, Directorate of Strategy Development, No: 15, Ankara/Turkey, As of 1st January 2017. (In Turkish: T.C. Enerji ve Tabii Kaynaklar Bakanlığı, Dünya ve Türkiye Enerji ve Tabii Kaynaklar Görünümü, Strateji Geliştirme Başkanlığı. No: 15. Ankara/Türkiye. 1 Ocak 2017)
13. Energy Situation in Turkey and the World, GAZBİR, (In Turkish: Dünyada ve Türkiye'de Enerji Durumu, GAZBİR). https://www.gazbir.org.tr/uploads/page/Dunya-ve-Turkiye-Enerji-Gorunumu.pdf. Accessed on 20 Dec 2019
14. BP Statistical Review of World Energy 2015
15. BP Statistical Review of World Energy 2020
16. Smil V (2017) https://ourworldindata.org/energy-production-and-changing-energy-sources
17. BP Statistical Review of World Energy 2017
18. SE4ALL database, IEA and World Bank. Release Date: December 20, 2013 and Last Updated: June 30, 2018
19. Kilic MK, Abut N, Canka Kilic F, Coban V, Kaya D (2017) Capacity increase and energy efficiency improvement studies in a biogas power generation plant. 7th International Conference of Ecosystems (ICE2017) Tirana, Albania, June 02.05.2017

20. U.S. Energy Information Administration I International Energy Outlook 2016. DOE/EIA-0484 (2016) I May 2016. https://www.eia.gov/outlooks/ieo/pdf/transportation.pdf
21. EPR, IEA, IRENA, UNSD, WB, WHO (2019) Tracking SDG 7: The Energy Progress Report 2019, Washington DC/USA
22. Kaya D, Öztürk HH (2014) Sanayide Enerji Yönetimi ve Enerji Verimliliği, Umuttepe Yayınları, Yayın No: 114, Mühendislik Dizisi: 12, Kocaeli/Türkiye

Chapter 5
Energy Performance Certificate

An Energy Performance Certificate (EPC) is a document, which contains minimum energy requirement of buildings, energy consumption classifications, insulating properties, and efficiency of heating/cooling systems. The purpose of getting EPC is using energy and energy sources efficiently at buildings, avoiding energy wastage, and protecting environment (see Figs. 5.1 and 5.2) [1, 2].

Energy performance certification ensures by virtue of rating original buildings (public, commercial, or residential) on how efficient (or inefficient) they are respecting the amount of energy needed to procure users with expected degrees of performance and well-being. The degree of efficiency can vary based on many factors such as the design of the building, regional climate; construction methods and materials; systems installed to provide air conditioning, heating, ventilating, or hot hygienic water; and the equipment and appliances needed to support the operations of the buildings and their users.

Apparently, certification is a complicated procedure, which obligates profound knowledge of building units. It also influences increasing recognition of the need to think of buildings as "integrated systems", rather only the sum of their sections.

Energy certification of buildings generally associates three main stages:

- A competent assessor carries out **the assessment of the energy performance** of a building by using a nominated methodology.
- **The issuance of a certificate** grading the building's energy performance that comprises, in some instances, information on possible advancements presumably bringing energy savings.
- **The communication of this information** to stakeholders by means of publication of the certificate.

Certification is often used due to the finishing of new buildings through proving conformity with building codes. In the event of existing buildings, certification is used to contrast similar buildings and to appraise the degree to which an older building fails

D. Kaya et al., *Energy Management and Energy Efficiency in Industry*,
Green Energy and Technology,
https://doi.org/10.1007/978-3-030-25995-2_5

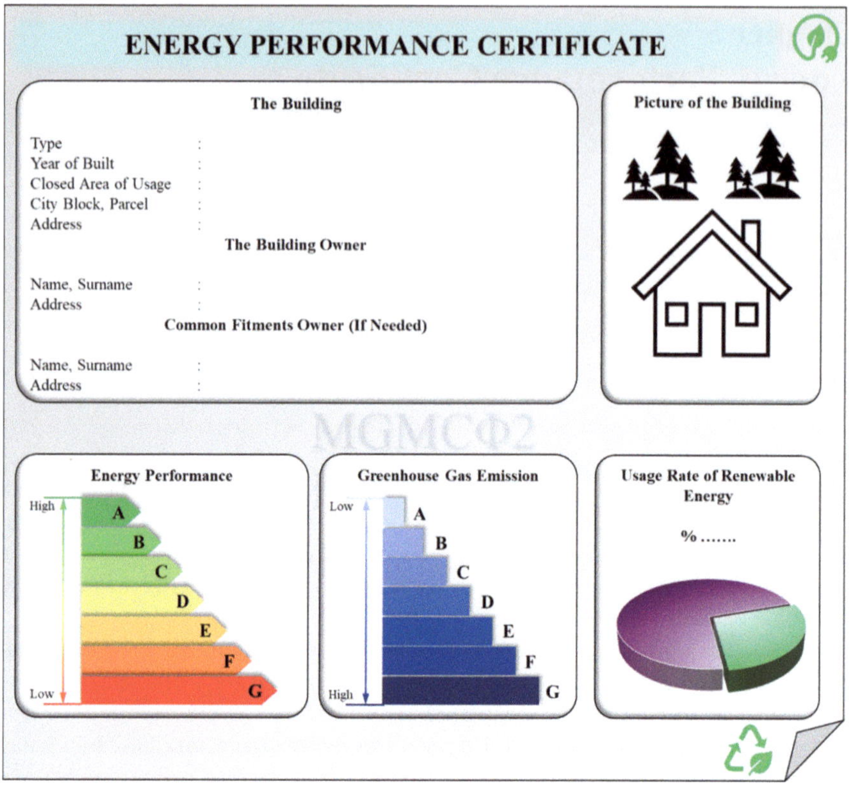

Fig. 5.1 An example of the energy performance certificate (EPC) (I)

codes that have been introduced since the time of its building. As much of the existing structure stock was built before energy efficiency became a center of government policy, certification of existing buildings can do more than procure ratings: it can describe precautions to develop energy performance [3].

Energy performance certificates are very important to all stakeholders in the construction industry. They supply a mechanism by which probable clients and hirers can contrast the energy efficiency of different buildings or the energy rating across a series of similar buildings. Certification besides contrasts existing buildings to recent building codes, ensuring a way to contrast existing and new buildings. In this respect, certificates are often preconditioned a very important piece of information in the course of purchasers are making decisions on property buys or leases for either existing or new buildings. Certificates can also be important to dealers and possessions possessors: customers/tenants might be captivated by the chance to save on energy bills by buying or renting a more efficient building. On the other hand, they may decide to buy/rent a less expensive building, knowing beforehand that it is less efficient but can be developed by means of upgrades described on the certificate.

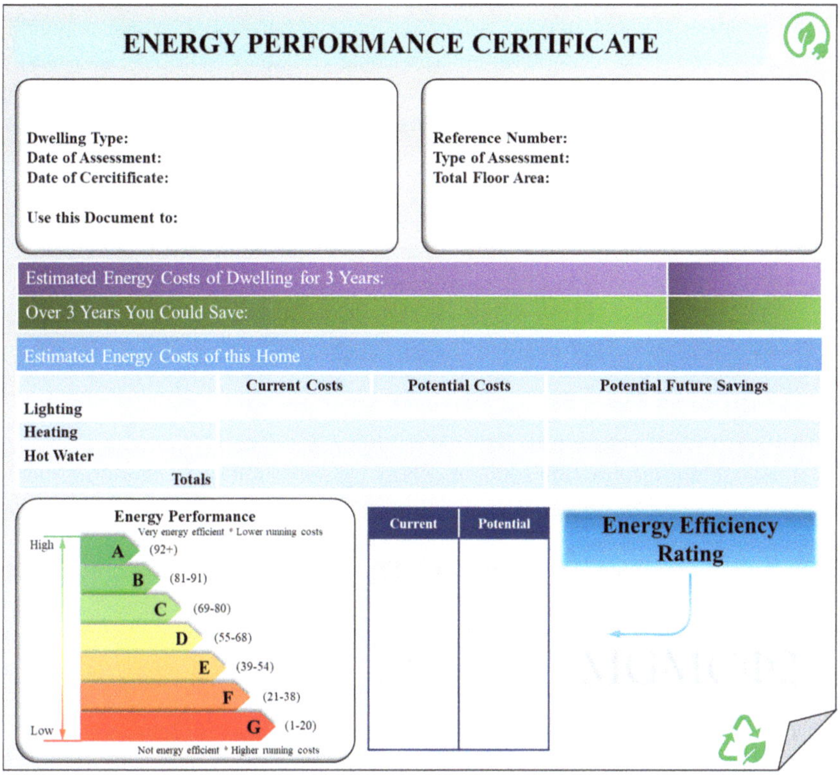

Fig. 5.2 An example of the energy performance certificate (EPC) (II)

5.1 Energy Performance Assessment

To prepare an energy certificate, it is first necessary to undertake an energy performance appraisal of the building's features and systems. This is realized by a skilled expertise who collects information on the building's features and parts, in addition to its energy systems and energy utilization. An appraisal generally comprises

- The area, form, and other details of the construction,
- The solar, thermal, and daylight specifications of the building envelope and its air conductance,
- Installation of space heating and hot water supply, comprising of their efficiency, control, and responsiveness,
- Ventilating, air-conditioning systems and controls, and constant lighting,
- Renewable energy sources and fuel,
- Systems of lighting and installed equipment and appliances.

This information is input into an authorized calculation model, which appraises the building's energy consumption under regional climatic conditions. The energy expertise submits the evaluation and results to a centralized system that carries out an automatic check of the appraisal outputs and procures the certificate (either in paper form or electronically). If government agencies or regional authorities execute the system, it commonly comprises a quality control. With this way, the certificate is circulated by a centralized administration system that can demonstrate the effectiveness of the certification plan. This aids to establish stakeholder trust and increases the reputation of the certificate.

5.2 Content of the Energy Identity Certificate

EPC must contain information such as energy consumption of the building, insulating properties, efficiency of heating/cooling and energy classification. The energy performance certificate is issued that contains as the following information for the existing buildings (which have total usage area 1000 m^2 or above in Turkey, as an example) and buildings that are recently taken into operation.

- Guideline about building,
- Information of organization and organizer,
- Usage area of building (m^2),
- The purpose of the use of the building,
- The amount of energy consumption (kWh/year) that cooling, heating, air conditioning, ventilation, and sanitary hot water,
- The amount of primary energy that being needed according to each energy consumption (kWh/year),
- Classification of annual primary energy consumption per usage area of the building according to a reference scale that is ranging from A to G,
- The annual amount of greenhouse gases per usage area by generated from final energy consumption (kg CO_2/m^2year),
- Classification of annual greenhouse gas emission per usage area of the building according to a reference scale that is ranging from A to G,
- Lighting energy consumption value of the building,
- Energy class according to primary energy consumption,
- CO_2 emission class according to final energy consumption,
- The rate of the renewable energy usage of the building.

5.3 Preparation of the Energy Identity Certificate

Information such as energy consumptions of building, insulating properties, efficiency of heating/cooling system, energy consumptions classify is kept on the EPC as a minimum. EPC is prepared according to the following principles:

- EPC is prepared according to the EN 15217 standards,
- EPC is valid for 10 years (from the preparing date). At the end of this period, the energy performance certificate is rearranged according to the report will be prepared,
- EPC is organized according to the format, which has been determined on the national regulations,
- EPC is organized by authority foundation that is authorized to give energy performance certificate and municipalities approve EPC. This document is an integral part of the occupancy permit for new buildings,
- EPC should be prepared for the entire building. In addition, EPC can be prepared for detached section or different using area as optionally,
- The following parameters are prepared as the attachments of EPC within scope of the project; method of obtaining surface areas that is used in calculations, energy conversion coefficient, converting final energy to primary energy and energy consumptions scale, final energy consumptions, conversion coefficients of greenhouse gas emission, and emission of carbon dioxide,
- The conversion coefficients of greenhouse gas emission are expressed in kg CO_2 per final energy consumption.

5.4 Use of the Energy Identity Certificate

A sample of EPC (energy performance certificate) is kept by the building owner, building manager, board of directors and/or energy manager, and another copy is kept hanging easily visible in the place at the entrance of the building. In addition, when the building or detached section are sold or rented out, EPC is given to buyer or renter by owner.

EPC is renewed according to this regulation if any implementation is made to change primary energy needs of the building. The EPC has to be implemented in every building except for some special situations in the buildings depending on the country.

5.5 Energy Label

Energy label is the document, which contains information on energy consumption levels of the energy-consuming equipment. An example of the energy label can be seen in Fig. 5.3.

An example of the Energy Label in Fig. 5.3 contains the followings:

I. The name or trademark of the manufacturer shall be written.
II. The manufacturer's model description will be specified.
III. Energy efficiency class of the device will be determined in accordance with the relevant regulation with regard to labeling of energy-related products. The appropriate letter will be written in line with the corresponding arrow mark.

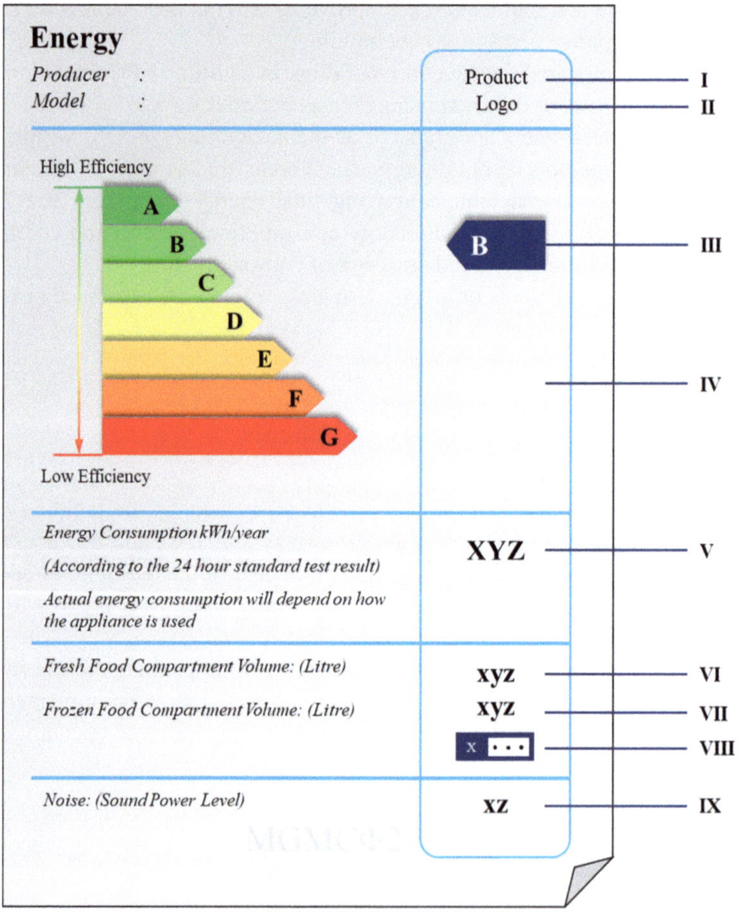

Fig. 5.3 An example of the energy label

IV. In case the European Community Council gives "Community environmental label award" to a product in accordance with the directive no. 880/92/EEC, the environmental award mark can be attached to the product, provided that the rules specified in the directive are complied with.
V. Energy consumption will be expressed in accordance with the standard specified in the relevant law and in kWh/year (24 h consumption × 365 will be specified).
VI. The net storage volume total of all food storage compartments that do not need to be specified with a star shall be written (operating temperature > –6 °C).
VII. The net storage volume total of all frozen food storage compartments that must be specified with a star shall be stated (operating temperature < –6 °C).
VIII. The star rating of the frozen food storage compartment will be indicated according to the specified standards. If a star is not required for this division, this line will be left blank.
IX. The noise level will be written in decibels.

Classification of EU energy efficiency label consists of seven groups based on annual energy consumption of a device [4]. Letter A shows minimum energy consumption class. If you buy an A class power tool, it will consume 45% less energy than average energy consumption. However, G class devices consume 25% more energy than average energy consumption. Thus, consumption of a power tool which has A, or B, or C energy class will be lower than the average energy consumption. In the future, all household electrical appliances will have to have labels that show energy consumption. In European Union countries, this classification was made for refrigerators and freezers.

The main purposes of energy labeling in the energy-consuming equipment are

- Providing information for the customer about the amount of energy consumption per year of the product (before the purchasing),
- Encouraging manufacturers to take precautions to reduce energy consumption of devices that they produce. In other words, to ensure the rational and efficient use of energy.

References

1. Kaya D, Öztürk HH (2014) Sanayide Enerji Yönetimi ve Enerji Verimliliği, Umuttepe Yayınları, Yayın No: 114, Mühendislik Dizisi: 12, Kocaeli/Türkiye
2. https://enerjiapi.etkb.gov.tr//Media/Dizin/SGB/Faaliyet_Raporlari/2019/496210-etkb_2019_yili_i%CC%87dare_faaliyet_raporu28.02.2020.pdf
3. Arkesteijn K, van Dijk D (2010) Energy performance certification for new and existing buildings, EC Cense P156, EC
4. IEA International Energy Agency (2010) Energy Performance Certification of Buildings. A policy tool to improve energy efficiency. Policy pathway. IEA Publications, 9, rue de la Fédération, 75739 Paris Cedex 15. Printed in France by Corlet, November 2010

Chapter 6
Energy Efficiency Services Sector

Energy efficiency services assist firms to implement new technologies and necessary precautions to have more efficient use of the energy by reducing the energy consumption and costs. Energy efficiency services comprise education of energy managers for the purpose of increase the energy efficiency, also preparing projects about energy audits, project implementation, and consultancy services. Energy efficiency sectors have economic, political, and cultural importance in many countries, and the range and diversity of these sectors increasing with each passing year. The development and strengthening of Energy Service Companies (ESCOs) mean also the improvement and growth of industrial sectors [1].

6.1 Energy Service Companies (ESCOs)

An ESCO is an energy service company, which procures a wide range of energy solutions consisting of designs and implementation of energy efficiency and energy savings projects, for the business processes betterment, energy conservation, energy infrastructure evaluating, energy supply and power production, and risk administration. Within the scope of the relevant legislation, the Ministry and the private companies that apply to the authorized institutions give an authority certificate. The companies that have a certificate of authorization offer the following services like organizing energy manager training, doing energy audits, preparing projects to increase efficiency and implementing projects, and consultancy. Energy service companies (ESCOs) have authorization certificate (Fig. 6.1), skilled staff, and suitable infrastructure to work in energy efficiency field.

D. Kaya et al., *Energy Management and Energy Efficiency in Industry*,
Green Energy and Technology,
https://doi.org/10.1007/978-3-030-25995-2_6

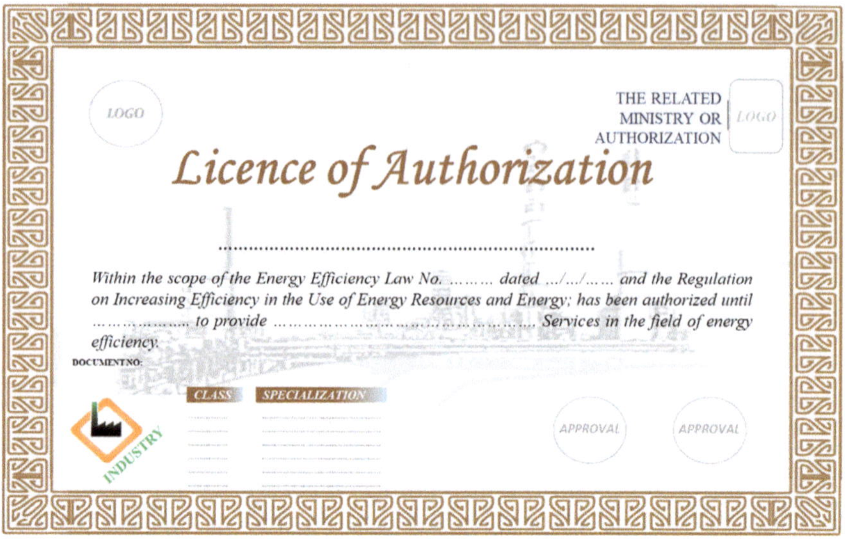

Fig. 6.1 An example of an authorization certificate for the industrial sector corporations [2]

6.1.1 Energy Efficiency Services

Education, survey, consultancy, and project services in the field of energy efficiency are being spread countrywide by giving authorization certificate to universities with specific qualifications, chamber of electrical engineers, chamber of mechanical engineers, and ESCOs. The legal entities that are given authorization certificates are defined as ESCOs and they can provide energy manager training for industrial enterprises and buildings, energy audit, project preparation, consulting for the implementation of project, and energy manager services for the buildings.

In case of having an agreement with the customer; ESCOs can prepare the projects for the implementation of the measures determined by the energy audit that has been realized earlier and these prepared projects are implemented in the customers' workplace by guaranteeing the energy-saving amounts. ESCOs provide services to increase energy efficiency. ESCOs provide project development, project implementation, project finance, and undertake maintenance repair costs (between 7 and 10 years periods) of projects in the final consumption sectors.

The main services provided by ESCOs are

- Determination and evaluation of energy-saving opportunities (energy audits),
- Preparation and implementation of the energy efficiency program according to the customer's needs,

- Management of all activities from the installation of the necessary systems to the monitoring of the project,
- Financing,
- Monitoring of the obtained savings rates,
- Supply the needs of education and maintenance/repair.

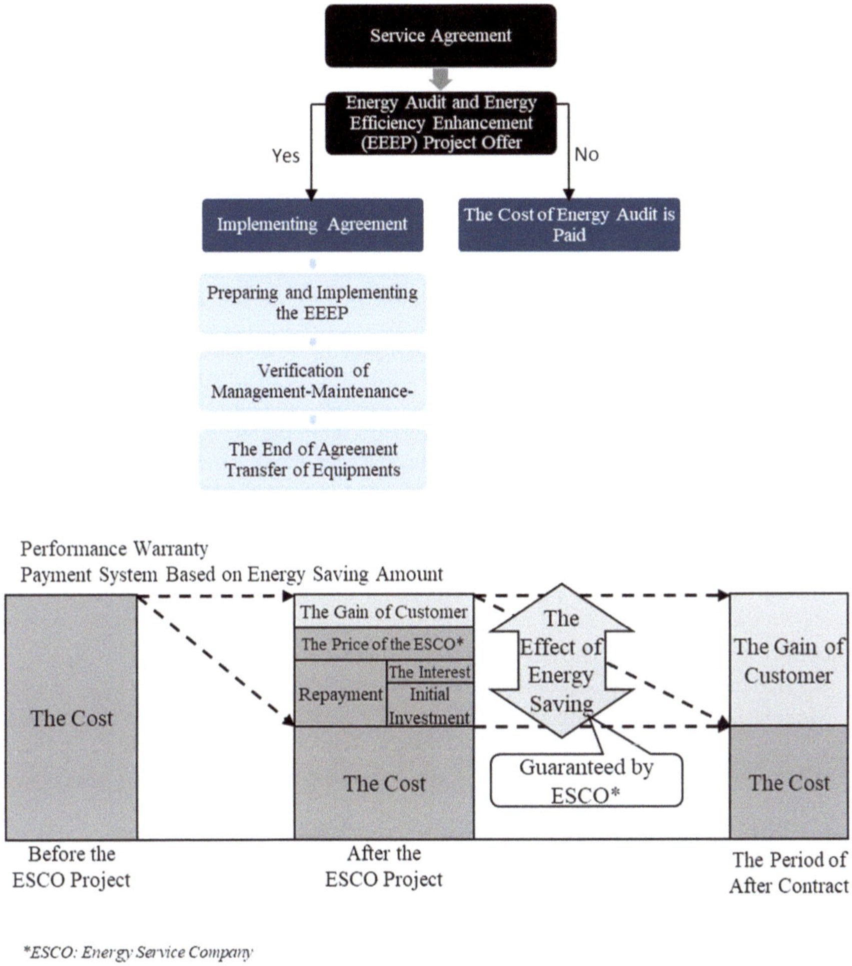

Fig. 6.2 Service models of ESCOs

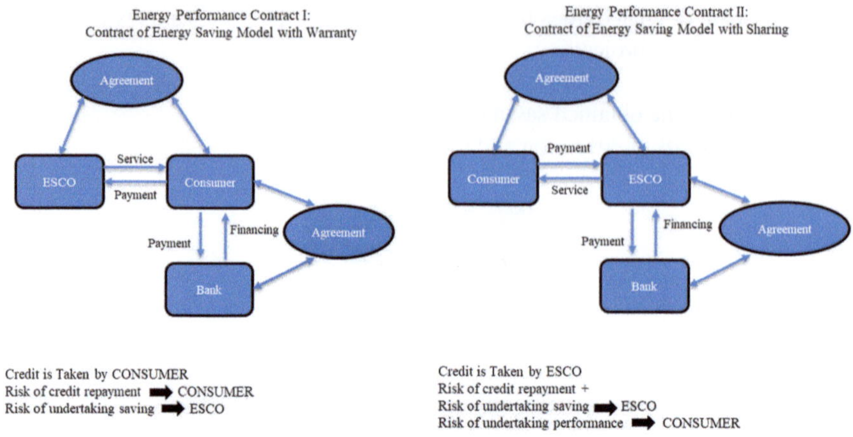

Fig. 6.3 Model of project financing for ESCOs

6.1.2 Energy Efficiency Service Models

Most of the ESCOs operating in many countries actualizing the followings in order to improve service standards, establishing national and international quality systems, monitoring the technological developments, establishing R&D units, creating the product range, attempting to establish foreign partnerships, tending to have qualified staff, and educating the staff.

The financing for investment can be provided by the client or ESCOs by using their own financial resources or can be provided by borrowing the necessary money amount from third parts. The savings provide the return on investment.

Service models of ESCOs can be seen in Fig. 6.2 [3].

Model of project financing for ESCOs can be seen in Fig. 6.3.

6.2 The Required Qualifications for the Measurements in ESCOs

Competencies to be required for the measurements in companies can be seen in Table 6.1.

Table 6.1 Competencies to be required for the measurements in companies

Flue gas	To be able to measure that of oxygen in the flue gas by doing that analysis of flue gas (including large flue), actual value of carbon monoxide (including high value), temperature of flue gas (including dry chamber temperature), velocity, particle, and CH_4. Measuring of Particle and CH_4 parameters are not considered as a mandatory criterion about giving an authority certificate
Monitoring heat/temperature	To be able to record that heat/temperature distribution as an image in order to detect heat losses on various surface areas
The heat transmission coefficient	To be able to measure or determine heat transmission coefficient of the constructional element, measuring humidity and temperature of wall surface
Liquid conductivity	To be able to measure electrical conductivity of boiler, feeding water, blowdown, raw water, etc. as μS/cm and (Total Dissolved Solids) TDS ppm units, measuring temperature values
Steam escapes	To be able to control existing steam traps in the steam systems, to be able to control or measure the amount of steam escapes
Temperature	To be able to measure temperature (including below or above 0 °C) of the fluid (liquid, air, etc.), granular material or similar topics with the contact device, to be able to measure the temperature that is difficult to reach in a place such as rotary furnace, moving surface, melting furnace, melted material with the contactless device
Flow	• To be able to measure particle, flow rate of clean fluid and fibrous pollution fluid (including high-temperature fluid) from outside the pipe without cutting pipe lines and without installing measuring equipment • To be able to measure the flow rate and/or various pressure values in any closed pipes and canal (including wide canal through which air and low-pressure gases pass by using pitot tubes or similar equipment) • To be able to measure flow rate the air and low-pressure gases through the canal (including air and gases which have high temperature) • To be able to measure the velocity or the flow rate of the air and gases at the intake of the air fans and at the outlet of the exhaust canal Solid particles, including high temperatures for the building sector, are not considered as a criterion for measuring the flow rate of fluids containing fiber-like impurities
Humidity	To be able to measure relative humidity and ambient temperatures of various areas
Pressure	To be able to measure high-and low-pressure values (including low-pressure values such as in furnace chamber etc.)

(continued)

Table 6.1 (continued)

Electric energy	To be able to measure electrical parameters (including medium voltage) such as voltage (V ("U" can also be used)), current (A), power factor (cos φ), power (kW, kVA, kVAr), energy consumptions (kWh, kVAh, kVArh), frequency (Hz), True RMS and harmonics (in the single-phase and three-phase systems). Measuring medium voltage is not obligatory for the building sectors
Lighting	To be able to measure lighting levels at the designated areas
Velocity and cycle	To be able to measure speed of rotating equipment such as motor, fan. To be able to measure feed rate of belt conveyor, fabric, and similar systems
Sound and noise	To be able to measure sound and noise levels at the designated areas
Recording	To be able to record long-term measurements and to be able to transfer these data to computer and similar devices if needed

References

1. https://en.wikipedia.org/wiki/Energy_service_company
2. https://ekosmart.com.tr/
3. Aksoy S, Çalıkoğlu E, Aras H, Karakoç N (2013) Energy management and policies, Rebuplic of Turkey Anadolu University, Publication No: 2787, Distance Education Faculty, Publication No:1745, 2013. (In Turkish: Aksoy S, Çalıkoğlu E, Aras H, Karakoç N, Enerji Yönetimi ve Politikaları, T.C. Anadolu Üniversitesi, Yayın No: 2787, Açık Öğretim Fakültesi, Yayın No: 1745, 2013)

Chapter 7
Measurement Techniques and Instruments

7.1 Measurement Techniques

Measurement is the process of comparing any size to a unit of its own type. *Measurement* is the process of investigating how many or how much amount of a given size is within a measured size.

Today, as industrial applications develop rapidly, the success of science, technological developments, and all other works depends on precise, accurate, and reliable measurements to be made. One of the most important factors in many studies, which is regarded as successful technological progress, the produced product must have high quality and be precise, durable, and reliable. A quality and precise production are also inseparable factors of the whole success because it brings durability and reliability of the product. The degree of prevalence of the ability to perform measurements with high accuracy and precision required in health, environment, scientific and technological research, industry, trade, defense, and similar fields is one of the fundamental elements that determine the technological level of countries. Measuring means efficiency and quality, and it ensures healthy and happy lives of communities [1].

7.2 The Quantities to Be Measured

The purpose of the measurement is to determine the magnitude of the quantity to be measured. To accomplish this, we need to know the structure of the quantity. The quantity to be measured can generally be divided into two parts:

(1) Physical quantities,
(2) Non-physical quantities.

D. Kaya et al., *Energy Management and Energy Efficiency in Industry*,
Green Energy and Technology,
https://doi.org/10.1007/978-3-030-25995-2_7

Force, pressure, length, mass, and temperature are examples of *physical quantities*. Quantities such as hardness, surface roughness are *non-physical quantities*, also called *industrial quantities*. Physical quantities are based on physical laws. Physical quantities are derived from these basic quantities by these laws. However, there is no scientific law based on non-physical quantities.

7.3 Measurement Methods

7.3.1 Direct Measurement

If the value of a searched dimension is directly compared with a true scale or can be read directly by a calibrated measuring device adapted to the measuring unit, such measurements are called direct measurements. For example, if a ruler measures a specific length, caliper, or any measuring instrument and the result can be read directly from the scale or the display, this type of measurement is directly measured. In the same way, the mass can be measured directly by scale and temperature can be measured directly with a thermometer. An equivalent reference quantity is required for each measurement and it is usually measured directly by comparison.

7.3.2 Indirect Measurement

If the value of the wanted quantity can be found indirectly according to the results of the direct measurements, based on specific laws with the wanted quantity, such measures are called indirect measurements. Indirect measurements can only be used when measuring physical quantities. Because there are specific physical laws between physical and base quantities, and the indirect measurements of results are determined in accordance with these laws.

Although direct measurement methods have been developed for many physical quantities such as force and pressure values, indirect measurement methods can also be used. As can be seen from this, indirect measurement methods for non-physical quantities cannot be discussed. The determination of heat transfer coefficients, expansion coefficients, modulus of elasticity, etc. of materials can be given as an example of the indirect measurement.

7.3.3 *Absolute Measurement*

If the value of a measurement quantity is based on the reference specified for the same magnitude, then this is called an absolute measurement. In fact, an absolute measurement has a comparison, but a direct measurement is applied.

At the same time, the accuracy of the fundamental constants (for example, light, speed, electron mass, etc.) in physics must be determined by a very high correctness of the measurement and based on the basic units. If the measurement of physical quantities can be defined by being reduced to a measure of the basic quantities, then this is called absolute measurement.

7.3.4 *Comparative Measurement*

If the physical quantity can be measured directly by comparison with a derived or base reference of the same kind, then this is called a comparative measurement. The accuracy of a comparative measure is lower than an absolute measure. Practically, a sensitive comparative measurement is less time-consuming than an absolute measurement. However, the accuracy must be guaranteed by an absolute measurement. Absolute measurement is not practical for industrial measurements but it is used to obtain references in advanced metrological studies.

7.4 The Properties of the Measurement Systems

7.4.1 *Error*

The measurement error is the actual value obtained by subtracting the actual value of the measured size from the measurement result. The actual value is a value corresponding to the definition of a particular size being considered and which can only be obtained by an ideal measurement. Since the actual values cannot be determined by the facts, the actual real value is used in practice. The traditional real value is the value given to a certain magnitude and assumed to have an appropriate uncertainty for the purpose considered.

The measurement error is limited to the resolution of the measuring device in the first instance. Resolution is defined as the smallest detectable difference between the display values of a display element. Because of the measurement, even by chance, the probability of reaching the absolute value is very low. For this reason, the aim is to ensure that measurement errors reach the lowest acceptable level for the purpose of measurement, not the elimination or removal of the whole. To achieve this goal, it is necessary to have sufficient knowledge about the source and type of measurement errors. *The measurement error* is the difference between the measured value and the actual value. The error that a measuring instrument makes during measurement is defined in three ways.

(1) *Absolute error*: The difference between the value read from the instrument and the actual value that the instrument should show.
(2) *Relative error*: The ratio of the absolute error to the desired value is called relative error.
(3) *Manufacturing error*: This value is found by dividing the absolute error by the last value on the tool dial.

7.4.2 Accuracy

What is expected from a good measurement is that it has an appropriate degree of accuracy for performing the measurement. The measurement accuracy is the degree of closeness between the measured value and the actual value. Accuracy is defined by measuring error or error limit.

The measurement error can be described numerically, since it is a quantitative concept. In many measurement reports, the error limit is $\pm$ 0.05%, which is defined as "the measurement accuracy is $\pm$ 0.05%", and the results obtained from the measurements will not exceed these limits. Since a value of $\pm$ 0.05% determines the limits of the measurement error, it should be regarded as an error value, not as an accuracy value. As the accuracy increases, a measurement with a $\pm$ 0.05% fault will have lower accuracy than a measurement with a fault $\pm$ 0.03%, as the measurement error will be reduced. Since accuracy is a qualitative concept, it must not be expressed numerically. When expressing the concept of accuracy, terms such as high accuracy or low accuracy or sufficient accuracy should be used, and the accuracy level should be avoided by using the figures.

7.4.3 Precision

Another problem with the definition of measurement accuracy is that precision is used instead of accuracy. These two terms are distinctly different from each other and this difference needs to be fully understood. If measurement result has accuracy, it means it has the precision having no tendency. Accuracy refers to the value of the measurement, its true value proximity or suitability, while precision defines the degree of agreement or adaptation within a group of measurements or devices. Precision is a condition for accuracy, but not sufficient condition. In other words, the accuracy of the reading value cannot be guaranteed with precision.

Precision is a qualitative definition of repeatability. In case of non-technical writing language, comparison of two measurements or information about the measurement/device allows to express by using comparative adjectives such as higher precision, lower precision. If the repeatability of a system is high, it can be said that the readings are heavily grouped, and the results have "low propagation". Low propagation is high precision. The accuracy of a measurement depends on the equipment and measurement technique.

7.4.4 Repeatability

Repeatability is the degree of closeness between successive measurement results of the same measured size under the same measurement conditions (same measurement method, same observer, same measuring instrument used under the same conditions, same position, same usage conditions, and short time interval) and quantitatively defined as the distribution of results.

Repeatability is also defined as the coherence and repeatability of successive measures or their closeness to each other and it depends on the random errors.

7.4.5 Measurement Uncertainty

Measurement uncertainty is a display that is associated with the measurement result and describes the distribution of the values that can correspond to the range of values in which the actual value is within the measured magnitude. This indicator can be, for example, the standard deviation (or certain folds thereof) or the half-width of the interval having a certain level of reliability.

Measurement uncertainty generally includes many components. All components of uncertainty, including those resulting from systematic effects, are components of this distribution, such as remediation and reference standards, measurement equipment, and components that come with the measured standards. Some of these components can be calculated from the statistical distribution of the measurement series results and are defined by the help of the experimental standard deviation. Other components that can be defined by standard deviation can also be calculated from the accepted probability distributions based on experience or other information.

7.4.6 Calibration

Under specified conditions, the sequence of operations that determines the relationship between the values indicated by the measuring system or the measuring system or the values indicated by the material measure and the known values of the measured size correspond to the calibration sequence. A simpler definition of calibration is the process of setting the accuracy level of a device with another device that has known that its accuracy level is exact.

7.5 Measurement Instruments

In energy surveys, energy-consuming equipment and systems are examined. These equipment and systems are mainly comprised of engines, furnaces, fans, steam plants, boilers and its equipment, lighting, electrical energy systems, harmonics, compensation and automation, air conditioning systems, compressed air systems, pumps, and insulation systems. In order to be able to perform these measurements mentioned in the energy audits, different measuring instruments are used. The measurement instruments used in these energy audits for the industry are given in Table 7.1 [1].

7.5.1 Electrical Measuring Instruments

The electrical values must be continuous and regular so that electricity can be used without damaging the facilities and consumers. It is necessary to know the connection of measuring instruments to evaluate the safety of the electric circuits and energy. For this purpose, electrical authorities should be trained well to ensure that measurements are finalized with certain values.

Electrical measuring instruments are devices that are comparing, indicating, writing, and counting of electrical values in their own values and units. The measurements are made by connecting the measuring instruments directly or with auxiliary units to the circuit. Electric current is used as direct current (DC) and alternating current (AC). In particular, in many measuring instruments, DC and AC values can also be measured with the same measuring instrument using auxiliary units.

7.5.1.1 Electrical Current Measuring Instruments

Since the electric current is the amount of charge passing through of a circuit part in the unit of time (internal variable), it is necessary to place a measuring device on the line through which the current flows so that the current can be measured. Devices that measure current intensity through an electric circuit are called ammeters (Fig. 7.1) [2]. "A" represents the symbol of the ammeters in electrical circuits.

The reason for the serial connection is that the entire current must pass ampere meter in order to measure the current through the receiver. The instrument is

Table 7.1 Measurement instruments used in industrial energy audits

• Thermal camera • Flue gas measuring instrument (Flue gas analyzers) • Electrical analysis devices • Ultrasonic flow meter • Ultrasonic sound level meter	• Decibel meter • Conductivity meter • Thermometer • Clamp ammeter • Humidity meter	• U-value meter • Lux meter (illuminance meter) • Earth (Ground) resistance meter • Air velocity meter • Tachometer

Fig. 7.1 Analog and pens ammeters [2]

connected in series with the receiver (load or device) so that the electric current from the receiver can also pass through the ammeter at the same time. This connection is called a serial connection. Ammeters are small measuring instruments with low internal resistances. The ammeter is manufactured with very little internal resistance in order not to affect the flow through the circuit. For this reason, the ammeters are connected to the electric circuits before electric application and in series. Because ammeters are series connected to the circuit, they must not have a limiting effect on the current like a load in the circuits they measure [2]. For this reason, the internal resistances of the ammeters are very small (0–1 Ω), and if they are accidentally connected in parallel, they may become unusable in a short period because of a very large current flow through them. If these devices are to be connected in parallel to the circuit, the coil of the instrument will be damaged because a large current will pass through it. No more current should be passed from the circuit of ammeters than the maximum value written on the indication. Connecting the amperemeter to the circuit can be seen in Fig. 7.2.

On the quadrant of the ammeters, the letter (A), which is the initial letter of the amperage, is written. Ammeters should accurately measure the value of current flow through the electric circuit. For this, the coil of the ammeter is made thick and thin. Ammeters according to the value to be measured are manufactured with

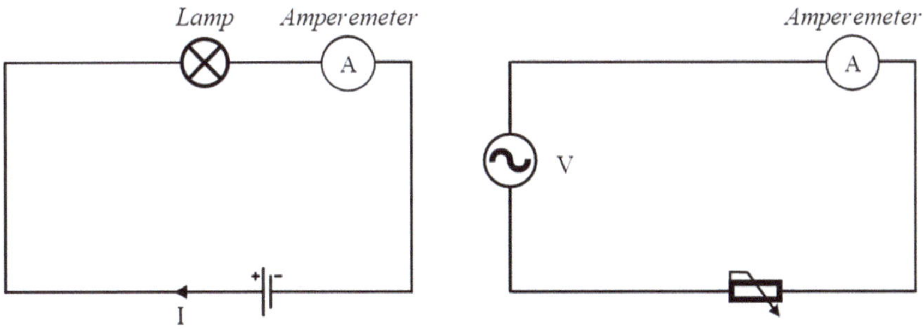

Fig. 7.2 Connecting the amperemeter to the circuit

a measurement range from the mA level to the kA level. Ammeters, milliammeter, microammeter, and kilo ammeter are named according to the value of the current they measure. These instruments measure the current magnitude and have variations as digital, analog, and clamp ammeters (two types of them can be seen in Fig. 7.1). Depending on whether the current to be measured is DC or AC, DC amperemeter or AC amperemeter should be used.

Ammeters operate according to the magnetic field's effect on the iron core. The structure is produced in two different types as soft and rotary iron. In the structure of soft iron measuring instruments, which are not widely used today, a permanent magnet coil and an iron core are disposed within the coil. The upper part of the iron core is connected to the body of the measuring instrument by means of an arc shown in Fig. 7.3. The measured value of the ammeter is read by means of the indicator attached to the iron core, and through the graduated ruler opposite this display. Measuring devices with rotary iron are produced in two different forms, namely, push and pull type.

Pull type measuring instruments are used with both direct and negative current. The quadrant of these measuring instruments is uniformly spaced as shown in Fig. 7.4.

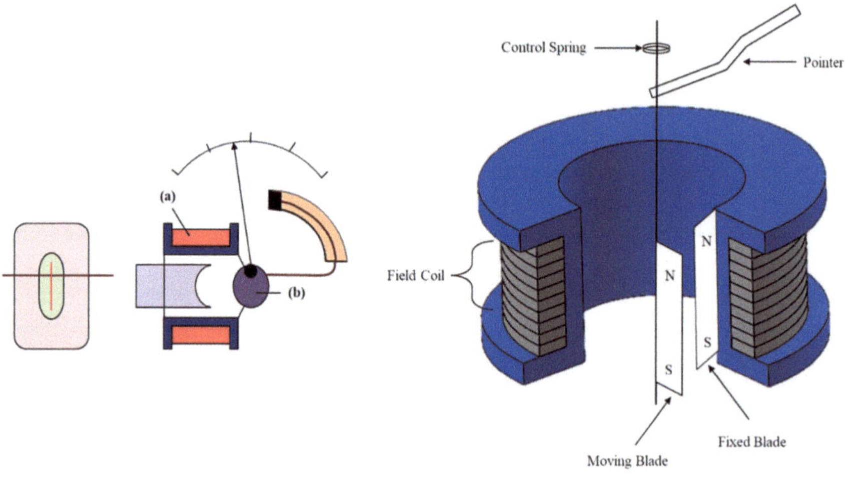

Fig. 7.3 **a** Pull type flat coil measuring instrument **b** Push type electromagnetic measuring instrument with round coil

Fig. 7.4 Ammeter and voltmeter dial graduation

The construction of the pull type measuring instruments has a flat coil as shown in Fig. 7.3a. Next to the hole of this coil, there is a moving plate (b), off the center. Plate (b) is prepared in accordance with coil (a). When a current is passed through the coil (a), the coil is magnetized and it pulls the soft plate (b) into it. (b) shows the value of the current flowing through the effect of pulling in the display connected to the same axis as the plate. In the case of puller type gauges, the drawing of the iron plate is not dependent on the direction of current flowing through the coil. In a pusher-type measuring device (Fig. 7.3b), a plate with a wide end and the other with a narrow end is secured by bending it parallel to the circumference of the plate (a) and fixing it into the coil. Plate (b), which is narrower and curved than plate (a), is connected to the moving axis of the tool.

7.5.1.2 Electrical Voltage Measuring Instruments

Voltage is defined as the potential difference between two electrically energized two-circuit points; the magnitude to be measured is an end-effector. In other words, in order to measure the stress on a circuit element, it is necessary to connect a measuring instrument parallel to the element ends. A voltage-measuring instrument is called a voltmeter. In other words, in an electric circuit, devices that measure the potential difference between two points are called voltmeters (Fig. 7.5).

In contrast to ammeters, the voltage is connected directly parallel to the junction between the two ends of the electrical circuit or voltage source to be measured (Fig. 7.6). Voltmeters are measuring instruments with large internal resistances. For this reason, the voltmeters are connected in parallel to the electric circuits. If this measuring instrument is to be connected to the circuit, a large voltage drop across the internal resistance will occur. Voltmeters are not damaged in series connection. In summary, the voltmeter voltage is connected in parallel to the ends of the element to be measured. The voltage of the voltmeter should not be subjected to more voltage

Fig. 7.5 Analog and digital panel type voltmeters

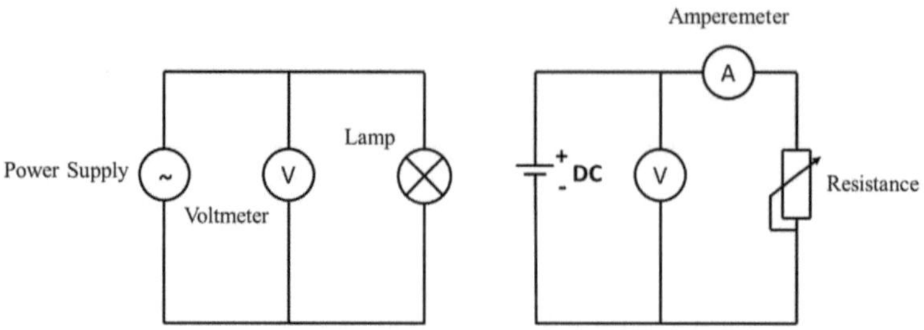

Fig. 7.6 Connecting voltmeter to the circuit

than the maximum value written on the indicator. Namely, the measuring capacity of the voltmeter to be connected to the circuit must be higher than the voltage to be applied.

When the voltmeters are connected to the circuit, they should not draw a current that is too high to reduce the source voltage. Voltmeters are wrapped very thinly on the coil so that the current drawn from the coil is small. The internal resistance of this measuring instrument is quite large compared to the ammeter. If the voltmeters are connected to the circuit in series, the total resistance of the circuit increases. A large part of the welding voltage falls on the coil of the measuring instrument. The difference voltage is applied to the receiver. Since the difference voltage is small, a low current amount flows through the receivers. This low current does not allow the current receiver to operate normally.

In electricity, the voltmeter sensitivity is the ratio of the total resistance of the voltmeter to its full-scale reading in volts, expressed in ohms per volt. For example, if the internal resistance of a 500 V voltmeter is 250,000 Ω, the accuracy of the instrument is 250,000/500 = 500 O/volt. The sensitivity per unit volt is 500 Ω. The greater the sensitivity of the gauge, the smaller the coil of the gauge works.

7.5.1.3 Instruments for Measuring Electrical Current, Voltage, and Resistance

A wide variety of measuring instruments have been developed to measure the sizes of electrical circuits and to test circuit elements. Measuring instruments measuring current, voltage, and resistance are called avometers (multimeters) (Fig. 7.7). When the AVO word is noted, it is to be seen that the initials form the Ampere-Volt-Ohm units. Because the avometer is a multipurpose measuring instrument, the other name is multimeter. Their structures are manufactured in two types, analog and digital, according to their construction. Depending on the model with these measuring instruments, variables such as current, voltage, resistance, and frequency can be measured accurately and alternatively, and tests such as transistor, capacitor, and continuity tests can be done successfully [2].

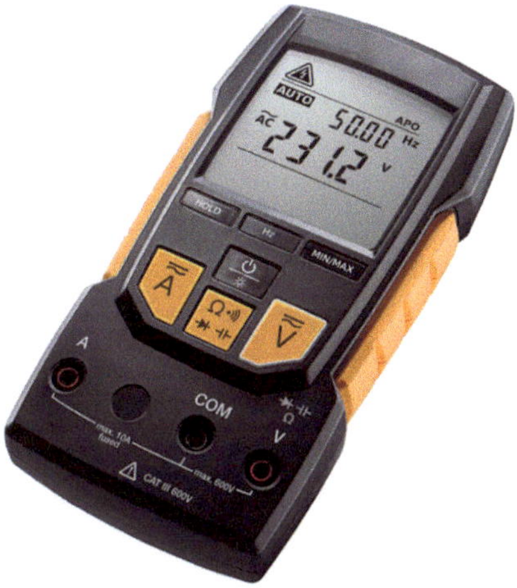

Fig. 7.7 Avometer (Multimeter) [2]

There are different types of multimeters that are manually selected or automatically selected in range. In a manually selected range, the user can determine the range. When measuring with multimeters, the magnitude to be measured is first selected. For example, voltmeter position is indicated by **V** if the voltage is measured, ammeter position indicated by **A** if the current is measured; if the resistance is to be measured, a suitable selection such as OHM position should be made. Then, the multiplier or upper limit is selected for each magnitude.

It is possible to find many types and models of analog and digital avometers according to the manufacturer. However, in principle. the measuring and reading process is almost the same. The user can understand the methods of the use of the avometer according to the explanatory notes from its user guide of the instrument.

Analog Avometers: For example, the use of the analog avometers given in Fig. 7.8 is explained in the following sections.

Selector Switch Locations (Commutator Key Locations)

Direct Voltage (V−): In the case of correct voltage measurements, the selector switch shall be brought to the appropriate level in this section. If the DCV value to be measured is not known as an estimate, the selector switch is first measured to the highest tier value. According to the value measured here, it is possible to pass to lower stages suitable for measurements that are more precise.

Alternating Voltage (V~): In the case of the alternating voltage measurements, the selector switch must be set to one of the appropriate levels.

Resistance (Ω): This is the resistance measurement step of avometer. If the instrument is to be used at this stage, care should be taken that there is no energy, as in the case of an ohmmeter in the circuit to be used. When the resistance measurement is to be made, the selector switch must be brought to the appropriate level from these levels.

Fig. 7.8 Analog avometer [2]

Direct current (A−): If the direct current value to be measured is unknown, the measurement starts from the highest step value.

Alternating current (A~): If the alternating current value to be measured is unknown, it starts to measure from the highest step value.

Probe Connection Terminals: The four-probe connection terminals on the top of the measuring instrument are as follows:

(−) Com: Common terminal is used in any measurement. What we measure is that one of the measurement probes has to be plugged into here. In DC measurements, it is a negative terminal. The probe cables of measuring instruments are usually black-red. The black cable is plugged into the com terminal.

Ω: The second terminal to be used is for resistance measurements. The red cable plugs into here.

+ A: This is the second terminal to be used for both alternating current and direct current measurements. The + sign indicates + terminal is in the direct current. When measuring current, the red probe is connected to this terminal.

+ ~ V: The second terminal to be used for both the alternating voltage and the direct voltage measurements. The + sign indicates the correct terminal is + terminal. When measuring voltage, the red probe is connected to this terminal.

Zero Adjustment Device: It is the zero adjustment device used in every measurement and step change to make accurate measurement in resistance measurements.

Digital Avometers: In recent years, because of the rapid development of technology, it is possible to come across many models of digital avometers, which are becoming increasingly popular, for reasons such as cost reduction and readability. Since the reading process is directly numerical on the LCD screen, their use is much easier and more practical than analog avometers. For the measurement, the user will be able to select the appropriate commutator stage and the size of the probe connection terminals is to be measured. The use of a sample digital multimeter shown in Fig. 7.9 is described in the following sections [2].

Selector Switch Locations (Commutator Key Locations)

Direct Voltage (DCV): When it is desired to measure the direct current voltage, the selector switch is brought to the appropriate stage in the switch DCV region. If the direct voltage value to be measured is not known as an estimate, it should be started at the highest level in the measurement process.

Alternating Voltage (ACV): In alternating voltage measurements, the selector switch is brought to the appropriate stage in the ACV region.

Fig. 7.9 Digital avometers (Digital multimeters) [2]

Direct Current 1 (DCA-1): In direct current measurements, the selector switch is brought to the appropriate stage in the DCA zone.

Direct Current 2 (DCA-2): The current value to be measured may be a larger value than that can be measured in the commutator region DCA-1. In this case, DCA-2 stage should be preferred. For two DCA measurement zones, the red probe is different from the plug connection terminal. This step was made for alternating current measurements in some measuring instruments.

Resistance Ω: Resistance is measured with the avometer, the selector switch is brought to the Ω region. Digital avometers do not multiply according to the step (level), as in analog ones. The levels here indicate the measurement limit. If the measured resistance value is in the kΩ range, the display shows “k”, if the measured resistance value is in the MΩ, the display shows “M” together with the resistance value.

When using combined instruments, the following factors should be considered:

- Status key: It is taken to the position (~) or (−) according to the type of current or voltage to be measured.
- The stage key is brought to the largest stage that can be measured.
- According to current or voltage value to be measured, connection of the measuring device is done properly.
- In the case of direct current measurement, if the tool is deflected reversely, the positions of the terminals are changed.
- If there is no deviation from the display when the stage key is set to the lowest value, it is checked whether there is a break in the circuit, whether the instrument is robust, or whether the current passing through the circuit is as small as the instrument cannot measure.

7.5.1.4 Electrical Power Measuring Instruments

The power used by electric loads in the direct and alternating current over time is defined as *electrical work* and it is measured by electricity meters. The work done in unit time is known as *power* and is measured by *Wattmeters*.

Since rectifiers obtain direct current electricity from alternating current electricity, direct current electricity generation, transmission, and distribution are not carried out. For this reason, the measurement of the work and the power of direct current electricity are also not carried out. For this reason, the work and power issues to be examined in this section will be mainly related to AC electricity.

Power Types: There is a phase angle between the current and the voltage of the receiver in the AC current receiver. The cosine value of this phase angle causes the receiver’s power to be different from the power that the receiver receives from the network. In the following vectors, this is shown by the receiver characteristic. For this reason, three separate power factors arise in inductive and capacitive receivers in the AC circuit. These are *visible, active,* and *reactive* forces.

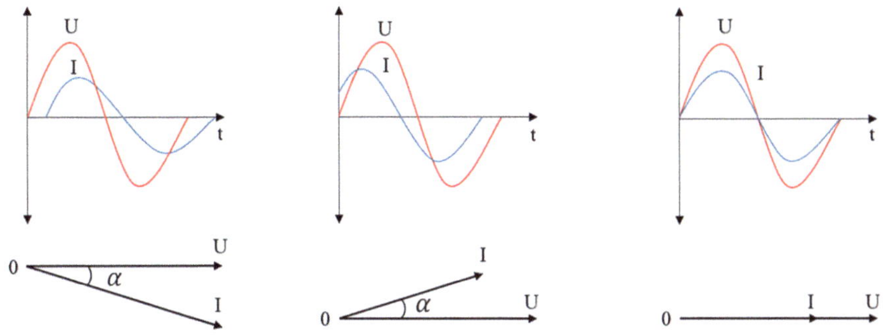

(a) Inductive Receiver *(b) Capacitive Receiver* *(c) Ohmic Receiver*

Fig. 7.10 Inductive, capacitive, and ohmic receiver sinusoidal curves and vector diagrams in alternating current

Depending on the curve and vector diagrams given in Fig. 7.10, inductive, capacitive, and ohmic loads are defined as follows:

- Inductive load: If the current on the load is behind the voltage, the load is called the inductive load.
- Capacitive load: If the stress on the load is behind the current, the load is called capacitive load.
- Resistive load: If the current on the load is a multiplier of the voltage and there is no phase difference between the current and the voltage, this type of load is called an ohmic (resistance type) load.

Apparent Power and Calculation: Apparent power is the power that the consumer draws from the network. It is indicated by the letter S. It is found by the formula: S = V x I. Unit (VA) is Volt Ampere. It equals the vector sum of the active and reactive powers taken by the consumer.

Where

S is the apparent power in Volt Ampere (VA).
V is the voltage in volts (V ("U" can also be used)).
I is the current in amps (A).

When we connect it to an alternating current circuit, the ammeter measures and displays the current and voltmeter voltage of the consumer. The apparent power of that consumer can be calculated by using the values we receive from these measuring instruments.

Example In the circuit given in Fig. 7.11, the value read from the amperemeter is I = 1.5 A, and the value read from the voltmeter is V = 220 V; accordingly calculate the apparent power of the load.

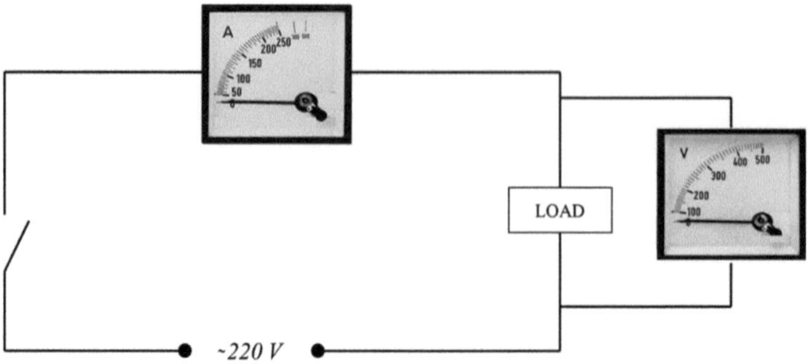

Fig. 7.11 A simple circuit diagram of apparent power calculation

Solution S = V × I = 220 × 1.5 = 330 VA.

Active Power (Real or True Power): The useful power is used to do the work on the load. It is indicated by the letter "P". It is found by the formula P = V × I × Cos φ. The unit is Watt (W).

Real or true power is the power that is used to do the work on the load.

$$P = V_{rms} \times I_{rms} \times \text{Cos}\ \varphi$$

P = The real power in watts [W]
V_{rms} = The rms voltage = $V_{peak}/\sqrt{2}$ in Volts [V]
I_{rms} = The rms current = $I_{peak}/\sqrt{2}$ in Amperes [A]
φ = The impedance phase angle, which means the phase difference between voltage and current.

Active Power Consumers: Active power is the useful power drawn by various electrical loads (motors, transformers, fluorescent lamps). Electricity is generated and distributed only as an alternative current. Consumers draw a current like I (ampere) from the network. For the convenience both physically and mathematically, this current drawn by consumers is considered to have two components, one theoretically active and the other reactive current.

The active power that the consumer brings to the market is made useful by the consumer. For example, in the case of heat-dissipating devices, the active power turns into thermal power, mechanical power in motor, lighting power in lamps. In other words, the active power of the active current can be transformed into useful power. However, the reactive power brought by the reactive current cannot be turned into useful power. Reactive power is only an alternative flux-related feature that has an undesirable effect on electrical installations. This affect generators, transformers, lines unnecessarily occupied, unnecessary loads, additional heat losses, and voltage drops. Different meters are available to measure active and reactive forces.

Active Power Calculation: The active power formula is expressed as:

P = V x I x Cos φ. Accordingly, in order for the active power of the device to be available, the current drawn by the consumer, the voltage on the consumer, and the power coefficient (Cos φ) between the current and the voltage must be known. We can measure these values with an ammeter, a voltmeter, and a cosine phi meter when we connect the circuit.

Example In the above circuit, the value read from ampere meter is 4 A, the value read from the voltmeter is 220 V, and the value read from cosine phi meter is 0.6. Accordingly, find the active power of the consumer.

Solution

$$V = 220\,V, \quad I = 4\,A, \quad Cos\varphi = 0.6, \quad P = ?$$
$$P = V \times I \times Cos\varphi = 220 \times 4 \times 0.60 = 528\,W$$

Example It is observed from the measuring instruments that a single-phase asynchronous motor draws 20 A current from the circuit during the loaded operation and the mains voltage is 220 V. From the motor label information, it is read that the value of Cos φ is 0.80. Find the visible and active powers of this engine.

Solution

$$V = 220\,V \quad S = V \times I = 220 \times 20 = 4400\,VA$$
$$I = 20\,A, \quad Cos\varphi = 0.80, \quad P = V \times I \times Cos\varphi = 220 \times 20 \times 0.80 = 3520\,W$$

A simple circuit diagram of active power calculation can be seen in Fig. 7.12.

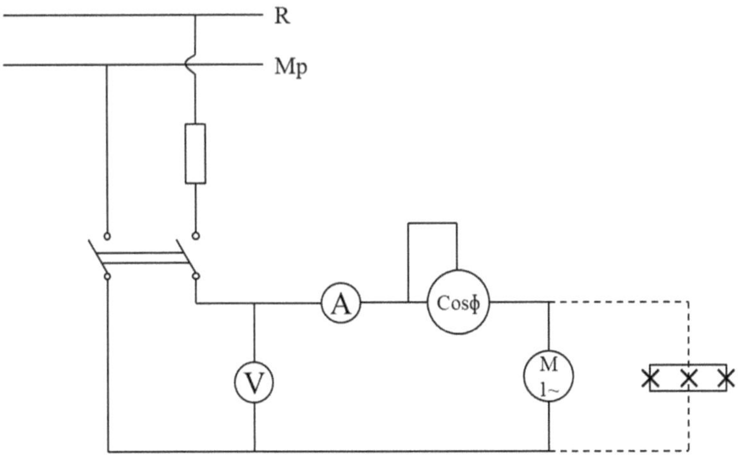

Fig. 7.12 A simple circuit diagram of active power calculation

Reactive Power: Reactive power is the power that is wasted and not used to do work on the load. It is a kind of blind power that does not work, but it is drawn from the source. In other words, the power required to excite the magnetic circuit in inductively charged circuits. It is indicated by the letter "Q". This force is not spent on the inductive load, it is only stored and sent back to the source. That is to say, reactive power is constantly exchanged between the source and the inductive load. This causes the current flowing through the conductors in the system to increase. It can be calculated as

$$\mathrm{Q} = \mathrm{V} \times \mathrm{I} \times \mathrm{Sin}\varphi \text{ equation. The unit (Volt} - \text{Ampere} - \text{Reactive) is VAr}$$

or

Reactive power is

$$\mathrm{Q} = \mathrm{V}_{\mathrm{rms}} \times \mathrm{I}_{\mathrm{rms}} \times \mathrm{Sin}\ \varphi$$

$\mathrm{Q} =$ The reactive power in volt-ampere-reactive [VAr]
$\mathrm{V}_{\mathrm{rms}} =$ The rms voltage $= \mathrm{V}_{\mathrm{peak}}/\sqrt{2}$ in Volts [V]
$\mathrm{I}_{\mathrm{rms}} =$ The rms current $= \mathrm{I}_{\mathrm{peak}}/\sqrt{2}$ in Amperes [A]
$\varphi =$ The impedance phase angle, which means phase difference between voltage and current.

Loads That Need Reactive Power: All electrical circuits with capacitors and/or coils need reactive power. When the capacitive circuits pull forward reactive power, the coiled circuits draw reactive power back. The loads that draw back reactive power are as follows:

- Low-excitation synchronous machines,
- Induction and arc furnaces,
- Transformers,
- Welding machines,
- Coils,
- Fluorescent lamp ballasts,
- Overhead lines,
- Sodium and mercury vapor lamp ballasts,
- Asynchronous machines,
- Neon lamp transformers,
- Rectifiers.

Reactive Power Calculation: The reactive power formula is $\mathrm{Q} = \mathrm{V} \times \mathrm{I} \times \mathrm{Sin}\ \varphi$. Accordingly, the current, voltage drop, and the power coefficient between current and voltage (Cos φ) must be known by the consumers in order to find the reactive power of the circuit.

If the power coefficient (Cos φ) is known, Sin φ can be found with the aid of a trigonometric table. Another method is to calculate reactive power if apparent and active powers are known. In other words, finding any two of P, S, and Q powers helps to find the other.

Example Find the active and reactive power of engine, which are given below:

Solution

$$V = 220V, \quad S = V \times I = 220 \times 20 = 4400 \text{ VA}$$
$$I = 20 \text{ A}, \quad P = V \times I \times \text{Cos } \varphi = 220 \times 20 \times 0.80 = 3520\text{W}$$
$$\text{Cos } \varphi = 0.80, \quad \varphi =\sim 36.87 \quad \text{Sin } 36.87 = 0.6$$
$$Q = V \times I \times \text{Sin } \varphi = 220 \times 20 \times 0.6 = 2640 \text{ VAr}$$

Calculation using power triangle:

$$Q = \sqrt{S^2 - P^2} = \sqrt{4400^2 - 3520^2} = \sqrt{6969600} = 2640 VAr$$

Power Vectors: Power triangles are utilized in the calculation of apparent, active (true, real), and reactive powers (Fig. 7.13). The power vector is defined as the vector of three different powers belonging to a consumer.

There is a 90° phase difference between the current flowing through or between the applied voltages in the inductive and capacitive loads. It is 90° ahead from the capacitive current voltage, 90° behind from the inductive current voltage. In ohmic circuits, current and voltage are in the same phase.

Active Power Measuring: The measuring instrument that allows us to measure the active power directly on the electric circuit is called *Wattmeter*. In order to measure the power at an electric circuit, the values of the current and voltage passing through that circuit must be known. In other words, the Wattmeter measures both the current value and the voltage value of the unit that is connected to and multiplying these two values gives the power value of the unit.

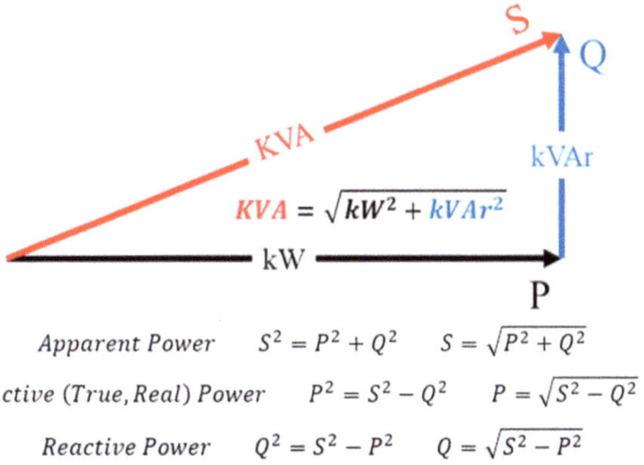

Fig. 7.13 The calculation of the Apparent Power, Active (True, Real) and Reactive Power and the power triangle

Wattmeters (Fig. 7.14a) measure power in the circuit, in terms of watt, kilowatts, and megawatt units [3]. Wattmeters are distinguished from other measuring instruments by means of one of the letters W, kW, MW written on the measuring instrument. Wattmeters are measuring instruments that combine amperemeter and voltmeter features. Wattmeters also have different structures designed for direct current and alternating current.

Structure of Wattmeter: The result of the multiplication of measured variables is displayed in analog or digital form on a Wattmeter. In the direct current circuits, the current and voltage are multiplied by the electric power in the circuit in Watts ($P = V \times I$). According to this connection, the current and voltage must be measured at the same time in order to be able to read the electric power as Watt in the measuring

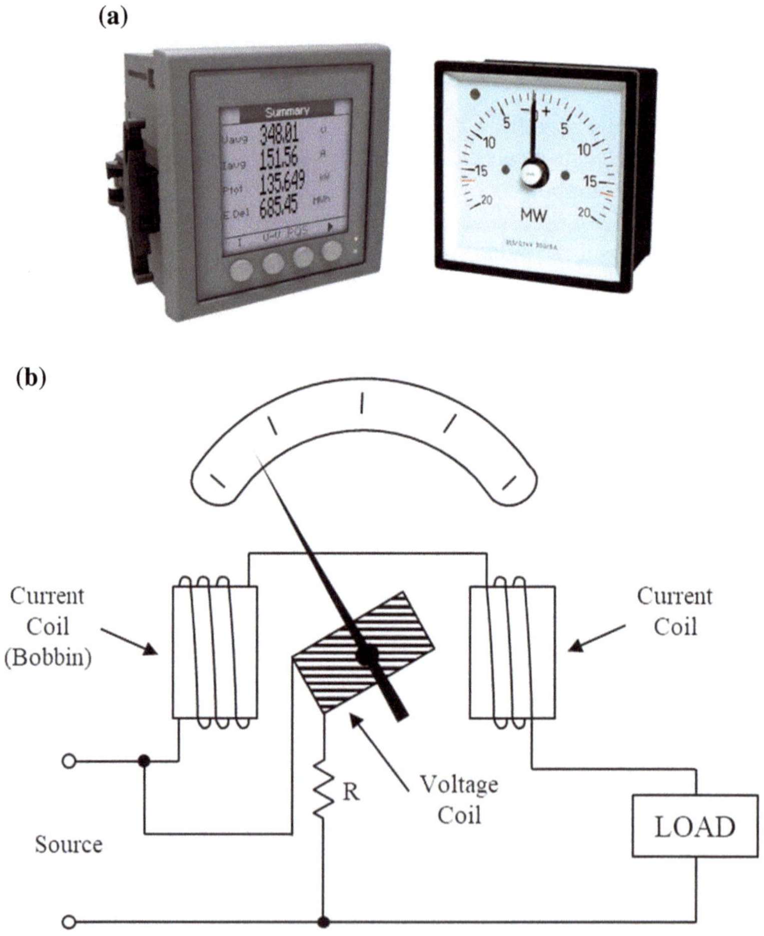

Fig. 7.14 **a** Analog and digital Wattmeters [3] **b** Internal structure of Wattmeter

instruments. To measure the current and voltage at the same time, there are two coils each for the measurement of the current and the voltage in the Wattmeter. The coils are named as voltage-measuring coil and the current-measuring coil. The current coil is connected to the circuit in series like an ammeter and voltage coil is connected to the circuit in parallel like a voltmeter.

Wattmeters are measuring instruments that have two coils, one of them is a current coil and the other is a voltage coil (Fig. 7.14b). The current coil is thick sectioned; it has low number of turn and shows amperemeter characteristics. It is connected in series with the receiver to measure the power. Voltage coil is made of thin-sectioned conductors that has high number of turn (multi-winding) and shows voltmeter characteristics.

The voltage coil is designed to move in wattmeters and a pointer is attached to it. In order to reduce their weight and reduce friction, some of the tension coil was reduced, and the fixed electronic resistance (pre-resistance) compensated the reduced resistance due to the limit. Since the voltage coils are connected parallel to the receiver whose power will be measured, the current passing through them and the magnetic field they generate are constant.

The current coil is heavier than the voltage coil and it is stationary. The current of the receiver (which its power is to be measured) passes through them. Since the current will change continuously depending on the state of the load, the repulsive magnetic field from the current coil will also change continuously. Depending on the strength of the changing magnetic field, the Wattmeter pointer and the voltage coil will move on the dial and indicate the power of the receiver.

When connecting the Wattmeters, if the power is to be measured in a large-scaled power measurement of a consumers, the current coil should be connected first, and if the power is to be measured in a small-scaled power measurement of a consumers, the current coil should be connected later (Fig. 7.15).

The Wattmeter connection diagram and internal structures of different types of Wattmeters are given in Fig. 7.16(I) and (II) [a) and b)].

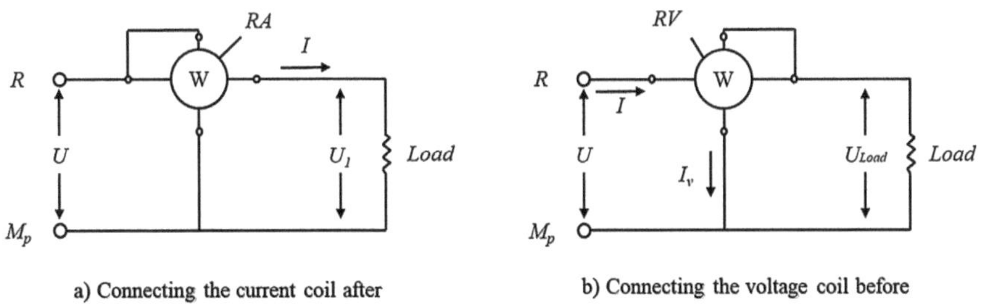

Fig. 7.15 An example of connecting current and voltage coil in Wattmeters

Wattmeter Types

Wattmeters are manufactured according to phase types; there are single-phase and three-phase Wattmeters. At the same time, with the development of technology, analog-produced Wattmeters leave their places to digital Wattmeters, but the working principles are the same. According to their structure, Wattmeters types are

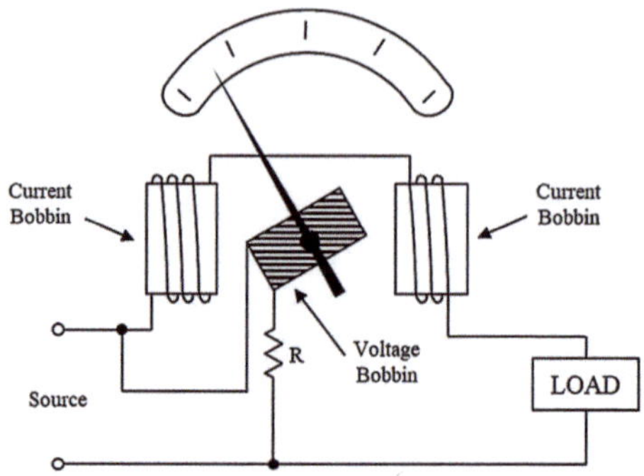

a) Electrodynamic Wattmeter

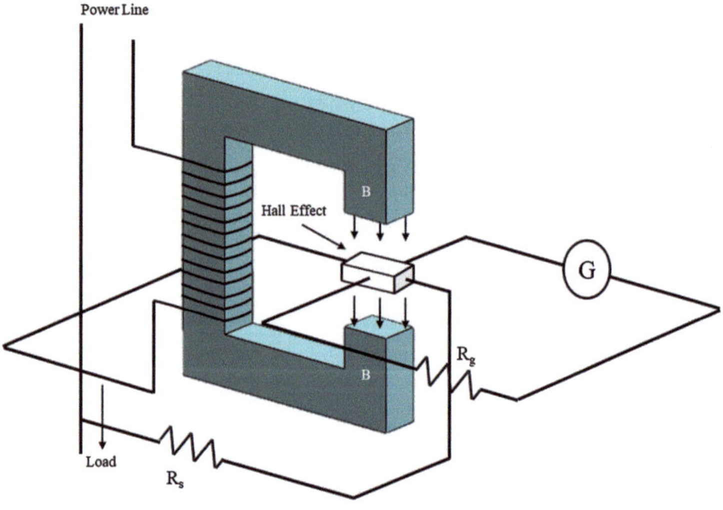

b) Hall Effect Wattmeter

Fig. 7.16 (I). Wattmeter connection diagram. (II). Internal structure of different types of Wattmeters [a) and b)]

- Electrodynamic Wattmeters,
- Hall effect Wattmeters,
- Induction Wattmeters,
- Electrostatic Wattmeters.

Wattmeter is manufactured in three different ways, namely, induction, electrostatic, and electrodynamic type in the measuring technique for the measurement of electrical power. The usage area of the induction and electrostatic types of Wattmeters has been reduced due to the fact that they are incorrectly measured, their power losses are excessive, and they only use in the alternating current. In practice, *electrodynamic Wattmeters* are widely used.

Single-phase Wattmeters: This requirement is met by connecting the current coil of the Wattmeter to the circuit that is made the measurement of power in series and the voltage coil in parallel. If small power is to be measured in the Wattmeter, the current coil should be connected later, if the large power is to be measured, the current coil should be connected first, and therefore, the measurement error can be reduced.

Three-Phase Wattmeters: Three current coils of three-phase Wattmeters are connected to three separate phases. As for voltage coils, they are connected to their neutral inputs by connecting their terminals to their respective phase inputs.

Power Measurement in Three-Phase Balanced Circuits: The power measurement in three-phase balanced circuits is done by the following methods:

Three-phase power measurement with one-phase Wattmeter in a balanced three-phase system: This method is used in balanced three-phase circuits. In three-phase and balanced-load systems, one-phase Wattmeter is connected to any of the phases of the three-phase receiver as one-phase circuits, since the power drawn from each phase is the same. With this method, only one-phase power is measured. The power obtained is multiplied by 3 to find the total power of the circuit. An example connection diagram is given in Fig. 7.17.

Three-phase power measurement with two Wattmeters in three-phase balanced loads (Fig. 7.18): This method is used in three-phase balanced three-line circuits. In the case of the system balanced and loaded, the power drawn from three-phase system can be measured with two Wattmeters or a single Wattmeter connected to Aron.

Power Measurement in Three-Phase Unbalanced Circuits: The power measurement method in three-phase unbalanced circuits can be done by the following methods:

The power measurement in three-phase unbalanced circuits with three Wattmeters with single-phase (Fig. 7.19): If the power of the system or the power of the receiver is unbalanced, each phase is connected to a single-phase Wattmeter to find the total power. This method can also be used for unbalanced loads. The total power is the arithmetic sum of the values read from the Wattmeter. In other words

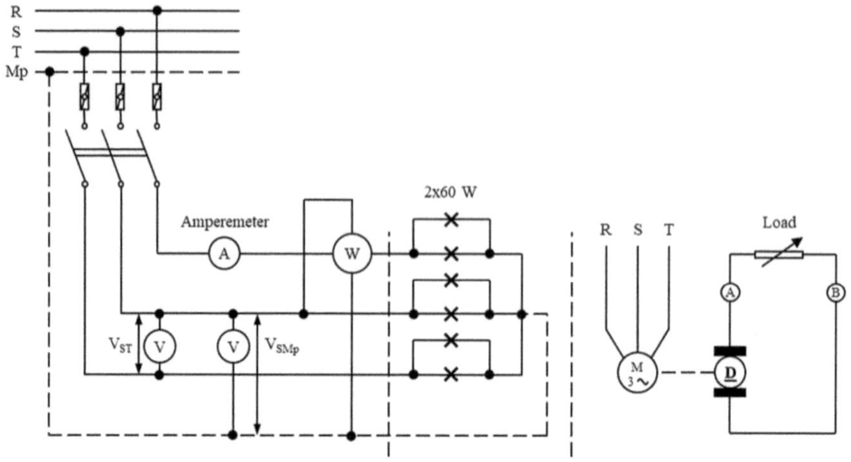

Fig. 7.17 Three-phase power measurement with one-phase Wattmeter in a balanced three-phase system

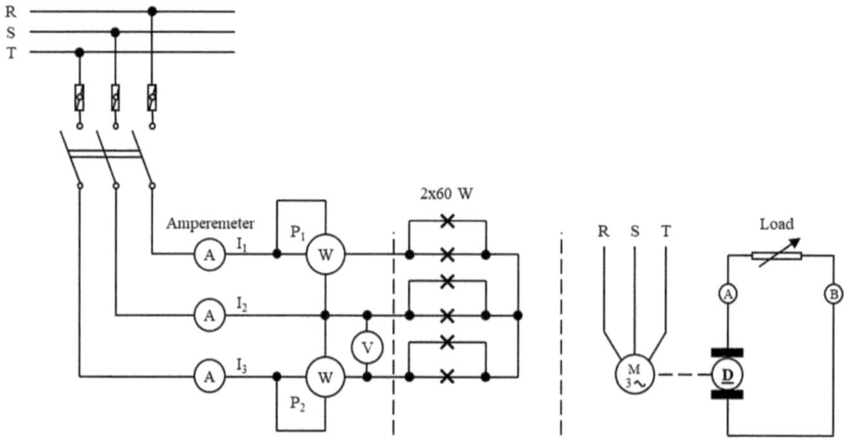

Fig. 7.18 Three-phase power measurement with two Wattmeters in three-phase balanced loads

$$P_{Total} = P_R + P_S + P_T$$

The three-Wattmeter method is not very useful due to the connection difficulty and the requirement of three Wattmeters, and therefore, the cost is higher.

Power measurement with a three-phase Wattmeter: Three current coils of the three-phase Wattmeter are connected to each individual phase one by one (Fig. 7.20). As for voltage coils, they are connected to their neutral inputs by connecting their terminals to their respective phase inputs. Measurement is not possible with analog and digital measuring instruments.

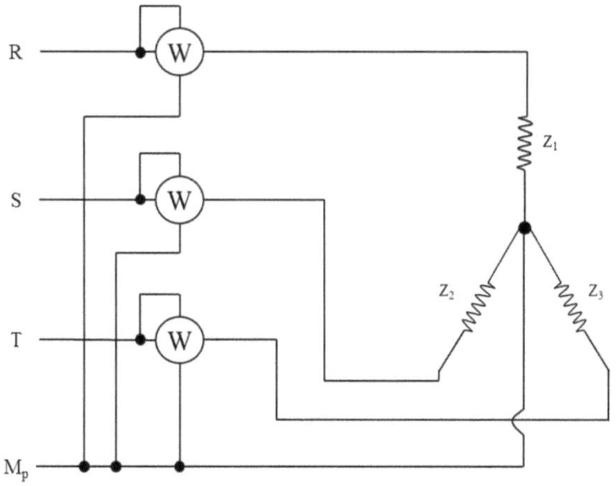

Fig. 7.19 The power measurement in three-phase unbalanced circuits with three Wattmeters with single-phase

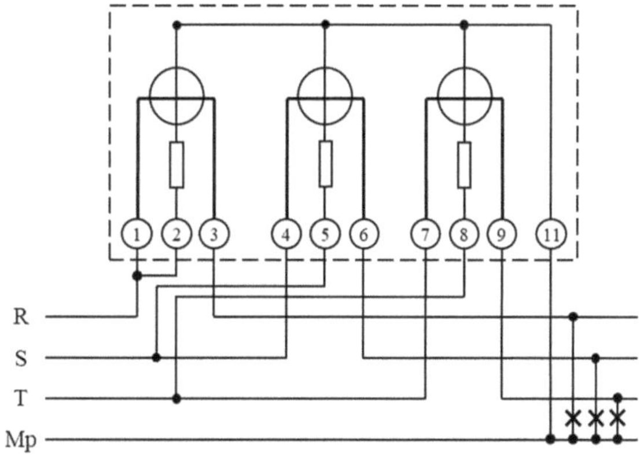

Fig. 7.20 Power measurement with a three-phase Wattmeter

Power Measurement with Aron Connection: Aron connected Wattmeters have two currents and two voltage coils. The current coils are connected to any two phases. As for voltage coils, they are connected between the phase to which its current coil is connected and the third phase to be idle. Figure 7.21 shows an internal wiring diagram of a Wattmeter connected to a three-phase Aron Connection.

If the power factor of the system (Cos φ) is less than 0.5, the total power is obtained by taking the difference of the measured powers in the Wattmeters. If it is greater than 0.5, the total power of the system is calculated by taking the sum of the powers measured in the Wattmeters. The connection made in this way is called an Aron Connection.

Reactive Power Measurement

Varmeter: The working power of the circuit (useful-active power) is measured by a one- and three-phase Wattmeters. The Wattmeter shows the multiplication of the alternating current by the fraction of current and voltage in the same phase. In the

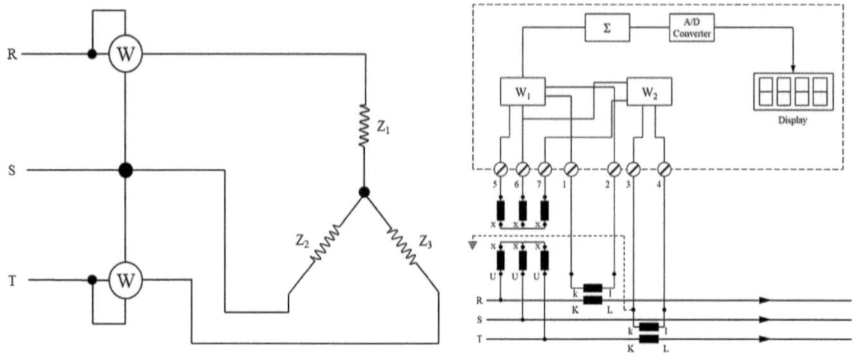

Fig. 7.21 The internal connection scheme of a three-phase Aron connected Wattmeter

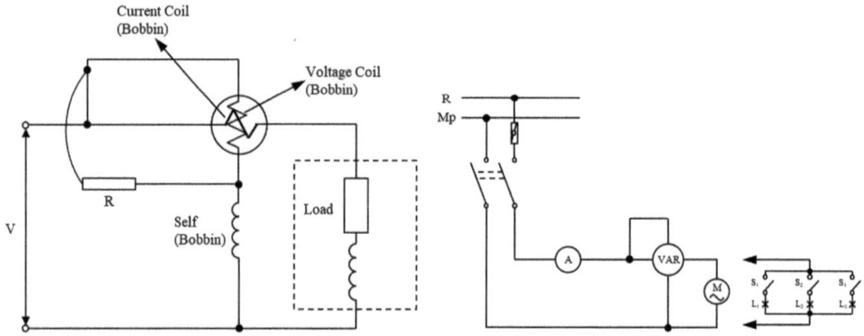

a) Structure of a varmeter

b) Connection of the Varmeter to the circuit

Fig. 7.22 Structure and connection of varmeter

inductive and capacitive states of the consumers, there is a reactive power (blind power = non-working power) in the circuit. The instrument, which measures the reactive power of the circuit, is called a varmeter.

Structure of Varmeter: Varmeters are manufactured by making small changes in the Wattmeters. This change is made by adding a series coil to the voltage coil, which is the moving coil, as shown in Fig. 7.22a. Thus, the current in the voltage coil is shifted by 90°. This is only a reactive power meter. Varmeter is connected to the circuit like a Wattmeter (Fig. 7.22b).

In order to measure the reactive power in single- and three-phase circuits, the varmeters are connected to the circuit as they are connected in the Wattmeter. In balanced systems in three-phase alternating current circuits, the reactive power of a line is measured and multiplied by 3 and the total power can be found.

Fig. 7.23 Digital and analog varmeters

Varmeter Types: The varmeters are produced in two types as in Wattmeters, single-phase and three-phase types, according to the phase types. With the development of technology, analog-produced varmeters leave their place to digital varmeters (Fig. 7.23) and the working principles are the same.

The points to Take into Consideration while Measuring Power: First, the current and voltage limits of the measuring instrument must be observed. Some Wattmeters have separate connection terminals according to current and voltage ranges. These are preferred according to the suitability to the circuit and the reading process is realized for these levels. Pay attention to which voltage is used (low voltage or high voltage). If it is used at low voltage, the phase terminals must be connected directly to R, S, T phases, respectively, since there are no voltage transformers.

The polarities of the voltages and currents of the phases must be connected correctly. For example, if the voltages are connected correctly and the polarity of one of the current traces is reversed, the Wattmeter will display an erroneous value proportional to the reactive power instead of the active power. To be sure that the Wattmeter is accurate, it must be ensured that the mounting is done in accordance with the wiring diagram. The procedure should not be carried out without reading the instructions for the use and the technical data for the installation.

7.5.1.5 Electrical Work (Energy) Measuring Instruments

The measuring device of the amount of produced or consumed electricity and/or the produced and consumed work in the circuits is described as the electricity meter (energy meter) (Fig. 7.24) [3]. Electricity meters measure the consumed energy in kilowatt-hours (kWh).

The electricity meters are used to determine the amount of produced or consumed electricity, and they collect the fixed or variable values of the power drawn from the circuit on the time axis. The Wattmeter is an instrument for measuring the electric power in watts of a given circuit. An electric meter (energy meter) is a device that measures the amount of electrical energy consumed by buildings or electrically powered equipment. The structures of an electric meter and Wattmeter are the same. The difference between them is the Wattmeter has the pointer (indicator) and the counter has a thumbwheel switch [3].

Fig. 7.24 Electricity meter [3]

In measuring devices, the system that returns with the effect of the current deviates by a certain angle. In electric meters, the rotating system makes a continuous rotational motion. The speed of this rotation is proportional to the power consumed. In these devices, instead of the counter moment, the effect of the braking torque (moment) is proportional to the speed of the moving system. The braking moment is provided by rotating the aluminum system between the magnet poles and an aluminum disc attached to the same shaft as the moving system.

In alternating current circuits, counters are used according to the induction principle. Induction counters are produced in two different ways, single and three phase. In addition to these counters, there are also dual timer, price, time, money, and current limiter types.

Structure and Operation: Induction counters with analog measurement consist of U- and G-shaped electromagnets, aluminum disc, and constant magnet. The aluminum disc is mounted so as to rotate freely between the electromagnet poles. Electric meters include a voltage and current coil, aluminum disc and a permanent magnet, similar to the G and U shape. The voltage coil is wound with a thin section on an iron core, with a very thin winding, and the current coil with a thin section with a small section on a separate iron core. In addition, the current coil has a further fused coil on the core, the ends of which are connected to an R resistance. The aluminum disc is placed so that it can rotate freely between the voltage coil and the current coil poles. Current is induced on this disk, which changes in terms of time and is located within an area. Therefore, there is no need for brushes and collectors to transmit current to the disk, as in direct current meters. Current to the disc is provided by induction. The current flowing through the disc creates a torque by the effect of the area in which it is located and thus the disc rotates. The transfer of the disk is transmitted to the numerator with an infinite screw. The energy consumed by the electricity meter reading is collected and measured. The rotational speed of the meter is directly proportional to the power it draws from the system.

The digital electricity meter consists of electronic components. On the digital counter, there are reset button, led signal outlets, call button, LCD panel, and optical communication interface. All counters have lock or seal parts in front of the terminal connection plate.

Single-Phase Meters:

The Structure: Single-phase induction counters measure the electricity amount (in kWh unit) consumed by devices such as televisions, lamps, ovens, motors, radios that work in single-phase alternating current. Figure 7.25 shows the internal structure of a single-phase induction counter.

The structure of the induction meters includes G- and U-shaped K_1 and K_2 electromagnets, M is the permanent magnet, and A is the aluminum disc. On the K_2 electromagnet, there is a thick section and a couple of winding on the current coil (St); and on the K_1 electromagnet, there is a thin section and a lot of winding on the voltage coil (Sp). On the K_2 electromagnet there is another coil (Spa), which is connected to the ends of the R resistance other than the current coil and consists of several windings. The aluminum disc A in Fig. 7.25 is mounted in a way that rotates between the poles of K_1 and K_2 electromagnets. The speed of the disc is transmitted to the Z speed counter with the help of an endless screw.

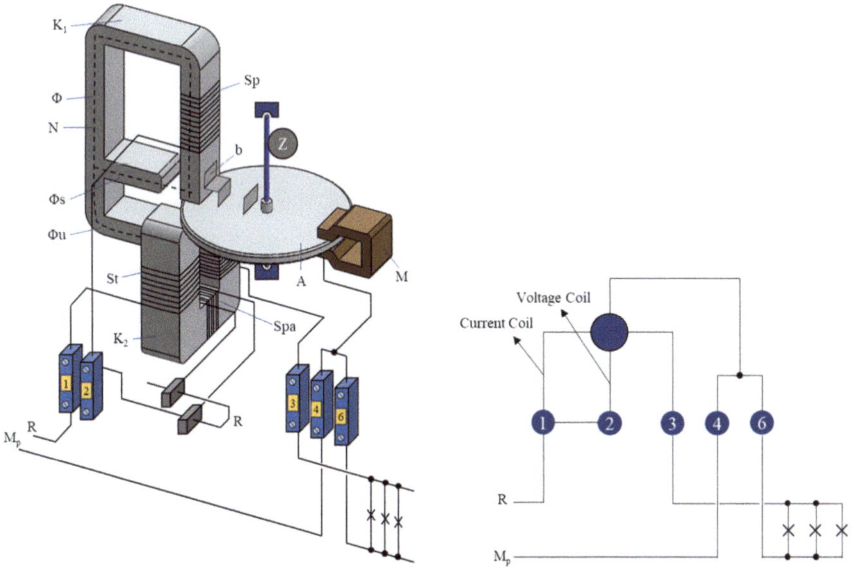

Fig. 7.25 Single-phase induction counter and connection scheme

The Operation of the Counter: As the current and voltage coils apply a force in proportion to the current that goes through the disk and the applied voltage, the rotational speed of the disc increases with the power drawn from the circuit. When the counters are connected to the circuit, the currents flowing through the current and voltage coils bring about the magnetic field in the coils. The bottom face of the disk is under the influence of the field created by the current coils and the top face is under the influence of the field created by the voltage coil. This resultant magnetic field, which flows through the geometric sum of the current and voltage coils, acts on the point of the disc in the air gap, allowing it to rotate continuously in a direction. The disc speed is transmitted to the counter with an infinite screw. The speed counter counts the speed of the disc. The thumbwheel switch writes a number after a certain cycle (rotation) of the disk. Thus, the electricity consumed by the counter is measured. The speed of the aluminum disc is dependent on the current that flows through the circuit. In order to make the correct measurement, the induction meters must have a phase difference of 90° between the magnetic fluxes of the current and voltage coils. This angle difference is possible if the voltage coil is made of pure inductive and the current coil is made of pure ohmic resistor. In practice, it is very difficult to find pure ohmic and pure inductive resistances. For this reason, the N component is connected in parallel to the K_j electromagnet to ensure that the magnetic flux generated by the voltage coil is 90° behind the mains voltage. The correct measurement is made with the M permanent magnet shown in Fig. 7.25. By this means, it is ensured that the disk rotation is proportional to the load current. If the load is to be deactivated while the meter is rotating, the f iron hook on the shaft and the (b) magnetic hook on the K_1 will stop the disc at the same place.

The Direction of Rotation of the Counters: The direction of rotation of aluminum discs in the counters is from the left to the right. This is indicated on the counter cover with the help of an arrow. A red mark is placed on the edge of the disc to see if the meter is working and to count the number of revolutions.

If the position of the conductor connected to the terminal of the current coil is changed in single-phase (monophonic) meters, the direction of the disc rotation in the counter also changes. In three-phase meters, if the inlet and outlet locations of at least two of the phases are changed, the direction of rotation of the aluminum disc changes.

The counters placed in their consumption places are usually read by staff once a month. The amount of energy consumption is calculated according to the difference between the last value read in the previous month and the last value read on the counter. The number of the rotation of the disc for one kWh of energy is required to be written on the labels of the counters. If a meter has 750 rotation/kWh written on it, the electrical energy consumption for each 750 rotations of disc is 1 kWh.

Control of Counters: A good counter should show actual consumption with the desired unit. Due to long time usage and some adverse factors, meters sometimes make erroneous measurements. The settings of the counters that make incorrect measurements are realized at the counter setting centers. It can be determined by a practical method whether a meter is measuring correctly or not. How this is carried out will be briefly explained with the example below.

Let us say you want to check whether a counter with 1200 rpm/kWh written on the label is written correctly or not. To do this, a timer is connected to a transceiver that its power has been known. Then, with the help of a clock or a chronometer, the number of rotations of the receiver in one minute is determined. For example, this receiver has 40 rotations per 2 min. On the label of counter, it is written 1200 rpm/kWh. If a 1000 W receiver was connected to the circuit of this counter, the receiver would make $1200/60 = 20$ rotations per minute. As the receiver connected to the counter circuit makes 40 rotations per 2 min, this counter makes accurate measurements.

If a clock detects the number of rotations of the disc or a timer is greater than the number of rotations that must be done per minute in accordance with the label value, it means that the counter is over-written. Otherwise, it is lower written. One of the methods to be used for controlling the meters is to calculate how much time it takes to reach a certain number of rotations of the disc.

Example You have connected a lamp in a 200 W power pack with 600 rotation/kWh printed on the label. 200 W This receiver takes three rotations per minute. Let us check whether the meter is correctly written in the direction of this information or not.

Solution According to the label value, the number of rotations per minute of the disc counter must be $600/60 = 10$ rpm. The number of rotations that the meter must make is calculated as follows.

If a 1000 W receiver has 10 rotations per minute on the counter,
The 200 W receiver is rotated X times in the same counter in 1 min.
$X = (200 \times 10)/1000 = 2$ rpm

A receiver in the 200 W power has two rotations per minute. According to our example of the counter disk, three rotations per minute, this counter is writing one more rotation per minute.

It is also possible to find the value of a receiver whose value is not known at the same time. For this, the receiver that has unknown value is connected to the counter circuit. The number of disc rotations in a given time zone is counted and the power of the receiver is calculated by the following method.

Example A consumer with an unknown power has six rotations per minute, counting 1200 rpm/kWh on the label. Calculate the power of the consumer.

Solution As the consumer in the 1 kW power has 1200 rotations per hour, the number of rotations per minute is $1200/60 = 20$.

If a 1000 W receiver has 20 rotations per minute,
The X (W) receiver has six revolutions in the same number of 1 min.
$X = (6 \times 1000)/20 = 300$ W

Counters are maintained and checked by authorized personnel at certain times, not more than 10 years. Today, along with the development of digital technology, electronic counters are produced besides induction counters.

Three-Phase Meters (Counter): According to the feature of the circuit, there are three-phase induction meters; these are manufactured as three-phase three-wire and three-phase four-wire of three-phase meters. Three-phase counters consist of two or three single-phase induction counters. Three-phase counters have no difference from the single-phase counters in terms of operation and characteristics. The only difference between three-phase counters and single-phase counters in structure is that each aluminum disk of the counter is mounted on a mile on the same axis. The internal structure of the three-phase three-wire induction counter and the connection diagram to the circuit are given in Fig. 7.26. Since there is a neutral line in the three-phase four-wire meters, these meters are connected to the balanced and unbalanced receivers and they measure the total energy of the three phases in kWh units.

Counters used in three-phase three-wire distribution systems have one disk. The two electromagnets that are shown in the figure affect this disc. Another counter type used in three-phase three-line circuits is a counter with two disks connected to the same disc. These counters, like the Wattmeters, measure the energy of the line belonging to each phase in kWh by connecting an Aron method to the circuit.

Connecting counters to the circuit: Counters must be connected to the electric circuit correctly so that the consumed electricity can be measured accurately. The current coils of the meters are connected in series and the voltage coil in parallel, as in the Wattmeter. The connection cover of each meter contains the connection diagram for the counter connection. The meters are connected to the electric circuits in three different ways.

Direct Connection: In low voltage networks and in circuits with low current values, counters are connected directly as shown in Fig. 7.27. In single-phase meters, the phase from the network enters the number 1 end of the meter. If it is neutral, it goes into the number 4 end. The meter is connected to the consumer by passing out the number 3 end the phase and number 5 end is the neutral.

Single-phase counters are produced with current values of 5, 10, 20, 30 A and 110–220 and 380 V, respectively.

Single-phase counters are referred as the current values of 5–10 and 30 A, and three-phase counters are referred to as 3×10, 3×30, 3×100, and 3×200 A meters. According to the state of the electricity enterprises, the current transformers (reducers) must be connected to the meters for current values above 50 A and the voltage transformers must be connected in the high voltage circuits.

Connecting with a Current Measuring Transformer: In low voltage circuits where the consumers draw excess current, the meters are connected to the secondary ends of a measuring transformer, as shown in Fig. 7.28a). If there is a lot of current in the low voltage network, the connection is made with the current measurement transformer. The secondary ends of the measuring transformer are connected to the current coil.

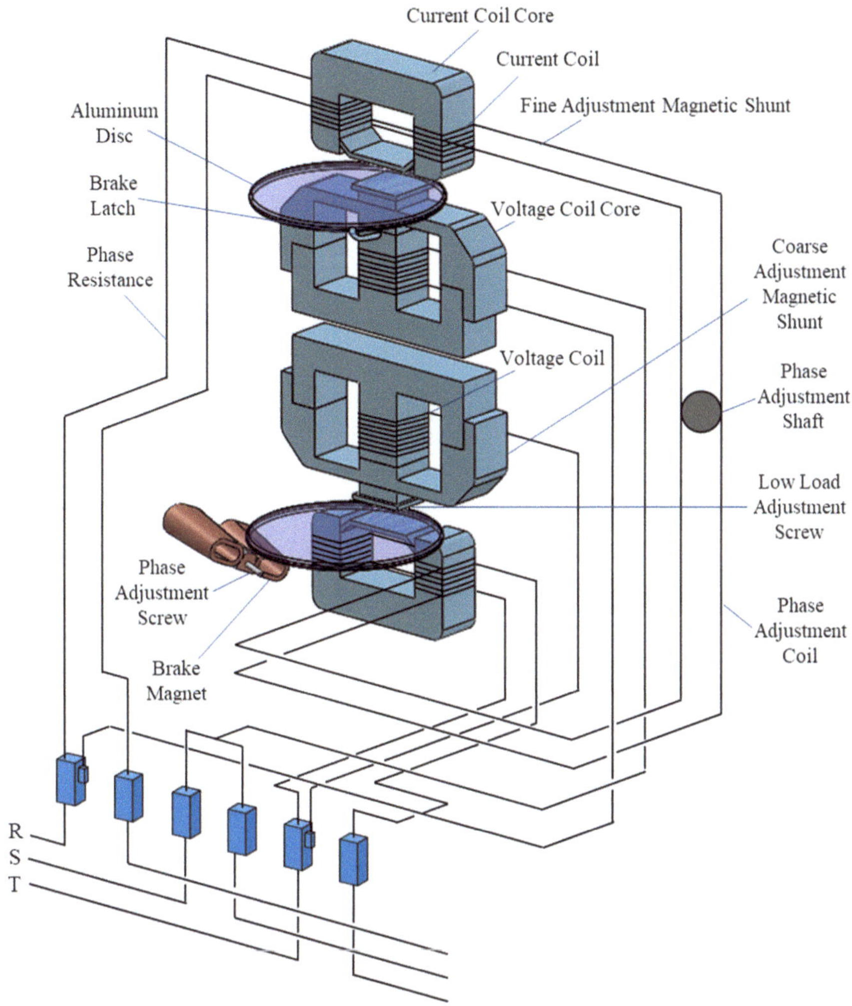

Fig. 7.26 Internal structure of three-phase three-wire induction counter

Connecting with Current and Voltage Measuring Transformer: In order to measure the electrical energy consumed in high voltage circuits, the measuring transformers are connected to the counters as shown in Fig. 7.28b); and therefore, the necessary measurements are made. High voltage is reduced for counter voltage. Since the high voltage value is reduced to the voltage value of 100–110 V for the meters, this meter is also isolated from the high voltage.

Electronic Meters and Connections: Electronic electricity meters designed for the use in electrical installations and residences are produced with single-phase and three-phase class 1 sensitivity. These counters are also called smart counters. Sensitivity

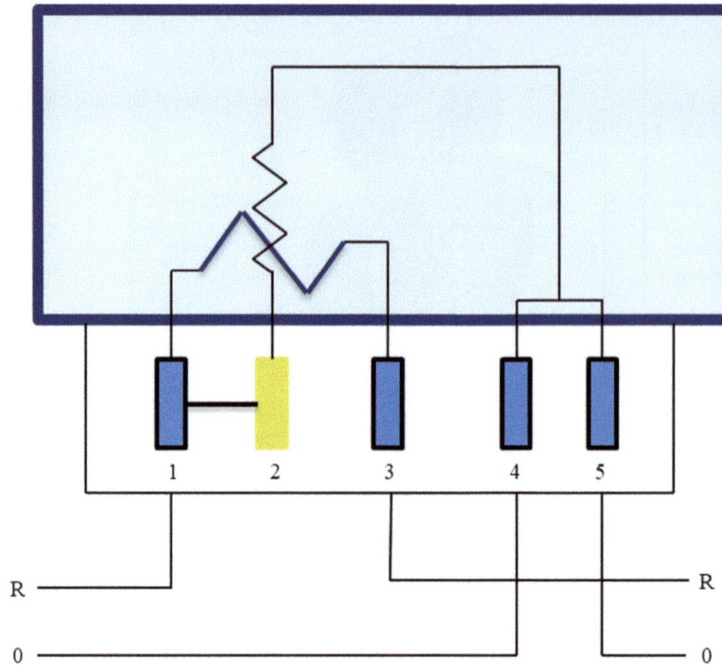

Fig. 7.27 Direct connection of the single-phase electricity meter to the circuit

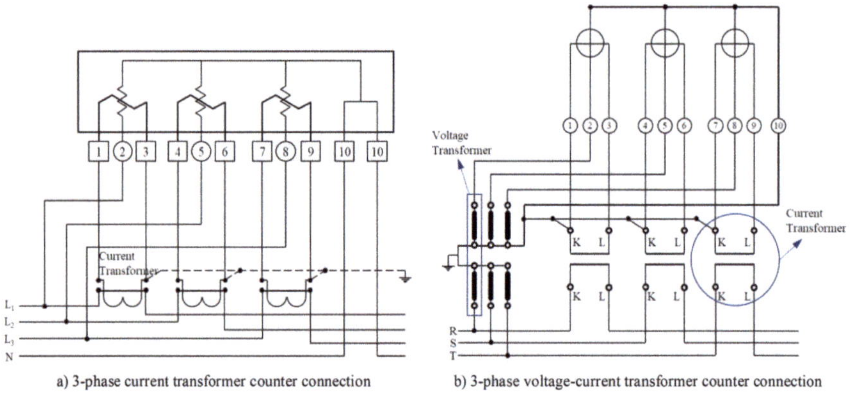

Fig. 7.28 Three-phase counter connections

classes have been reduced to 0.5. Today, it is necessary to use electronic counters in subscriptions of new structures and buildings. The widespread use of electronic meters makes a great contribution to both energy producers and subscribers as well as the economy of the country.

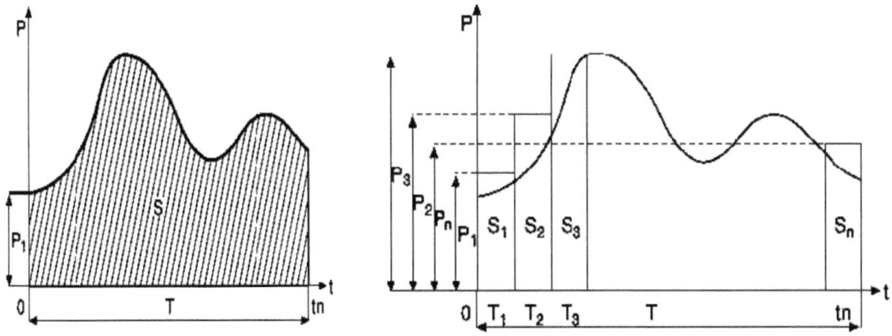

Fig. 7.29 The change of the power depending on time duration

The counter (like Wattmeter) measures the power of the circuit to which it is connected, and records these power values, which change over time, as in kWh or MWh (the calculation can be realized by multiplying power and time). For example, depending on the duration, the power drawn by the receiver should be as given in the graph on the right in Fig. 7.29. The energy that this receiver draws between 0 and t_n (within time T) is equal to the combed S area. The combed area (S) can be divided into small pieces as shown on the left in Fig. 7.29. In other words, the sum of the small parts gives the combed area S.

Where

$$S = S_1 + S_2 + S_3 + \cdots + S_n$$
$$S_1 = P_1 \times T_1$$
$$S_2 = P_2 \times T_2$$
$$S_3 = P_3 \times T_3$$
$$S_n = P_n \times T_n$$

If A uses instead of S

$$A = P_1 \times T_1 + P_2 \times T_2 + P_3 \times T_3 + \cdots + P_n \times T_n \rightarrow A = P \times t$$

where A—total energy amount (kWh), P—power drawn (kW), and T—power drawn time (hour).

To provide the tariff categorizations determined in the Electricity Tariff Regulations (varies by country), electronic meters usually (Fig. 7.30) can divide a day into 4 separate, eight separate time slots for weekly [3]. Also, Saturday and Sunday days to the main purpose of the description is to spread the electricity consumption in a balanced way in 24 h and to make the consumer conscious about electricity consumption. Therefore, the user can control consumption by checking the counter information; and since every time zone will be priced differently, the scheduled electricity usage will direct the consumer to consume electricity when it is cheap.

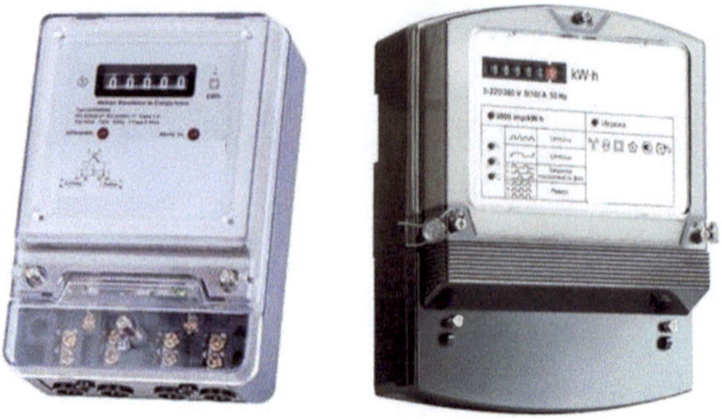

Fig. 7.30 Single- and three-phase electronic counters [3]

The differences between the mechanical counter/meter and the electronic counter are

- The measurement system of mechanical and electronic meters is different.
- There is no problem in the electronic meters due to mechanical parts.
- There is no or very little risk of using illegal electricity since electronic meters have records of the date and time of the opening of the terminals and a body cover.
- Energy information can be followed when retroactive consumption information is kept on electronic meters.
- It is possible to pay less money against the same energy consumed by subscribing to a multi-tariff system in electronic meters.

Energy Saving with Electronic Meters: Since mechanical meters are priced at a single tariff, the monthly expenditure is fixed. If the electronic meter is purchased and subscribed to the multi-tariff system, the amount of energy consumed in the electronic meter (depending on the time zone that electricity is used) is less than the mechanical meter. In order to save the money, it is necessary to subscribe to the multi-tariff system by applying only to the related institution, because only purchasing the electronic meter is not enough to benefit from these advantages.

Multi-Tariff Systems: According to when customers (who subscribe to multi-tariff system) use their electricity, the electricity meters (counters) for consumers have these tariffs as

$(06^{00}–17^{00})$ *daytime tariff,*
$(17^{00}–22^{00})$ *peak time tariff,*
$(22^{00}–06^{00})$ *nighttime tariff.*

- *The daytime tariff is normally about 5% cheaper.*
- *The peak time tariff is normally about 50% more expensive.*
- *The nighttime tariff is normally about 50% cheaper.*

In electronic meters, the billing system is transferred to the information computer that is received from the subscriber. Then the information is transferred to the data processing center and it is invoiced and sent to the subscriber.

Structure and Connection of the Electronic Counter: There are active and reactive electronic counters in the market. In addition, today in three-phase systems, three-phase combo meters are used (active-reactive-capacitive). The connection of electronic electricity meters is similar in mechanical meters. In addition, the terminal connections have the same characteristics as the mechanical meters.

Technical Features: The basic functions of electronic meters are given below:

Measuring Functions:

- In general, there are four tariffs for total active energy measurements.
- Total reactive-capacitive energy measurements can be made in four tariffs.
- Time zones and tariffs can be programmed.
- Twelve different time slots can be selected during the day, and one of the four types can be selected in the specified time zone.
- Additionally, 32 holiday days, 8 separate days, 8 different weeks, and 12 different monthly programs can be realized.

Billing Function:

- Billing indices can be read automatically.
- The start of the billing is at 00:00 on the first day of every month.
- It can be programmed on any day when requested.
- It can store information for the last 12 months.
- It has a storage capacity of 12 months based on four tariffs as total active, reactive, capacitive energies.

- Self-Diagnosis: The meter automatically checks the circuits continuously and displays the fault menu on the screen whether there is a malfunction or not. These faults include memory failure, real-time clock fault, clock pin weak fault, system pin weak fault; top cover and terminal cover open faults.
- Electric Terminal and Top Cover Opening, Closing, and Saving Feature: When opening the top cover or terminal cover, it records the first opening date for each month and the total number of openings in that month. The cover opening date and the last 12 months of the unit information are stored in memory.
- Communication with Counter, Reading with Optical Port: The electronic meters have optical ports on them. The meter can be read on both the computer and the handheld device with the optical port interface without any intervention. It can also be read in the dark with the Backlight system, which illuminates the LCD screen when the power is off.
- Reading via RS 485: Thanks to the RS 485 output, RS 232 can be read directly with the help of the reading program by connecting to the computer via the RS 85 adapter. Even though there is no obligation to study remote reading in future periods, many companies have already prepared the substructure of this system for the meters, owing to this data output. This will remove the function of the assigned staff for the meter reading operation for the future applications. This is also seen as the most important factor in reducing illegal electricity usage.

- Backup Power Supply: Provides continuous feed of real-time clock and calendar by taking advantage of uninterrupted power supply consisting of electronic counter, super capacitor, and battery in power failure. With this system, time and date/calendar information can be securely protected in case of power failure.
- Led Signal and Pulse Outputs: There are LEDs flashing on the counter.
- Summer/Winter Automatic Time Clock Setting: Electronic timers automatically adjust summer/winter time application for 16 years from production date.
- Time Programming: It can be set to 5, 10, 15, 30, 45, 60, and 120 min. 6 months of information is stored in the memory. A daily time can be divided into a maximum of 12-time slots and each time slot can be matched and selected one of the four tariffs. Furthermore, weekly time programming (WP), monthly time programming (MP), and holiday time programming (HP) can be performed besides the daily time programming.
- Operation and Function Controls: When the meter is installed and energized, the LCD screen appears.
- Electric Terminal Cover Open/Close Control: After screwing in electric terminal cover, the lock icons on the LCD screen start flashing. This icon disappears when reading with the optical port.
- Auto Display: Auto-Display mode switches on when the meter starts to work. Every piece of information appears on the screen every 5 s and the following information is displayed on the screen after that.
- Recording of the Phase and Current Break: Many of the electronic meters record the dates and times of the phase break beginning and ending separately for each phase and the total break time in minutes. It performs the same operations in case of a current break and reverses current formation.
- Compliance with Existing Panels: Electronic counters, as required by the conditions in any country, and other electrical panels and equipment in use are manufactured according to the country's standard sizes. It is possible to use in the same panel with mechanical meters.

Factors to be Considered While Using Electronic Meters:

- Do not run your meter for more than 10 s at the highest current.
- Do not touch the parts with electricity.
- During the dismantling and fitting of the meter, electricity must be cut off.
- Installation of meter should be done by authorized institutions or technicians by taking into account the connection scheme as a reference.
- Do not expose your meter to direct sunlight.
- Keep your meter protected against the impact.
- Protect your meter from water and excessive moisture.
- Do not clean the outer surface of the meter with solvents. In addition, plastic parts and its derivatives of the meter should be handled with care.
- Only clean the windscreen with a dry cloth if it is necessary.
- Do not interfere with the terminal cover.

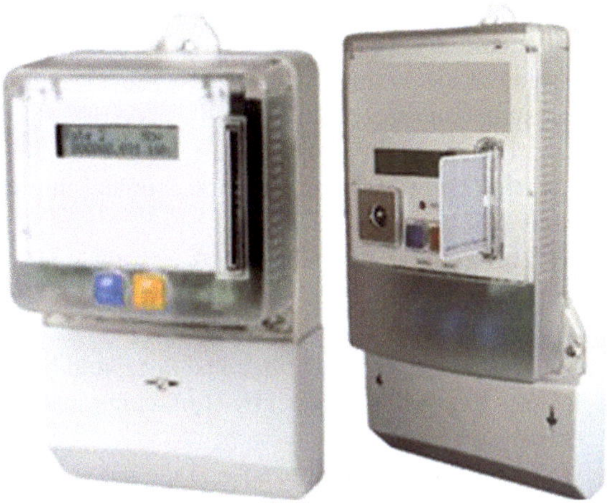

Fig. 7.31 Single- and three-phase electrical meters with card [3]

- There are two seals in the meter. Do not interfere with these points.
- This instrument is a measuring instrument. Please be careful when transporting.
- Do not use your meter except for the instructions given in the instruction manual. Otherwise, the company will not take any responsibility.

Meters with Card: Because of the physical structures of the electromechanical meters, it is not possible to accurately measure and record the consumption by making illegal transactions easily on these meters. However, with card counters, these situations are prevented.

Meters with card (Fig. 7.31) are electrical meters designed for prepaid use. In other words, a system allows the amount of electricity consumption as much as it is desired to be consumed by being loaded on the card and mounted on the meter [3].

In counter with card, the microprocessor reads the data from the smart card, stores the measured data by analyzing, and records other counter information in the non-volatile memory. It also transfers measurements and other data to the card. It regulates the operation of the LCD, controls the circuit breaker and the horn. It provides the genuine real-time clock for more than seven tariffs and four list tariffs and keeps a record of usage time.

Distribution companies want to incorporate the sub-systems that will reduce the time spent from the reading of the meters until the collection to the management systems. Thus, they want to reduce annual operating costs, secure collections, shorten the current collection periods, and reduce financial losses to a minimum. Card counters eliminate collection and reading problems. Other benefits of card counters can be summarized as follows:

- Directs the consumer with multiple tariff applications.
- Contributes to conscious consumer formation.
- Provides faster and more accurate data reading.
- Allows detecting and correcting network problems more quickly.

- Counter readers do not have to go to the meter.
- Automatic, fast, and healthy data reading and recording are ensured successfully.
- Instead of electromechanical measurement, more precise and reliable electronic measurements are made.
- By utilizing the data saved in the memory of the meter, it facilitates the detection of illegal interventions in the counters. It ensures the minimization of such interventions.
- With the creation of healthy databases, the consumption can be tracked periodically, individually, and in total. It also makes available to realize some different plans, successfully.
- Network problems can be handled faster and healthier.
- It saves time and personnel by eliminating mistakes in meter reading and billing operations.
- Unlawful uses are minimized, and the income increase can be provided.
- It provides the regulation of cash flow with credit or prepaid sales.
- Possible mishaps that may occur to counter readers (meter readers) are not in question.
- Customer service quality is increased.

7.5.1.6 Power Factor Measurement

In alternating current circuits, the angle φ between the grid voltage applied to the circuit and the circuit current varies depending on the ohmic, inductive, and capacitive resistances in the circuit. The whole process of boosting the small power coefficient of the circuit load to a greater value is called *the correction of power coefficient* or *compensation.*

Power Factor: The power factor is defined by the ratio between active power and apparent power in an alternating current circuit.

$$\cos\varphi = \frac{\mathrm{P}}{\mathrm{S}} = \frac{\mathrm{W}}{\mathrm{V} \times \mathrm{A}} \tag{7.1}$$

In power measuring in AC circuits, if the circuit is inductive or capacitive, there is a phase difference between current and voltage in such circuits. This difference is indicated by the angle and gives the cosine power coefficient or power factor of this angle.

Power Factor Measurement: At the plants, the reactive energy drawn from the system (apart from the active energy) is measured by the relevant meters and related institution collects the price. Reactive power is unavailable power and it is possible to reduce this power. To prevent power loss, the power factor must be measured. The power coefficient can be measured directly or indirectly. Instruments that are directly measuring power coefficient are called *cosine phi meter* or *power factor meter*. However, the power coefficient can also be calculated from the active and reactive meter readings.

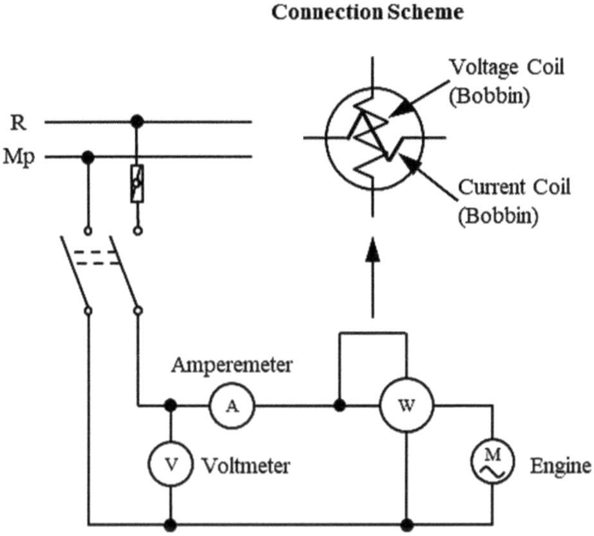

Fig. 7.32 Measuring the power factor by using the amperemeter, voltmeter, and Wattmeter

The power coefficient is measured in two ways:

(1) Measuring the power factor by using the amperemeter, voltmeter, and Wattmeter (Fig. 7.32).
(2) Measuring the power factor by using the cosine phi meter (Fig. 7.33).

The power factor is determined by reading the counter as follows. The values of the active energy meter (kWh) and the reactive energy meter (kVArh) at normal operating conditions and the longest possible time interval are taken at the beginning and the end of the working day. Thus, the active energy consumption during working

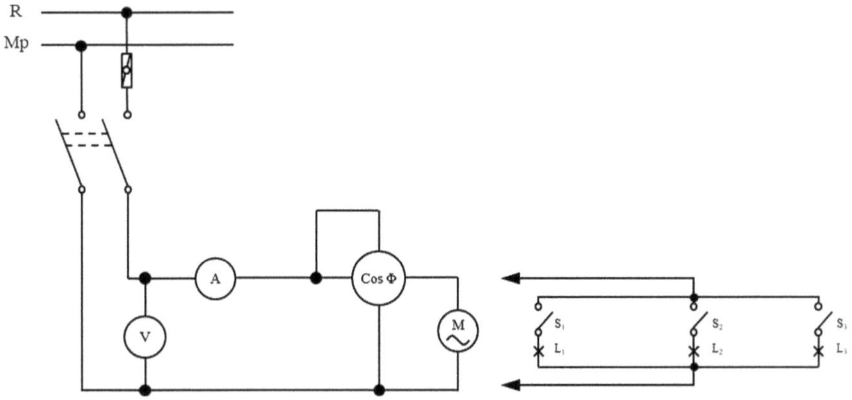

Fig. 7.33 Measuring the power factor by using the cosine phi meter in single-phase circuit

hours is determined from the $E_a = L_e - L_i$ relationship. Where E_a shows the active energy consumption during working hours, L_e is the energy value that is read at the end of the workday, and L_i is the energy value that is read at the beginning of workday.

The reactive energy that is taken during working hours is determined from $E_r = L_f - L_i$ relation. Where, E_r is the reactive energy taken at work hours, L_f is the energy read at the end of the working day, and L_i is the energy read at the beginning of the working day. The power factor is calculated from the values of E_a and E_r as it can be seen the following equation:

$$\cos\varphi = \frac{E_a}{\sqrt{E_a^2 + E_r^2}} \tag{7.2}$$

The time interval between L_e and L_i and Cos φ and the reactive power can be calculated as

$$kW = \frac{E_a}{T} \qquad kVAr = kW \times \tan\varphi$$

Measuring with the method of Amperemeter, Voltmeter, and Wattmeter in Single-Phase Circuits: Active power is the useful power that turns it into work on the receiver. It is calculated by the formula $P = V \times I \times \mathrm{Cos}\ \varphi$. The measured current, voltage, and active power are used in the following equation, and therefore, the power factor is calculated as

$$\cos\varphi = \frac{P}{V \times I} \tag{7.3}$$

In Three-Phase Circuits: The power factor can be determined by two methods. It is calculated according to the following equations for balanced and unbalanced loads. It is calculated according to the values read from Wattmeter.

(1) *For balanced loads*; P value can be found by $P_{Total} = 3P_1$ formula. Then power coefficient will be calculated from the equation $\mathrm{Cos}\ \varphi = (P/V \times I)$.
(2) *For unbalanced loads;* P value can be found by $P_{Total} = P_R + P_S + P_T$ formula. Then, power factor will be calculated from the equation: $\mathrm{Cos}\ \varphi = (P/V \times I)$.

Connection diagram for finding the power factor with the help of amperemeter, voltmeter, and Wattmeter in three-phase balanced loads can be seen in Fig. 7.34.

Measuring with Cos phi Meter: The methods given above for measuring the value of power factor are not preferred by the enterprises. The value of the power factor must be measured directly. Instruments that directly measure the power factor are called the *cosine phi meters*.

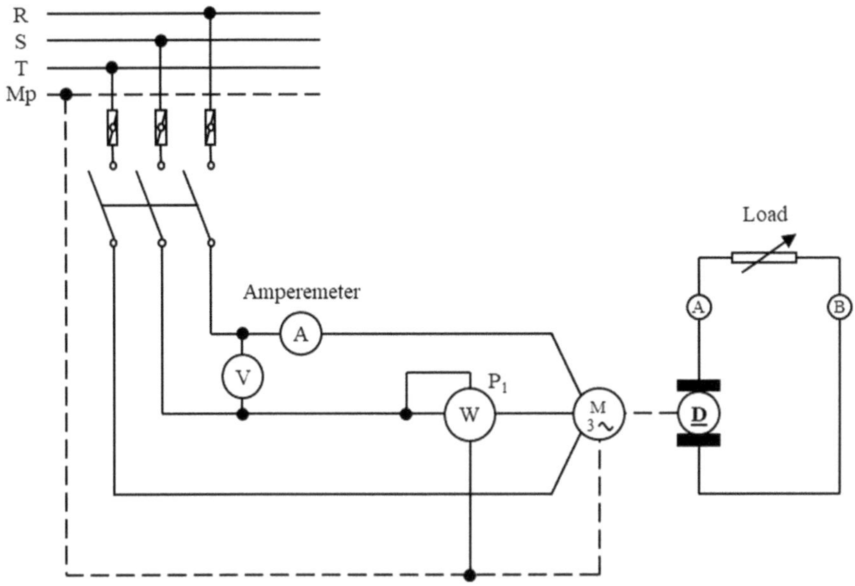

Fig. 7.34 Connection diagram for finding the power factor with the help of amperemeter, voltmeter, and Wattmeter in three-phase balanced loads

The Structure of Cosine phi meter: Single-phase cosine phi meters (Fig. 7.35) are placed so that the two voltage coils, whose magnetic axes are perpendicular to each other, can move diagonally into the fixed current coil, as it is the case with electrodynamic Wattmeters. One of the voltage coils is connected to an ohmic resistance, the inductive reactance in series. The number of turn (number of winding) and wire diameters of the two coils are the same. The current passing through the ohmic resistance connected coil and the voltage are in the same phase, the current passing through the coil connected in series with the inductive reactance is 90° behind the voltage. The phase difference between the currents and voltages of the two cross coils is made 90° by means of resistance and coil.

When the ohmic load is connected to the circuit (power factor is 1), the current passing through the current coil is in the same phase as the voltage. The current passing through the voltage coil 1 is in the same phase as the current passing through the current coil. The current through coil 2 is 90° behind them. The rotational force is generated by the resultant magnetic fields formed by these two currents. This force affects only coil 1 and rotates the transverse coil until the axis of this coil is at the current axis 90°, indicating the number 1 on the pointer scale. In the case of ohmic load, there is no force acting on coil 2. Because the current passing in ohmic load is also ohmic. The ohmic current also has no inductive and capacitive components. When the circuit is fully inductive (power factor is zero), the current passing through the current coils is in the same phase, the current passing through coil 1 is 90° ahead of them. In this case, coil 2 turns until the current coil is perpendicular to the axis and the pointer indicates zero. In other words, $\cos \varphi = 0$.

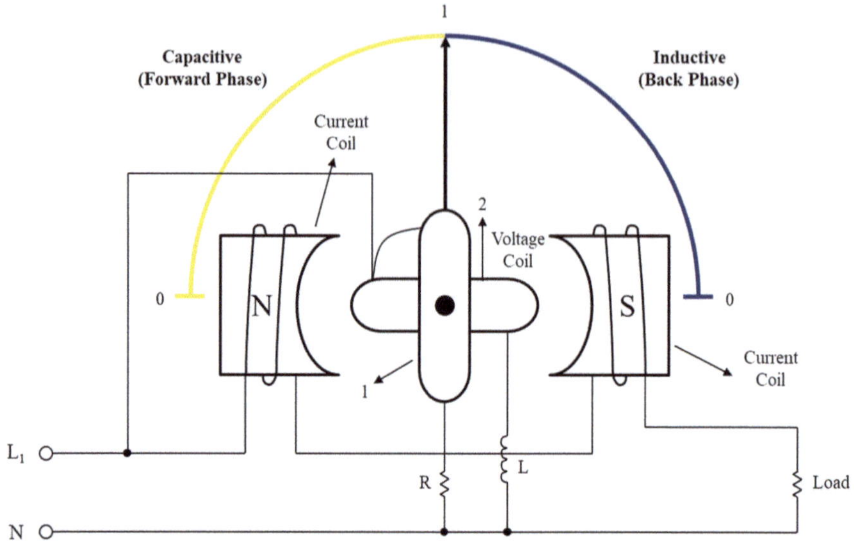

Fig. 7.35 Internal structure of a single-phase cosine phi meter

When the circuit is fully capacitive (power factor (coefficient) is zero), the opposite situation happens when the circuit is fully inductive and coil 1 rotates 180°, this time the indicator shows zero on the capacitive side. Power coefficient (factor) is between 0 and 1; the phase difference between the voltage coils and the current coils is in such a way so that they can give the power coefficient. The right side of the scale shows the inductive load state and the left side shows the capacitive load state.

Power coefficient is between 0 and 1; the phase difference between the voltage coils and the current coils is in such a way so that they can give the power coefficient. The right side of the scale shows the inductive load status and the left side shows the capacitive load status.

It is placed between moving voltage coils 120°—phase difference. The voltage coils are connected to the two other terminals outside the phase to which the current coil is connected. The stationary coils carry the line current. If the instrument is connected to a three-phase load, the pointer shows the power factor on the scale. Three-phase cosine phi meter (see Fig. 7.36) current coils are connected in series to the load. One of the ends of the voltage coils are connected to the other phases. The measurements give a directly power factor value.

Types of Cosine phi meters: Cosine phi meters are manufactured in two types, single-phase and three-phase types according to the phase diagram. They are produced as analog and digital types according to the production type (see Fig. 7.37).

As shown in the single-phase and three-phase digital cosine phi meter connection diagram (Fig. 7.38), a suitable current must be connected to the network via a current

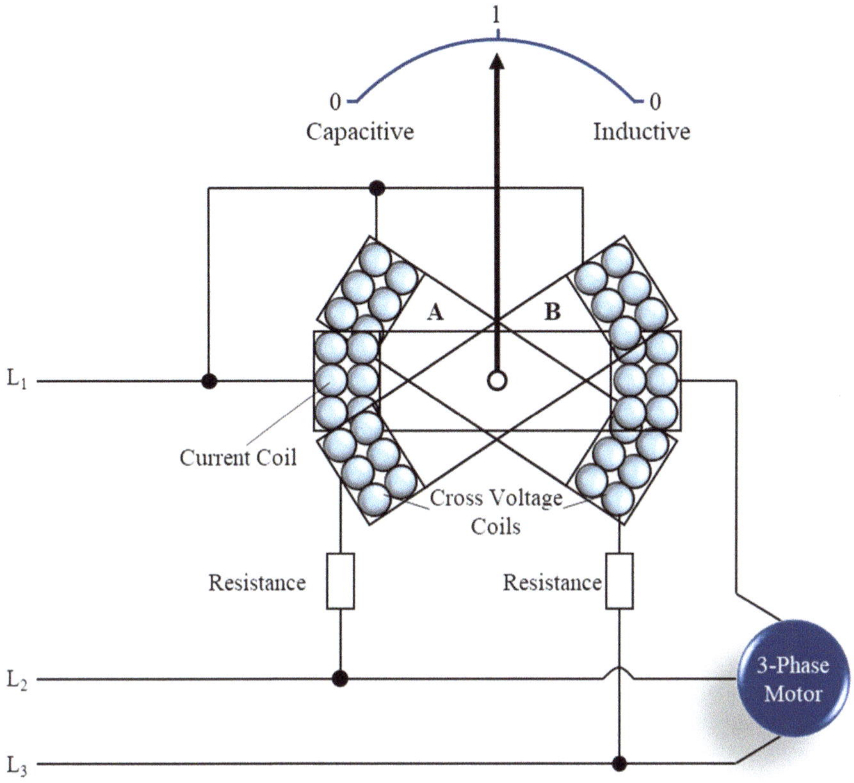

Fig. 7.36 Internal structure of three-phase cosine phi meter

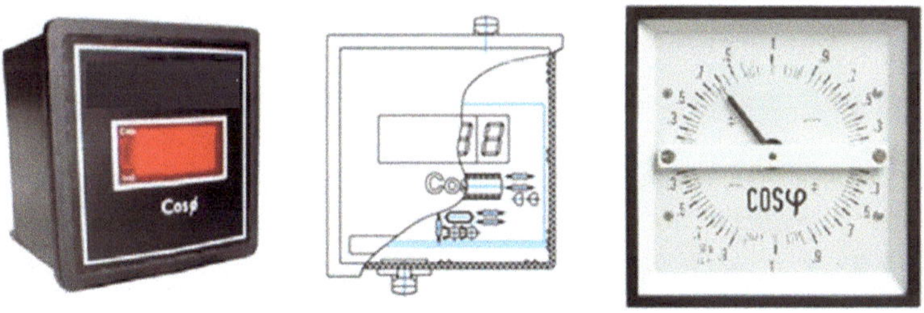

Fig. 7.37 Analog and digital cosine phi meters

transformer. When connection is made, care must be taken to ensure that the supply voltage is taken from the phase to which the current transformer is connected (U_n is supply voltage).

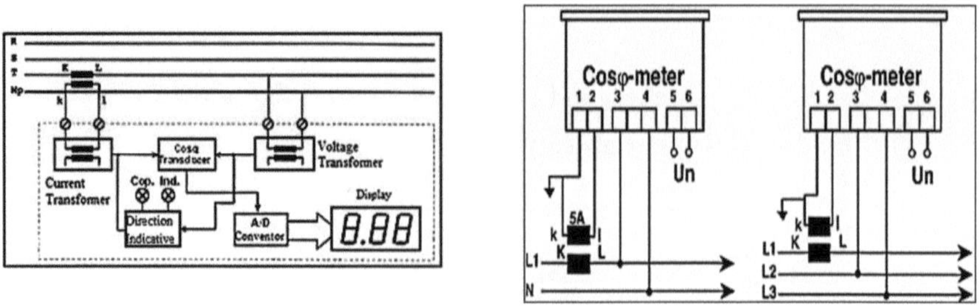

Fig. 7.38 Connections of cosine phi meters to the circuit

7.5.2 Force Measurement

The concept of force or weight measurement is the most common topic even in non-industrial notions. Indeed, the measurement of these quantities is an important part of the productions. The following quantities have to be measured in many industrial areas such as storage filling, bagging, dosing, quality controlling, screwing, power, and level measuring. Force or weight measurement is a process that requires great care. The concepts of mass, force, and weight are defined as follows:

Mass (It is symbolized "m"): Mass is a dimensionless quantity that represents the amount of matter in an object or a particle. In other words, it can be defined as the amount of substance in that object. This amount remains constant even if the object is taken anywhere. In the International System (SI), the standard unit of mass is the kilogram (kg). The mass of an object can be calculated by using the force and the acceleration. Mass is not the same as weight, they are different.

Force (It is symbolized "F"): A force can be described as any interaction that, when unopposed, will change the motion of an object. A force can cause an object with mass to change its velocity, which includes beginning moving from a state of stationary, for example, to accelerate. As it is well known, a force has both a numerical magnitude and direction, and an application point that makes it a vector quantity. The force is associated with mass through Newton's second law of motion. Newton's second law says that when a constant force acts on a massive body, it causes it to accelerate, for example, to change its velocity, at a constant rate. With a constant mass, $F = m \times a$ (Force = mass × acceleration). Force, acceleration, and velocity are vector quantities. Each has a magnitude and a direction.

Weight (It is symbolized "W"): It is a quantity representing the force exerted on a particle or an object by an acceleration field, especially the gravitational field of the Earth at the surface. The constant acceleration due to gravity is written as g, and therefore, Newton's Second Law can be written as

$$F = m \times g$$

Mass Units Conversion (EN-45501 Balance Standard) can be seen in Table 7.2.

Table 7.2 Mass Units Conversion (EN-45501 Balance Standard)

Table of Weight Unit Conversion-EN 45501 Balance Standard							
Ton Mg	kg	g	UK Ton	US Ton	cwt	lb	oz
1	1000	10^6	0.9832	1.1011	19.66	2.205×10^3	3.527×10^4
10^{-3}	1	1000	9.832×10^{-4}	1.101×10^{-3}	1.966×10^{-2}	2.2046	35.274
10^{-6}	10^{-3}	1	9.832×10^{-7}	1.101×10^{-6}	1.966×10^{-5}	2.204×10^{-3}	3.527×10^{-2}
1.017	1017	1.017×10^6	1	1.12	20	2240	35840
0.9081	908.1	9.081×10^5	0.8928	1	17.856	2000	32000
5.085×10^{-2}	50.85	5.085×10^4	0.05	0.0560	1	112	1792
4.536×10^{-4}	0.4536	453.6	4.46×10^{-4}	5×10^{-4}	8.92×10^{-3}	1	16
2.835×10^{-5}	2.835×10^{-2}	28.349	2.79×10^{-5}	3.125×10^{-5}	5.580×10^{-4}	6.25×10^{-2}	1

7.5.2.1 Mechanical Measurement Methods

Force measurement methods can be divided into two groups as direct comparison and indirect comparison. In the direct comparison method, an unknown force is directly compared with the gravitational force acting on a known mass. A simple analytical balance is an example of this method. The indirect comparison method requires the use of tuned masses or transducers.

- Arm-balance methods,
- Force-balance method,
- Hydraulic pressure measurement,
- Acceleration measurement,
- Elastic elements (elastic materials).

7.5.2.2 Load Cell

The Structure and Types of a Strain Gauge: Strain gauge is an element whose resistance changes when the length of the wire changes. The strain gauge is connected to a steel cylinder that extends when stretched and shortens when compressed. The strain can be measured by using several methods, but the most common is with a strain gauge. A strain gauge's electrical resistance varies in proportion to the amount of strain in the device. The most commonly used strain gauge is the bonded metallic strain gauge. The metallic strain gauge consists of a very fine wire or in a widespread manner, metallic foil arranged in a grid pattern. The grid pattern enlarges the highest amount of metallic wire or foil subject to strain in the parallel direction. The grid is bonded to a thin backing called the carrier, attached directly to the test sample. Therefore, the strain experienced by the test sample is transferred directly to the strain gauge that responds with a linear change in electrical resistance.

The resistance value is proportional to the length of the wire made of the strain gauge. It is possible to determine the load on the load cell by measuring the resistance of the strain gauge. Force cannot be applied directly in practice. The strain gauge is usually attached to structure element that is under the tension. The device that measures the shape change must be very light and responsive so that the object to which it is attached is able to accurately show the deformity. Furthermore, it should not take into account the shape changes of the undesired side of the shape change. To achieve this, there is a concept called cross sensitivity or indicator factor.

Load Cell Structure: The load cell is the transducer used to convert a stress (strain) or load into an electrical signal. This transformation is indirect and takes place in two stages. A strain or load changes the shape of a strain gauge. The strain gauge measures the deformation (caused by strain) as an electrical signal. Because strain changes the effective electrical resistance of the wire. Four strain gauges are used in load cells. These are linked in the form of the Wheatstone bridge (Fig. 7.39). When a stress or load is applied to the bridge, the output voltage will be a voltage proportional to the applied stress or load. The electrical voltage output is usually several millivolts and must be amplified with an amplifier before it can be used. The converter output is associated with an algorithm to calculate the force applied to the converter. Load cells have a total of four leads, two inputs and two outputs, as shown in Fig. 7.39. In some load cells, in addition to these ends (terminals), there are two other terminals. These are the (+) sense and (-) sense terminals. These terminals are connected to the same location as the input ends, as shown in Fig. 7.39b. Its function is to determine whether there are any breaks in the load cell connections.

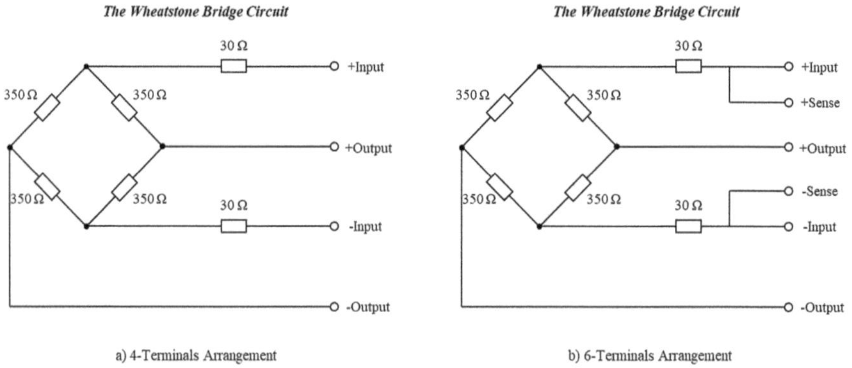

Fig. 7.39 The internal structure of load cell

Terms Related to Load Cells

Full display output voltage: It is the output value of the load cell under full load. Its unit is mV/V. For example, if this value is 2 mV/V and the supply voltage is 10 V, the output voltage under full load is 20 mV.

Supply (Excitation) Voltage: The supply voltage required for the operation of the load cell. Generally, 10 V is selected for this value, and its maximum value is 15 V.

Input Resistance: The resistance measured between the input terminals of the load cell.

Output Resistance: The resistance is measured between output terminals of load cell.

Hysteresis: The difference between load cell output readings for the same applied load; one reading obtained by increasing the load from zero and the other by decreasing the load from rated output. It is described as percentage of full display output voltage.

Nonlinearity: The maximum deviation from the calibration curve drawn between the output in unloaded state and the output values taken under the full load. It is defined as the percentage of the full indicator value.

Balanced Temperature Range: The temperature range at which the load cell is balanced to maintain the output and zero balance.

Safe Temperature Range: The temperature range at which the load cell can operate without permanent changes in yield characteristics.

Insulation Resistance: The resistance measured between the bridge circuit and the building material of the load cell.

Zero Balance: The voltage value that is taken from the outlet of the load cell when it is in the unloaded state. Usually, it is defined as the percentage of the full display output voltage.

Types of Load Cell: Load cells are produced in various designs depending on their capacity. (Table 7.3 and Fig. 7.40).

Beam-type load cells (Fig. 7.40a) have been developed in electronic weight and force measurement applications that operate according to shear force principle for the use in industrial environments. This type of load cells are used where places have low weight, especially platform scales, bagging, and dosing machines.

Table 7.3 Types of the load cells

Type of load cell	Capacity
Beam type	20 kg–200 kg
S beam type	500 kg–2 tons
Shear type	200 kg–5 tons
Platform type	6 kg–600 kg
Compression type	10 tons–200 tons

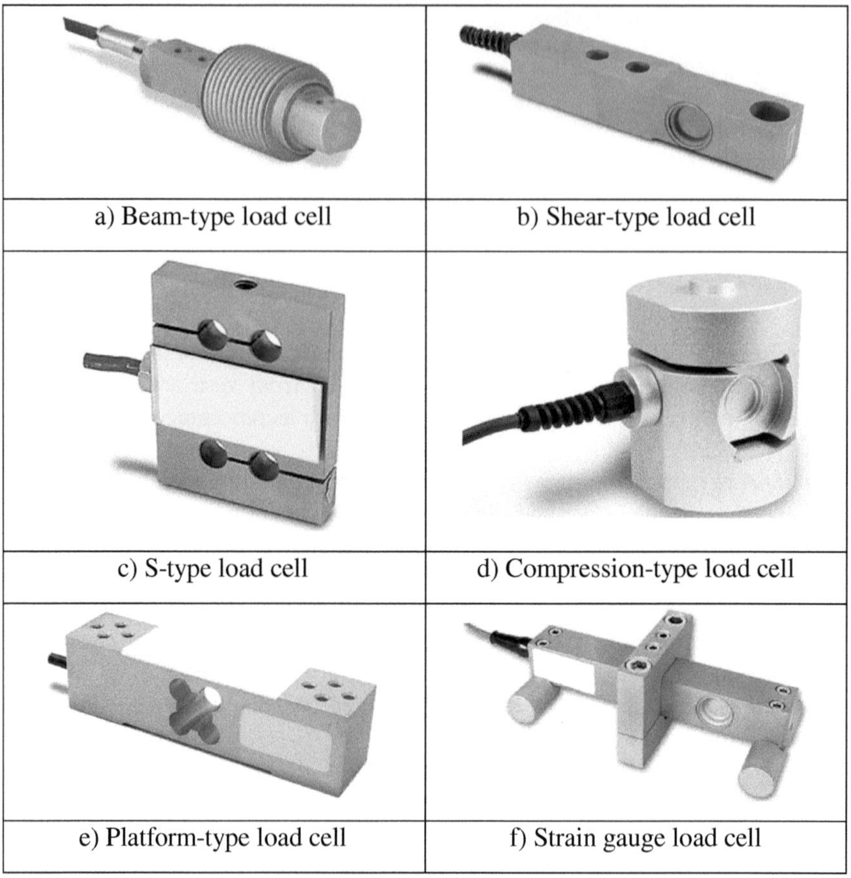

Fig. 7.40 Types of load cells

Shear-type load cells (Fig. 7.40b) are designed for high-precision electronic force and weight measuring systems that operate according to shear force principle. Shear-type load cells have been developed for use in the direction of pressure and to find application areas particularly for platform scales, process measurements, and mechanical electronic transformations.

S-type load cells (Fig. 7.40c) were developed for force measurement applications in the direction of pushing and pulling with the principle of shear force. It is used especially in the translation of band scales and mechanical scales into electronic.

Compression-type load cells (Fig. 7.40d) have been developed for the use in high sensitivity and high-capacity electronic weight and force measurement applications in the direction of shearing and compression. They have been developed for load applications in the push direction such as tank and wagon weighing systems and very high-capacity weighbridges.

Platform-type load cells (Fig. 7.40e) are used in low capacity, single-load platform weighbridges where the loading point can vary. It is used intensively in trading scales, industrial parts counting scales, bagging, and dosing machines and in scientific fields.

Table 7.4 Measuring force by the load cell

Process steps	Guidelines
• Install the load cell mechanically on the measuring device • Connect the terminals of the load cell electrically • Check the strength of the load cell electrically • Carry out a measurement by applying various forces to the load cell • Connect the load cell to the amplifier electrically • Prepare the final report	• Make mechanical connection by using the connection elements of the load cell and taking into account the direction of force • Make sure that the mounting surface is parallel to the ground • Do not forget that the terminals that have a resistance of 400 Ω are the input terminals and the terminals that have a resistance of 350 Ω are the output terminals • If you cannot see between 500 and 200 Ω between terminals, the tool is defective • Install the terminals of the load cell with respect to the directions to the input of the amplifier • Note the values by applying various forces whose value is known and not exceeding the capacity of the load cell • Connect the measuring device to the output of the amplifier and reset the output when there is no force • Report your actions and the challenges that you have encountered

The strain gauge load cells (Fig. 7.40f) are mounted on steel ropes, which are immobilized on the elevator or winches for overload protection purpose. It detects the tension of the load at the steel rope. It is used at 2 or 4-rope systems.

The tension-sensing roller is used to detect the loads on the rollers by attaching to the shafts of the rollers. It provides ease of assembly with its own built-in ball. It is used to perceive the tension of the material passing through the cylinder.

Measuring Force by the Load Cell: The process steps and recommendations related to the load cell and force measurement are given in Table 7.4.

7.5.3 Pressure Measurement

7.5.3.1 Pressure

The pressure is the amount of perpendicular force acting on a surface per unit area. Because of their weight, solids, liquids, and gases apply a force to the surface. Regardless of the source of the force, which is applied perpendicular to the surface of an object per unit area on which that force is distributed is called as pressure (its symbol is P), the force acting perpendicularly to the whole surface is called the pressure force (its symbol is F).

The effect of the air column on this surface to the unit area of a surface is called the pressure force. The force acting perpendicularly to the unit area: weight, force, or gas molecules (because of their mobility) may be the sum of the impacts of impacts on the surface. The pressure is calculated by dividing the total amount of force by the surface area, acting on the surface.

$$P = \frac{F}{A} = \text{Pressure} = \frac{\text{Force}}{\text{Area}} = \text{Pa} = \frac{N}{m^2} \tag{7.4}$$

where

Pa: Pascal, which is the SI unit for pressure.

7.5.3.2 Pressure Units

Many units are used to define the extent of pressure. In the International System of Units (SI) (Such as Meters, Kilograms, Seconds), the pressure unit is Pascal (Pa).

The force is equal to the multiplication of mass and acceleration. If the mass unit is kg, the acceleration unit is m/s^2, the force unit is N = Newton, $N = kg \times m/s^2$. In the CGS (Centimeters, Gram, and Seconds) unit system, the pressure unit is defined as follows:

$$\begin{aligned}\text{Pressure} = \frac{\text{Force}}{\text{Area}} &= \frac{\text{dyn}\left[\left(g \times cm/s^2\right)\right]}{cm^2} \\ &= (\text{Dyne per Square Centimetre}) \times 10^6 = \text{bar}\end{aligned} \tag{7.5}$$

Conversion of pressure units can be seen in Table 7.5.

Since dyne per square centimeter is a very small pressure unit (1dyn/cm^2=0.1 Pa (N/m^2), so 10^6 times of this unit is generally used, which is bar (10^5 Pa), as a common pressure unit. The pound-force per square inch (psi) is the traditional unit of pressure in the imperial and US customary systems. Another commonly used pressure unit is the *atmosphere*. An atmosphere is the pressure of a 76 cm high mercury column to the unit area. Since the specific gravity of the mercury is 13.6 g/cm^3, the atmosphere is equal to 1.033 kg/cm^2. Pressure may also be stated in terms of standard atmospheric pressure; the atmosphere (atm) is equal to this pressure, and the *torr* is described as 1/760 of this. Manometric units such as the centimeter of water, millimeter of mercury, and inch of mercury are used to define pressures in terms of the height of column of a specific fluid in a manometer.

Table 7.5 Conversion of pressure units

	Pascal	Bar	Standard atmosphere	Kilogram force per cm^2	Pound force per $inch^2$	mm water column	mm mercury column	Inch water column	Inch mercury column
	Pa	bar	atm	kgf/cm^2	$Lbf/inch^2$ (PSI)	mmH_2o	mmHg (Torr)	$inchH_2O$	inchHg
Pa	1	10^{-5}	9.86923×10^{-6}	1.01972×10^{-5}	1.45038×10^{-4}	0.101971737	7.50064×10^{-3}	4.01463×10^{-3}	2.95300×10^{-4}
bar	10^5	1	9.86923×10^{-1}	1.01972	14.5038	1.01972×10^{-2}	750.064	401.463	29.53
atm	101325	1.01325	1	1.03323	14.6959	10332.287	760	406.782	29.9212
kgf/cm^2	98066.5	0.980665	0 .967841	1	14.2233	9999.69	735.561	393.701	28.959
$Lbf/inch^2$ (PSI)	6894.76	0.0689476	0.068046	0.070307	1	703.069	51.7151	27.6799	2.03602
mmH_2O	9.80663	9.80663×10^{-5}	9.6784×10^{-5}	1.00003×10^{-4}	1.42233×10^{-5}	1	7.35558×10^{-2}	0.039370	2.89589×10^{-3}
mmHg (Torr)	133.322	0.00133322	0.00131579	0.00135951	0.0193367	13.5951	1	0.535238	0.0393699
$inchH_2O$	249.089	0.00249089	0.00245832	0.00254	0.0361273	25.400	1.86833	1	0.0735559
inchHg	3386.39	0.0338639	0.033421	0.0345316	0.491154	345.316	25.4001	13 .5951	1

7.5.3.3 Pressure Measurement

A variety of techniques has been developed for pressure measurement. Devices used for pressure measurement are classified in Table 7.6.

Differential Elements: Differential elements are used to see the difference between two different pressures in the pressure measured system. It is usually used to see the difference in inlet and outlet pressure differences in the filters and the difference between the outlet and return pressures in heating and cooling systems. Figure 7.41 shows the internal structure of the pressure measurement system with differential elements.

The pressure gauge has two inputs. These are marked as "+" and "-". During installation, the positive pressure must first be given to the system. The pressure measured on the scale is the difference between the two pressures. These devices are called differential pressure gauges. This measurement is done by Bourdon tube manometer or diaphragm pressure gauge. These devices are filled with damping fluid (usually glycerin) when measuring the pressure at the test points under high dynamic load with impact and vibration.

Bellow-Type Pressure Gauge: Bellows-type differential pressure gauges (Fig. 7.42) are used to measure pressure values between 0 and 1000 Pa. The bellows will be damaged at higher pressures. The bellows are designed as a material in which the copper alloy is pressed into each other in the form of a thin corrugated sheet.

When pressure is applied inside the bellows, the bellows will expand and lengthen. This elongation value is related to pressure. The increase in the amount of pressure is

Table 7.6 Classification of pressure gauges

Liquid Column Manometer (Pressure gauge)	• U -Tube (Type) Manometer • Well-Type Manometer • Inclined Manometer • Micro Manometer • Barometer • Bell Type Manometer • Circular Balanced Type Manometer
Elastic Element Manometer (Pressure gauge)	• Bourdon Manometer • Diaphragm • Capsules • Bellows
Piezoelectric Manometer (Pressure gauge)	
Bridgman Manometer (Pressure gauge)	
Low Pressure Manometer (Pressure gauge)	• McLeod Device • Knudsen Device • Ionization Device

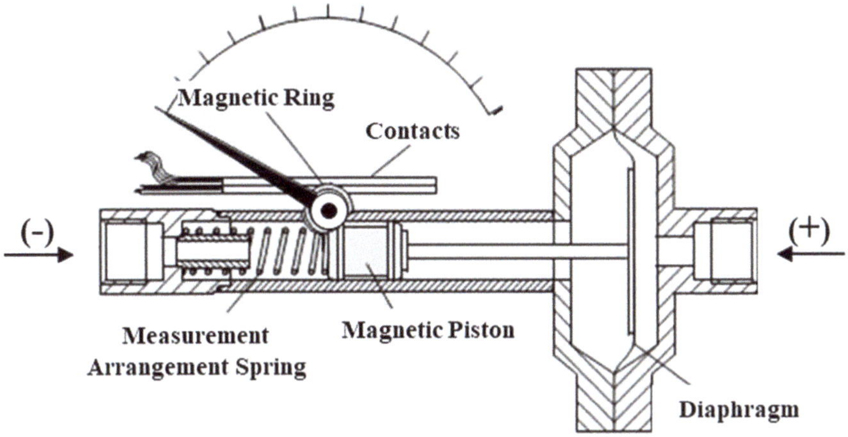

Fig. 7.41 Pressure measurement system with differential elements

$$P = \frac{d}{A}\lambda$$

P= Applied Pressure [N/m^2]
d= Extension Amount of Bellow [m]
A= Bellow Cross-Sectional Area [m^2]
λ=Strain [N/m]

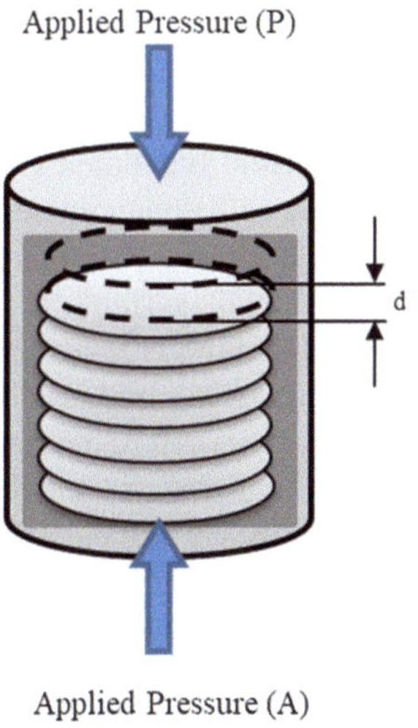

Fig. 7.42 Operation principle of the bellows-type pressure gauge

perceived as the distance change in the bellows. It is possible to read this elongation value as a pressure value, which is changed on the screen with various pressure values.

Diaphragm Manometer (Pressure Gauge): Diaphragm: It is a thin metal sheet and is supported in bearings along its circumference. It is usually made of stainless steel or brass. When pressure is applied, the diaphragm makes a swelling. Movement caused by swelling of the diaphragm displaces a shaft or a mechanical device. This displacement is converted to electrical data by various sensors and the applied pressure is measured (Fig. 7.43).

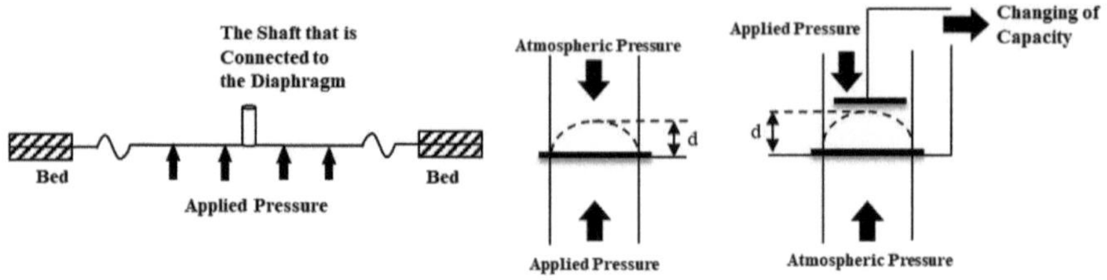

a) Internal structure b) Working principle

Fig. 7.43 Internal structure and working principle of diaphragm pressure gauge

Figure 7.43 shows a diaphragm and a conductive plate located at a distance from the diaphragm. Since there is air between the plate and the installed sheet as insulator, a capacity measurement takes place here. When pressure is applied to the diaphragm, there is a swelling in the diaphragm and thus a displacement up to "d". This displacement amount also changes the capacity evenly. In other words, the change in the applied pressure means that the capacity changes. A linear scaling between the lowest and highest values in the display allows the pressure to be measured in terms of capacity or pressure.

Manometer: Instruments used to measure the pressure of gas or liquid fluids are generally called manometers. The manometer measures the working pressure in the system according to atmospheric pressure. Manometers are used in both hydraulic and pneumatic systems. It is used to measure the pressure of the fluid in hydraulic systems and the pressure of the air in pneumatic systems. Manometers are usually connected to the pressure line. In the system, pressure can be controlled in places where it is desired to measure the pressure in these systems. Manometers are widely used in industrial areas and ships with adverse conditions and shaking.

The mercury column manometer is the simplest type of the manometer. It is used extensively to measure fluid pressures continuously in the laboratory. This type of manometer includes U-type, well-type, and inclined-type manometers (Fig. 7.44). By measuring the heights of the appropriate liquids placed inside them, the desired pressure or pressure differences can be found directly.

U-Type Manometers: A suitable fluid is put into the transparent tube such as U-shaped glass or plastic to the middle level. To measure the pressure that is applied to one of the manometer's column of the pressure gauge, while the other end of the pressure gauge is open to the atmosphere, P_a and the fluid heights in the manometer columns are balanced at different levels. If the total difference between these levels is h, the density of the fluid transmitting pressure is ρ_f, the density of the liquid in the manometer is ρ_m, and the acceleration of gravity is g, then the following equation is valid for the equilibrium of pressures in the two columns.

$$P_a + \rho_m \times g \times h = P + \rho_f \times g \times h \quad \rightarrow \quad P - P_a = g \times h \times (\rho_m - \rho_f) \quad (7.6)$$

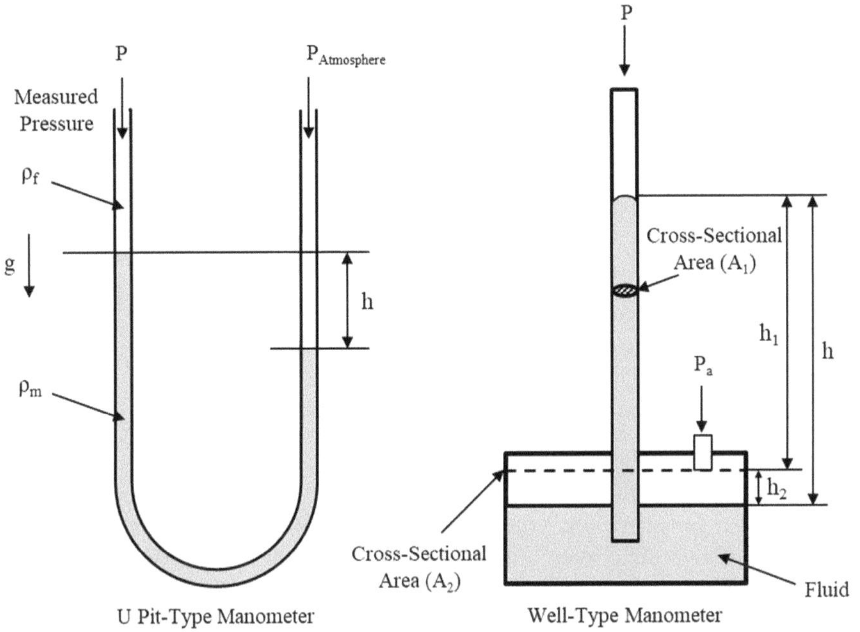

Fig. 7.44 Mercury column manometers

In U-type manometer, the height (h) can be measured with ± 2 mm sensitivity with the naked eye. Usually, pure water, alcohol, and mercury are used in the manometers. The densities of these fluids can be obtained with a sensitivity of 0.005%. As it can be seen, the density of the pressure gauge fluid has little effect on the result. In order to increase the sensitivity, reading must be done very carefully. In addition, the length of the connecting tube, which transfers the pressure, should not be too long.

Usage Areas of the U-Type Manometer: U-tube manometers are not much used in the industry. U-tube manometers are used for the calibration of other instruments used in the industry (Fig. 7.45). It is also used for experimental studies based on observation in laboratory environments. The biggest drawback is that they have to make the measurements at close range. This requires that there is an attendant who immediately informs the pressure changes of the manometer in the applications that require a quick response. Such usage difficulties limit the use of U-tube manometers in the industry.

As shown in Fig. 7.45, any type of pressure measurement can be made by using the U-type manometer. This makes the U-type manometer an indispensable measuring device of the calibration tables. Nevertheless, the poisonous property of mercury should not be ignored. Mercury vapor must not be inhaled. Minimizing reading error during measurement will also be the main calibration rule.

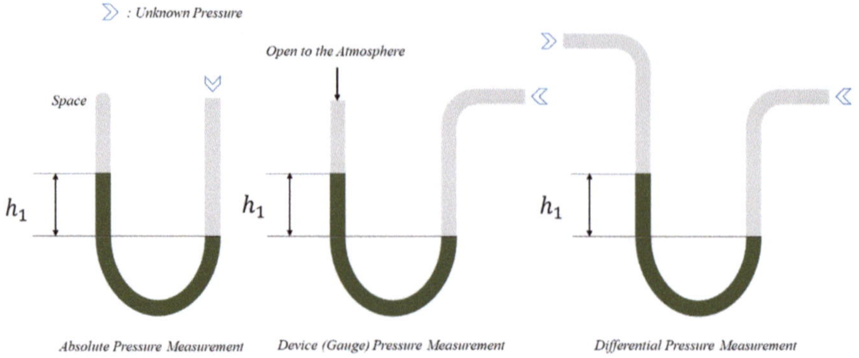

Fig. 7.45 Basic pressure measurements with U-type manometer

General Purpose Manometers: The structure and measurement sensitivity are designed in accordance with normal conditions. They are used to measure the pressure values of all non-corrosive liquids and gases of copper alloys. General purpose manometer types and usage areas are given in Table 7.7.

Bourdon Tube Manometers: They are used for pressure measurement in liquid and gas fluids. However, they are not used in media with high viscosity, in crystallized environments, or in liquid media, which reacts with copper alloys.

Pressure gauges that using the Bourdon tube are the most used and inexpensive devices for static pressure measurements. As its working principle, the shape change

Table 7.7 General purpose of manometers, their types, and areas of the usage

Manometertypes	Usage areas
Hydraulic type manometer	It is used in places that have high stability and high sensitivity is required
Glycerin-filled manometer	It is used to measure the pressure values of the liquids and gases that are non-corrosive of copper alloys. Due to its glycerin filling, it is used especially in places with dynamic pressure, in places with sudden pressure rise and drop occurred and in jolty environments
Stainless hydraulic type manometer	It is used in abrasive environments such as screw connection alloys and in environments where corrosion is unwanted
Diaphragm type manometer	The diaphragm type pressure gauge is used especially in places where the liquid or gas flowing is not allowed to enter the manometer. For example, in dirty liquids, in liquids that have the characteristic of freezing and crystallizing. They are also used where hygiene is of great importance, in applications that bacteria formation is not desired in terms of health concern
Contact type manometer	It is used in places requiring automatic pressure control

of elastic element under pressure is used for the measurements. The Bourdon tube is mostly a C-shaped pipe with an elliptical cross section. One end of the pipe is fixed and the other is free. When pressure is applied to this pipe, an elastic shape change occurs at the end of the pipe in the direction of the arrow (on the right side of the Fig. 7.46). The change of the idle end of the pipe is transmitted to the needle (pointer) as a rotation by spring and gear mechanisms (Figs. 7.46). General purpose manometers can be seen in Fig. 7.47 [1].

Electric Contact Manometer: The electrical circuit is automatically switched on or off at the values, set by the user. By pressing the button on the front of the manometer, the contact value is set to the desired value on the scale. Electric contact manometers (Fig. 7.48) are manufactured as standard production, with nominal diameters of 100 and 160 mm.

There are types of shock absorbers with high-frequency dynamic pressure, sudden pressure rise-drop and showing glycerin-filling feature for jolty environments. The contacts are designed as single or double contact as standard. Contact material is made of a very delicate and special alloy. Therefore, high current must not be passed over the contacts. The current that will pass over the contacts must be 0.4 A maximum. Therefore, electric contact manometers should not be directly connected to the electrical circuit. Miniature relays must be connected to the circuit in order for the contacts to operate effectively and long lasting.

The contacts are designed as in the characteristic feature of nonmagnetic standard. In the case of vibrating areas, the contacts can be switched on and off due to the

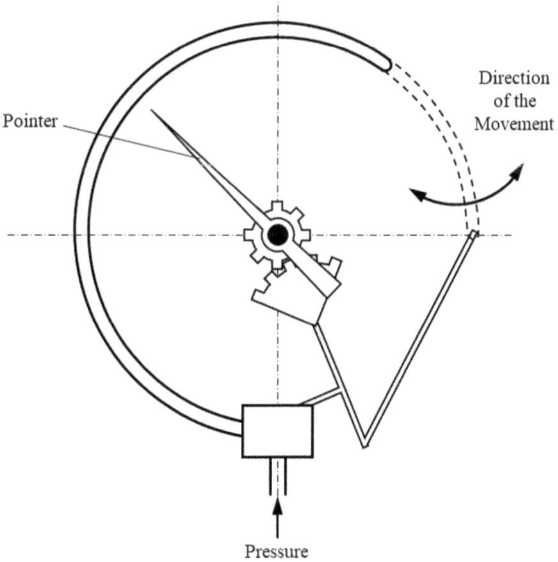

Fig. 7.46 Bourdon tube, its working movement, and internal structure of Bourdon tube manometer (a simple representation)

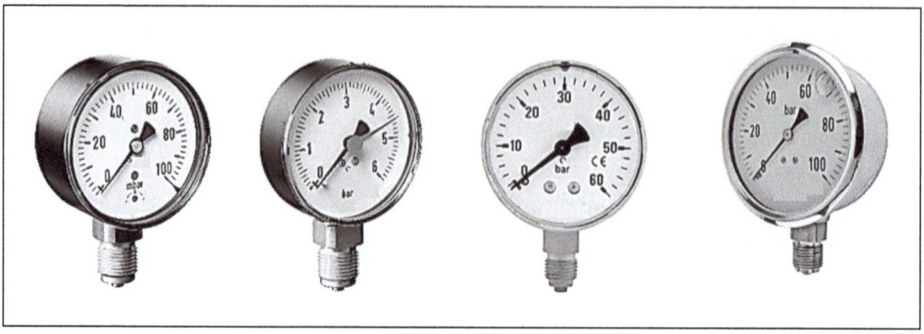

Fig. 7.47 General purpose manometers [1]

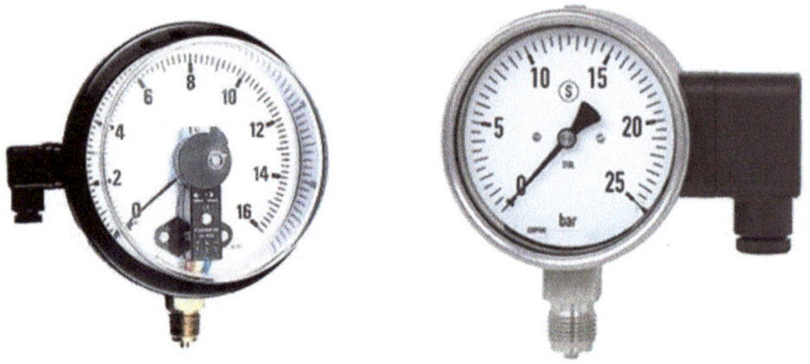

Fig. 7.48 Contact manometers

vibration. This causes the contacts to wear in a short time. In the case of magnetic types, these unwanted on and off can be avoided depending on the magnet force. In such cases, the relays with vibration breakers can also be used. Due to the magnets in the contact lever in magnet types, depending on the setting state of the magnet, they can cause an early "off" and late "on" at the rate of 2–5% according to the set pressure value. Electric contact manometers: they may be of a type with stainless contacts and damper contacts.

Diaphragm Manometers (Pressure Gauges): They consist of manometer and diaphragm group. The diaphragm group is connected to the connection mouth of the manometer. In diaphragm pressure gauges, the pressure or liquid that is measured does not enter the pressure gauge. The pressure is transmitted linearly to the manometer through the diaphragm group. Due to this feature, they are used in places where the measured liquid or gas does not want to enter the manometer (for example, in liquids and gases containing dirty, foreign particles, in liquids that show freezing or crystallization features).

Diaphragm manometers are designed in tongue, screw, and flange connection, capillary tube and vibration-cutting diaphragm types (Fig. 7.49) [4]. In particular, at high pressures, in refrigerated liquids, tongue diaphragm pressure gauges are recommended for cold and dirty liquids. Diaphragm pressure gauges are more sensitive to

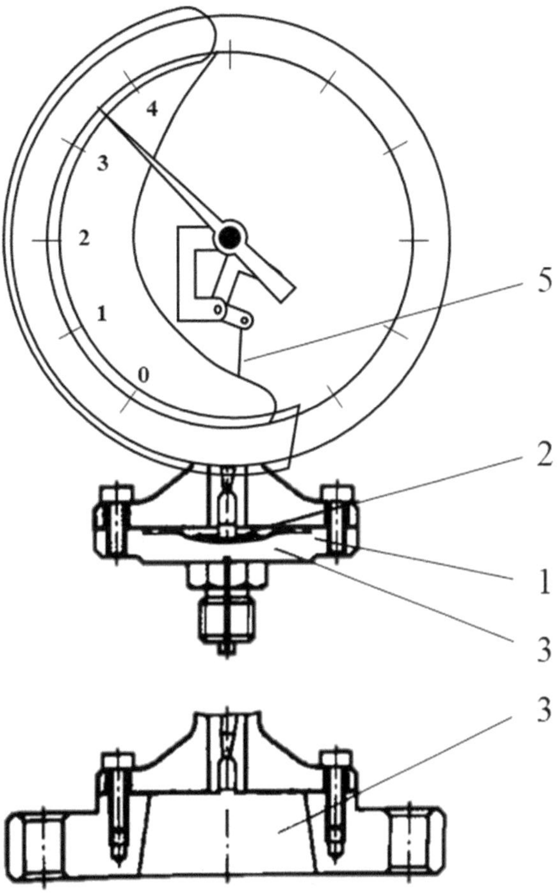

Fig. 7.49 Structure and working principle of diaphragm manometer. *Source:* https://www.makinaegitimi.com/manometreler-ve-manometre-cesitleri/ [4]

the vibration than Bourdon tube pressure measuring devices. They are also more suitable for dirty and high viscosity fluids. As shown in Fig. 7.49, a concentric diaphragm, tensioned between the two flanges, (1) divides the pressure vessel into two separate sections. The pressure chamber (2) is connected to the outside and pressurized by atmospheric pressure. The chamber (3) is pressurized by the working pressure in the measurement position and forms the measuring section. The differential pressure between the pressure in the chamber (2) and the pressure (3) in the measuring section causes the diaphragm to deteriorate in one direction. This deterioration is transmitted by a pushing lever (5) to the display mechanism and a pointer to the correct position on the scale.

Calibration of the manometer: The temperature of the measured material and the environment is important. Manometers have a calibration temperature of 20 °C

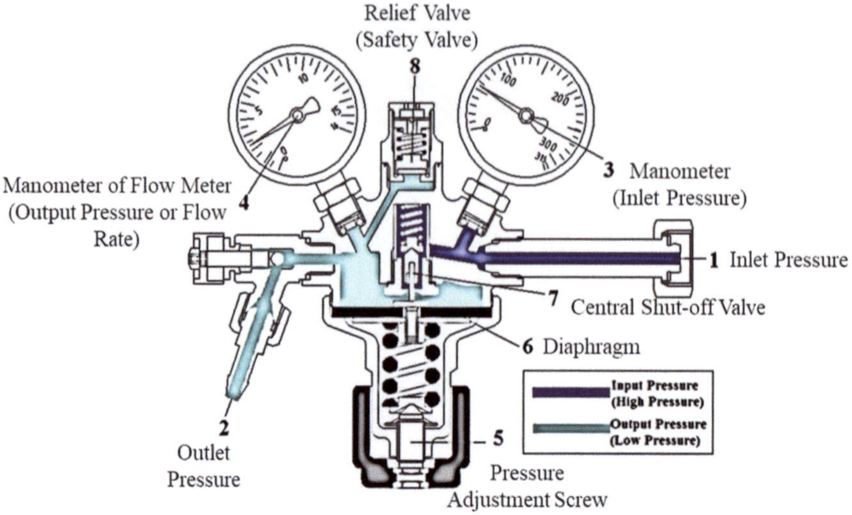

Fig. 7.50 Connecting manometer to the system

according to DIN 16005. For every 10 °C that exceeding this temperature, an error of 0.3% of the scale value may occur due to the temperature. This temperature is the temperature of the manometer; it is not the temperature of the material being measured. In diaphragm pressure gauges, this ratio may vary according to the liquid fill.

The device to be calibrated is disassembled and brought to the suitable laboratory and then it is waited in the calibration environment, at +20 °C for 4 h. For any ± 10 °C exceeding this degree ± 0.3% error may be made. Then, physical control is performed (to check whether there is broken part or crack, etc.). In addition, the mounting teeth are checked, and the tooth bottoms are cleaned with a soft wire brush (copper) and then it is mounted on the device to be calibrated.

Connecting manometer to the system can be seen in Fig. 7.50.

Calibrations are made in special laboratories with precision measuring instruments. The test manometers are usually used where a very precise measurement is required in the control of manometers in the plant. It is recommended not to exceed 3/4 of the scale in the measurements. Test manometers were calibrated at 90° upright. There may be deviations in the positions other than this position. Test manometers are carried in specially protected bags. They should never be subjected to impact or fall. If the manometers tested are within the error limits as much as the amount of error allowed, the tested manometer is within the error limits. Test Manometers should be tested at least once a year in laboratories. For frequent use, this period should be kept short. It is necessary to test the manometer immediately in cases of impact and fall.

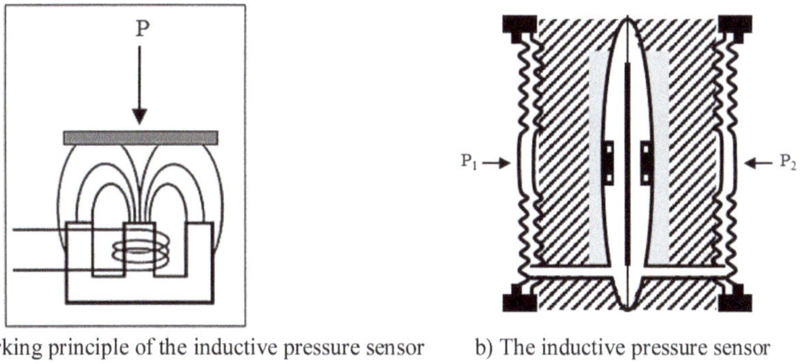

Fig. 7.51 The inductive pressure sensor and its working principle

Pressure Sensors

Inductive Pressure Sensors: Inductive based pressure sensors (Fig. 7.51) use the effect of changing the self-inductance of a coil on the pressure-reducing metallic diaphragm.

The basic approach in inductive pressure sensors is the change of the inductance of the coil. To achieve this, the installed mechanical structures may vary. Such pressure sensors for special purposes are expensive. It is especially used to minimize the temperature interaction.

Capacitive Pressure Sensors: Capacitive pressure gauges are generally used for differential pressure measurement. The effect of temperature is well adjusted. It is expensive. The principle of operation is based on the placement of the elastic diaphragm and an electrode above or against the opposite wall. This structure forms the condenser. If pressure is applied to the diaphragm, the distance between the electrodes decreases and the capacitance of the capacitor changes. The space between the electrodes can be filled with liquid or air.

The working principle of capacitive pressure sensor can be seen in Fig. 7.52. Capillary holes provide connection to the atmosphere. In other words, there is a

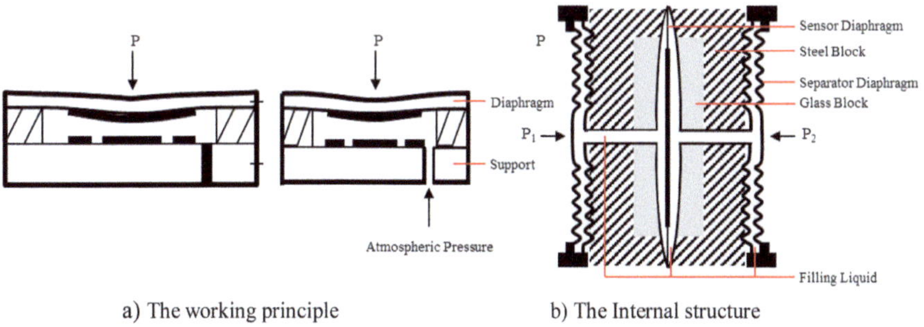

Fig. 7.52 The working principle of capacitive pressure sensor

special differential pressure measurement where atmospheric pressure is also taken into account. Differences in humidity may affect measured values. In a closed system, one of the functions of the reference electrode is to adjust the temperature effects. In particular, in the case where the inner compartment is closed, the air inside becomes hot, expands and increases the plate distance. This effect can also be adjusted with the aid of reference electrodes, and the capacitive absolute pressure sensor can be used for this purpose, which is an effective and inexpensive alternative for differential pressure sensors.

In differential pressure measurements, systems with two capacitors of the same type are usually used (Fig. 7.52b). When the pressure changes, the capacity also changes in the opposite direction. For example, one's capacity increases, while the other one decreases. This type of pressure sensor is differential capacitor. If there is a diaphragm inside and two separating diaphragms (from the medium/fluid), a two-compartment version, if there are only two open diaphragms, it is a single-compartment version.

Quartz Electrostatic Pressure Sensors: Dynamic pressure sensors operate on a *piezoelectric effect*. The piezoelectric feature is the ability of the material to change the electric field or electrical potential as a result of mechanical pressure applied to certain materials (especially crystals and certain crystals, such as bone) (Fig. 7.53). This effect is directly related to the change in polarization density in the material. If the material is not short-circuited, the applied voltage generates a voltage in the material. The word "piezo" is derived from Greek and it means "compress" and "apply pressure".

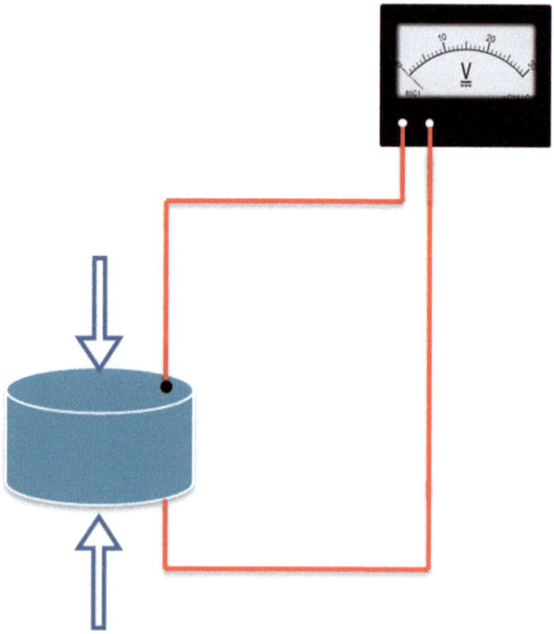

Fig. 7.53 Voltage generation in piezoelectric deformed disk

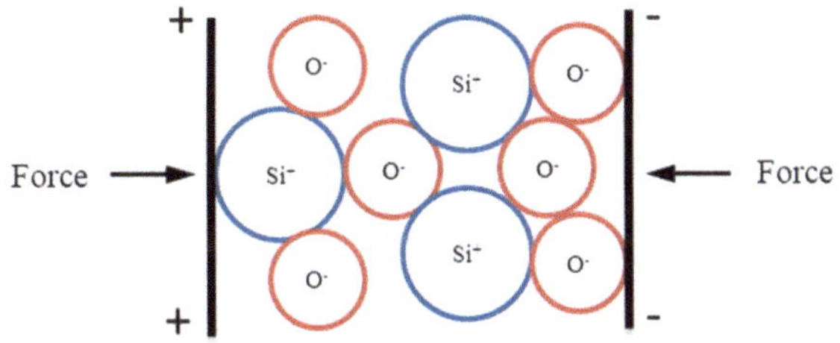

Fig. 7.54 Structure of the quartz crystal

Piezoelectric materials can also show the opposite characteristics. In other words, they are the materials that show "direct piezoelectric effect" (they produce electricity when voltage is applied). They also show reverse piezoelectric effect (they also show stress-strain production, as a result of applied electric field). That means the piezoelectric effect is a bi-directional effect on materials. In general, a piezoelectric material is placed between the two electrodes of the condenser as the insulating element. In response to the tension caused by bending the material, a tension occurs between the electrodes, or a tension in the material, i.e., a bending-twist, occurs as a result of the tension applied to the electrodes. For example, lead zirconate titanate crystals may deform from their original size to about 0.1%. This effect can be used in applications such as "sound creation and sensation", "high voltage generation", "electronic frequency creation", "microbalance", and "superfine focus of optical transducers".

Quartz materials can give a linear output in a very high-frequency range, such as 400 kHz. Such a material can measure high frequency but small amplitude fluctuations over large static pressure values. Monitoring and control of pump pressure, hydraulic and pneumatic pressure lines in industry; investigation of flow-induced vibrations; cavitation, water hammer, pulsation, acoustic measurements; aviation tests; valve dynamics, explosive and weapon tests; internal combustion engine tests can be done by using these sensors.

When the piezoelectric elements are under an external force, an electric charge is formed on their reciprocal surfaces. The large circles shown in Fig. 7.54 indicate silicon atoms and the smaller ones indicate oxygen atoms.

Natural or processed quartz crystal is one of the most sensitive and stable piezoelectric materials. In addition to natural materials, materials such as polycrystalline and piezo ceramic produced by high technologies can also show piezoelectric properties when they are exposed to high electric field. These crystals produce a very high load charge. Thanks to these features, they are especially used for measuring low amplitude signals. The design, which is based on compressive force, shows high durability. Thanks to this feature, it is used in high-frequency pressure and force measurements. The negative feature is that it is sensitive to temperature changes.

Piezo-Resistance Type Pressure Sensors: Some industrial pressure sensors use the piezo resistance feature of silicon. The piezo resistance element converts the pressure directly into resistance. The resistance change can be converted into voltage by the bellows. It is ideal for pressure values between 0–1.5 psi and 0–200,000 psi.

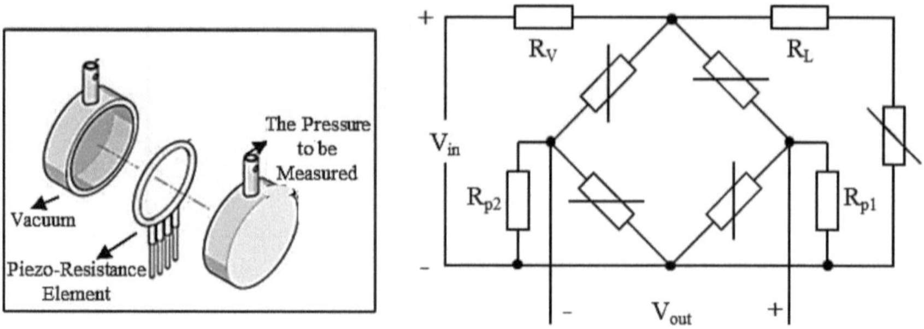

a) Piezo-resistance pressure sensor b) Bridge connection of piezoelectric sensor.

Fig. 7.55 Piezo electric pressure sensor

The piezoelectric sensors operate by stimulating the pressure applied to the diaphragm by the bending of the bridge (Fig. 7.55a, b). The change of the voltage at the bridge terminals is read by turning it to the desired value by means of an electronic board. The main advantages of piezoelectric sensors are that they have no mechanical parts and do not require frequent adjustment. With the help of electronic cards, high accuracy and repeatable measurements can be made. Thanks to their small size and compact structure, they can be used in every process. They have a wide range of output information to be used with various measurement and control instruments.

Figure 7.56 shows the internal structure of a piezoelectric sensor. The sensor is placed in a ceramic protector in order to protect it from any mechanical impact, physical or chemical effects that may occur in any way. Then, it is fitted to a metal structure suitable for the process connection as seen in Fig. 7.57. By inserting the appropriate electronic card into the input structure of the device to work in conjunction with the sensor, the piezoelectric sensor is placed in the stainless tube with which the connector is connected with the electronic card. Various types and shapes of pressure sensors are usually designed steel body to resist to impact. The pressure application inputs can contain connection elements that is of various size.

Differential Pressure Sensor: t is defined as the abbreviation of Differential Pressure Cell (DP/Cell). They are generally used for flow measurement. The device measures the pressure difference between two separate sources and shows it as the standard output size (Fig. 7.58). Electronic differential pressure sensor is designed for differential pressure measurement, so any type of pressure measurement can be used. It is used for measurement both level and fluid. As the measurement uses the differential pressure indicator, the usage area is large. Differential pressure sensors have two inputs, and these are high and low pressures (Fig. 7.58). High and low pressures are connected to the respective arms. Through the diaphragm type sensor inside, it takes the difference between the two-pressure applied to the arms.

The measurement output is usually 4–20 mA current and can be adjusted. Some types express the output directly in numerical form. What kind of pressure is to be measured for connection to the system is important. Both wet and dry types are available.

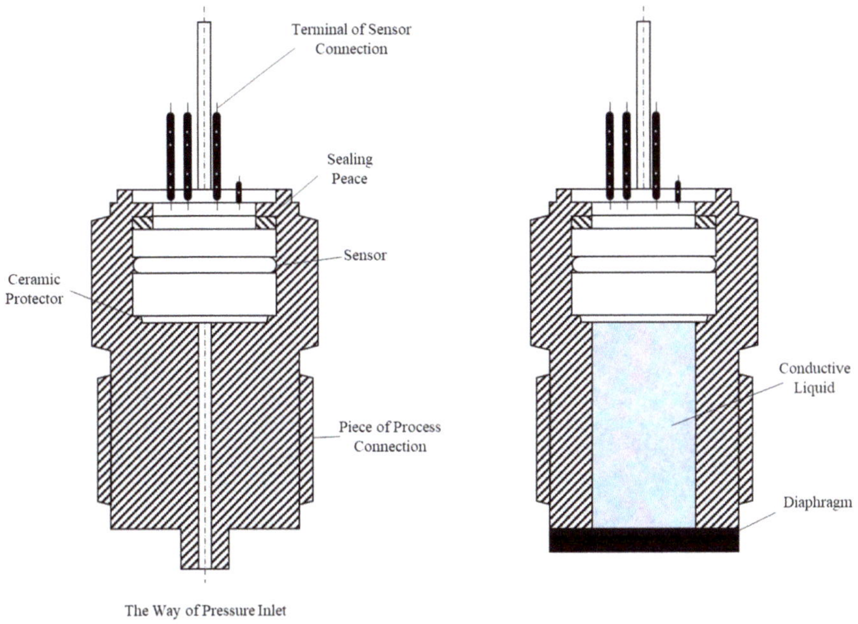

Fig. 7.56 Internal structure of piezoelectric sensor

Fig. 7.57 Piezoelectric pressure sensors

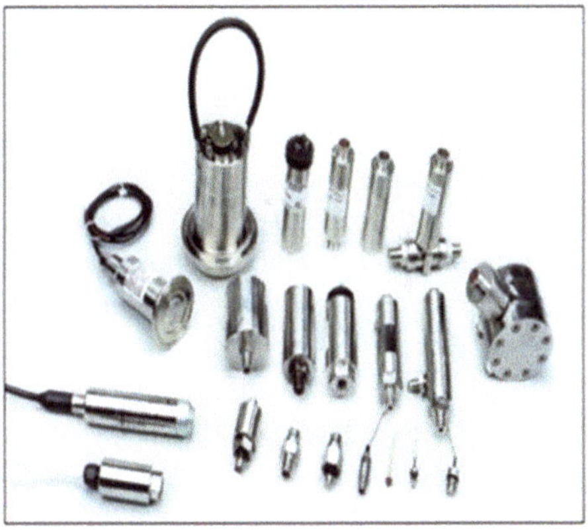

While the differential pressure sensor is connected to the system, low and high pressures are connected to the corresponding pressure inlets. This type of pressure measurement is often used for liquid flow or liquid level measurement. Zero and top settings are adjusted according to the lowest and highest values of the level. In absolute pressure measurement (Fig. 7.59a), the low-pressure inlet is closed, preventing exposure to any pressure. Thus, the measured value will be absolute pressure. Such measurements are barometric measurements. It can be used to measure atmospheric pressure. It is also used in situations where atmospheric pressure varies.

Fig. 7.58 Differential pressure sensor

For the calibration of the differential pressure sensors, the apparatus given in Fig. 7.59b is prepared. The standard test gauge output is connected to the high-pressure inlet. The low-pressure input will be open to the atmosphere. The sensor supply is set to 24 VDC. The regulator is set for the air supply, which is 1.5 kg/cm^2.

7.5.4 Flow Rate Measurement

In engineering and science, the flow rate can be described as the amount of fluid passing through the unit section of a channel or a pipe in the unit time (generally, a minute or an hour). The determination of the flow rates of fluids flowing in the channel has an important place in energy calculations. The measurements of velocities or flow rates of fluids flowing through any of the sectioned channels or tubes are important for the examination of many biological, meteorological phenomena, for the various processes of the industry and for carrying out various experiments in laboratories.

Different devices and methods have been developed in order to measure velocity and flow rate or flow events in general. They have various advantages and disadvantages. For example, changing the diaphragm is easier than changing the nozzle or Venturi. On the other hand, the flow rate coefficient of nozzle is more precise than the diaphragm, which is a reason for preference. In any flow measurement process, various factors, including sensitivity, capacity range, and price, play a role in the selection of the measuring device and method. Depending on the desired property and price, a wide range of measurement tools is used in flow rate measurement.

In closed channels there are three flow meters that are used to measure mass or volume flow rate based on the principle of measuring the pressure difference occurring in the flow by forming a section narrowing in the channel. These are Venturi meter, nozzle, and diaphragm (Fig. 7.60).

The measuring accuracy of the Venturi meters is very high. Unlike other flowmeters, the pressure loss in the device can be recovered to a large extent. The diameter of the narrow section in the Venturi meter does not deteriorate and the device does not hold residues.

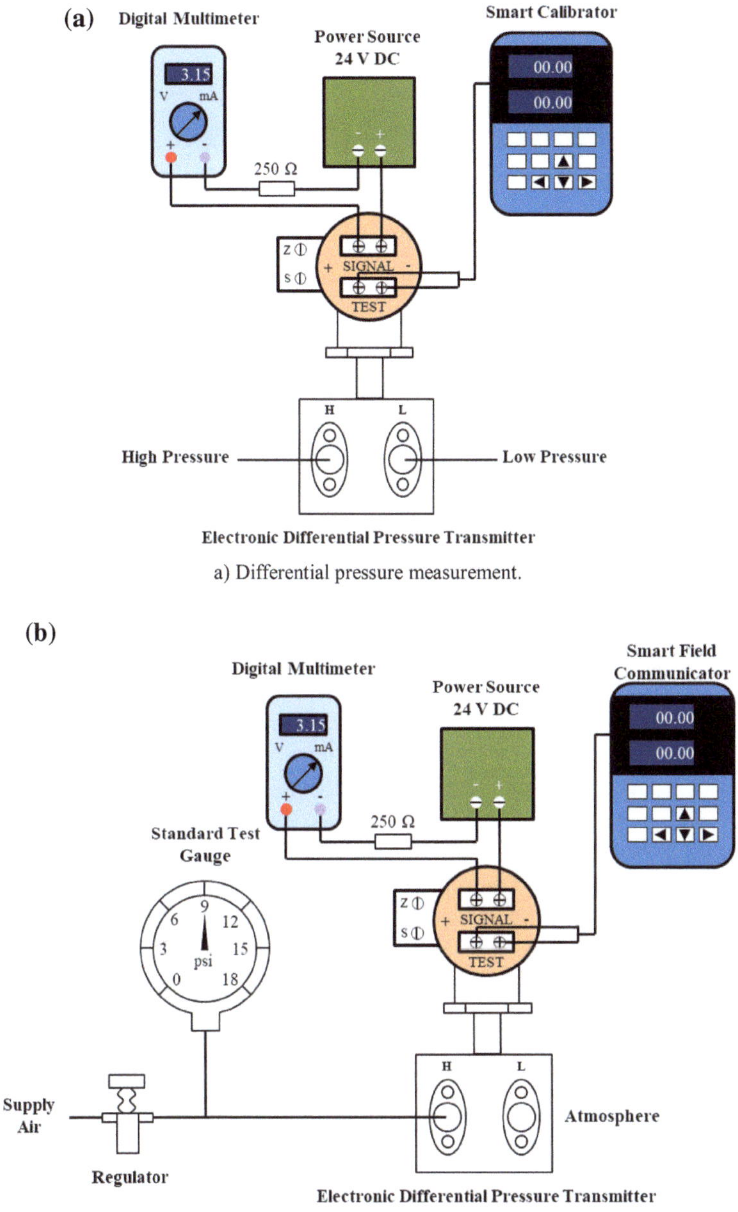

a) Differential pressure measurement.

b) Gauge pressure measurement.

Fig. 7.59 Pressure measurement with the differential pressure sensor

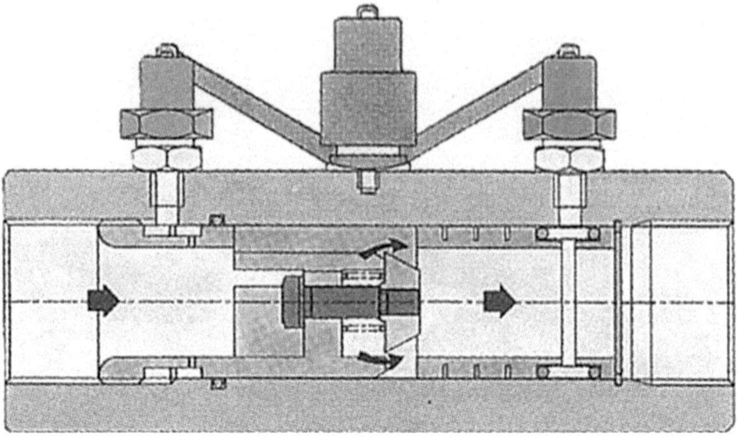

Fig. 7.60 Working principle of a flowmeter based on cross-sectional narrowing

In *the nozzles*, the pressure loss can be recovered at a certain rate, not even the same amounts as in the Venturi. Their lengths are not as long as the Venturi meters. A nozzle is resistant to erosion and does not hold sediment. Manufacture of the nozzles are more difficult and expensive than diaphragm. However, it is cheaper and easier than Venturi meter.

The design of *diaphragms* is very simple and inexpensive. Due to their small footprint, they can be installed anywhere on the piping. It is also easy to install and remove. They are not suitable for large capacities. As their strength is not high, they may deteriorate in shape due to impacts in flow. If standard diaphragms are used to measure the flow of contaminated fluids (dirty particulate liquids or very moist and dusty gases), the accumulation of particles and dust on both sides of the diaphragm can cause erroneous measurements. The use of slit-type diaphragms is recommended for the flow measurement of contaminated fluids.

Flow Rate Measurement with Articulated Wing: It is possible to measure the flow of liquids or gases by means of a wing suspended from a horizontal axis. The movement of this wing is proportional to the volumetric flow rate of the passing fluid.

Flow Rate Measurement by Centrifugal Effect: A pressure difference is created in the fluids flowing in a curved channel due to the effect of the centrifugal force and the different velocity between the inner part of the channel and the outer part. By means of the pressure difference measured at the inner and outer part of the duct, the flow within this channel can be determined. This method is particularly suitable for installations that are difficult to assemble and use liquids such as diaphragm, nozzle, and Venturi meter.

Vortex Generating Flow Meter: As seen in Fig. 7.61, an object with flow-disturbing angular shape placed in a smooth flow causes vortex formation in this flow. The

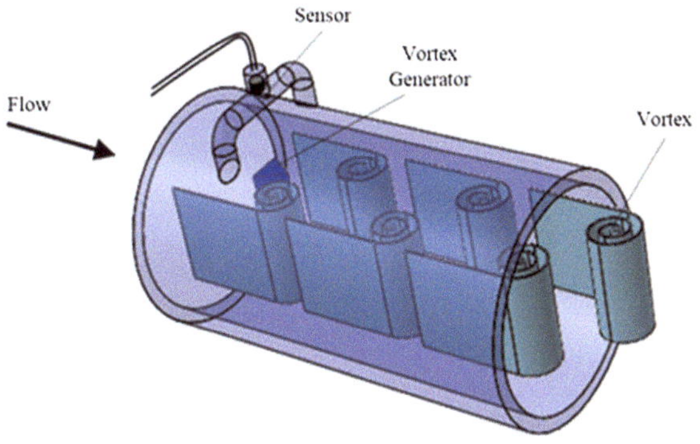

Fig. 7.61 Vortex flow meter

frequency of the resulting vortex is directly proportional to the velocity of the flow or flow rate. Pressure changes in a vortexed flow can be measured by pressure sensors placed in the flow.

Vortex flow meter; they operate according to the flow measurement principle that calculate the volumetric size of the substance passing through without looking for any properties in liquids. Vortex flow meters, which generally can be found in many applications, especially in the flow measurement of steam, compressed air, gases, and liquids. They are among the most preferred flow meters due to their versatility.

7.5.4.1 Flowmeter

The simplest and most sensitive method of flow measurement is the weighing of the mass of fluid flowing in a given time. Although atmospheric conditions are very easy for non-volatile liquids, special precautions should be taken when using this method for volatile liquids and gases.

The determination of the amount of liquid or gas fluids in a given mass or volume in a container is the most precise and simple method of measuring flow. This method is usually used for the calibration of other flow meters; it can also be used where the flow rate is small, the flow is discontinuous, and high accuracy results are needed. Because of their simplicity, these systems are usually large and slow.

The volume of the fluid passing through the flow line and deposited in a tank is indicated by ΔV, and the elapsed time is measured as Δt, the flow rate can be determined directly by $\dot{V} = \Delta V/\Delta t$.

The working principle of two different flowmeters with volumetric sweeping is shown in Fig. 7.62. In the flowmeter shown in Fig. 7.62.a), the blades placed on the rotatable piston with springs behind them come into contact with the walls. As the cylinder with eccentric rotates in the body, it shows flow rate of the fluid that passing through the wings. The system is independent of fluid viscosity, and its measurement sensitivity (accuracy of the measurement) is $\pm$ 0.5%.

Another example of the volumetric flowmeters is the turbine type flowmeter (Fig. 7.63). It can be used to measure the flow rate of liquids and gases.

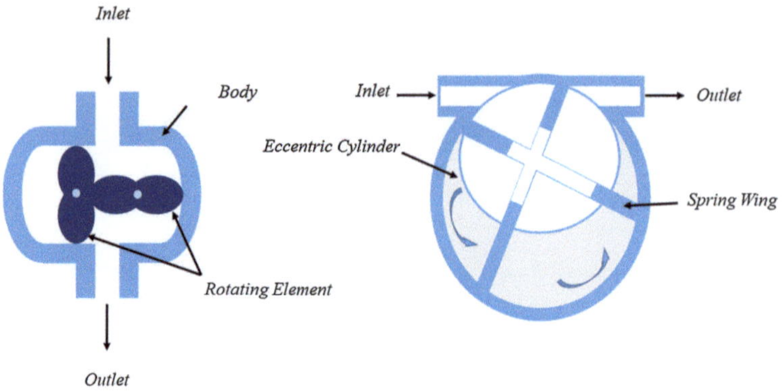

(a) Rotary vane flowmeter. **(b) Flow meter with two drive elements**

Fig. 7.62 Swept volume flow meters

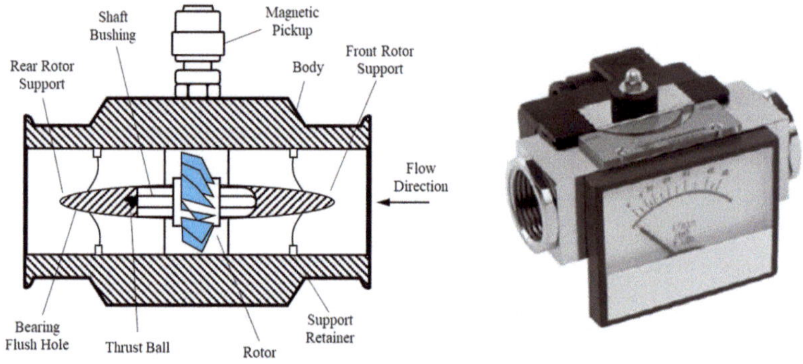

Fig. 7.63 A type of turbine flowmeters (the simple structure)

Capacities are between 3–12,000 m^3/h and their sensitivities are $\pm$ 2%. This type of flowmeters can be installed in horizontal and vertical pipes.

A different type of turbine flowmeter is shown in Fig. 7.64. A permanent magnet is placed in the turbine, and as the blades rotate, voltage pulses are generated in the electrical winding in the fixed body. These pulses can be measured with a frequency meter and the instantaneous velocity or flow rate can be found in terms of the electrical signal.

Flow Measurement

Flow Measurements: *Flow* is the continuous movement of a fluid through a line. The fluids can be in the form of three substances (solid, liquid, gas) [1]. For flow measurement, the following sizes and units are used:

- Volumetric flow rate (m^3/s): It is the volume of fluid flowing per unit of time.
- Mass flow rate (kg/s): It is the mass of fluid or a substance flowing a cross-sectional area in the system per unit of time.

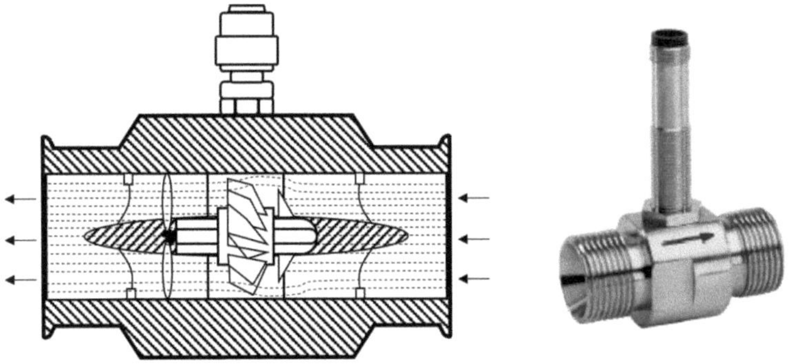

Fig. 7.64 Turbine flow (or velocity) meter giving off voltage pulses

- Velocity (m/s): The path taken by the fluid in the unit time. It can be also described as the speed of fluid in a given direction.

Some measurement methods used in flow measurement are [5–8]

- Volumetric flow measurement: In volumetric flow measurement, helical screw flow meter, rotary flow meter, turbine flowmeter, impeller system flow meter, and disk counters are used.
- Mass measurement with force balance and variable area flowmeter (rotameter) methods are used for mass flow measurement.
- Flow measurement with velocity measurement method: Velocity measurement with Pitot tube, velocity measurement with anemometer, velocity measurement with reduction effect, Venturi pipe velocity measurement, orifice flow measurement, nozzle flow measurement methods are used in flow measurement by using velocity measurement methods.
- Electromagnetic flow measurements.
- Ultrasonic flow measurements.

In order to be able to control an industrial process, it is necessary to know the amount of material entering and leaving the process. In the case of substances being fluid, it is important to measure the flow rate in a pipe or duct. Some measuring devices used for this purpose are

- Direct mass or volume measurement devices,
- Variable height measuring devices,
- Area meters,
- Current meters,
- Positive displacement devices,
- Magnetic meters.

In measuring with flow meters, the measuring element is immersed in the fluid; the element rotates according to the velocity of the fluid and reads the velocity of the fluid from the meter. *Positive displacement meters*, including various metering pumps, operate according to the same principles as rotary and piston pumps.

The magnetic flow meters operate according to the principle of creating an electrical potential through the movement of a conductive fluid through an externally generated magnetic field. According to Faraday's law, the voltage generated is directly proportional to the velocity of the fluid. Commercial magnetic flow meters measure the velocity of all fluids except hydrocarbons (their electrical conductivity is low).

The most commonly used devices for flow measurement are variable height gauges and field measurement devices. Variable height measuring devices include Venturi meter, orifice meter, and Pitot tubes; as for field meters, rotameters can be mentioned accordingly.

Electromagnetic flowmeters: These devices measure the flow according to the principle of measuring the magnetic field formed by the fluids in the pipe. Electromagnetic flow measurement is based on Faraday's law. The electromagnetic flowmeter uses liquid as a conductor. As shown in Fig. 7.65, the electromagnetic flowmeter has two coils connected in series. When the coil is energized, a magnetic field is formed with the current effect around the liquid. Two conductive electrodes are placed perpendicular to the liquid flow and magnetic field to the pipe edge. The velocity change of the liquid fluid in the pipe acts on the magnetic field and transmits the change to the electrodes. The electrodes send signals to the measuring instrument. The change in the flow rate shows a linear change with the signal received by the meter.

The disc meter given in Fig. 7.66 can be used for volumetric flow measurement [9]. In such meters, similar to the meters used in homes, the disc takes a certain volume of fluid from the inlet at each full rotation, making a wobble movement. Depending on the flow rate of the fluid or the amount of flow flowing, the disk, whose movement is increased or decreased, transmits this amount to the gear system on the upper part and each oscillation of the disc is processed in the gear wheels by the amount of fluid delivered by the disk. The gear ratios of the gearwheels and the topmost numerator indicate the volume delivered by the disc in the unit of time.

Venturi meter: As seen in Fig. 7.67, the Venturi meter is a measuring device that has a narrowing and expanding flow region and determines the flow rate of the flow

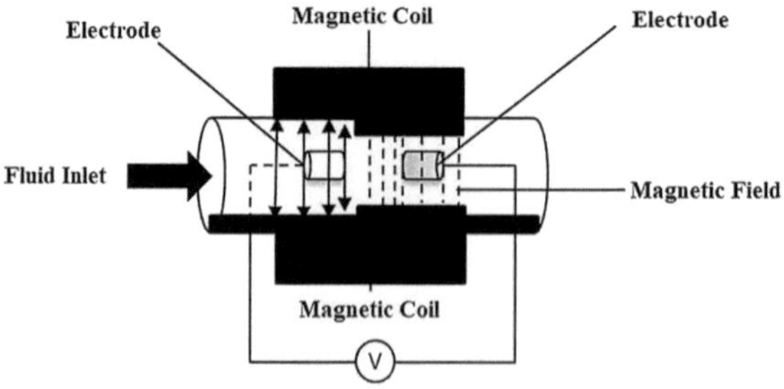

Fig. 7.65 Electromagnetic flowmeter

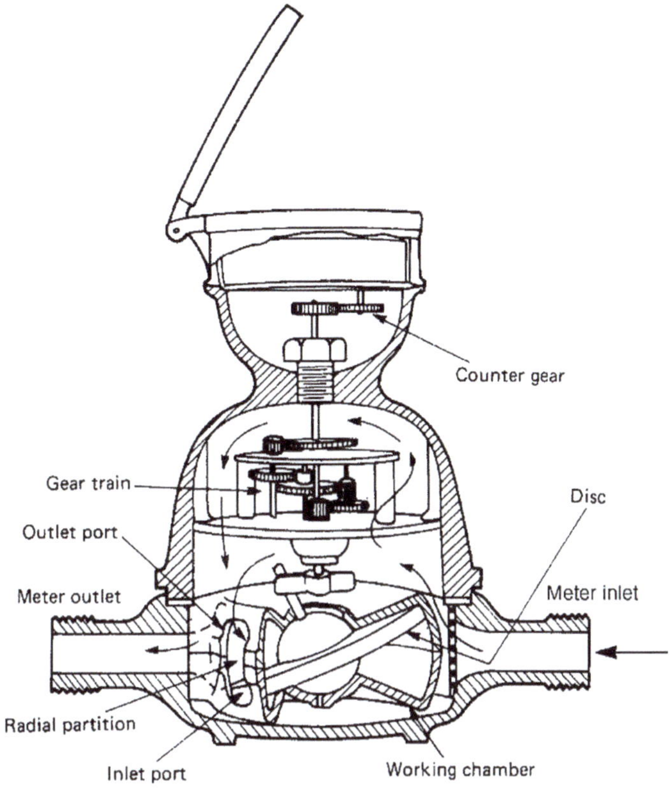

Fig. 7.66 Nutating disc flowmeter [9]

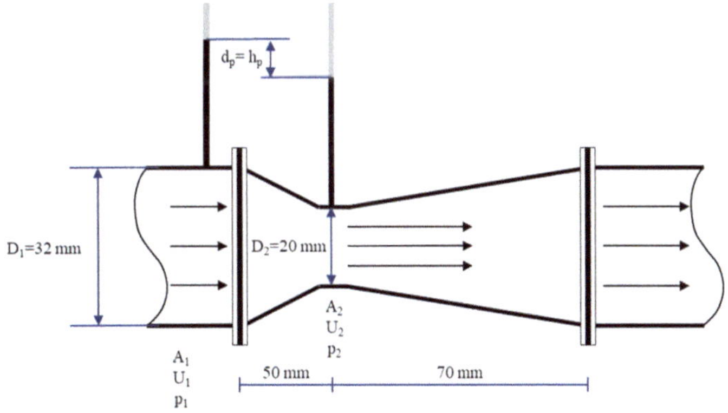

Fig. 7.67 Venturi meter

in pipes and ducts. Venturi meter is a measuring instrument used for measuring fluid flow, open at both ends, narrow at both ends toward the middle of the cross section and in the form of a thin tube. It takes its name from the Italian physicist Giovanni Battista Venturi [10].

In the narrowest section of the tube, which thins the throat, the fluid velocity has the largest value while the fluid pressure decreases (Fig. 7.68). A pressure difference occurs between the large section of the tube and its narrow section. This pressure difference is measured by the "closed manometer" connected to the input and output of the Venturi meter. By means of the pressure values obtained from this measurement, the flow rate of the fluid flowing through the tube is determined.

A schematic view of a Venturi meter is shown in Fig. 7.69. The current enters the flanged part A, which is cut in a conical shape, and passes through the throat B and exits the long section C with a conical cut. The input current (upper side flow) is regulated at the connection point of the cylindrical and conical portion as it passes through a circular ring (D) having small openings (holes) (E); It is called "piezometer (pressure gauge)", which consists of the part D and E. The pressure of

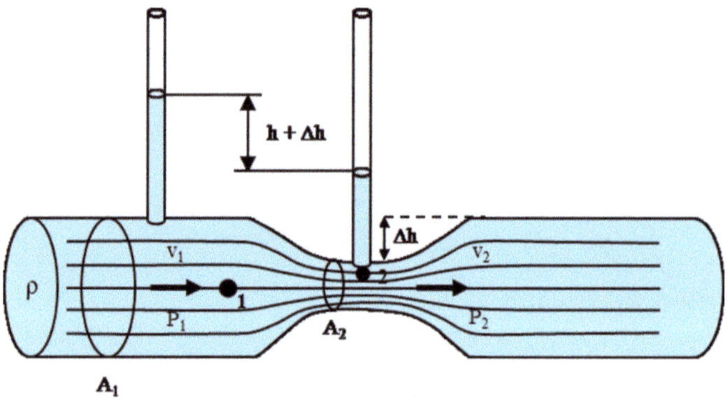

Fig. 7.68 Fluid velocity and pressure in Venturi meter

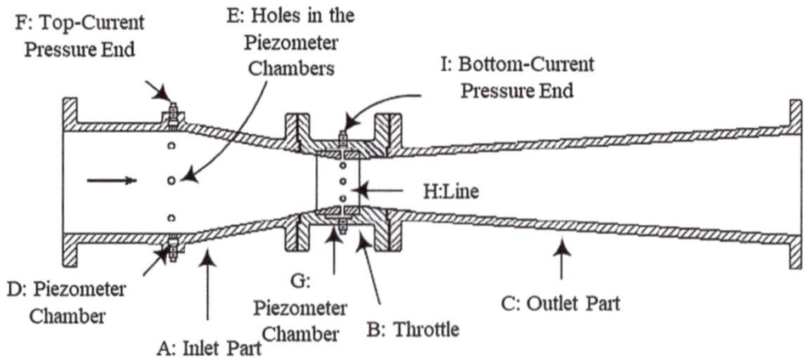

Fig. 7.69 The parts of the Venturi meter

the input current is measured in plug F. A second piezometer is located in the throat with G-holes and H-holes; the holes are made and processed very precisely. The plug I controls the pressure in the throat. A suitable pressure gauge is connected between the plugs F and I to measure the pressure difference between the input and output currents.

In the Venturi meter, the velocity increases , the pressure decreases at the inlet cone. The pressure drop in the inlet cone allows the flow rate to be measured throughout the system. Then the velocity decreases and the current (flow) to the output of the cone C returns to its original pressure. In order to completely recover the dropping pressure, the angle of the conicity in C is kept small, to prevent boundary layer separation and minimize friction.

Venturi meters are suitable for measurement of gases, but they are also suitable for certain liquids, especially water. For incompressible fluids, the fundamental equation of the Venturi meter is subtracted from the Bernoulli equation. The Bernoulli equation is written between the two-pressure points D and G. It is accepted that there is no friction, the meter is standing horizontally and there is no pump. Due to the conservation of energy, the energy equation is valid. This equation, which takes the Bernoulli equation for continuous, frictionless, and incompressible flow, is as follows:

$$\frac{\mathrm{P}_1}{\rho} + \frac{\mathrm{v}_1^2}{2} + \mathrm{g} \times \mathrm{h}_1 = \frac{\mathrm{P}_2}{\rho} + \frac{\mathrm{v}_2^2}{2} + \mathrm{g} \times \mathrm{h}_2 = \text{Constant} \tag{7.7}$$

The Bernoulli equation refers to the relationship between velocity, pressure and elevation in frictionless flows in a horizontal system, $h_1 = h_2$ and the equation becomes the following state.

$$\frac{\mathrm{P}_1}{\rho} + \frac{\mathrm{v}_1^2}{2} = \frac{\mathrm{P}_2}{\rho} + \frac{\mathrm{v}_2^2}{2} = \text{Constant} \tag{7.8}$$

In addition, the continuity equation is valid due to the conservation of the mass and the inlet mass flow rate ($\dot{m}_1$) must be equal to the outlet flow rate ($\dot{m}_2$):

$$\rho_1 \times \mathrm{v}_1 \times \mathrm{A}_1 = \rho_2 \times \mathrm{v}_2 \times \mathrm{A}_2 \tag{7.9}$$

As the density for the incompressible flows is constant ($\rho = \rho_1 = \rho_2$), the continuity equation becomes the following expression.

$$\mathrm{v}_1 \times \mathrm{A}_1 = \mathrm{v}_2 \times \mathrm{A}_2 \tag{7.10}$$

The definitions to the right and left of the above equation give the volumetric flow rate. Accordingly, the volumetric flow rate is also constant in incompressible flows.

$$\dot{V}_1 = \dot{V}_2 = \dot{V} = \text{Constant} \tag{7.11}$$

With the help of the equation of continuity, the following equation can be obtained from the energy equation for the velocity in the narrow section:

$$\mathrm{v} = \sqrt{\frac{2\Delta \mathrm{P}}{\rho\left(1-\beta^{4}\right)}} \tag{7.12}$$

Here, β is the ratio of narrow section diameter to pipe diameter ($\beta = d/D$) and ΔP is the pressure difference in both sections. The pressure difference can be defined and measured.

$$\Delta \mathrm{P} = \mathrm{P}_1 - \mathrm{P}_2 = \rho \times \mathrm{g}(\mathrm{h}_1 - \mathrm{h}_2) \tag{7.13}$$

Furthermore, the calculated velocity is greater than the actual value because friction and local losses are not taken into account. If this is taken into account with a C_d coefficient, the following correlation is obtained for the volumetric flow rate:

$$\dot{V} = \mathrm{A}_0 \times \mathrm{C}_\mathrm{d}\sqrt{\frac{2\mathrm{g}(\mathrm{h}_1 - \mathrm{h}_2)}{\left(1-\beta^{4}\right)}} \tag{7.14}$$

where A_0—is the area of the narrowest section and h—is the manometric static pressure height. The flow rate coefficient is $C_d < 1$ continuously and for only the frictionless current is $C_d = 1$.

Orifice Meter: It is essentially the same as the Venturi meter. Venturi meters have some drawbacks in applications, they are expensive, they take up much space and the ratio of throat diameter to pipe diameter cannot be changed. The highest measurable flow rate is constant in a Venturi meter and manometer system. In this case, when the flow range changes, the throat diameter remains too big or too small and the result is not accurate. The orifice meters do not contain these negatives.

Figure 7.70 shows a standard sharp-edged orifice. The orifice plate, which is very well machined, perforated, and attached between the two flanges, is mounted to the pipe in a central position. The opening in the plate may be grooved on the lower current side. One of the pressure tips is on the upper side of the orifice and the other on the lower side; these ends are connected to a manometer or a suitable pressure gauge. The locations of the pressure tips may vary, but the coefficient of the meter changes depending on the position of these tips.

As seen in Fig. 7.70, the orifice meter is a measuring device that has a sudden shrinking and expanding flow region and determines the volumetric flow rate of the flow by measuring the flow velocity in pipes and ducts. Compared to Venturi, the geometry of the orifice is simpler, easier to manufacture, cheaper, take up less space. In contrast, it causes more permanent pressure loss than Venturi. In principle, the theory described above for Venturi also applies to the orifice. However, the flow movement is more complex than the Venturi. After the narrow section exit, the flow movement is slightly narrowed. In addition, vortices are formed before and after

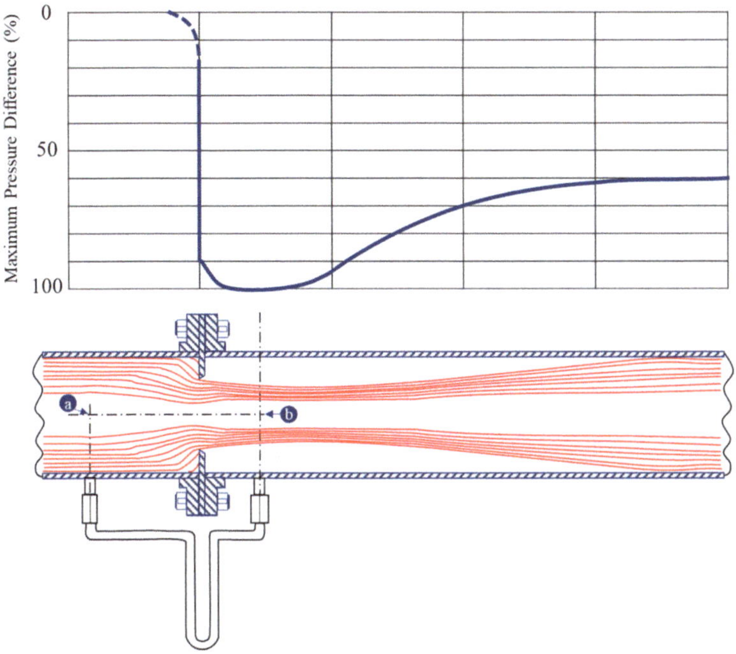

Fig. 7.70 Orifice meter

the narrow section. Therefore, the pressure drop cannot be recovered again, and a significant amount of pressure loss occurs. The correlation in the Venturi meter for the flow calculation can also be used here. However, the flow coefficient is much smaller than the C_d.

In the orifice meter, with the decrease of the cross-sectional area, as the fluid passes through the orifice, the velocity increases and the pressure drop between the ends is measured with a manometer. The relation between the decrease in the pressure height and the increase in the velocity height is obtained from the Bernoulli equation. A significant problem is encountered in the orifice meter, which is not seen in the venturimeter. Since the orifice is sharp, the fluid flow is separated from the downstream side of the orifice plate and a free-flowing jet stream is formed. The jet stream is not under control of the pipe walls and its area varies with the ratio of the orifice opening to the orifice pressure tip distance. The area at any point (e.g., at the downstream pressure end) cannot be easily determined. The velocity of the jet at this point cannot be defined by a simple correlation with the diameter of the orifice. Orifice coefficients are more experimental than Venturi.

Flow measurement according to the differential pressure generated by the orifice plate method: The orifice plate is made of stainless steel and has an outer diameter close to the diameter of the pipe where it will be placed (Fig. 7.71). The inner diameter of the fluid passing through it is such that it creates a differential pressure depending on the physical properties.

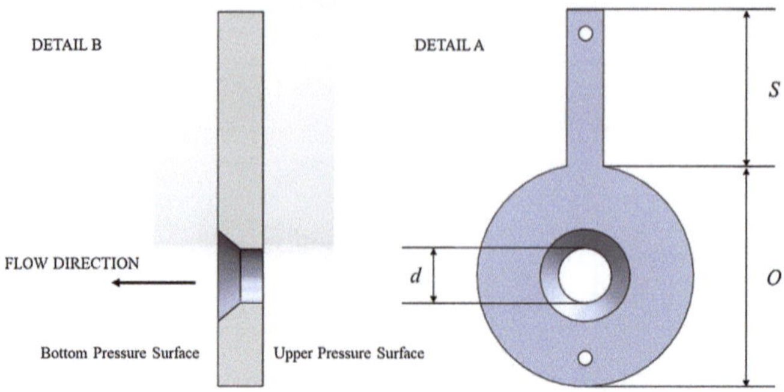

Fig. 7.71 Orifice plate and its structure

Orifice plate and its operation can be seen in Fig. 7.72.

The material used in the inner and outer diameters on the orifice plates and the line in which the fluid flows are described in Tag No. When the orifice plate is mounted between one or two flanges, the inlet of the fluid is at high pressure and the outlet is at a lower pressure. This pressure difference will vary with the ratio of the fluid passing through it. This ratio is not linear and shows logarithmic behavior. At the beginning of the pressure difference, fluid flow is almost doubled.

Placing the orifice plate on the fluid line can be seen in Figs. 7.73 and 7.74.

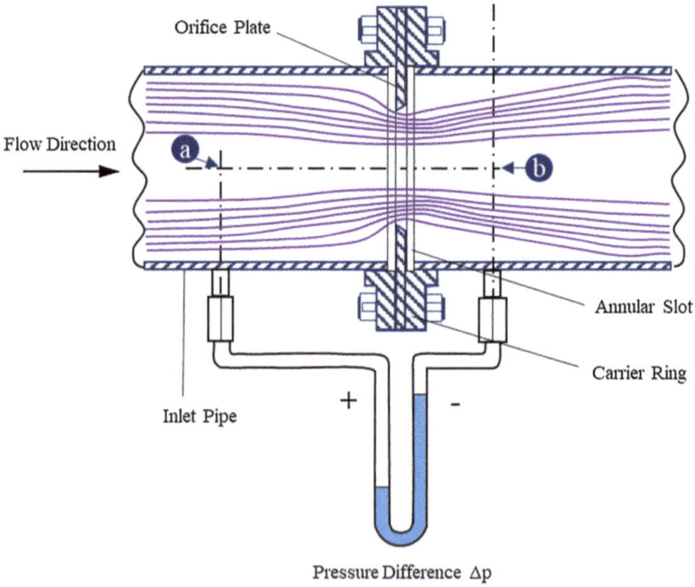

Fig. 7.72 Orifice plate and its operation

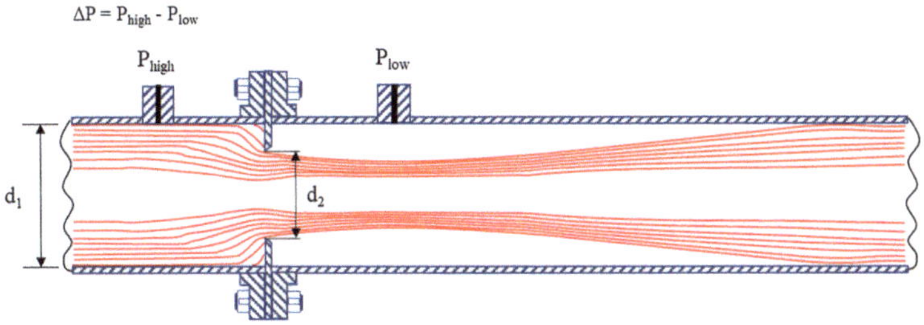

Fig. 7.73 Placing the orifice plate on the flow line

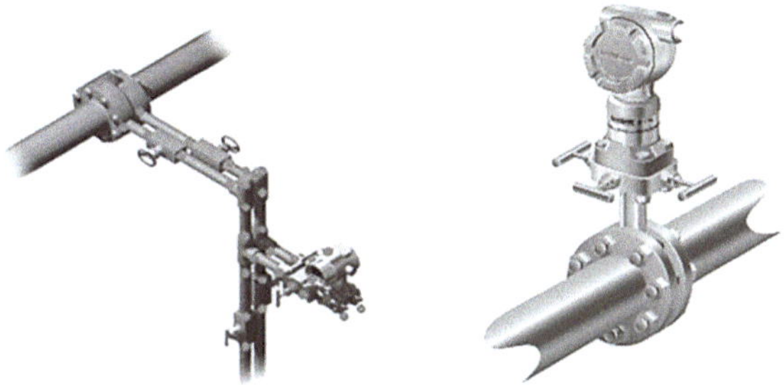

Fig. 7.74 Placing the orifice plate on the flow line

The orifice plate is placed in the flow line as seen in Fig. 7.74. The pipe is divided into two and the two flanges are welded (Fig. 7.75). The orifice plate is placed between the flanges. Two graphite gaskets are placed to prevent fluid leakage from the edges at high pressures. Finally, the bolts are crossed through the two flange holes and tightened with the nut to connect the pipeline.

Connecting the orifice plate to the electronic differential pressure sensor: After the orifice plate is placed between the flanges on the pipeline, 1/2 inch carbon steel pipes are connected to the holes on the flanges and connected to the sensor to be measured. The high-pressure outlet of this pipe is connected to the "High" side of the sensor and the low-pressure side to the "Low" side (Fig. 7.76) [7].

The manifold used here performs many functions, such as the zero setting of the sensor or the isolation of the sensor from the steam when disassembled for maintenance. The sensor sends a value to an indicator or controller that is mounted on the panel via a connection method called two wires. In other words, the standard 4–20 mA current that will transmit both 24 VDC supply and process variables in current form sends values to the indicator or controller via these two cables. The operator can change the flow passage by opening or closing the pneumatic control valve on the Loop according to the reading value. This can be done by hand, or by typing a set value in the controller, it can automatically connect the output in the pipe around this set. In other words, it can leave operations to the controller.

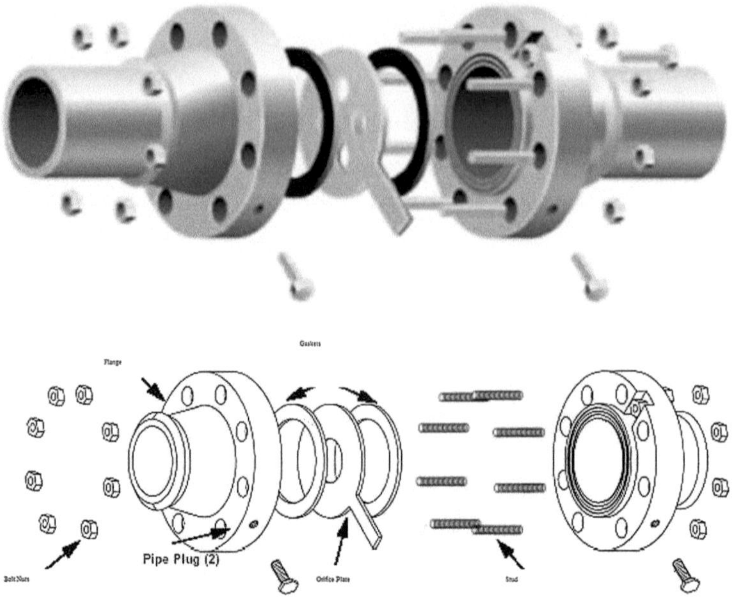

Fig. 7.75 Placing the orifice plate on the pipeline between the flanges

Rotameter: It is a device that shows the flow rate of liquid or gas fluid directly. This device consists of a vertically scaled vertical tube in the direction of flow of the cross-sectional area and a floating element which can remain constant at different heights in proportion to the flow (Fig. 7.77) [11]. It determines the state of the floating element, the force of the fluid, and the resistance force of the flow from bottom to top.

Superiority of rotameters are high accuracy, durability, and ease of use. It is especially used in flow measurement of liquid and gas flow in small diameter pipes. The rotameter consists of a conical measuring tube and a float that can move freely up and down in the tube. Unlike the previously described Venturi and orifices, the flow measurement is based on the resistive forces due to the external current acting on an object. When the buoyant force acting on the immersed body and the resistance force are balanced by its own weight, the body remains constant at a certain height in the flow. This height gives the measured flow rate according to the calibration result. For the flow to be measured according to the height of the submerged object in the tube, the inner surface of the tube is of a certain degree tapered. In addition to this conicity, the density and viscosity of the fluid, the coefficient of resistance of the body, and the weight of the body are in great importance. The measurement tube is placed vertically with the narrow end at the bottom. The fluid to be measured enters the tube from the lower point, passes around the float, rises up and exits from the top. When there is no flow in the rotameter, the float stops at the bottom of the measuring tube. This is where the largest diameter of the float almost corresponds to

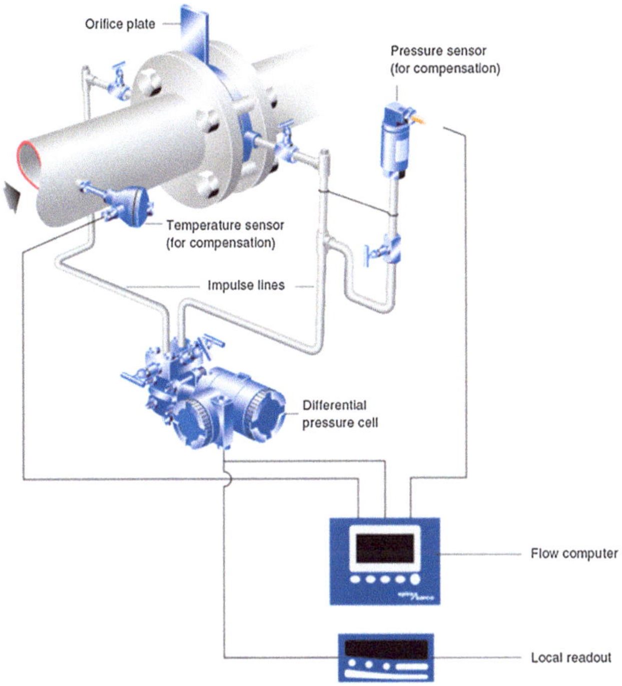

Fig. 7.76 Connecting the orifice plate to the electronic differential pressure sensor. *Source* https://www.spiraxsarco.com/learn-about-steam/flowmetering/types-of-steam-flowmeter [7]

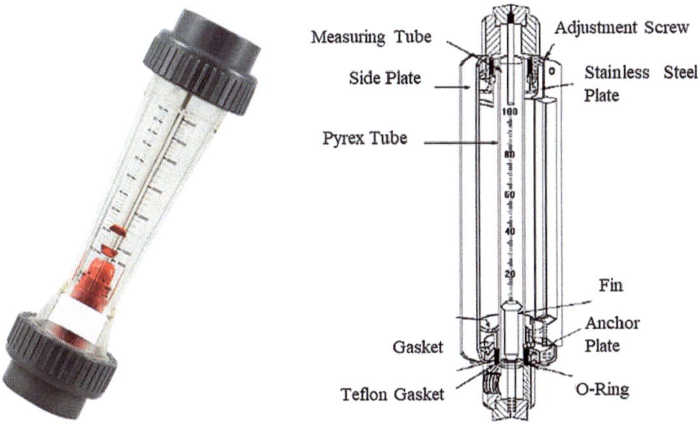

Fig. 7.77 Rotameter*. **Source:* https://www.pce-instruments.com/f/t/us/main.htm [11]

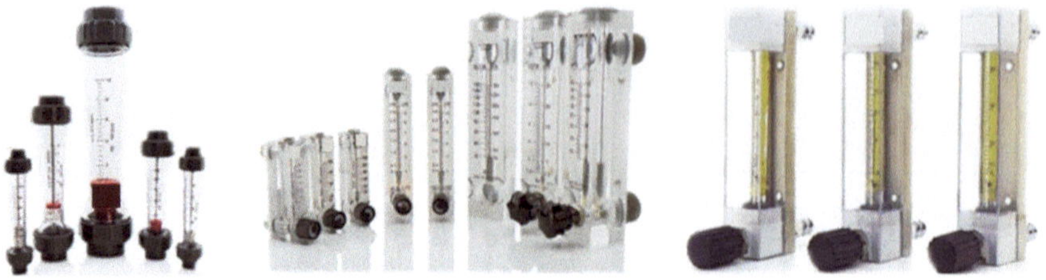

(a) Plastic tube (b) Glass tube with regulated valve (c) Low volume metal tube

Fig. 7.78 Rotameters in different designs

the diameter of the tube. When there is no flow, the buoyant force relieves the float, but since its density is greater than that of the liquid, it is not enough to raise the float. When the fluid enters the measuring tube, an annular opening with an annular section between the float and the tube increases. The buoy is raised until the buoyancy force and the resistance force resulting from the flow are balanced with the weight force. After this equilibrium of the buoy, any increase in flow can cause the buoy to a higher position and a decrease to a lower position. Each position of the float corresponds to a flow rate. The flow rate at which the position corresponds to the flow rate is determined in advance by known measurements and marked with a scale on the rotameter (calibration). Then, the desired flow rates are measured in the tube.

Changing the flow rate through a fixed area in the orifice, nozzle, or Venturi causes a variable pressure drop. This depends on the velocity of the flow. In the measurement instruments defined as field meters, the pressure drop is constant or very close to the constant; the current velocity changes the area through which the current passes. The rotameter, which is the most important area meter, is a graded and upward expanding glass tube (Fig. 7.78). The fluid moves up (vertically) in the tube. At this time, there is a standard floating particle in the suspended state. This particle rises as the velocity of the current increases; when the velocity is constant, the place of the particle, for example, its height in the tube, remains constant. When the velocity of the current decreases, the particle descends in the tube. The level at which the particle is located is read from the graduated glass tube and the flow rate of the current is found from the individual particle calibration curve data of each flowmeter.

Rotameter: unlike measuring devices which have a constant geometry such as orifice meter, Venturi meter, and Pitot tube and read the pressure difference which is proportional to the square root of the volumetric flow rate, it reads the volumetric flow rate which will create a constant pressure difference. The rotameters have a very variable geometry, i.e., they are variable area type flowmeters. The rotameter consists of a conical measuring tube and a float that can move freely up and down in the tube. The fluid raises the float with the pressure it creates and reads the value on the rotameter.

Rotameters are the most commonly used in flow measurements in practice, composed of a vertical transparent tapered pipe and an element which can play in fluid motion. The fluid enters through the narrow section of the cone and exits from its large cross section. When the fluid flow rate is zero, the moving element is at the bottom of the conical section. When the flow begins, the tensile forces acting on the moving element and the resistance forces are equilibrated to a certain height. The height at which the movable element is in equilibrium with the aid of the division in the walls of the tapered transparent pipe determines the flow rate of the flowing fluid.

Flow Measurement with Turbine Meter: Turbine type flowmeters are devices that measure volumetric flow with a rotatable turbine rotor made of stainless steel inside a non-magnetic body. The flow exerts a force on the blades of the turbine rotor. This applied force assures the rotation of the blades and turns the linear velocity of the fluid into the angular velocity. The turbine type flowmeters are based on the principle that the rotational speed of the turbine is proportional to the average fluid velocity, and therefore to the volume flow rate of the fluid. The turbine meter is sometimes referred to as a propeller meter and is a freely rotating propeller connected to a pipeline. A typical design is seen in Fig. 7.79. In the rotor flow direction, rectifier fixed blades are present, and the speed is measured by electrical or magnetic signal elevations (pulses) generated by a hole placed on a point on the rotor. The rotational speed of the propeller is approximately proportional to the velocity of the fluid.

The most important advantage of turbine meters is that each signal can be measured easily as a response to a very small increase in fluid flow. Liquid type turbine meters are made in two blades and the fixed number of signals is measured in a 5:1 flow stage with a sensitivity of $\pm$ 0.25% for each flow unit. Gas meters are made of multi-blades and their precision is $\pm$ 1%.

Turbine meters are highly individual devices and must be calibrated. For this purpose, manufacturers have prepared calibration curves for specific flow ranges. In addition, flow rate calibrations can be made with the help of the settings of the digital

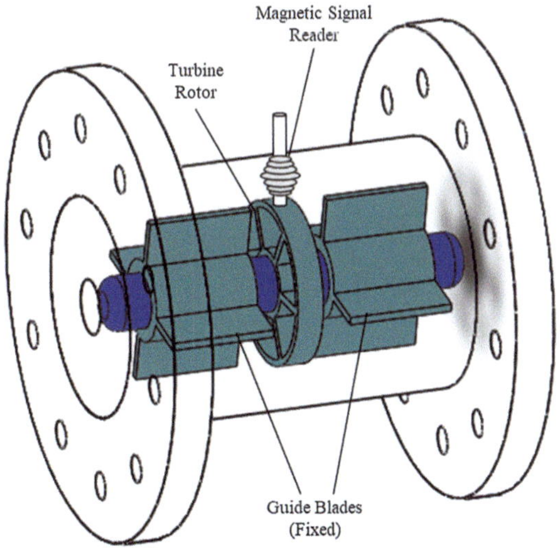

Fig. 7.79 Turbine meter

display devices. Turbine meters are also used for free flow measurements such as ocean currents and wind.

Pitot Tube: It is used to measure the local velocity along a current line. It consists of a tube opposite the direction of flow in a pipe. In the flow pipes, the pressure in the center of the pipe is higher than the pressure on the walls of the pipe. It operates according to the principle of comparing the two pressures that it has. The negativity of the Pitot tube is that it does not give direct average velocity and the readings for gases are very small.

The diagram for the use of the Pitot tube is given in Fig. 7.80. The opening of the tube (a) is perpendicular to the current direction and it is parallel to the stationary tube (b). These two tubes can be connected to the ends of a manometer to measure small pressure differences. The stationary tube measures the stationary pressure (p_0) because there is no velocity component perpendicular to the open part. The point B of the impact-resistant tube (a) is the point at which the AB current line is terminated (stationary point). The static pressure p_s is measured here. The pressure gauge measures the pressure difference p_s-p_0.

The Pitot or Pitot tube measures the total pressure created by a fluid, and therefore, the velocity of the fluid. In particular, the Pitot tube, which is commonly used in aircraft, is used to transform dynamic pressure with the static system. He was named after the French engineer Henri Pitot.

Connecting Pitot tube on the manometer can be seen in Fig. 7.81 [12].

The Pitot tube is placed in regions where its current is regular and is not affected by external factors. The Pitot tube is placed parallel and in the opposite direction to

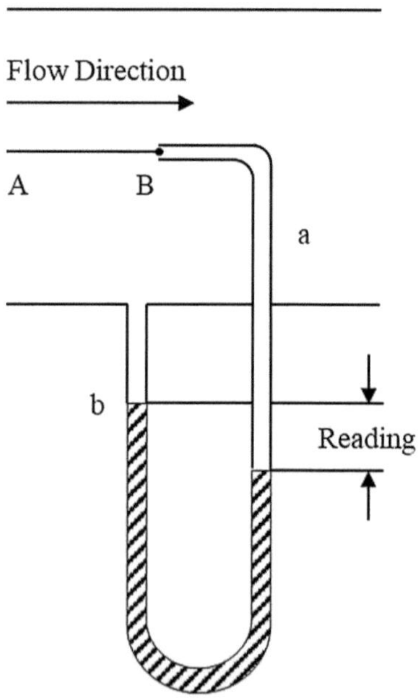

Fig. 7.80 Working principle of Pitot tube

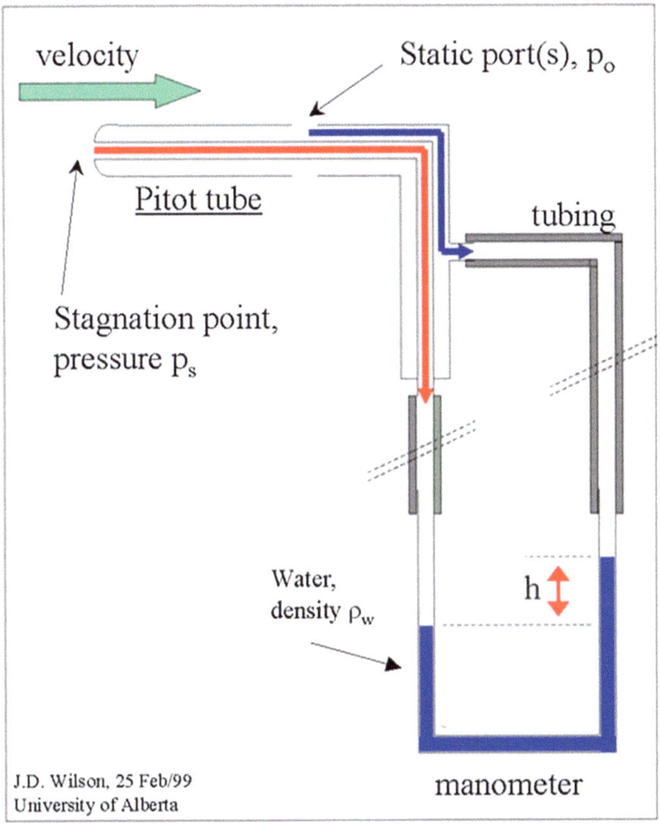

Fig. 7.81 Connecting Pitot tube on the manometer. *Source:* Dr. J.D. Wilson University of Alberta, 25 Feb/99 [12]

the direction of movement of the fluid. As the fluid passes through the Pitot tube, due to the relative motion of the device and the fluid, it measures the combined pressure of the dynamic pressure generated by the fluid (including the pitot tube) on the entire surface, e.g. the total pressure. The total pressure is also called Pitot pressure and ram pressure in some sources. Pitot tube is a part of the Pitot-static system in aircraft (Fig. 7.82). The total pressure that comes from the Pitot tube causes to expand the sensitive diaphragm in the system. The static pressure from the static holes allows the diaphragm to contract at the rate of static pressure. The combination of these two movements gives dynamic pressure and this information is reflected in the scale of the speed clock. This can be formulated as

Total pressure (p_t) = Static pressure (p_s) + Dynamic pressure [$p_d = (1/2\ \rho \times V^2)$] = Constant

$$p_t = p_s + \left(\frac{\rho \times v^2}{2}\right) \tag{7.15}$$

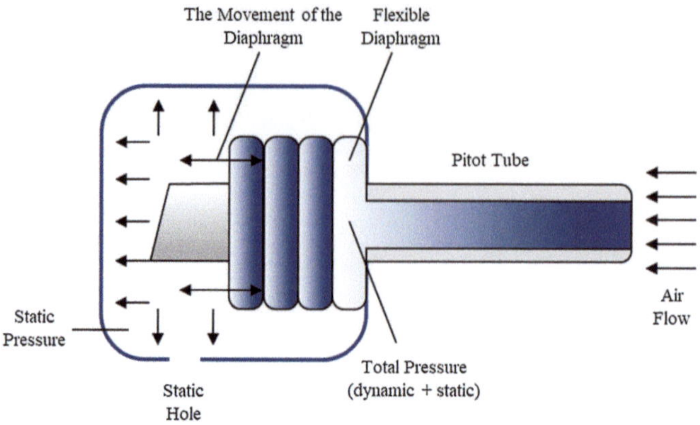

Fig. 7.82 Pitot-static system

If velocity is drawn here, the following equation is obtained:

$$v = \sqrt{\frac{2(p_t - p_s)}{\rho}} \tag{7.16}$$

7.5.5 Velocity Measurement

Velocity measurement can be done with devices using different methods in various fields of industry. Hinged channel anemometers have a wing on the shaft in an enclosure. As the air passes through the device, it exerts a force on the wing due to the flow. A spring and magnetic damping element apply force to the deflector in the opposite direction of movement. The instantaneous value of the velocity can be seen on the display of the device. This device is used to measure the velocity of the laboratory hoods by determining the air velocities in the intake and return duct emitters and grids in the determination of air movements in the hall.

7.5.5.1 Anemometer

Velocity measuring devices are generally called anemometers. An anemometer is a simple instrument that displays the velocity of a rotating element of the air flow, showing the speed of rotation [1]. Anemometers are generally designed based on the following principles:

- The change in pressure caused by the velocity,
- Cooling of hot objects in air flow,
- Changes in velocity of sound due to velocity and pressure change,
- Drift effect of air.

Anemometer: Anemometers are generally designed using the drag effect of air. For example, the anemometer consisting of rotating hemispherical cups (Fig. 7.83) operates on the principle that the wind force is greater than the pit (hollow) face and the bump (hump) face. As a result of these forces, the valves rotate about the vertical axis. By means of an established mechanism, the rotational energy is converted to electrical effect, in other words by means of an electric current, to measure the wind speed even at a remote station. Such anemometers measure speeds between 7 and 160 km/h. However, the sudden decrease in wind speed may not immediately affect the rotational speed of the flaps. Therefore, the resulting error can reach up to 30%. In other words, this kind of anemometer cannot show the exact speed changes. Some hemispherical cup type anemometers also have data accumulation capability. Technical characteristics of a hemispherical cup type anemometer capable of measuring very low air currents are given in Table 7.8. Operation principle of the hemispherical cup type anemometer can be seen in Fig. 7.84 [1].

The hemispherical cup type anemometers consist of a light turbine that rotates easily by air movement and the gear mechanism that transmits the rotational movement of the turbine to the dial indicating the path traveled by the turbine during a given period of time (Fig. 7.85). Such anemometers can be made in different sizes. The most commonly used ones are in the diameters of 75, 100, 150 mm. Especially at low speeds, the effect of friction on the device becomes important. There is a gear mechanism that increases the rotational speed to prevent deceleration due to friction.

Fig. 7.83 Hemispherical cup anemometer

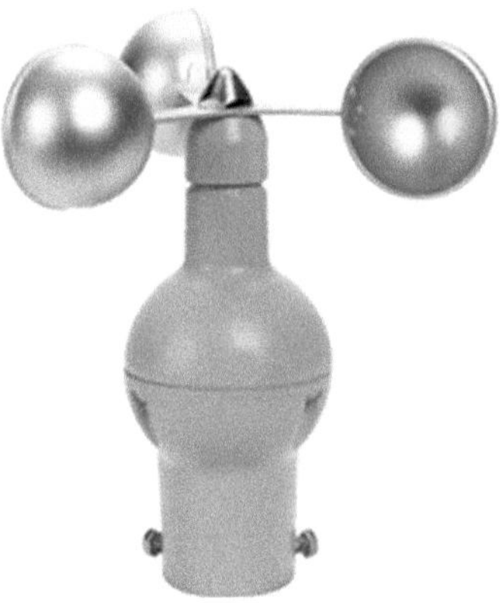

Table 7.8 Technical specifications of a hemispherical cup type anemometer

Technical specifications	Value
Measuring range	0.9–35 m/s
Measuring stability	0.1 m/s
Measurement accuracy	±2%
Bucket (Cup) diameter	70 mm
Display screen	28 × 19 mm LCD
Power supply	4 AAA Battery (1.5 V)
Mass	180 g
Dimensions	190 × 40 × 32 mm
Operating temperature	0–50 °C
Relative humidity	< 80%

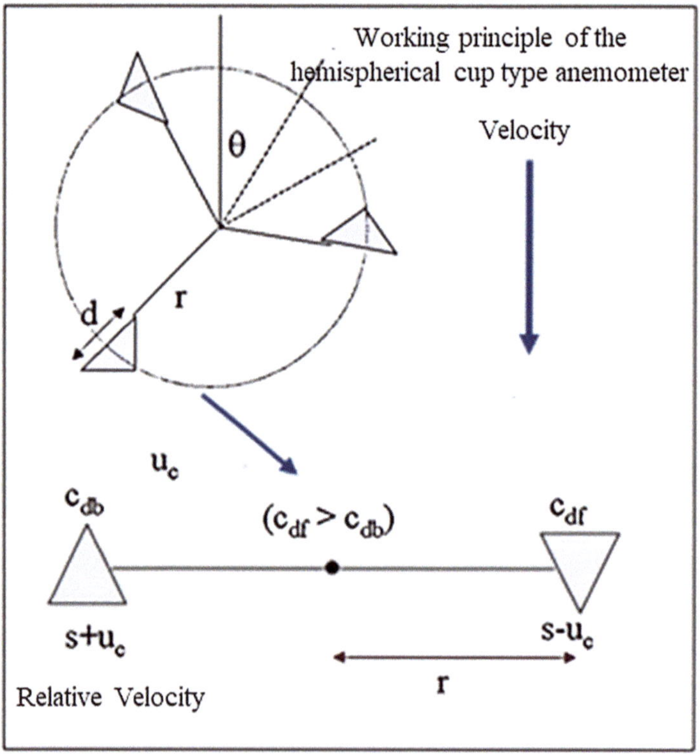

Fig. 7.84 Operation principle of the hemispherical cup type anemometer [1]

Therefore, the corrections are added at low speed values and at high speed values. For the medium speed values, the correction required is at the lowest level.

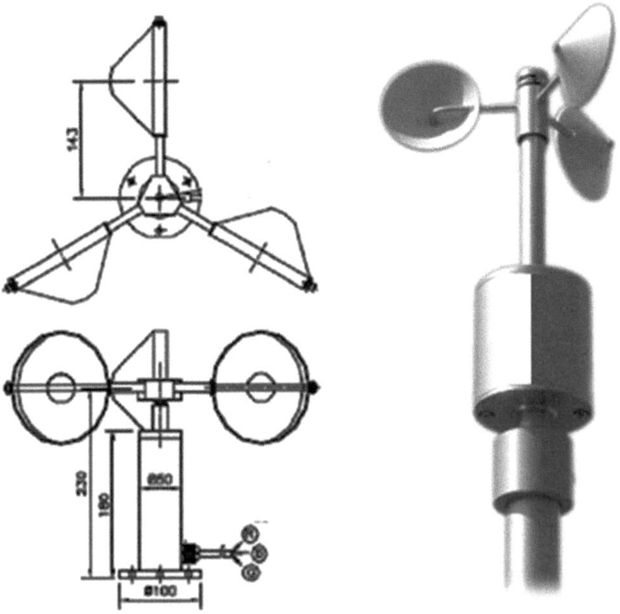

Fig. 7.85 Structure of the hemispherical cup type anemometer

Vane Anemometer: It is a kind of windmill. If the speed is too low, the bearing is made with rhinestone. Vane anemometers are used to measure wind speeds in the range of 1–40 km/h (Fig. 7.86) [2]. Vane anemometers are designed to measure and record other climate variables such as temperature, relative humidity, etc. The characteristics of a measuring system with vane anemometer, which can collect data, are given in Table 7.9.

Operation principle of the vane (propeller) anemometer can be seen in Fig. 7.87.

Thermal or Hot Wire Anemometer:

Anemometers operating on the principle of heat loss are used in the measurement of wind speeds that are less than 7 km/h. Such anemometers are called hot wire anemometers (Fig. 7.88). Hot wire anemometer consists of a heated wire, an electrical power supply, and a device for measuring current. When the heat loss increases with airflow, the temperature of the wire decreases. As a result, the falling resistance causes more current to flow. Thus, by producing more heat, a balance is created against heat loss. From the measurement of the current flow, the velocity of the air flow is determined.

Working principle of the hot wire anemometer: When an electrically heated wire is placed in a gas stream, heat is transferred from the wire to the gas and hence the temperature of the wire decreases. Depending on the wire temperature decreasing, the resistance of the wire also changes. This change in wire resistance is a measure

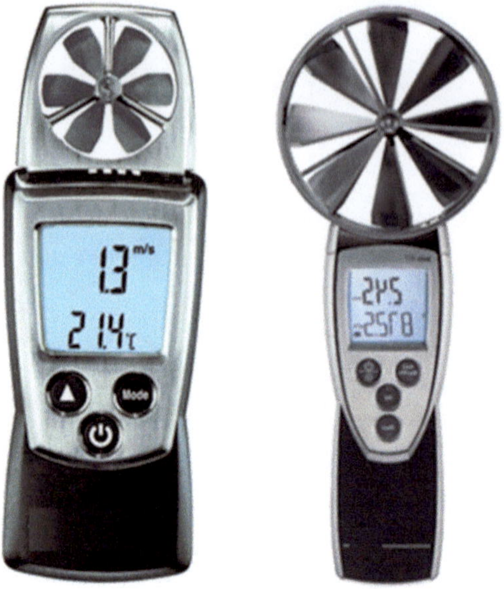

Fig. 7.86 Vane anemometers [2]

Table 7.9 An example of the vane anemometer technical data

Technical data	NTC (negative temperature coefficient)	Vane	Volumetric flow rate	
Measurement range	0…+ 50 °C	+0.3… + 20 m/s	0… + 99999 m^3/h	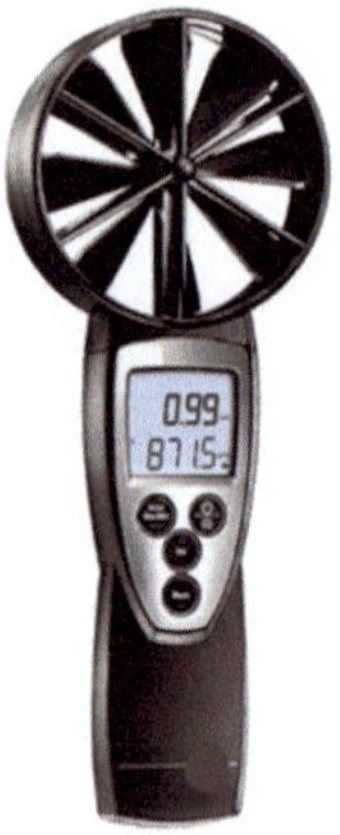
Sensitivity ±1 digit	±0.5 °C	± (0.1 m/s + 1.5% measurement value)		
Resolution	0.1 °C	0.01 m/s	0.1 m^3/h (0… + 99.9 m^3/h) 1 m^3/h (+100… + 99999 m^3/h)	

of the flow rate (Fig. 7.89b). The conductor wires are placed in a ceramic body. The leads from the conductive wires are soldered and connected to one of the legs of the Wheatstone bridge to measure the change in wire resistance. There are two methods to measure flow velocity using anemometer:

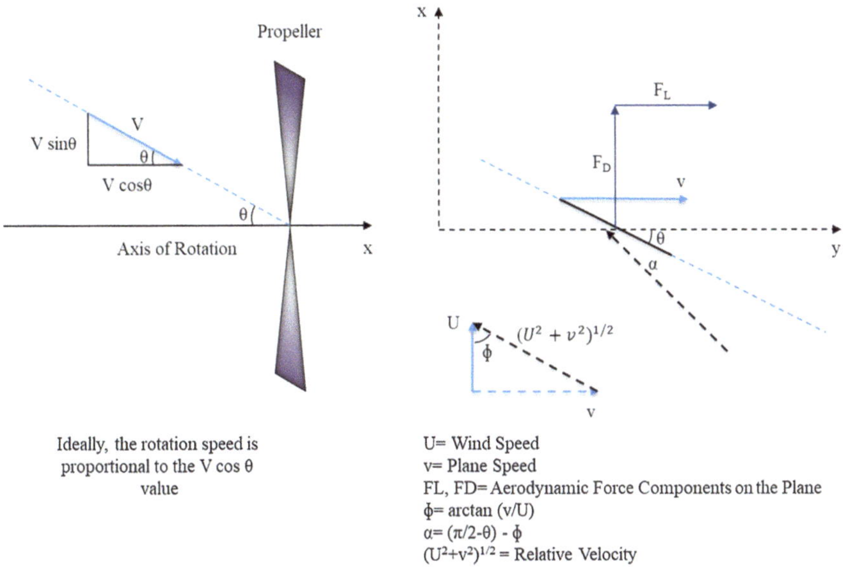

Fig. 7.87 Operation principle of the vane (propeller) anemometer

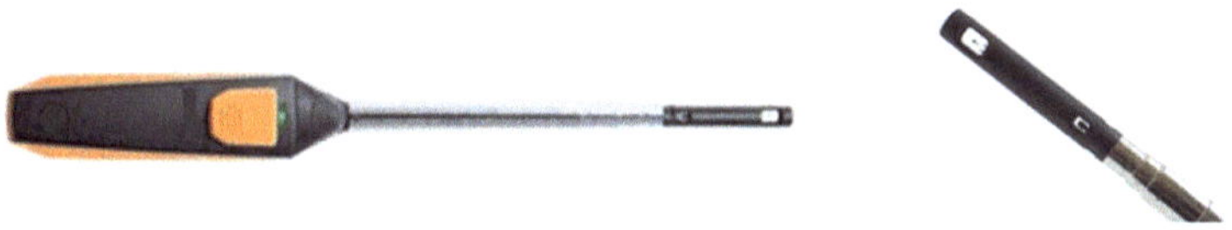

Fig. 7.88 Hot wire anemometers [2]

(1) Constant current method,
(2) Constant temperature method.

In *the constant current method*, the bridge arrangement with the anemometer is shown in the diagram given in Fig. 7.89c. The anemometer is held in the gas stream to measure the flow rate. A constant current is passed through the detection (sensation) wires. In other words, the voltage passing through the bridge circuit is kept constant. Due to the gas flow, heat transfer takes place from the sensing wire to the flowing gas. In this case, the change of the detection wire temperature causes the wire resistance to change. The change of wire resistance is a measure of the flow rate. This causes a deviation in the current meter, which is initially in the zero position. When calibrated, this deflection in the flowmeter becomes a measure of the gas flow rate.

In *the constant temperature method,* the bridge arrangement with the anemometer is shown in the diagram given in Fig. 7.89d. The anemometer is held in the gas stream to measure the flow rate. The current is passed through the wires. Due to the gas flow, heat transfer from the sensing wire to the gas stream takes place. In this case, as the temperature of the wire changes, its resistance also changes. The

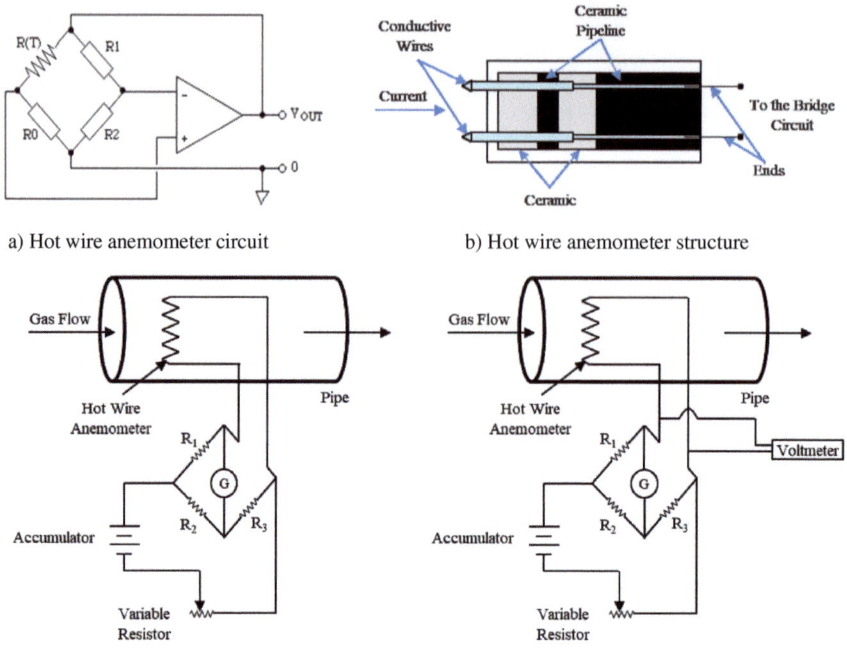

a) Hot wire anemometer circuit b) Hot wire anemometer structure

c) Constant current method d) Constant temperature method

Fig. 7.89 Working principle of the hot wire anemometer

principle of this method is to maintain the temperature and resistance of the detection wire at a constant level. Therefore, the current passing through the sensing wire is increased in order to bring the wire to its temperature and resistance in the provincial state. The electrical current required for bringing the wire to its initial resistance and temperature becomes a measure of the flow velocity.

The thermal or hot wire anemometer comprises a heated resistance temperature device, thermocouple (thermo-element) or thermistor placed at the end of a probe as a sensitive element (Fig. 7.90). Such devices are designed for easy and direct detection of air velocity at a point in the flow area. The electrically heated sensing element of the probe placed in this flux area cools in proportion to the velocity of the fluid passing through it. A thermal anemometer consisting of a responsive element and an electronic circuit can evaluate the flow rate directly from the sensitive element and can show the flow rate directly as analog or numerical.

These types of anemometers operating according to the principle of heat loss do not change the current during measurement as they are used for measuring airflow in narrow volumes. The main advantages of thermal anemometers are the dynamic working ranges that are large and they can measure very low speeds. With the same measuring device, air speeds from 0.1 m/s up to 50 m/s can be measured. If the sensitive element becomes dirty, it should be kept clean since it will change the calibration

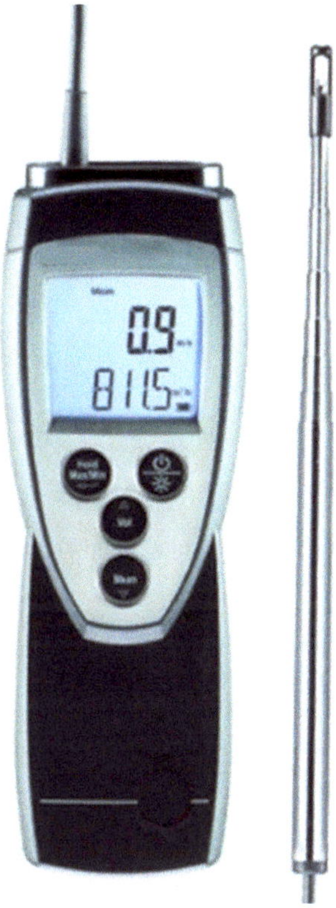

Fig. 7.90 A hot wire anemometer [2]

of the device. Thermal anemometers are widely used in heating, ventilation, and air conditioning applications.

Ultrasonic anemometer: Another method for measuring air velocity is realized by measuring the speed of sound in the air. Because the sound velocity in the stagnant air is constant, any deviation in this velocity value becomes a measure of the wind speed. Ultrasonic anemometers operating according to this principle are shown in Fig. 7.91.

An example of the main specifications of ultrasonic anemometers can be seen in Table 7.10.

7.5.6 Temperature Measurement

Molecules, which are the smallest grains of objects, are constantly moving or vibrating due to the heat energy, they have in their mass. This vibration of molecules is formed by the movements in the solids, in short movements, longer in the liquids by the displacement movements of the particles and in the gases in the continuous and

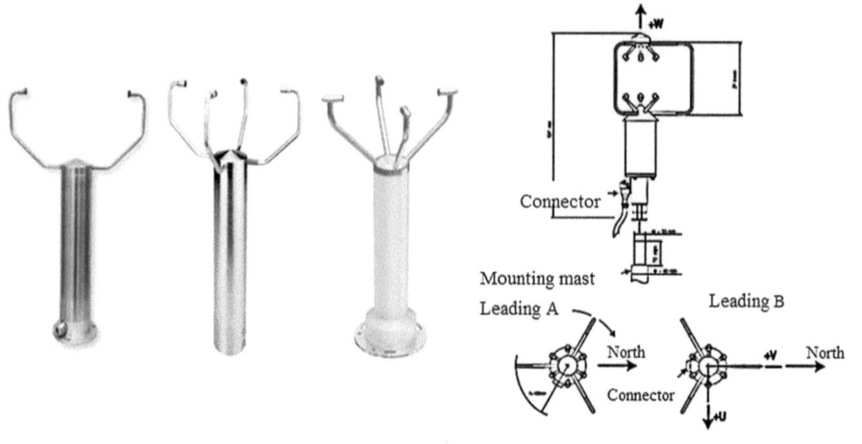

Fig. 7.91 Ultrasonic anemometers

Table 7.10 An example of the main specifications of ultrasonic anemometers

Wind speed and direction	Measuring range: 0.01–65 m/s
	Measurement accuracy: ± 0.1
Operating voltage	12–24 V AC/DC
Power consumption	3 VA
Temperature	Measuring range: −40–70 °C
	Measurement accuracy: ± 0.5
Protection	IP 65 (against dust and water spray)

mixed directions. According to the ability of the substance in these three states, while the shape of the solids does not change, the liquids take the shape of the container in which they form, and the gases can cover the volume into which they are filled in a short period of time. The intensity of these movements of the molecules increases in proportion to the increase in the energy from the external environment. Conversely, as the energy decreases, the movement of the molecules decreases and eventually stops at a certain energy level. This level is called absolute zero degree and it is shown as −273 °C or 0 K. The total amount of energy that an object has in its mass is called heat [1].

This heat energy, which provides the molecular movements or vibrations in the bodies, cannot be directly detected or measured. When the sum of the energy of an object increases in its mass, in other words, the increasing energy is dispersed into the molecules in the matter and the proportion of energy falling into each of the molecules increases. The increase in energy in each molecule also increases the kinetic motion energy of the molecules, in other words increases their vibration. These increasing molecular vibrations have an effect on the environment in the form of electromagnetic waves. This effect is called as the *temperature.*

Temperature is a measure of the average thermal or heat energy of the particles in a substance and it does not depend on the number of particles in an object, since it is an average measurement. Considering what has been described so far, *heat* is a potential energy present in the bodies, and the *temperature* is the kinetics of this energy or the effect of the energy. Because of these characteristics, the concepts of heat and temperature, which are closely related to each other and directly proportional to each other are completely different in terms of their properties.

In physics, the temperature is measured by the thermometer and determined by the degree (°) (usually Celsius (°C) or Fahrenheit (°F), which is the unit of temperature). Heat is not determined directly, but its appearance is determined by the help of the temperature, and its unit is indicated by the Joule (J).

Temperature refers to the thermal state associated with the orientation to heat transfer of a substance. It is a feature that defines the condition of a substance that is taken or given heat. It is a measure of the molecular kinetic energy of the matter. *The temperature* is a physical indication of the heat energy in terms of degree or intensity.

Heat is a kind of energy. Heat is present with the movements or vibrations of the molecules that make up the substance. The nature of the heat is not dependent on the external movements of the object but depends on the movements of the molecules that make up the body. *Heat* is a form of transfer of energy caused by a temperature difference. In some instances, the molecular energy stored in the body is energy that can be transferred by conduction, convection, and radiation in some cases. Heat is a physical energy that causes an object to increase in temperature, lengthen its length, expand, melt, evaporate, and do work. The size of heat is indicated by its intensity and amount. The temperature indicates the intensity of the heat. The amount of heat required to increase the temperature of a unit mass material by one degree under certain pressure and volume conditions is called specific heat. The specific heat varies slightly depending on the temperature. Especially, the temperature of the specific heat in gas exchange is important. *Sensible heat* is the heat that changes the temperature of a substance when it is given or received. *The latent heat* is the heat that the materials take or give at a constant temperature during its physical state changes. Three different temperatures are defined for air-water vapor mixture:

(1) Dry bulb temperature,
(2) Wet bulb temperature,
(3) Dew point temperature.

Dry Bulb Temperature: The dry bulb temperature, which is the reference temperature for specifying the general air temperature, is the temperature detected by a thermometer that is in thermal equilibrium with air. The dry bulb temperature, which is a measure of the temperature of the sensible heat in the air, is the temperature of the air-water vapor mixture in the stagnant state and determines the saturation partial pressure of the water vapor. The dry bulb temperature is the true thermodynamic temperature.

Wet Bulb Temperature: *The wet bulb temperature* is the temperature in a liquid or solidified water that evaporates into the air and which adheres to the air at the same temperature, adiabatic saturation. It can be said that the wet bulb temperature is the adiabatic saturation temperature. The wet bulb temperature is the temperature felt by the human skin in the wet state exposed to the air flow. If the air is not initially saturated, heat evaporation occurs from the air. Water vapor mixes with air and absolute humidity rises. This process continues until the air is saturated. When the air is cooled, the sensible heat is converted to latent heat. The cooling process continues until the saturation point. If the air is too dry at first, excessive cooling takes place. The temperature that limits the saturation of air is the wet bulb temperature.

The psychrometric wet bulb temperature is measured by means of a thermometer covered with a wick that is continuously wetted with distilled water. When the unsaturated air stream passes over the chamber of the thermometer, water evaporates from the moist wick. In other words, the temperature of the thermometer chamber falls below the air temperature. Heat transfer from the air to the wetted wick continues until a certain equilibrium is established. The temperature measured when equilibrium occurs is *the psychrometric wet bulb temperature.* The difference between *thermodynamic wet bulb temperature* (adiabatic saturation temperature) and *psychrometric wet bulb temperature* is quite low. For any condition of the moist air, there is such a temperature value that the liquid (or solid) water at this temperature and pressure saturated this air that taken into account. In the adiabatic saturation process, the air is saturated with water, and the air is at the same temperature. This saturation temperature is called *thermodynamic wet bulb temperature.*

The psychrometric wet bulb temperature is the steady-state temperature determined by evaporation of water from the wet wick in the thermometer. The temperature value is read after the thermometer is released to the airflow. With the evaporation of water in the wick, the wick temperature decreases. In this case, the temperature read on the thermometer is the wet bulb temperature. The wet thermometer reads a temperature that is lower than the dry bulb temperature of the air. The dry bulb temperature is low as the ambient air is drier, which means that the evaporation rate is greater. The temperature read at the wet bulb thermometer varies considerably depending on the following two factors:

(1) The initial temperature of the water used to keep the filter wet,
(2) Heat exchange between the wick and the medium by radiation.

Dew Point Temperature: When the moist air is cooled sufficiently, the water vapor in the air condenses. Climatic events such as rain, snow, fog, and dew occur depending on this principle. The temperature at which the condensation occurs is called the dew point temperature. When the air conditions are below the saturation line in the psychrometric diagram, the dew point temperature is lower than the dry bulb temperature. In the case of saturation, dew and dry bulb temperatures are equal.

If the moist air is cooled under constant absolute humidity and constant atmospheric pressure, it cannot carry all of the water vapor in the air after a certain temperature. The water vapor, which the air cannot carry, is condensed and separated from the air. The temperature at the time the condensation starts is called *the dew point temperature.* When the air is cooled, the moisture content does not change initially. Because the cold air is capable of holding less moisture than warm air, its relative humidity increases. However, if the cooling process takes place below the dew point temperature, the absolute humidity remains unchanged until the air cools at the saturation point where the water begins to condense. The dew point temperature is defined as the temperature of the saturated air with the same vapor pressure as the moist air. The dew point temperature can also be defined as the temperature of the air-water vapor mixture when the air is condensed to the outside of the air when the air is cooled in constant pressure and constant water vapor content.

The dew point temperature (t_d) is defined as the temperature of the saturated air when the humid air with absolute humidity *(w)* is equal to the absolute humidity of the saturated air (w_s) at the same *P* pressure. Since the dew point temperature is a function of the water vapor partial pressure in the air, it can be calculated from the following equations.

At temperatures between −60 and 0 °C:

$$t_d = -60.45 + 7.0322[\ln(P_w)] + 0.3700[\ln(P_w)]^2 \quad (7.17)$$

At temperatures in the range between 0–70 °C:

$$t_d = -35.957 - 1.8726[\ln(P_w)] + 1.1689[\ln(P_w)]^2 \quad (7.18)$$

where t_d = dew point temperature (°C) and P_w = water vapor partial pressure (Pa).

7.5.6.1 Temperature Scales

As a temperature scale in climate science, different scales such as Celsius (°C), Fahrenheit (°F) or Kelvin (K) are used (Table 7.11 (a), (b)). At normal atmospheric pressure (101325 Pa), the temperature points where the water freezes and boils are referenced (Fig. 7.92). The thermometer is calibrated between two fixed points. These are the freezing point and boiling point of water. At normal atmospheric pressure (760 mm mercury pressure), the distance between these two points is divided into 100 equal parts in the Celsius thermometer. Each of these shows a Celsius (1 °C).

In the Fahrenheit scale, this range is divided into 180 equal parts. Each of these shows a Fahrenheit (1 °F). In this measurement, the freezing and boiling point of the water are determined to be 32 °F and 212 °F, respectively. In the Réaumur scale, these points are called 0 °R and 80 °R. The intermediate distance (spacing) is divided into 80 parts. Since mercury freezes at −39 °C, it is not suitable for measuring very low temperatures. These are filled with colored alcohol with a low freezing point. The lowest temperature that can be reached is absolute zero, which is −273.15 °C. Another scale starting from absolute zero is the Kelvin scale. Absolute zero is the temperature at which the molecular movement stops.

As temperature scales, Celsius (Centigrade) is used in many parts of the world, Fahrenheit (Fahrenheit) and Réaumur are used in some countries, mainly in England and America. The transformation of these temperature scales is given in Table 7.11 (a, b). Today, two common temperature scales are used, namely, Celsius and Fahrenheit.

Table 7.11 (a) Temperature scales and the features. (b) Temperature scale formulae and equations

Features	Temperature Scales			
	Celsius (°C)	Fahrenheit (°F)	Réaumur (°r or °Re)	Kelvin (K)
Boiling point of water	100	212	80	373
Freezing point of water	0	32	0	273
Absolute zero	– 273.15	– 459.67	– 218.52	0
Average room temperature	20–25	68–77		293–298
Human average body temperature	37	98.6	–	310
Conversion	°C = [(°F–32) × (5/9)]	°F = [(9/5 × °C) + 32]	°Re = [(4/5) × °C]	K = °C + 273.15

From	To	Formulae
Réaumur	**Celsius**	°C = °Re × 1.25
Réaumur	Fahrenheit	°F = °Re × 2.25 + 32
Rankine	Kelvin	K = °R × 5/9
Rankine	**Réaumur**	(°Re = (°R – 491.67) × 4/9

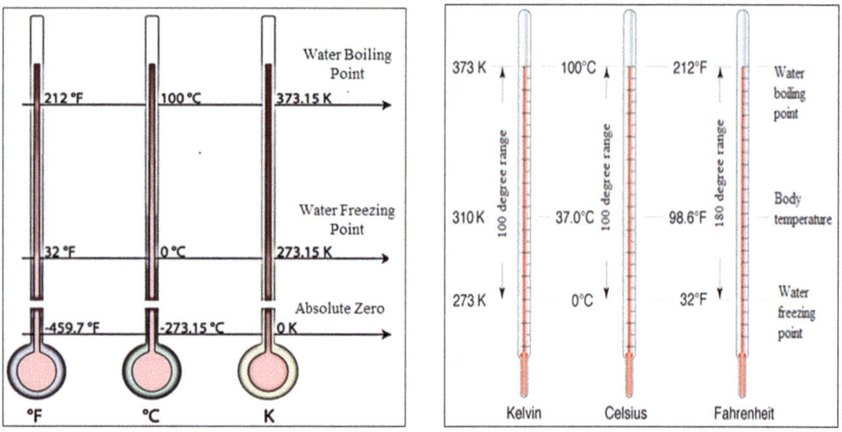

Fig. 7.92 Temperature scales

The Celsius temperature scale was defined by Anders Celsius (1701–1744) and the boiling point of the water at the pressure of 1 atmosphere (1.01325 bar) was defined as 100 degrees (°C), freezing point was defined as 0 °C. Later, Carrolus Linneaeus (1707–1778) reversed this scale, accordingly, it changed the freezing point of water to 0 °C and the boiling point to 100 °C at 1 atmosphere pressure. This measurement system is now called as Celsius (°C). As it can be seen from Fig. 7.92, the freezing point of the water in Celsius grading was accepted to be 0 °C and the boiling point to be 100 °C and the Celsius scale was obtained by dividing this point by 100 equal parts.

A further temperature scale was defined by the German physicist Gabriel D. Fahrenheit (1686–1736), which in Fahrenheit the freezing point of water is 32 °F and the boiling point is 212 °F. This temperature scale is called degrees Fahrenheit (°F). In this way, the freezing point of water in 1 atmosphere equals 32 °F.

In engineering applications, the point at which the kinetic energy of atoms and molecules is usually reset is taken as the basis for the measurement, the value corresponding to degrees Celsius is −273.15. The degree starting from this point is called the degree Kelvin (K) which goes to the same size as the Celsius. The basic SI unit is degrees K. In the Fahrenheit (°F) scale, the same point corresponds to −459.67 °F. The scale starting from this point and going with the degree Fahrenheit size is called degree Rankine (°R). These four temperature scales can be converted to each other (Table 7.11 (a) and (b)).

7.5.6.2 Temperature Measurement

The temperature is the most commonly measured and used quantity. Temperature can be defined as a measure of the kinetic energy of atomic and molecular movements, reflected in the macro world. According to a reference system, temperature measuring devices are called thermometers or temperature meters. Thermometer means instrument used to measure temperature (from ancient Greek: *thermos* is temperature and *métron* is measure). In a thermometer, the property that used to evaluate the temperature is called *thermometric property*. Length, volume, pressure, electrical resistance, potential difference, color change, and radiation intensities of surfaces are thermometric properties. Depending on these thermometric properties, different types of temperature gauges have been developed.

Thermometer Types: A thermometer is a measuring instrument used to measure the temperature of an object. It is the most widely used temperature measuring device. The expansion difference is used to determine the temperature. The sensitivity of the thermometer is related to the calibration sensitivity. The error rate of a thermometer does not exceed 0.1 °C. For measurement of temperature, different types of thermometers are used by taking into consideration factors such as expected accuracy and temperature measurement range (Table 7.12 (a) and (b)).

Thermometers used for climatic measurements are designed by using the properties of different materials in the face of heat. A thermometer design takes into account one of the following:

- The expansion under the influence of heat,
- Water vapor depends on temperature,
- Changing the resistance of the conductors with heating,
- Thermoelectric effect.

Table 7.12 (a) Classification of thermometers. (b) Comparison between different types of Thermometers

Contact Thermometers
(1) Expansion thermometers • Liquid expansion • Gas thermometers • Bimetallic thermometers
(2) Thermistors
(3) Electric resistance thermometers (Nickel, Copper, Platinum, Tungsten)
(4) Thermoelectric thermometers (Thermocouple, thermoelement)
(5) Liquid crystal thermometers
(6) Combined circuit thermometers
Non-Contact Thermometers
(1) Optical thermometers
(2) Radiation thermometers

S. No	Thermometer Type	Range	Thermometric Property	Remarks
1	Liquid Thermometer	– 190 °C to 750 °C	Thermal expansion of a liquid	Simple and convenient but small range
2	Gas Thermometer	– 268 °C to 1600 °C	Thermal expansion of the given mass of a gas	Very sensitive, accurate and large range but bulky and slow
3	Resistance Thermometer	– 272 °C to 1200 °C	Variation of electric resistance with temperature	Wide rage, accurate but not suitable for rapid changes
4	Thermoelectric Thermometers	– 250 °C to 3000 °C	Seebeck effect of a thermocouple	Wide rage, quick but complicate
5	Radiation Thermometers	Above 3000 °C	Seebeck effect of thermocouples, thermopiles etc.	Measures the temperature or inaccessible objects (Stars) but not direct reading
6	Vapor Pressure Thermometers	– 272 °C to –193 °C	Saturation pressure of a pure liquid	High precision and sensitivity for low temperatures, but scale is non linear

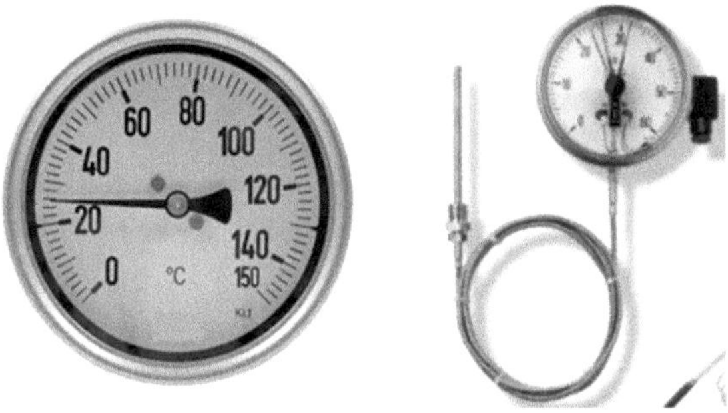

Fig. 7.93 Pressure thermometers

Pressure Thermometers: They have similar properties with ideal gas thermometers. The difference between these types of thermometers is that the reservoir (sensor part) is filled with a liquid (propyl alcohol), gas (nitrogen), or liquid/vapor (freon 22, propyl, methyl chloride, acetone, ethyl benzene). Pressure thermometers are made in three different designs:

(1) Gas filled,
(2) Liquid filled,
(3) Liquid/gas filled.

Pressure thermometers (Fig. 7.93) are similar to the *Bourdon manometer*. It is based on measuring the pressure generated by thermal expansion of the fluid in the reservoir with temperature. These thermometers are also called fluid expansion thermometers. Their capillary tubes can be 20 m.

Such thermometers operate according to the principle of measuring the pressure resulting from the thermal expansion of the viscosity in a closed container, as shown in Fig. 7.94 [13]. For this reason, pressure thermometers are also called *fluid expansion thermometers.*

Pressure thermometers are widely used in industrial temperature measurements because they are economical and sensitive to other temperature meters. The distance between the measuring point and the reading point can reach up to 60 m. They can be used between −150 °C and +600 °C, and they can also be used with special precautions at temperatures between 100 °C and 1000 °C.

Liquid/Steam Filled Pressure Thermometers: These thermometers are filled with a fluid that is easily evaporable, whose pressures are smaller than their critical pressures. Working ranges range from −50 °C to +300 °C, temperature measurement is made by using the principle of changing the vapor pressure of the fluid stream that placed inside them. The most important advantage of liquid/steam-filled thermometers is that they use a smaller reservoir. For example, this type of thermometer with a capillary tube length of 0.6 m to 25 m has a diameter of 16 mm and a length of 78 mm.

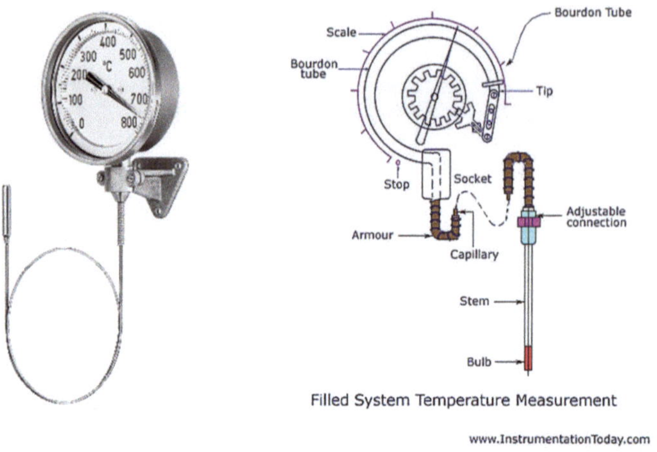

Fig. 7.94 Structural components of the pressure thermometer [13]. *Source*: http://www.instrumentationtoday.com/

Liquid Filled Pressure Thermometers: It is possible to make a liquid-filled thermometer by filling the pressure thermometers with propylene alcohol, silicone, and similar liquids. The capillary tube varies between 0.6 and 10 m. The temperature range is between −150 and + 300 °C. The measuring scales are linear with a great approximation. Practically, in liquids, there must be a good match between the temperature and the measured temperature range, the volume of the reservoir and the volume of the spring system measuring a volumetric expansion.

Bimetallic Thermometers: They work in accordance with the principle that the thermal expansions of different metals change with temperature (Fig. 7.95). The two-metal adhered on top of each other, due to the different thermal expansion, are

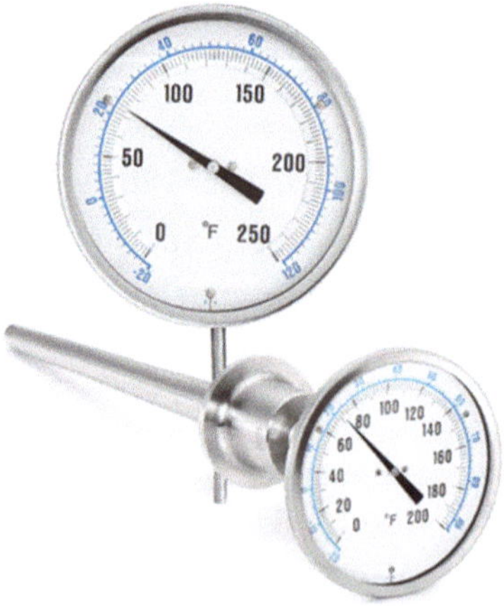

Fig. 7.95 Examples of bimetallic thermometers

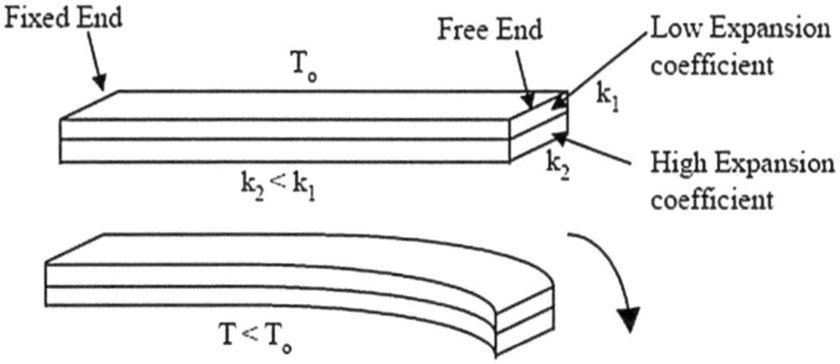

Fig. 7.96 Operating principle of the bimetallic thermometer

tilted by the effect of temperature. (Fig. 7.96). The temperature value can be measured by determining this level of bending, which depends on the temperature. It is based on the linear elongation of the rigid bodies with temperature. At a temperature of 0 °C the length of a solid rod is l_0, at a temperature of t, the length of a solid rod is $l_t = l_0 (1 + \alpha t)$. Where α (1/K) is the mean linear thermal extension coefficient between the 0 and t temperatures of the bar.

Bimetallic materials are the materials used in thermostat construction by using two different metals with different expansion coefficients. Bimetallic materials (brass (copper/zinc alloy), stainless steel) are also used for temperature control.

Bimetal thermometers used in industry are designed by helical winding. One end of the helix is placed in the sheath and the other end is connected to the needle. The temperature expands end rotates the pointer in a circular motion. As the temperature increases, a widening helical metal rotates the pointer by moving in a circular manner. In bimetal thermometers, unequal expansion of two separate metal sheets, which are joined together and encircled into a narrow strip, is used to move a pointer around a circular structure (Fig. 7.97).

They are used to measure the temperature that cannot be measured by mercury and alcohol thermometers. With such thermometers, high temperatures up to 1600 °C can be measured. Their sensitivity is in the range of ± 1–3%. Their most important advantages can be listed as being rugged, easy to use, cheap, and lack of power requirement. On the other hand, such thermometers are not suitable for low temperatures and they are not very sensitive. At low temperatures, the expansion and contraction of metals does not occur properly.

Thermistors: Thermistors (Fig. 7.98) are heat sensitive resistors and are generally designed from semiconductor materials such as ceramics. They work according to the principle of change of temperature and resistance. The resistances of semiconductor materials show a great change with temperature. As the temperature increases, the resistance decreases. The legs of the thermistor element made of semiconductor material are calibrated by connecting to a bridge circuit.

A thermistor is made from copper, manganese, nickel, cobalt, and lithium oxides. These oxides are mixed in appropriate proportions, given the desired shape and recrystallized by heating. Thus, a ceramic structure with the desired resistance

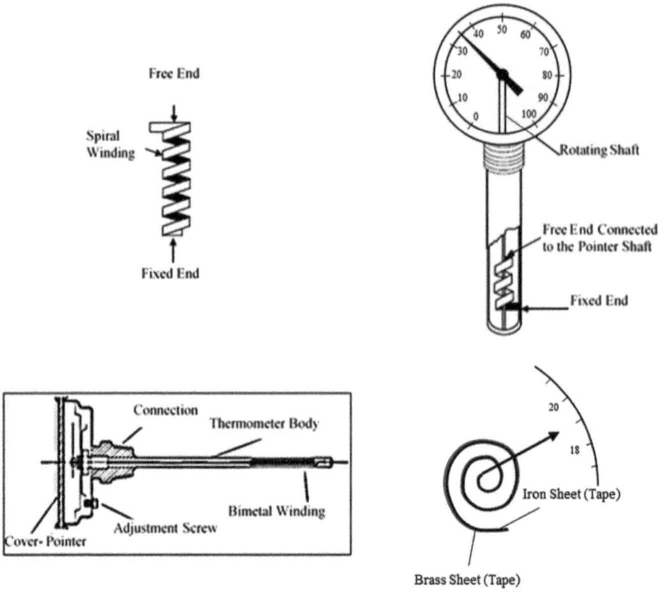

Fig. 7.97 Structural components of the bimetal thermometer

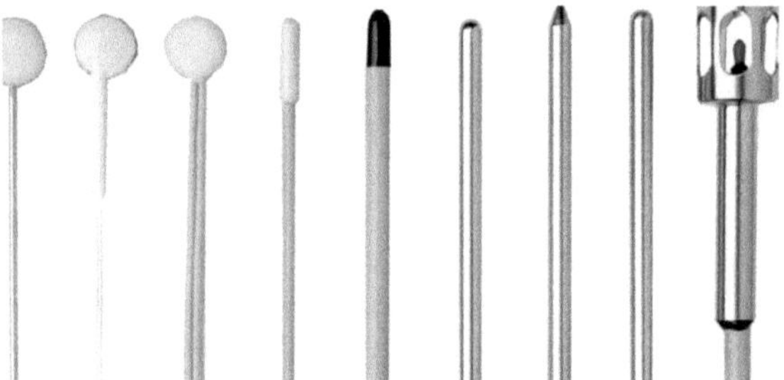

Fig. 7.98 Various types of thermistors

temperature characteristic is obtained. The thermistor is a temperature-sensitive variable resistance designed from ceramic-like semiconductor material. They are designed in different shapes and sizes ranging from a few ohms to higher ohms. Their designs may vary in size, from very small bead structure, membrane structure, thin chip or tiered structure, to wide road structure (Fig. 7.99).

They are very useful in dynamic temperature measurements. The advantages and disadvantages of thermistors are given in Table 7.13. The main advantages of temperature measurement method with the thermistor can be listed as easy to use,

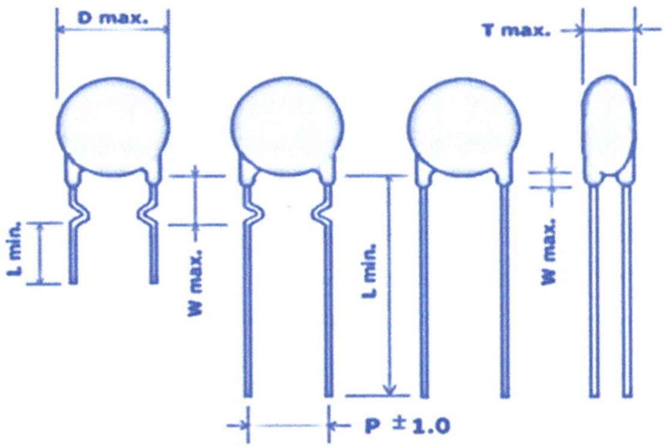

Fig. 7.99 Examples of the thermistor design

Table 7.13 The advantages and disadvantages of thermistors

Advantages	Disadvantages
• Measurement sensitivity is high • The measurement accuracy is 0.01 °C • Resistant to electrical and mechanical stresses • The measuring range is quite wide and ranges from –100 °C to 300 °C • They can easily adapt to the bridge circuits • The costs are quite low • Measuring velocity values are high • They are in small size	– The change in resistance value with temperature is not linear – It has an internal warming effect caused by much lower current levels than metal sensors – The temperature shows an exponential change with resistance

convenience, sensitivity, and speed. When the appropriate calibration is performed, the temperature can be measured by the thermistors with a sensitivity of 0.01 °C. Since such devices are made of semiconductor material, they are not suitable for use above 300 °C. Special precautions should be taken for high temperatures. Thermistors are very sensitive. They are 100 times more sensitive than electrical resistance thermometers (RTD: Resistance Temperature Detector) and 1000 times more than thermocouples. They can detect a small temperature change and they are very quick. They can detect sensitive temperature control and they can also detect small temperature differences in unit time.

Electrical Resistance Thermometers: Electrical resistance thermometers (RTDs, they are also called as Resistance Temperature Detectors) are devices that measure temperature with electrical effects. The principle of temperature measurement is based on the principle of temperature variation of electrical resistances of some resistance elements (materials such as platinum, nickel, tungsten). As the temperature increases, the resistance also increases. Resistance thermometers operate on the principle that the temperature of the sensitive metal element is mostly changed by electrical resistance.

An electrical resistance thermometer can be seen in Fig. 7.100.

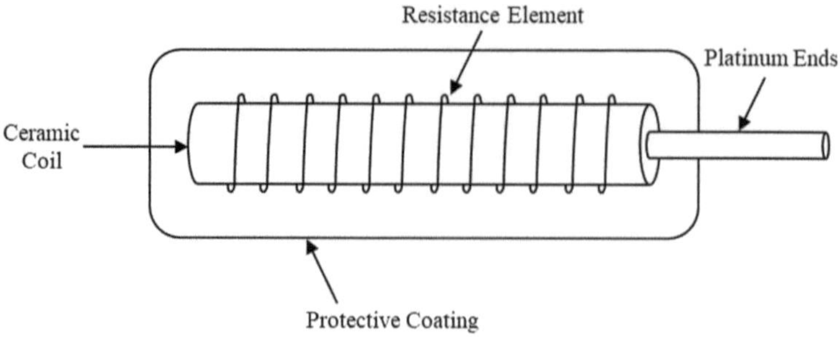

Fig. 7.100 An electrical resistance thermometer

This type of thermometer can be measured in the range of −260 °C/ +750 °C. Their sensitivity is around ± 0.01 °C. Prices of resistance thermometers are more expensive than thermocouples and response times are longer. Resistance thermometers provide the best results when used in slow-changing temperature measurements. Resistance thermometers measure the change in conductor or semiconductor resistance with temperature change. Platinum, nickel, and copper resistance are the most commonly used metals in thermometers. Metal sensing element can be made of platinum, rhodium-iron, nickel, nickel-iron, tungsten, or copper. Such devices have simple circuit structures, high linearity, sensitivity, and stability. The choice of these materials is made by taking into consideration the temperature range, corrosion protection, mechanical stability, and cost criteria to be used. The most used materials and temperature ranges in these thermometers are given in Table 7.14.

Thermoelectric Thermometers: Thermoelectric thermometers are often referred to as thermocouples, or thermo-elements (Fig. 7.101). The most widely used electrical temperature measurement methods are thermometers. With this type of thermometers, any liquid, solid, and gas temperatures between −85 and +1820 °C can be measured easily.

When one end of the two conductive wires made of different material is attached to each other, a voltage at the mV level occurs between the free ends, depending on the temperature of the end. Temperature measurements are made by using this thermal voltage. These types are the most widely used electrical temperature measurement methods. All kinds of liquids, solids, and gas temperatures can be easily measured

Table 7.14 Materials commonly used in resistance thermometers

Material	$\propto$ (1/°C)
Nickel	0.0067
Copper	0.0043
Platinum	0.00392
Gold	0.004
Tungsten	0.0048

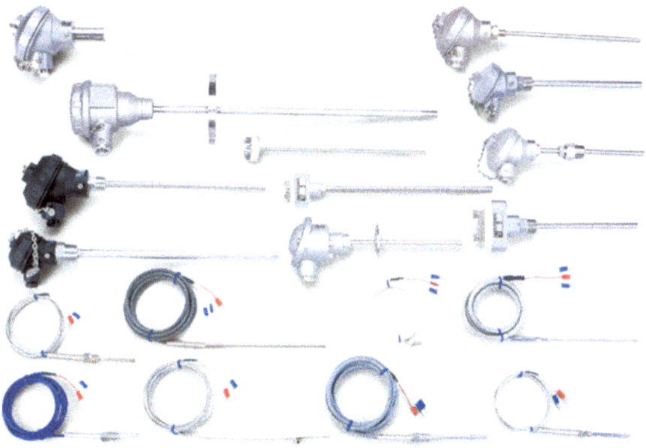

Fig. 7.101 Different types of thermocouples

between −185 °C and +1820 °C by using thermocouples. In order to form a thermocouple circuit, there must be at least two end points in this circuit. If the temperature is certain at this end, which is called the reference point, the temperature at the other end is determined by using the thermoelectric properties of the material used. Usually, a thermocouple circuit can be made with any two conductive wires. In the selection of thermocouple materials used in the applications, price, temperature range, corrosion resistance are all considered carefully.

The most commonly used method for temperature measurements are thermoelectric sensors known as thermal pairs. The structure of thermal pairs is quite simple. Two wires of different metals are twisted together (Fig. 7.102). When different temperatures are applied to the ends, the thermoelectric electromotive force (emf) occurs in the device, the magnitude of the emf is dependent on the temperature difference. This allows the use of a thermocouple as a thermometer in a limited temperature zone. One of the two joints, which is called hot or the measuring joint is exposed to the temperature to be measured and the other end, called a cold or reference, is left at the reference temperature of known value. The generated emf is measured with a suitable millivoltmeter connected to the circuit. It can be connected to electronic indicators with a capacity to measure temperature.

Thermocouples are usually made of copper-nickel chrome-plated wire. Chrome-plated wire is an alloy containing 50–60% of copper and 40–50% of nickel. The following factors are taken into account when selecting materials for thermocouples:

- Ability to resist temperatures when used,
- Immunity to degradation and oxidation, ensuring the sensitivity of thermo-electric properties in continuous use,
- Linear features.

Thermocouple Design Materials: The American National Standards Institute (ANSI) has issued standards for standard thermocouples and design methods. Since the voltage (emf) values of the wires formed according to these standards are known, cheap temperature measurement systems could be developed. The main standard

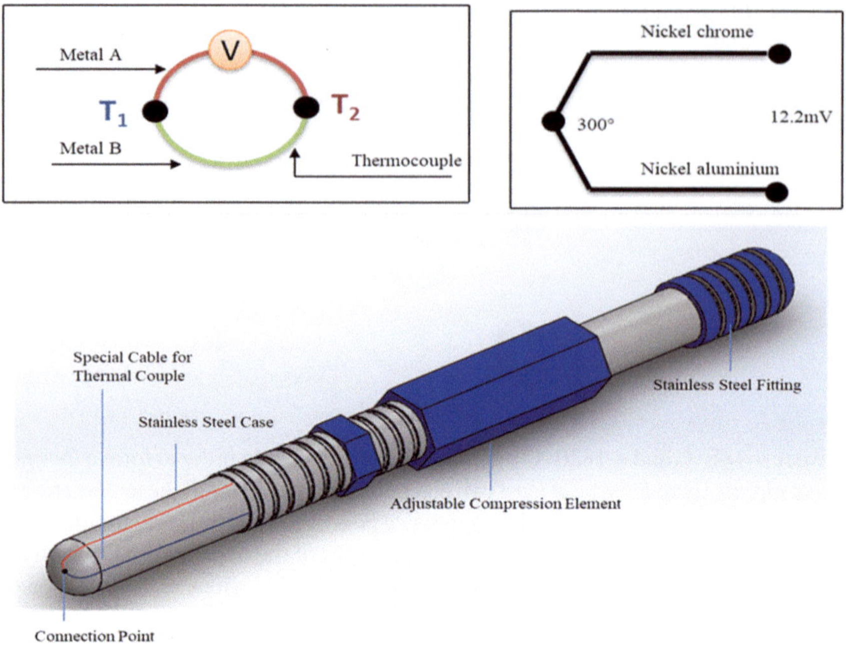

Fig. 7.102 Structural components of the thermocouple

thermocouples are given in Table 7.15. These have the corresponding standard voltage values (reference value 0 °C). For measurement of temperature with thermocouple, a variety of hand-held or laboratory-type measurement and control devices have been developed. Most of these devices have fixed temperature reference points or use a thermometer type thermometer to measure the ambient temperature and reference temperature. The fact that such devices are cheap causes them to be used without being calibrated. As with any measuring instrument used in engineering applications, the handheld thermocouple must be calibrated.

Table 7.15 Standard thermocouple components

Type	Type of wire material	
	Positive (+)	Negative (−)
S	Platinum	Platinum and 10% Rhodium
B	Platinum and 13% Rhodium	Platinum and 6% Rhodium
R	Platinum	Platinum and 13% Rhodium
J	Iron	Constantan (55% Copper and 45% Nickel)
T	Copper	Constantan (55% Copper and 45% Nickel)
K	Chromel (90% Ni and 10% Cr)	Alumel (94% Ni, 3% Ma, 2% Al and 1% Si)
E	Chromel	Constantan

Table 7.16 Commonly used thermocouple materials and their temperature ranges

Type	Symbol code	Temperature range (°C)
Copper—Constantan	(T) Blue/Red	– 185 and +370
Chromel—Constantan	(E) Purple/Red	0 and 550
Iron—Constantan	(J) White/Red	0 and 800
Platinum—Platinum/Rhodium	(R), (S) Black/Red	700 and 1480
Chromel—Alumel	(K) Yellow/Red	300 and 1100

Another important factor is that the voltage-temperature curves entered into such hand-held devices are usually for a specific temperature zone. Otherwise, these devices may give very incorrect results. The most commonly used alloys in thermocouples are constantan (copper-nickel), chromel (nickel-chrome), alumel (nickel-aluminum), nicrosil (nickel-chromium silicon), and nisil (nickel-silicon).

Commonly used thermocouple materials and their temperature ranges can be seen in Table 7.16.

Radiation Thermometers: The measurement of the temperature with the help of radiation is generally based on the principle of the determination of the energy output from the radiation source. In many manufacturing processes, measuring the surface temperature without touching the product is an important process in terms of quality and efficiency. Depending on the product produced and the technology applied, the need for non-contact temperature measurements varies. The selection of equipment for temperature measurement without touching the target is mainly depending on the features of the application, the accuracy, reliability, and price of the device. There are mainly two groups of measuring devices for contactless measurement, these are

(1) Optical pyrometers,
(2) Infrared radiation thermometers.

Optical pyrometers measure the temperature in the visible radiation spectrum between 0.4–0.7 μm wavelengths (especially 0.655 μm). As for infrared radiation thermometers, they are used to measure temperatures in the visible spectrum between 0.7–80 μm wavelengths (in particular 0.655 μm).

Apart from the traditionally used optical thermometers, there are two different technologies available; these are light separator sliding type and disappearing filament type pyrometers. As for infrared thermometers, they have two types, single wavelength (single color) and multiple wavelength (multi-color). Some basic properties of the temperature measurement need to be known in order to determine the technology to be used for a specific purpose.

Advantages and disadvantages of optical pyrometers can be seen in Table 7.17.

The Operating Principle of Optical Pyrometer: The principle of operation is based on measuring the change of the visible radiation wavelength emitted from the bodies by measuring the temperature (Fig. 7.103). To measure the temperature, the radiation emitted by the objects is compared to the electrically heated lamp filament. By

Table 7.17 Advantages and disadvantages of optical pyrometers

Usage Areas	Advantage	Disadvantage
• Melted metals • Hot substances • Flue (stack) temperature • Heated materials	• The physical contact of the device is not necessary to measure the temperature • The accuracy is high ±5 °C • An image proportional to the size of the heat source is obtained • The distance between the heat source and the device is not important • Using the device is very simple	– Temperatures higher than 700 °C can be measured if the temperature source is illuminated for measuring purposes – Not used for continuous monitoring and control since it is used manually

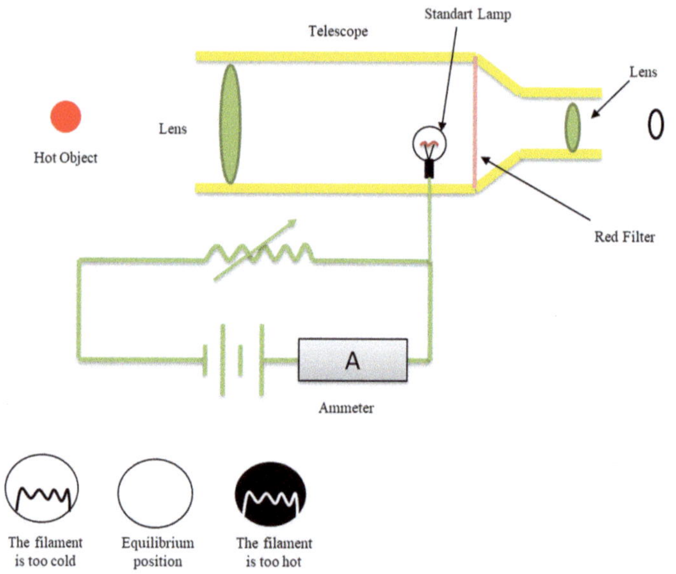

Fig. 7.103 The operating principle of the optical pyrometer

changing the current intensity given to the filament, it is ensured that the temperature of the material to be measured is brought to the same color. The flow through the filament was calibrated according to the temperature and the scale of the milliampere meter on the device was scaled as temperature.

Traditionally used optical pyrometers measure the brightness of the visible spectrum. By utilizing the natural characteristic of the human eye, the intensity of unknown radiation in the wavelength of 0.655 μm from a hot source is compared with the known brightness of the calibrated lamp in the interior of the device. Because the

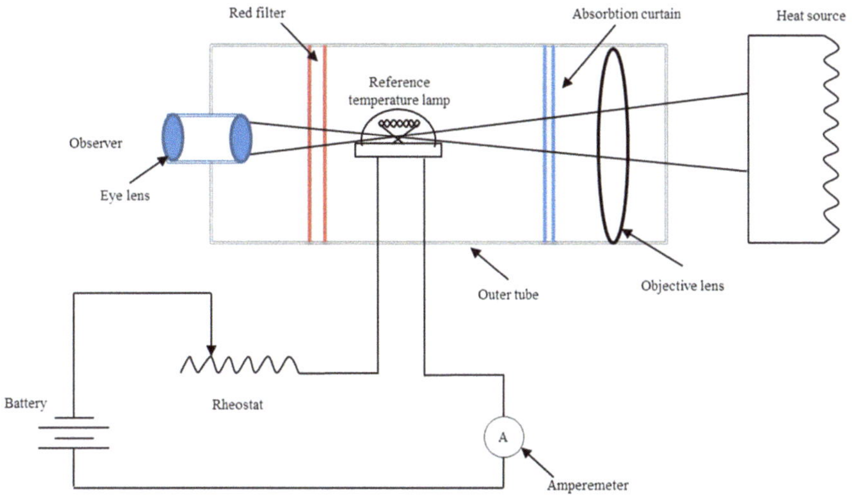

Fig. 7.104 Main parts of optical pyrometers

human eye perceives the reflected energy as glow. This level of luminosity is somewhat more effective than those detected by infrared thermometers, with bright or non-widespread targets. Lightweight and portable, this type of temperature measuring devices are more effective for certain applications. The temperature values measured by optical pyrometers are more accurate when compared with the black body temperature source. However, as with infrared thermometers, optical pyrometers are also affected by the spreading characteristic of the target. Therefore, for non-black target temperature measurements, the spreading temperature needs to be compensated.

The main parts of optical pyrometers are given in Fig. 7.104. The optical pyrometer has an eye lens at one end and an objective lens at the other end. The power supply (battery), the set resistance (rheostat), and the multimeter to measure current are connected to the reference temperature bulb. Between the objective lenses and the reference temperature bulb, the absorption screen is placed. The absorber is used to extend the temperature range which can be measured by the device. The red filter between the eyepiece and the lamp allows only a narrow wavelength range of about 0.65 μm.

Infrared Radiation Thermometers: In places where it is not possible to reach, in particular, measuring the temperature of moving objects is only possible without contact. Main features that are effective in non-contact temperature measurements are object and environmental conditions, lens and optical system, IR detector, indicator, and output characteristics.

Infrared radiation thermometers are widely used in temperature measurements. With this type of thermometer, the temperature values of approximately −20 and 4000 °C can be measured. Their designs look like a simple hand gun with a laser plug (Fig. 7.105). In industrial applications, process control can be carried out effectively by conducting temperature measurement on time.

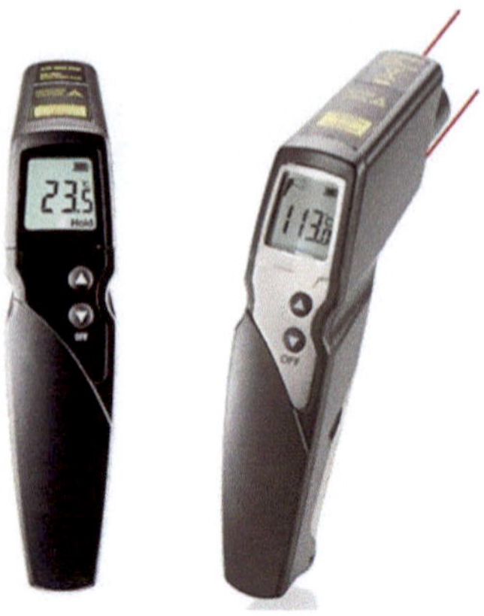

Fig. 7.105 Infrared radiation thermometers [2]

Advantages of Infrared Radiation Thermometers: Infrared radiation thermometers have many advantages compared to other touch type thermometers since their temperature does not come into contact with the measured object. These advantages are

Non-contact and clean measurement: They provide easy and clean measurement where materials are soft, wet, and inaccessible.

Small, moving, or very hot objects: Infrared thermometers are much more useful than contact thermometers in measuring the temperatures of small and moving objects, because they only perceive the energy emitted from the objects. Objects with temperatures up to 3000 °C can be easily measured over long distances.

Hard-to-reach objects: The temperature of objects that are in view and difficult to access can be measured very remotely.

Safety: Infrared thermometers can work safely in unsafe and difficult places where people cannot enter.

Measurement rate: Infrared measurements are much faster than other contact measurements. They do many readings per second and give the results precisely. It takes a long time to make the same measurements with contact thermometers.

Repeatability and accuracy: Infrared thermometers do not lose their accuracy and sensitivity as the temperature does not come into contact with the measured objects. Their repeatability is very high. They serve for many years without any problem or damage. Continuous control of the production process by means of infrared measurements ensures that production errors are reduced and as a result, the product quality increases. In addition, the identification of possible problems at the initial stage with infrared thermometers prevents unexpected abrupt developments and ensures better maintenance times and better programming of the required materials. As a

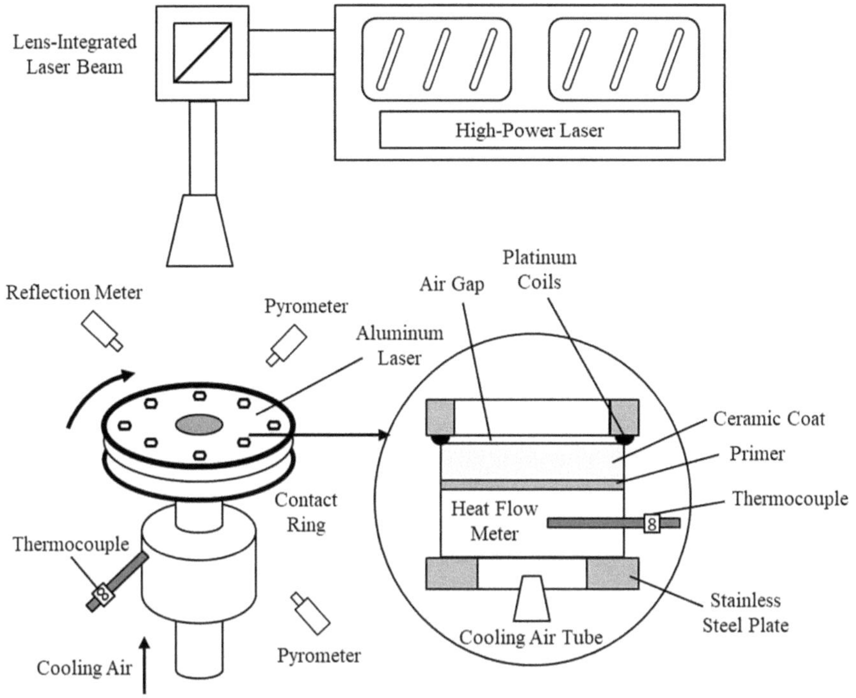

Fig. 7.106 Major components of the infrared radiation thermometer

result, maintenance costs and production losses are significantly reduced. In addition, infrared thermometers used in insulation lines play an important role in reducing energy losses.

Working Principle of Infrared Radiation Thermometers: The basic design principles of all infrared radiation thermometers are the same (Fig. 7.106). Using infrared radiation filters of different properties, the optical lens system focuses on an infrared sensor that converts energy into an electrical signal. This electrical signal is mainly manually compensated from the emissivity feature. There is an analog signal output via the linearizer and extender in the device. The characteristics of this signal are mainly: 1–5 VDC or 4–20 mA. There are also electronic units that convert the analog output of the device into digital signals that can be transmitted at high speed in order to increase the data acquisition rate. The temperature compensating electronic units used to ensure the temperature change inside the infrared radiation thermometer do not affect the output of the device. Some applications in very hot environments may require water jackets to cool the infrared sensor.

Infrared thermometers generally operate between very wide or very narrow wavelengths. Essentially, infrared thermometers operating over a wide wavelength range have a wider temperature range than infrared thermometers running between narrow wavelengths. The most important drawback of infrared thermometers operating in the wavelength range is that they are significantly affected by the spreading value. The spreading value is a variable property and takes different values at different wavelengths. Infrared thermometers operating in the wavelength range of 3–5 μm are as follows: It is 0.6 at a wavelength of 3 μm and 0.9 at a wavelength of 5 μm. The target temperature may not be determined correctly without knowing the spread value correctly. This limitation does not apply to infrared thermometers operating in the narrow wavelength range. Because, in the narrow wavelength range used, the spread value will not change substantially. Therefore, the spreading value of the target must be determined as it affects the temperature measurement significantly. Infrared radiation thermometers operating in the narrow wavelength range are slightly more expensive than infrared thermometers operating in the wide wavelength range, since sensitive electronic units are required to adapt low energy levels determined in these wavelength ranges. In addition, because of the reduced energy measured by infrared thermometers operating in the narrow wavelength range, temperatures less than 250 °C cannot be measured.

Single-color infrared thermometers are the most commonly used temperature measuring devices for non-contact temperature measurement. These measuring devices measure the temperature in just one piece of the infrared spectrum. When this approach is applied to narrow or wide wavelength ranges, monochromatic infrared thermometers are general purpose measuring devices, including wide wavelength ranges. They can be used for certain applications where a very narrow part of the infrared spectrum is important.

Multi-color infrared thermometers can measure temperature in two or more sections of the infrared spectrum or wavelengths. Theoretically, if the measurement is made at different wavelengths very close to each other, if the spreading values are at the same values at each wavelength, since the spreading value is removed from the equation, the spreading value does not affect the temperature measurement very much. This theory is based on the assumption that the emission values are the same at each wavelength. If the spreading values are the same at each wavelength, the multi-color infrared thermometers measure the correct temperature. In practice, this will not always be true with dynamically changing goals. Therefore, significant errors may occur in temperature measurement. Despite this drawback, multi-color infrared thermometers can be useful in many processing applications. The multi-color infrared thermometer can be used to measure the temperature of a target that does not fill the entire visual field. This feature provides important practical benefits for measuring the temperature in processes where only a certain part of the target can be seen because of the obstruction of the visual field.

Infrared thermometers can also be combined with other technologies. For example, fiber optic sensors allow the control of devices placed at a location away from the temperature of the target to be measured. In addition, special probes made from materials such as sapphire can be used to measure the temperature of unreachable targets with conventional sensor heads to measure. This feature is useful for measuring the temperature of a target in a reactor or vacuum chamber.

When measuring with many of the infrared thermometers, it is necessary to estimate the emissivity of the target by the user in order to accurately determine the temperature of the target. Recent developments in infrared thermometers are the incorporation of pulse laser technology with monochrome infrared thermometers operating in the narrow wavelength range to automatically determine the emissivity values for accurate determination of target temperature. In this technology, the pulse laser emits energy in the same narrow wavelength range as measured by the device. The infrared sensor in the device measures the amount of laser energy projected by the target. The microprocessor converts this additional energy to compensate for the emissivity. Because of this feature, this measuring device accurately measures the target temperature with an accuracy of ± 3 °C with the corrected spread value. Pulse laser technology is very efficient with common targets.

Process control systems using multiple sensor heads are another option. This arrangement allows for simultaneous data collection in many regions. Infrared temperature measurement can be made due to radiation or reflected energy emitted from other sources. This measuring device can be used to measure the radiated back radiation and to detect its total effect on temperature measurement. In the furnace application, the back radiation emitted from the hot walls can be compensated in order to more accurately determine the target temperature. This technology ensures that the target temperature is accurately controlled, effectively controlling the process.

Each optical device has a visual angle (field of view) (Fig. 7.107). Before the temperature is measured, the user should check that it is at a suitable distance from the object. The field of view can be adjusted from the V-shaped projection on the top of the device or when showing the laser circle shape. If there is another obstacle

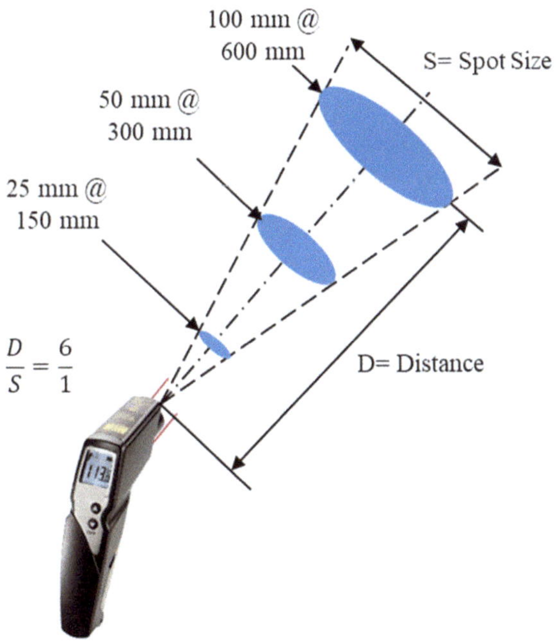

Fig. 7.107 The field of view of the infrared radiation thermometer

in front of the object to be measured, this must be removed. The Y object should be removed to measure the temperature of the X object, and the thermometer should be brought closer to measure the temperature of the Y object.

Infrared radiation energy is emitted as a result of atomic and molecular vibration. When the object is cold, these vibrations are relatively slow, and the sum of the resulting energy is relatively small. When the subject is warmer, the frequency of the vibration increases, and the sum of the infrared energy increases significantly.

The frequency of the vibration is related to the wavelength of the energy produced. High frequency provides short wavelength energy. In general, for many materials, the infrared radiation energy is not emitted at a single frequency or wavelength. It has a distribution pattern similar to a trapezoid bell curve above a certain wavelength level.

The reason for the measured temperature is not simply the sum of the infrared radiation emitted by an object. The emissivity value that should be considered is an important variable. The emissivity value of an object is affected by the material, surface conditions, reflectivity, and opacity properties. This can make the temperature measurement process very complicated at times. In order for an infrared radiation thermometer to be applied properly, it is necessary to know the emissivity value and its properties. The emissivity value is between 0.000–1.000.

Infrared Radiation Camera (Infrared Thermography): They work according to the principle of thermal radiation. The temperature distribution on the surface of the object can be measured in a non-contact manner. Radiation emissivity varies with temperature. The thermal radiation emissivity coefficient at different temperatures takes place on the surfaces in different colors. The temperature is determined according to the color change.

Any object in nature emits an energy. Visible radiation is the best-known form of electromagnetic energy. When looking at objects, the color spectrum ranging from red to purple is seen. The main difference between these colors is the wavelength. All beings present in nature emit electromagnetic energy, also called thermal radiation, which varies in intensity at different wavelengths, depending on the temperature they have. There are two wavelength ranges in the infrared range that are just above the red color. These are "Medium Infrared" and "Far Infrared" ranges, respectively.

It is known that all objects have the ability to absorb energy from other sources, except for the ability to emit thermal radiation. Examples of this are that objects heated by the sun in the day emitting their thermal energy into their environment during the night. The wavelength of the emitted energy and the total amount of emitted energy depends on the temperature of the object. The human body and many objects we encounter in our environment have a temperature of around 30 °C. An important part of the energy emitted from these objects is located in the far infrared range. As the imaging method, it is the imaging system based on the invisible infrared energy (heat) and the colors and shapes which are formed according to the infrared radiation energy of the general structure of the image.

Usually they are used for security purposes. However, it is possible to use a wide variety of sectors, particularly, their importance has increased with the development of military techniques like temperature-guided missile, night vision systems and so on. In the electricity sector, they are used to determine electrical problems. Also, they are used for temperature analysis in installations and buildings in the energy sector. In the architectural field, they are also used to determine the mold moisture or cracks formed under the plaster for the determination of metal fatigue in steel structures.

They detect the heat of objects with infrared radiation sensors. Gives an image that is formed between black/white or colorful (red hot, black cold), black/red. Infrared temperature measuring devices consist of a camera that scans for thermal radiation, and a screen portion showing the thermal image (Fig. 7.108a).

Infrared radiation thermography cameras produce invisible infrared radiation images and offer precise and non-contact temperature measurement. Almost all objects have increased temperature before they become corrupted, which makes infrared cameras a valuable diagnostic tool in a wide range of applications. While industry is trying to increase production efficiency, manage energy, improve production quality, and improve work reliability, new application areas for infrared cameras continue to emerge.

Areas of Use: Thermal imaging has many usage areas. For example, it can be used in military field, identification of the area of fire, person identification, vehicle, and aircraft recognition. In addition, thermal imaging can be used in many areas such as

(a)

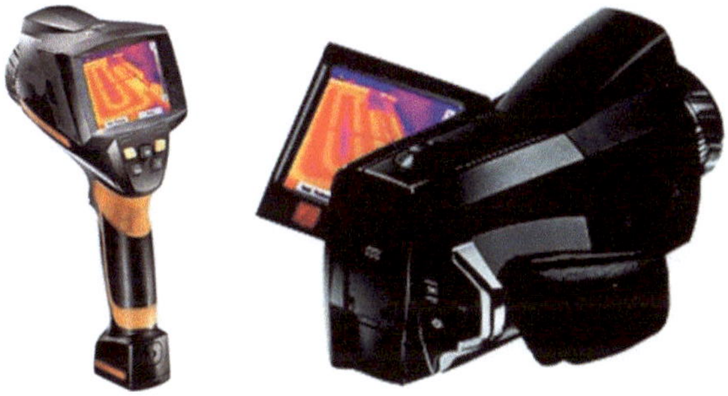

(b)

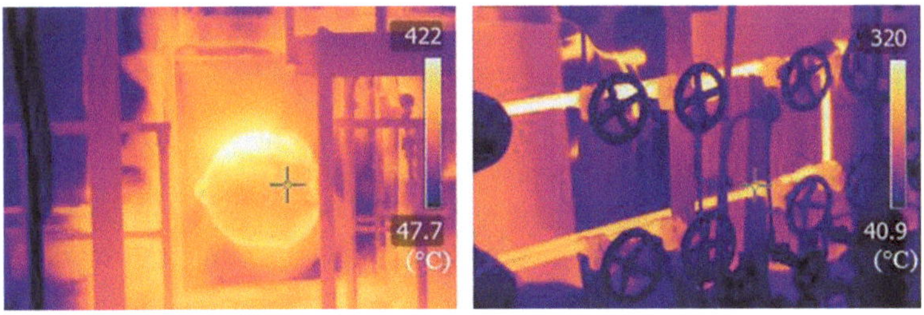

Fig. 7.108 **a** Infrared radiation cameras [2]. **b** Temperature measurements by using thermal cameras

analysis of energy transmission lines, ventilation systems analysis, building construction, heat sealing tests. Other areas of use can be listed as follows: Medical imaging; veterinary practices; night vision systems; research projects; process control; military defense and security purposes; chemical imaging; volcano examination systems; state monitoring systems.

It is very difficult to measure the temperatures occurring in the tool-chip interface and the sliding zone. Different methods have been developed by the researchers to measure the temperatures that occur during the cutting process. Temperatures at tool-chip interface are generally measured by conduction and radiation. However, with these methods, tool-chip interface temperatures cannot be measured with the desired completeness. Factors such as accurate determination of temperature, small area affected by temperature and very high temperatures in the insert area make temperature measurements difficult.

The principle of taking photos of heat with the thermal camera is based on the use of the thermal imager photography technique. The workpiece, shavings, and side surfaces of the cutting tool are photographed throughout the vertical cutting process. The photo is subsequently calibrated for temperature distribution. This method is very useful for measuring the temperatures on the sawdust, tool, and workpiece. Thus, it helps to find the source of the problem easily by removing the hotspots with a light color and cold spots with a dark color. Heat and digital photos of the problem point are placed side by side, and therefore, reporting process can be executed. It is possible with thermal cameras to prevent unexpected situations, reduce production losses, eliminate problems without further damage, control larger areas in a shorter period, analyze the integrity of the heat treatment and analyze for emerging problems.

This field, which is called infrared thermography, thermal imaging, thermographic imaging, thermal video, is a part of the science of infrared imaging. Thermographic cameras detect irradiations in the infrared range of the electromagnetic spectrum (about 900 nm–14000 nm), making these invisible radiations visible. All objects emit more or less infrared radiation because of their body temperature. Through thermography, these invisible rays can be made visible. The amount of radiation emitted by the object will increase in temperature as it increases. Therefore, thermography makes it possible to detect different temperatures. High temperature objects can easily be detected in front of cold objects. Therefore, warm-blooded creatures, such as humans, can easily be identified by the thermal imager outside. Thanks to these features, thermal imaging cameras have found a great use especially in military fields. The temperature measurement with thermal cameras can be seen in Fig. 7.108b.

Liquid Crystal Thermometers: In addition to solid, liquid, and gas phases, some organic substances in nature also have a phase state where they have solid and liquid properties simultaneously. This phase state is the liquid crystal phase state. They are solid in structure but are liquid in appearance. These structures are called liquid crystals. They act according to the arrangement of their molecules. These molecules are arranged in three different ways:

(1) Smectic: Molecules are arranged on a horizontal and vertical line.
(2) Nematic: Molecules are listed in a vertical line only.

(3) Cholesteric: The Nematic type is arranged horizontally for each layer, but spirally in order to jump in the vertical direction.

Many commercial fats and proteins and fats used in the industry are liquid crystals. The color of liquid crystals varies from red to purple. Color change with temperature is reversible. It is used for temperature measurement and image acquisition using liquid crystals. Liquid crystal temperature can be observed and photographed by applying it on the object to be measured. In the art, the liquid is applied to the crystal body and the temperature distribution on the body becomes visible. To protect against external influences, the liquid crystal is coated with polyvinyl alcohol. In liquid crystal second type temperature meters, the infrared rays emitted from the objects are dropped onto a plate which is in contact with the liquid crystal and absorbs the infrared rays. By observing the liquid crystal portion, the temperature zones can be conveniently monitored.

Temperature Measurement based on Shape Change: They are also called as Seger temperature gauges (Seger cones). This method of the measurement is based on the deterioration of the shapes of certain geometric bodies that are prepared from various material mixtures at certain temperatures. Especially in the soil (ceramic) industry, in the measurement of the temperatures inside the firing kilns, thermo-cones or pyramids are commonly used instead of thermo-elements, pyrometers. Measuring sensitivities are not high.

Temperature Measurement with Color Change: This method utilizes color changes of various mineral salts at certain temperatures, which means it is based on measuring the temperature at the values corresponding to these color change points. These temperature meters are available as adhesive strips. Such temperature gauges are disposable. They are economical, they are applied to electronic circuits, electric motors, aircraft engines, sensitive drugs. Overheating in the elements can be determined during handling, transportation and storage, or at which point it is damaged due to temperature.

7.5.7 Radiation Measurement

In our physical world, we encounter a wide range of frequencies ranging from the very low frequency of the electric waves produced by the power transmission lines to the very high frequency of the gamma rays emitted from the atomic nucleus. The frequency and wavelength of an electromagnetic wave depends on the source of that wave. This wide frequency range of electromagnetic waves forms the electromagnetic spectrum (Fig. 7.109). Most of the solar energy measurements on Earth are limited to visible and near infrared wavelengths that contain approximately equal amounts of energy.

Three different measurement systems are used for radiation measurements:

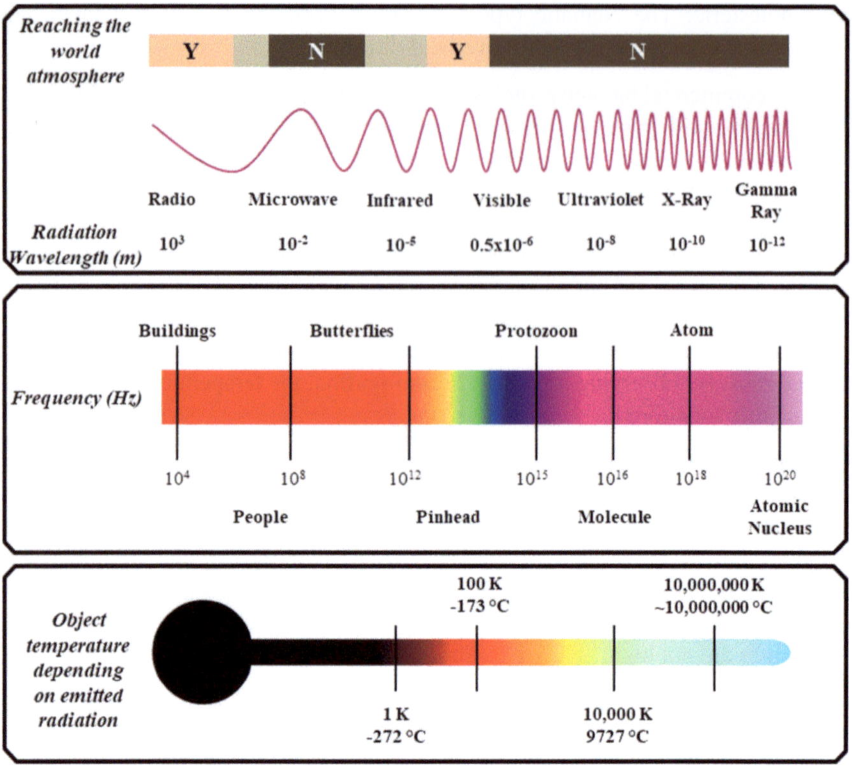

Fig. 7.109 Electromagnetic spectrum

(1) *Radiometric measurements*: In the specific wavelength range, W/m^2
(2) *Photometric measurements*: For illumination, lm/m^2 or lx,
(3) *Quantum measurements*: In determining photon flow density (usually 400–700 nm), μmol/sm^2. 1 molar photon flow in 1 s in 1 square meter is specified as 1 *Einstein.*

In the production of light sources and engineering, radiometric and photometric units are used. Radiometric units are used to determine the radiation energy (W/m^2). From any source of radiation, the energy coming from the unit area on a flat surface is defined as the energy of the radiation.

In the measurement of light and radiation there are two parallel systems known as radiometric and photometric quantities (Fig. 7.110). Each size in a system corresponds to a size in the other system.

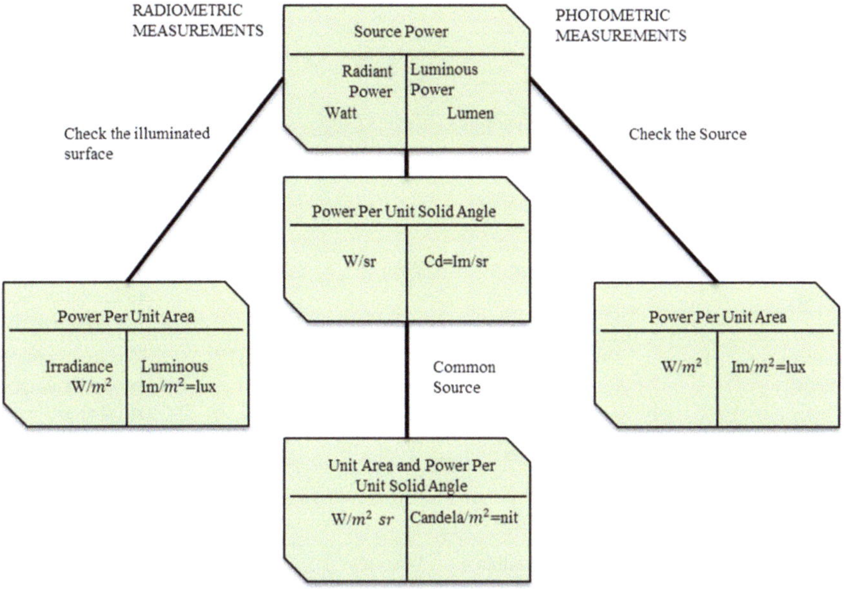

Fig. 7.110 Radiometric and photometric quantities

7.5.7.1 Radiometric Measurements

Radiometry is the science of measuring electromagnetic radiation. Radiometric measurements include infrared, visible and ultraviolet radiation measurements of the electromagnetic radiation spectrum, between x-ray radiation and radio waves, within a certain wavelength range. Although, in practice, it appears to be generally limited to IR (infrared), VIS (visible) and UV (ultraviolet) regions, theoretically all regions of the electromagnetic spectrum are within the subject of radiometry.

Radiometric SI Units can be seen in Table 7.18.

Net Radiometer: It is used to measure short and long wave solar radiation separately (Fig. 7.111) [14]. The net radiometer is also used to measure the net radiation with shortwave radiation and surface temperature.

7.5.7.2 Photometric Measurements

Photometry is a branch of science related to the light response of the human eye. Photometry is the science of measuring visible light with different units, depending on the sensitivity of the human eye. In other words, photometry is the name of the field for measuring radiant energy only for the visible region of the electromagnetic spectrum. In doing so, it refers to the eye, the perception in the brain, but not to measuring instruments as in radiometry. It is a quantitative branch of science based

Table 7.18 Radiometric SI units

Quantity	Symbol	Unit	Symbol	Dimension	Note
Radiant energy	Q_e	Joule	J	ML^2T^{-2}	It is a kind of energy
Radiant flux	Φ_e	Watt	W	ML^2T^{-3}	The radiation energy per unit time is the radiation power
Spectral power	$\Phi_{e\lambda}$	Watt/meter	W/m	MLT^{-3}	It is the radiation power per wavelength
Radiant intensity	I_e	Watt/steradian	W/sr	ML^2T^{-3}	The unit is the power value per solid angle
Spectral intensity	$I_{e\lambda}$	Watt/(steradian × meter)	W/(sr × m)	MLT^{-3}	It is the radiation intensity per wavelength
Radiance	L_e	Watt/(steradian × square meters)	W/(sr × m^2)	MT^{-3}	The source projection area and power values per solid angle
Spectral radiance	$L_{e\lambda}$ or $L_{e\nu}$	Watt/(steradian × cubic meters) or Watt/(steradian × square meters × Hertz)	W/(sr × m^3) or W/(sr × m^2 × Hz)	$ML^{-1}T^{-3}$ or MT^{-2}	It is usually measured in conjunction with the surface area, with wavelength or frequency W/(sr × m^2 × nm)
Radiant flux intensity	E_e	Watt/square meters	W/m^2	MT^{-3}	This is the amount of power per the unit surface area
Spectral irradiance	$E_{e\lambda}$ or $E_{e\nu}$	Watt/cubic meters or Watt/(square meters × Hertz)	W/m^3 or W/(m^2 × Hz)	$ML^{-1}T^{-3}$ or MT^{-2}	It is per wavelength or per frequency
Radiant emittance	M_e	Watt/square meters	W/m^2	MT^{-3}	It is the amount of power emitted from a surface

(continued)

Table 7.18 (continued)

Quantity	Symbol	Unit	Symbol	Dimension	Note
Radiosity	J_e or $J_{e\lambda}$	Watt/square meters	W/m^2	MT^{-3}	It is the amount of power emitted and reflected from a surface
Spectral radiant emittance	$M_{e\lambda}$ or $M_{e\upsilon}$	Watt/cubic meters or Watt/(square meters × Hertz)	W/m^3 or W/(m^2 × Hz)	ML^{-1}T^{-3} or MT^{-2}	It is the amount of power emitted from a surface per wavelength or frequency
Radiant exposure	H_e	Joule/square meters	J/m^2	MT^{-2}	Radiant energy delivered to a given area
Radiant energy density	ω_e	Joule/cubic meters	J/m^3	ML^{-1}T^{-2}	
Wavelength	λ	Nanometer	nm	L	Nanometer = 10^{-9} m or = mμ

Fig. 7.111 The net radiometer [14]. *Source:* https://www.hukseflux.com/

on the statistical model of light's human response. It is concerned with careful detection of light in controlled conditions. Photometry includes any light intensity and illumination energy measurements in the wavelength range of 380–760 nm that can be detected by the human eye. Photometry is the measurement of light quantities. The actual quantities taken into account in photometry are the light intensity of the source, the intensity of illumination and the luminous flux. *Photometry* is a light measurement science that is measured by considering the light created by light in the human eye. *Radiometry* is the science of measuring the absolute power of radiation energy, including light.

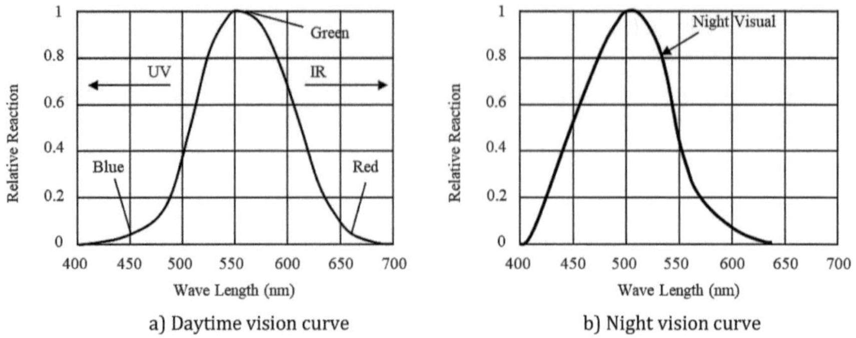

Fig. 7.112 Daytime and night vision curves

The International Commission on Illumination (CIE) determined the daytime vision curve by measuring the responses of the light-regulated eyes of selected sample groups (Fig. 7.112a). This curve indicates that the human eye reacts too much against the green color and is less sensitive to red and purple color. The eye reacts differently in dark conditions where it is difficult to determine color characteristics. The night vision curve (Fig. 7.112b) was determined based on the response of the eye in dark conditions.

Photometric Quantities: Although the human eye response is considered in photometric quantities, the radiometric quantities are not completely defined depending on the absolute power value. The radiation energy outside the visible spectrum does not fully contribute to photometric quantities. Some photometric quantities, radiometric responses and units are given in Table 7.19.

Photometric Units: Measurement of the effects of electromagnetic radiation has gained importance in the late eighteenth century. Measurement methods have changed, and different terms have emerged depending on the purpose of study. In order to develop radiometric units related to total energy and power, the total heating

Table 7.19 Photometric quantities

Photometric	Radiometric equivalent	Symbol	Unit
Luminous flux	Radiant flux (Power)	ϕ_V	Lumens (lm)
Luminous intensity	Radiant intensity	I_V	Candela (lm/sr)
Luminous exitance or emittance	Radiant exitance or Emittance	M_V	Lux (lm/m^2)
Illuminance	Irradiance	E_V	Lux (lm/m^2)
Luminance	Radiance	L_V	Nit = [lm/(m^2 sr)] = [cd/m^2]

effect of infrared radiation was measured by thermometers. In order to identify the photometric units, the human eye was used as a sensor and the reactive properties of the eye were taken into consideration. Many different units have been developed for photometric measurements (Table 7.20).

Photometric Measurement Methods: Photometric measurements are performed by light sensing devices. Some types of such devices produce an electrical signal when exposed to light. Simple applications of this technology are controlled by the on-off switches, depending on the ambient light conditions. Light meters are used to measure the total amount of light coming to a point.

More complex methods of photometric measurements are used in the illumination industry. For the measurement of directional light flux produced by the lamps, spherical photometers are used. This type of photometer consists of a large diameter sphere placed in the center of the lamp. By rotating the photocell around the lamp inside the inner axis, the output which can be composed of all sides is measured.

Table 7.20 Derived photometric units (SI)

Quantity	Symbol	Unit	Symbol	Note
Luminous energy	Q_V	Lumen × second	lm × s	It is also called the as Talbot
Luminous flux	Φ_V	Lumen	Lm	Also known as luminous power
Luminous intensity	I_V	Candela (lm/sr)	cd	The amount of luminous flux per solid angle. Candela is a light intensity emitted by a monochromatic radiation source with a frequency of 5.4×10^{14} Hz with a radiant intensity of (1/683) W/sr
Luminance	L_V	The amount of candela per square meter	cd/m^2	It is also called as nit
Luminous exitance or emittance	M_V	lux (lm/m^2)	lx	It is luminous flux emitted from a surface
Illuminance	E_V	lux (lm/m^2)	lx	It is used for light coming to the surface (It is light density falling upon a surface)
Luminous exposure	H_V	lux × second	lx × s	
Luminous energy density	ω_V	(lumen × second)/m^3	(lm × s)/m^3	
Luminous efficiency	η	Lumen/Watt	lm/W	The ratio of the luminous flux to the radiation flux

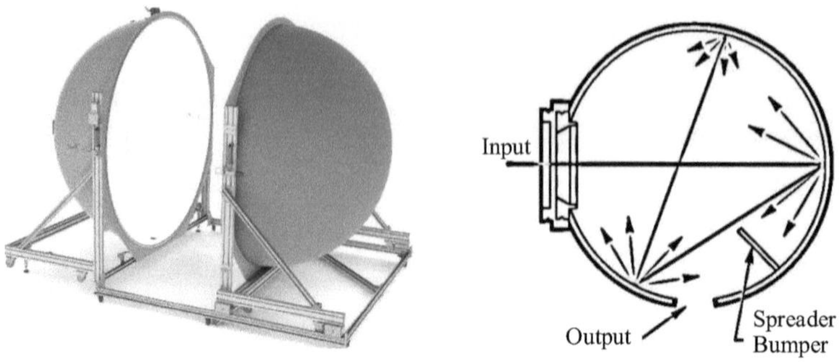

Fig. 7.113 Ulbricht sphere

Luminous flux measurement: The luminous flux is measured by the Ulbricht sphere. The Ulbricht sphere allows direct measurement of the light flux of the light source. The Ulbricht sphere is composed of a sphere with a measuring window that is painted with a non-selective solid paint and close to the reflecting factor 1 (Fig. 7.113). Inside the sphere, there is a screen painted with the same solid paint, which prevents the rays coming out of the light source from coming into the measuring window directly. The reason is that the inner surface of the sphere is painted with solid paint, in order to create an even light level in the measurements, to provide a proper reflection.

With a sphere measuring 1 m in diameter, a sensitive and heat compensated photocell, a specially designed precision measuring unit, special arrangements for attaching different lamps and Ulbricht sphere measuring units consisting of three different types of (standard) lamps, the following measurements are in accordance with international standards and can be done in a responsive format:

- Luminous flux of any type of light source [lm],
- Light efficiency of all types of light sources [lm/W],
- Standard rendering of all types of luminaires (luminaires).

Spectrophotometric Measurements: Spectrophotometric measurements are a method of measuring based on the principle of measuring the changes of a wave of atoms of a substance on a wavelength. Electromagnetic waves or visible light can be used for measurement. The measuring instrument is also called a spectrometer. There are varieties such as infrared spectra measurement, ultraviolet spectral measurement, electromagnetic spectrum measurement, x-ray spectral measurement, or visible light spectra measurement according to the wavelength used.

Photometry is the measurement of light, while spectrophotometry is the measurement of light intensity at the selected wavelength. Spectrophotometry is a method of analysis by measuring the light intensity absorbed by colored substances, a reagent and colored substances. In spectrophotometers, the wavelength is changed to absorb or measure the wavelength. A single beam UV-Visible Spectrometer's schematic diagram is given in Fig. 7.114.

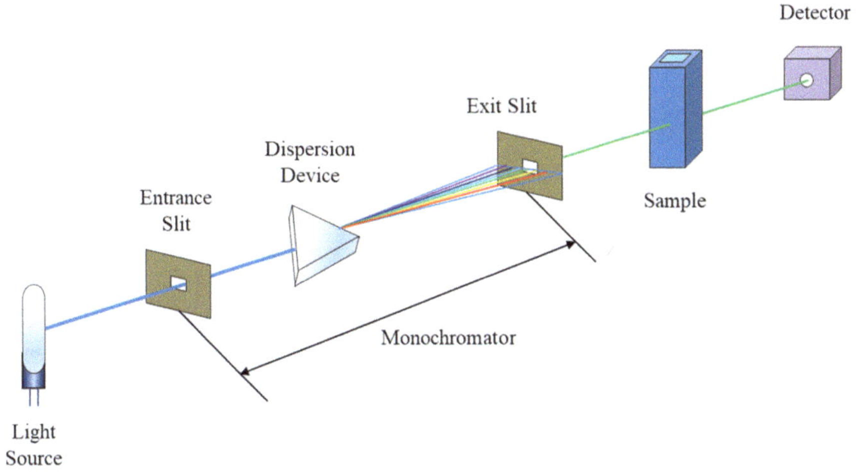

Fig. 7.114 A single beam UV-Visible Spectrometer's schematic diagram

Spectrophotometric analyzes are used to perform quantitative and qualitative analyzes of chemical and biological substances. Basic principle is the measurement of the light-absorbing properties of the substance or its substance derivatives.

Solar Radiation Measurement: Solar radiation is a short wavelength of radiation emitted from the sun in electromagnetic waves [14]. The part of this radiation that reaches the ground as a result of processes such as absorption, dispersion, and reflection in the atmosphere is divided into two:

(1) **Direct radiation**: In the sun, the radiation that reaches the earth in parallel rays is called direct solar radiation.
(2) **Diffuse radiation**: The part of the solar radiation that has entered the atmosphere from the sun is reflected in various ways by the clouds, dry air, and dust molecules, and after it is dispersed, it is called diffuse radiation.

The sum of direct and diffuse radiation reaching the earth is called total solar radiation. Solar radiation intensity is measured by three different methods:

(1) **Thermal resistance**: Solar energy is absorbed by the black painted disc whose resistance changes with temperature. The amount of sunbathing is measured by measuring the resistance change in the disc.
(2) **Thermoelectricity**: Solar radiation is collected at the connection point of two different types of metals connected together. Under the influence of heat, a voltage of mV is produced at the connection point of the metals. The voltage increases as the radiation intensity increases. Thus, the radiation is measured by measuring the voltage.

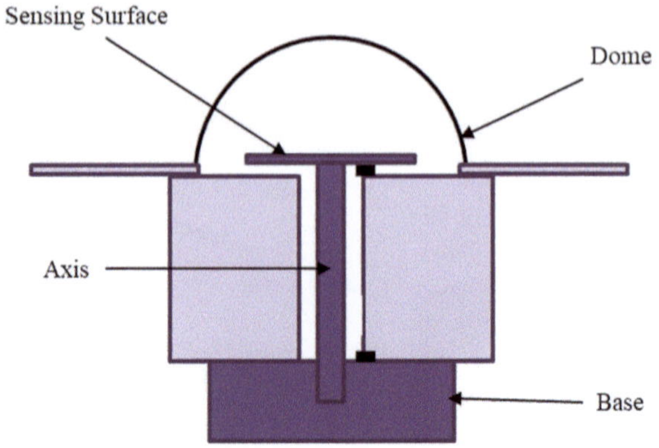

Fig. 7.115 The structural components of the Pyranometer

(3) **Photoelectric**: Light-sensitive photoelectric materials generate voltage with solar radiation. By measuring the produced voltage, the total radiation is measured.

Pyranometers: A Pyranometer is used to measure total solar radiation energy. Pyranometers (Fig. 7.115) are generally developed on the basis of radiation absorption and reflection. There is a black part and a white part at the top. The black part absorbs radiation and its temperature rises, the resulting temperature difference is measured.

Pyranometers are developed electronically and replaced actinometer and actinograph. Although pyranometers are electronic, contamination of their glass due to external factors which negatively affects the measurement results. Therefore, they should be checked regularly and their glass cleaned.

Pyranometers are used in solar energy, meteorological and agricultural applications. In recent years, pyranometers are developed with new designs, for example, a silicon photovoltaic sensor has been placed in a small head with cosine correction. Current outputs are directly related to solar radiation. From solar radiation, solar energy (J) per unit time (s), per unit area (m^2) is measured (W/m^2).

The properties of a Pyranometer can be seen in Table 7.21.

Table 7.21 The properties of a pyranometer

Features
Spectral response range: 305–2800 nm
Sensitivity: 7–15 $\mu V/W/m^2$
Temperature range: (−40)–(+80) °C
Measuring range: 0–2000 W/m^2
Temperature sensitivity: <±0.1%/°C
Response time: 1 s
Calibration traceability: WRR

Table 7.22 An example of the features of a spectroradiometer

Features	Value
Wavelength range	360–770 nm
Resolution	0.9 nm
Measuring angle	0.14–1°
Smallest measuring area diameter	1.15 mm
Measuring range	3–80.000 cd/m^2 for Illuminant A
Accuracy (for Illuminant A):	±1–2% x:±0.0015 y:±0.002
Repeatability:	±0.1–1% ±0.0002
Polarization error	<5%
Spectral response	±0.3 nm
Deposition time	40 ms–60 s
Mass	±5 kg

Spectroradiometer: *Spectrometer applications.* It is based on the measurement and interpretation of the electromagnetic radiation absorbed or emitted during the transition of atoms, molecules, or ions from one energy level to another. Spectroradiometric devices are the devices in which radiometric calibrations of spectrometric measurements are provided. Spectroradiometers such as satellite and other remote sensing sensors are based on quantitative measurements of radiation, brightness/reflection or transmission (Table 7.22). The main basis in spectro-radiometric methods is that the objects have their own reflection values in the electromagnetic regions. This reflection value is due to its chemical structure which gives the object features such as color, texture, brightness, and appearance.

The spectro-radiometric methods are based on the measurement of the reflected energy from any object in the reflection, radiation, or glare (flaming) values. Here, solar or artificial radiation can be used as the source of radiation. Spectroradiometers are systems using a photo-multiplexor or IR sensor. The radiation from the light source is divided into wavelengths in the distribution system and recorded at a high resolution of 1–10 nm by the photo-multiplexor or IR recorder. Spectroradiometers are devices capable of measuring reflections at wavelengths of 375–2500 nm. These devices have a spectral resolution of 3.5 nm at 700 nm. Each spectral step is 1.6 nm. The wavelength accuracy of the radiometer is ± 1. The spectroradiometer converts the 16-bit analog data collected in 512 channels into numerical terms. Spectralon, which is made of gypsum block, is used as a white reference in the calibration of the device. Detection of the radiation is done with 1°, 2°, 3°, 3.5°, 4°, 5°, 7.5°, 8°, 10° and 25° lenses or contact probe and plant probe. Batteries and chargers designed for use in the field, energy supply adapter for laboratory work, spot light used as light

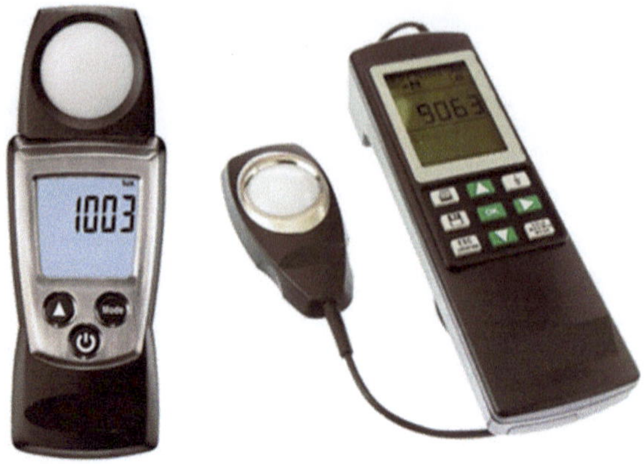

Fig. 7.116 Lux meter: Light intensity meter [2]

source, angle adjuster water assembly and tripod stand apparatus. An interface is used for data collection and original software is used for processing.

Spectroradiometers are applications with very little history. Yet these applications are not used as a standard analysis method, such as spectrometer or photometer applications. However, the studies suggest that it can be used as a standard method in the near future. The development of nanotechnology and the availability of these devices in smaller, portable and field usable applications make it necessary to conduct further research on spectro-radiometry applications.

When compared to analytical methods, spectro-radiometric applications have various advantages and disadvantages. The most important of these advantages is that it is portable in the field with all its apparatuses. In this way, sampling and analysis can be done at the same time, in a short time and in the field. With these devices, analysis can be done on live plants and without damaging the plant leaf or tissue. It is very fast compared to traditional analysis methods. It is possible to obtain hundreds of digits with a few seconds of measurement. With these values, it can be estimated for more than one variable or component. It is possible to reach appropriate modeling and statistical results by using these figures. No specific study is needed to prepare samples for analysis in these systems.

Lux Meter: The luminous intensity is measured with the lux meter (Fig. 7.116) [2]. Lux meter consists of a highly sensitive milli-amperemeter (galvanometer) calibrated with a photoelectric battery that converts light energy into electrical energy. The lux meter's filter system is made so that its spectral sensitivity is the same as that of the human eye.

Some chemical metals are very sensitive to light. Most sensitive metals to light are (such as) potassium, sodium, cesium, lithium, and rubidium. When light falls on these metals, the electrical event begins and an electromotor force (emf) is formed. This is called as photo-electricity. This event is created by the photoelectric battery or photocell. Most of the time, this photoelectric battery and galvanometer are placed in the same portable box and a fixed photocell lux meter is made. The lux meters used in the application are very diverse and generally made in portable type. With the same tool, there is a level switch on the tool to measure the luminous intensity at different levels. While reading the value, it should be read from the appropriate scale partition. In some lux meters, the dial is single-sectioned because there are no tap switches. When the measurement area of this type of luxury meter wants to be expanded, the photoelectric eye is screened. These curtains reduce the luminous intensity by 1/5 or 1/10. Thus, the measuring area of the tool is enlarged 5 or 10 times. Amplifier lux meters are used to measure the luminous intensity at very low values.

7.5.8 Analysis of Flue Gases

Emissions are defined as air pollutants emitted from the plant to the atmosphere by burning fuel, etc., and other mechanical processes of substances such as synthesis, decomposition, evaporation, and similar processes like, agglomeration, separation, and transportation.

The main reasons for the flue gas emission measurement are

- To control combustion efficiency,
- To increase the efficiency of boilers,
- Saving fuel and money,
- To reduce combustion emissions.

7.5.8.1 Preparations Before the Measurements

The preliminary preparations before the measurement are as follows:

- Before the measurement, the facility to be measured should be visited with the customer authorities, and the facility officials should be consulted on the sampling plan of the experimental studies planned to be carried out at the facility and final agreement should be reached on the measurement numbers and measurement points.
- Before starting work, information on occupational safety and health practices carried out at the facility and explosive, flammable and combustible materials, if any, should be obtained. In order to reach the name and contact information of the occupational safety officer of the facility should be obtained in case of an accident.
- Safety measures must be taken according to the measuring position.
- If the sampling points are equipped with a platform, the platform width must be sufficiently wide to allow measurements.

- During the experiments, observations of environmental factors should be made and situations that may have an impact on the measurement results should be recorded.
- All kinds of human and machine activities should be prevented in areas that may affect the measurements.

7.5.8.2 Determination of Sampling Points

The flat stack or duct length should be as much as 7 hydraulic diameter sizes, and the sampling plane should be set as 5 hydraulic diameter distances. If the flue or duct opens to the open air at the end of the straight line, the distance to the flue outlet must be 5 hydraulic diameters. The velocity profile must represent the laminar flow profile. In the laminar flow profile, the velocity value should be the highest in the middle and lower toward the edges due to the frictions.

The following factors are taken into account when determining the sampling location:

- Before starting the experiment, the sampling points should be determined, and necessary examinations are made.
- Sampling points are determined in accordance with the relevant standards (For example, in Turkey, there is one, which is TS ISO 9096 "Fixed Source Emissions—Mass Determination of Granular Material by Hand").
- The sampling point must be on a flat stack or duct.
- The sampling point should be at a sufficient distance from the effects of the waste gas that distort the characteristics of the flow (fans, valves, elbows (turning), contraction or expansion points on the flow line of the waste gas).
- There should be no factors that may block or divert the gas flow.

7.5.8.3 Flue Gas Measurement Methods

Flue gas measurement methods can be divided into two groups:

(1) Direct measurement with measuring instruments: Combustion gases, total volatile organic compounds, etc.
(2) Take a sample from a stack and analyze in a laboratory: Dust, heavy metals, fluorine, chlorine, PAH (Polycyclic Aromatic Hydrocarbons), dioxin-furan.

The devices used in the flue gas measurements generally operate according to the following measurement principles:

- Electrochemical cell method,
- UV (ultraviolet) radiation method,
- IR (infrared) radiation method.

Electrochemical Cell Method: Electrochemical cells are used to measure the electrical current generated by the electrochemical reaction of a selected gas component in a sensing electrode. The gas component to be determined reaches the battery through a semipermeable membrane. The reaction rate is proportional to the concentration of the gas component. When an oxidizing electrode is used in the process, electrons are also released to the sensing electrode by electrochemical reaction. The generation of electrons in this electrode in proportion to the concentration of the respective gas component results in a negative potential difference equal to the counter electrode. The electron current obtained by this method is measured by a suitable electronic device. The selectivity of the battery is determined by the semipermeable membrane, electrolyte, electrode materials, and retarding potential. With the devices working according to the Electrochemical Cell Method, O_2, CO, NO, SO_2, flue gas temperature, soot quantity, flue gas extraction, differential pressure, velocity measurements can be made as a result of the use of solid, liquid, and gas fuels.

Infrared Radiation Principle: For the measurement by infrared (NDIR: Non-Dispersive Infra-Red) method, the measurement gas is passed through an infrared radiation medium in the appropriate frequency band (as much as monochromatic as possible) to determine the concentration of a gas component that has the ability to absorb in a given infrared frequency band. The energy loss resulting from the absorption in proportion to the concentration of the gas component of interest in the measuring gas is evaluated by comparing it to the reference medium without any absorption energy loss.

Ultraviolet Radiation Principle: The same system is used in the ultraviolet (NDUV: Non-Dispersive Ultra Violet) method. It works only in the ultraviolet frequency region instead of infrared. Another method is the measurement of paramagnetic properties of some gases. In a strong magnetic field, the measuring gas is passed at a certain velocity over a heated resistance at a certain temperature, the value of which changes against the temperature change. The change in cooling due to turbulence proportional to the concentration of the paramagnetic gas component in the measuring gas via the resistance is measured by a Wheatstone bridge and evaluated by a suitable electronic device.

7.5.8.4 Flue Gas Measurement Units

ppm: The number of parts per million “ppm” represents a ratio, which is similar to “%”. For example, when we say 4%, we express four parts in a hundred pieces. Similarly, when it is said “250 ppm CO (carbon monoxide)”, it means there are 250 CO in one million gas parts.

ppm is independent of pressure and temperature and is used for low concentrations. If higher concentrations are present, the measurement is carried out in %. The conversion of two units can be made as follows:

10,000 ppm = 1%
1000 ppm = 0.1%
100 ppm = 0.01%
10 ppm = 0.001%
1 ppm = 0.0001%

mg/Nm³: In mg/Nm³ normal volume (normal cubic meter) is taken as reference and the harmful gas is given in milligrams (mg). In other words, the unit refers to the mass of gas in the unit of mg in air per cubic meter. Since this unit is dependent on pressure and temperature, the volume is taken as a reference in standard conditions (15 °C, 1013 mbar). However, since the volumetric ratios in the flue gas vary according to the amount of oxygen, this information is not taken as standard. For this reason, the measured values should be converted to a specific oxygen volume (O_2 reference).

7.5.8.5 Flue Gas Measuring Instruments

Devices working with electrochemical cell principle (Fig. 7.117) must have degassing, gas conditioning, and gas measuring sections [2]. In this section, the gas sample is taken with a probe and a pump on the conditioning unit via a heated line. After removing the remaining particles from the two-stage fine filter at the inlet of the conditioning unit, the water vapor in it is condensed by cooling and separated

Fig. 7.117 The flue gas measurement unit [2]

in condensed form. It automatically measures, writes and stores the gas components within the specified concentration limits.

With devices operating according to the principle of electrochemical cells, there are some measurements that can be made resulting from the use of solid, liquid, and gaseous fuels like CO_2, CO, NO, SO_2, flue gas temperature, soot, flue gas extraction, differential pressure, and velocity measurements. With such devices, CO_2, the combustion efficiency can be calculated. Since there is generally no carbon dioxide measurement cell, the percentage of carbon dioxide is determined according to the percentage of oxygen and fuel type, automatically.

The following factors should be taken into account when measuring combustion gas:

- There must be no air entry into the sampling system or gas leakage out of the system.
- Equipment should be made of suitable materials to avoid any chemical reactions between the gas and the pipes, hoses and other materials which are used in the system.
- The sampling line should be kept as short as possible.
- A refrigerant must be added to the sampling line to cool the hot gases.
- The flue gases must be submitted to the analysis system after being filtered to remove particulate matter such as dust and soot.
- The plant must operate at full capacity during measurements.
- At least three measurements should be made at the maximum load in the facilities operating in continuous regime.
- It is important to make sure that the values of the gases to be measured are checked on the screen and there should be no fluctuation in the values (they should be constant).
- The results should be taken from the printer while the values are constant.
- Periodic calibrations of the devices must be monitored and archived, and calibration certificates made by the manufacturer should be provided if calibrations are not made in the country.
- The flue gas measuring device can be verified with traceable reference gases.
- After the device has passed to the measurement position, it is necessary to check the appropriate connection hoses once again and check for leakage.
- After the readings on the screen are stabilized in combustion gas measurements, at least 3 measurements will be made according to Industrial Air Quality Protection Regulation. Furthermore, the number of measurements can be increased according to the type of measurement requested by the plant.
- The sampling line must be heated to prevent the loss of sulfur dioxide and to avoid false readings.
- Velocity, temperature, and humidity measurements should also be carried out in order to perform mass flow calculations with the tests performed within the scope of gas measurement.

- Moisture measurements can be made according to the dry-wet temperature principle, moisture measurement apparatus, and gravimetrically according to EPA Method 4 (Determination of Moisture Content in Stack Gases).

7.5.8.6 Carbon Monoxide (CO) Measurement

The amount of carbon monoxide (CO) can be measured by selecting gas with the same flue gas analyzer. Carbon monoxide is a colorless and odorless, poisonous gas. If its concentration is too high, it combines with hemoglobin in human blood to form carboxyhemoglobin and prevent oxygen from passing into the blood. For example, if you breathe air in an environment with 700 ppm CO, this causes people to breathe air in this environment to die within 3 h. The maximum permissible working area should have the value of 50 ppm. For example, in combustion products of fuel oil boilers, this value is between 80 and 150 ppm.

7.5.8.7 Carbon Dioxide (CO_2) Measurement

The amount of carbon dioxide (CO_2) gas present in the flue gases is also measured by the flue gas analyzer. The amount of CO_2 is measured from the center of the flue gas flow where the flue gas temperature is measured. This is the highest point in terms of temperature and CO_2, the lowest oxygen concentration. In gas furnaces, CO_2 should be as high as possible and the CO amount should be less than 1000 ppm in undiluted flue gas.

References

1. Kaya D, Öztürk HH (2014) Sanayide Enerji Yönetimi ve Enerji Verimliliği, Umuttepe Yayınları, Yayın No: 114, Mühendislik Dizisi: 12, Kocaeli/Türkiye
2. https://www.testo.com/tr-TR/
3. https://www.se.com/tr/tr/
4. https://www.makinaegitimi.com/manometreler-ve-manometre-cesitleri/
5. https://www.investopedia.com/ask/answers/030415/what-functional-difference-between-gdp-and-gnp.asp
6. https://www.nmbtc.com/white-papers/fan-efficiency-and-increasingly-important-selection-criteria/
7. Spirax Sarco: https://www.spiraxsarco.com/learn-about-steam/flowmetering/types-of-steam-flowmeter
8. T.C. Millî Eğitim Bakanliği. Elektrik Elektronik Teknolojisi, Endüstriyel Sayaçlar, 523EO0078, Ankara, 2011
9. Fowles G, Boyes WH (2009) Instrumentation Reference Book, 4th Edition, Edited by Walt Boyes. Chapter 6: Measurement of Flow. Butterworth-Heinemann is an imprint of Elsevier 30 Corporate Drive, Suite 400, Linacre House, Jordan Hill, Oxford OX2 8DP, UK. Hardcover ISBN: 9780750683081, eBook ISBN: 9780080941882, Published Date: 11th December 2009, Burlington, MA 01803, USA

10. https://commons.wikimedia.org/wiki/File:Venturi5.svg
11. https://www.pce-instruments.com/f/t/us/main.htm
12. Wilson JD, Connecting Pitot tube on the manometer. University of Alberta, 25 Feb/99
13. http://www.instrumentationtoday.com/
14. https://www.hukseflux.com/

Chapter 8
Fuels and Combustion

8.1 Types of Fuel

Any material that releases energy (heat) when a change occurs in its physical and chemical structure is generally called fuel. Generally, substances that give heat when burned are called fuel. Fuels can be stored in order to generate energy. They can be used to generate work when necessary, to provide energy. In order for matters to be defined as fuel, they must have features such as not harmful to health when burned, abundant in nature, not posing a danger and also, being cheap. Fuels are generally divided into three groups:

(1) **Solid fuels**: They are divided into three groups.

- **Natural solid fuels**: The most important solid fuels are natural solid fuels known as fossil coals. These coals according to the formation stages are; anthracite, hard coal, brown coal, lignite coal, peat follows the order.
- **Artificial solid fuels**: Coke and wood coal obtained from natural solid fuels.
- **Biomass solid fuels**: Biologically derived fuels, such as wood and biomass briquettes/pellets, obtained from industrial waste processing forest and agricultural products.

(2) **Liquid fuels**: They are obtained from crude oil and coal tar.

- **Natural liquid fuels**: They are produced by distillation of crude oil.
- **Artificial liquid fuels**: They are divided into 3 groups according to the source they are produced from:
  - Fuels produced by distillation from fuels such as coal,
  - Liquid fuels produced by cracking heavy petroleum products,
  - Liquid fuels produced by synthesis process.

(3) **Gaseous fuels**: The gas obtained as a result of the distillation of fossil coals is the gases in the structure of small molecule hydrocarbon mixture which is

D. Kaya et al., *Energy Management and Energy Efficiency in Industry*,
Green Energy and Technology,
https://doi.org/10.1007/978-3-030-25995-2_8

the product of petroleum distillation and natural ground gases. Gaseous fuel is produced by fermentation (biogas) or gasification of organic materials.

Liquid fuels in general are divided into three: (1) petroleum-based fuels, (2) alcohol, and (3) oils. Liquid fuels have the following advantages compared to solid fuels:

- The energy they give per unit mass or volume is very high.
- Combustion is more efficient in liquid fuels. For this reason, the energy contained in liquid fuels is quickly converted to heat energy.
- Liquid fuels are easier to transport, store, and quantify.
- Due to high thermal values, storage volumes are small.
- They can be easily mixed with air and can be operated with a smaller excess air coefficient (EAC). Elements such as smoke, air ducts, ventilators, and aspirators can be designed in smaller capacities and sizes.
- The change in fuel quality in liquid fuels is less.
- Initial investment and subsequent operating costs of liquid fuel combustion systems are lower.
- Automatic control of liquid fuel-burning systems is easier.
- Boilers burning liquid fuel start up faster and respond to load changes faster.
- Liquid fuels do not leave ash after burning. Liquid fuel systems operate cleanly and emissions are less harmful to the environment.

8.1.1 Fossil Fuels

Fossil fuels are fuels such as coal, oil, and natural gas. Most of the energy we use today comes from fossil fuels. For millions of years, fossil fuels have been formed with the decay of plants and animals (Fig. 8.1). Fossil fuels can be drilled (drilled) or excavated. Currently, these fuels are formed underground with heat and pressure. However, the rate at which fossil fuels are produced is much lower than their consumption rate. Therefore, fossil fuels can be considered as non-renewable energy sources in a short time (Fig. 8.2). In particular, population growth, urbanization, and industrialization lead to an increase in the energy requirements of these fuels. In addition to the depletion of fossil fuels and the steady increase in prices, the damage they cause to the environment as a result of their combustion and their impact on human health is also important.

Advantages and disadvantages of fossil fuels can be seen in Table 8.1 [1–3].

8.1.1.1 Fossil Coals

Coal is a mineral and a rock made up of mostly carbon, hydrogen, and oxygen and has a minor amount of sulfur and nitrogen with a different chemical and physical

Fig. 8.1 Formation of fossil fuels

structure. Other ingredients include ash-forming inorganic compounds and mineral substances. Some coals melt and become plastic when heated.

As a result of the processes, tar, liquor, and various gases can be obtained. Natural solid fuels: anthracite, hard coal, brown coal, lignite coal, and peat fuels are generally called fossil coal (Fig. 8.3).

Lignite is a type of coal, also known as brown coal, which has the lowest rank in the coal rankings, almost all of which is used as fuel in thermal power plants. Thermal value (calorific value) of lignite is low, ash and moisture contents are high. However, it is an energy raw material that is frequently used, especially in developed countries such as Germany, the United States of America, since it is abundant in the earth's crust.

Lignite coal and brown coal, which is an advanced type of lignite, are not as rich in thermal value as the others. They are often used where they are removed. Brown coal is similar to hard coal in its external appearance, but can be distinguished from hard coal in the following characteristics:

The line drawn with hard coal on the unglazed porcelain is black and the one drawn with brown coal is brown.

A If these two types of coal are pulverized and boiled with concentrated NaOH (sodium hydroxide) solution, hard coal cannot paint the solution, but brown coal dyes brown.

Fig. 8.2 Formation of fossil fuels by the effect of time, pressure, and heat

Chemical Structure of Fossil Coals

Fossil coals, which constitute the main part of natural solid fuels, are composed of organic and inorganic compounds. The organic part forms the combustible part of the coal and the inorganic part forms the ash of the coal.

The organic part of the coal is formed as a result of the aliphatic, heterocyclic, and carbocyclic compounds in plants converted into humin material by biological processes and the humin substance undergoes a condensation reaction under ambient conditions such as high pressure and long-term temperature (Fig. 8.4).

Coal is not a hydrocarbon. Because, in its structure, there are organically bound oxygen, nitrogen, sulfur, such molecules that contain heteroatoms. Oxygen, according to the formation cycle of coal can be found as hydroxyl, carbonyl, carboxyl oxygen, or in the form of further coals, in the form of heterocyclic carbon-oxygen rings or ether bridges.

Table 8.1 Advantages and disadvantages of fossil fuels

Advantages	Disadvantages
• The energy of all fossil fuels comes from the sun. In this respect, each of the fossil fuels is mainly derived from solar energy • The design of power plants using fossil fuels is simple • Fossil fuels can be transported by pipelines in gas and liquid forms • A lot of petroleum-derived fuels are used in the automotive sector • Fossil fuels are easier to extract and process. Therefore, they are cheaper than other non-traditional types of energy • Easy to transport and distribute anywhere • Too much electricity can be produced from fossil fuels on their own • Power plants using fossil fuels can be easily designed in almost any area Therefore, it is possible to store as much fuel as possible in power stations • Their costs are comparatively low because their reserves are large and easy to access	• The leakage of certain fossil fuels such as natural gas and crude oil during transport can cause significant environmental damage • Since they take a long time to occur, they are non-renewable • Due to uncertainties in the future reserves, fossil fuel prices are gradually increasing • The use of crude oil can cause environmental pollution and poisoning, such as damage from marine oil. Since crude oil contains harmful chemicals, it causes air pollution when burned • Excessive fuel is required for coal-fired power plants. It is therefore necessary to transport coal in a uniform manner

Fig. 8.3 Fossil coal

Nitrogen in the coal comes from the compounds of plant such as alkaloid, protein, and chlorophyll, and is generally found in large heterocyclic molecules. Sulfur, on the other hand, passes through the coal from plant proteins. In coals with sulfur content above 1%, some of the sulfur may also be present as inorganic pyrite sulfur. The hydrogen is present in both the aromatic molecule hydrogen and the alicyclic and aliphatic olefin hydrogen. In the composition of coal ash, the source of coal: silica, aluminum, iron, calcium, magnesium, and salts of alkali metals. For example, the highest percentage of these substances in the ash of coal extracted in the USA is as follows;

Fig. 8.4 Thermal (calorific) value increase in fossil coals

- SiO_2 (Silica): 30–60
- Al_2O_3 (Aluminum oxide): 10–14
- Fe_2O_3 (Iron-III oxide): 3–30
- CaO (Calcium oxide): 1–20
- MgO (Magnesium oxide): 0.5–4
- TiO_2 (Titanium dioxide): 0.5–3
- Alkali metal oxides: 1–4

The majority of fossil coals, which are natural solid fuels, consist of an organic network of high molecular weight, cyclic compounds made up of six rings, and inorganic compounds which are introduced by diffusion between these networks. The organic part forms the combustible part of the coal and the inorganic part forms the ash of the coal.

Classification of Coals: In addition to the carbonization process and depositing, moisture content, ash and volatile matter content, constant carbon content, sulfur, and mineral matter contents, coals are very diverse in terms of geological, petrographic, physical, chemical, and thermal properties. This necessitated the classification of coals in many countries on the basis of similar properties and close values.

In the production, use, and technology of coal, advanced countries have first developed a classification according to the characteristics of their own coal, and have developed common standards for an international general classification. In the international classification of different types of coals according to their intended use, first, in 1957, the International Coal Board, which was formed by members from various countries, and was supported by the International Standards Organization (ISO) carried out studies on samples taken from many countries and made a general classification (Table 8.2) [3]. Coals according to this classification, based on their thermal value, volatile content, constant carbon content, coking, and caking properties, coal is divided into two groups as hard and brown coals:

Table 8.2 International general coal classification

Brown coals	Hard coals
• Lower bituminous coals: Its thermal (calorific) values are between 4165 and 5700 kcal/kg. They do not exhibit agglomeration properties • Lignite: Its thermal (calorific) value is less than 4165 kcal/kg. It shows agglomeration properties	• Coking coals: It is suitable for coke production using blast furnaces • Non-coking coals: (a) Anthracite (b) Bituminous coals

(1) Hard coals: Its thermal value on wet and ashless (ash-free) bases is more than 5700 kcal/kg. They are subdivided according to volatile content, thermal value, and coking properties.
(2) Brown coals: The thermal value on wet and ashless basis is lower than 5700 kcal/kg. They are subdivided according to the total moisture content and thermal value.

Another classification criterion accepted in the international coal classification is the coal rank classification (carbonization degree classification). In this classification, carbon content is considered as the main variable. In high rank (high carbonization) coals, the volatile matter content and low rank (low carbonization) coals were classified according to heating values. Table 8.3 shows the classification and properties of coal rank, and the properties of coals in general classification is given in Table 8.4. The classified coals contain moisture, volatile matter, fixed carbon, and ash amounts which are determined by rough analysis, sulfur, hydrogen, carbon, nitrogen, and oxygen amounts are determined by detailed analysis and heating values are determined by calorimeter.

Table 8.3 Coal properties in different rank

Rank	Volatile content (mass, %)		Carbon content (mass, %)		Moisture content (mass, %)	Heating value (MJ/kg)
	Wet	Ash-free	Wet	Ash-free		
(1) Lignite	69	44	76	62	52–30	19.31–14.66
(2) Semi-bituminous	52	40	80	71	30–12	26.74–19.31
(3) Bituminous	–	–	–	–	–	–
(a) High volatile-B	50	29	86	76	15–2	26.74–24.42
(b) High volatile-C	–	–	–	–	–	–
(c) High volatile-A	49	31	88	78	5–1	32.56
(d) Medium volatile	31	22	91	86	5–1	32.56
(e) Low volatile	22	14	91	86	5–1	32.56
(4) Anthracite	14	2	99	91	5–1	32.56

Table 8.4 Properties of coals in general classification

Lignite	Semi-bituminous coal	Bituminous coal	Anthracite
Brown	Black	Dark black	Glossy black
Fragile, rapid powder crumbling	Breaking into fine parts and powder crumbling by oxidation or as a result of fracture	Bloc fracture	Lenticular refraction
Solid (massive), woody, or uniform clay texture	Massive	Banded and compact	Hard and durable
Volatile matter and moisture content is high	Volatile material and moisture content is higher than bituminous coal	Volatile material and moisture content is low	Volatile material and moisture content is low
Heating (calorific) value <4610 kcal/kg	Heating (calorific) value, 4610–6390 kcal/kg	Heating (calorific) value, 5390–7700 kcal/kg	Heating (calorific) value >7000 kcal/kg
Constant carbon content is low	Constant carbon is lower than bituminous coal	Constant carbon content is high	Constant carbon content is high

8.1.1.2 Oil

In Latin, the word petroleum is composed of the words Petra meaning stone and oleum meaning oil (Petra oleum = Petrol). The term petroleum refers to crude in its natural state and extracted from the ground. Oil is a liquid with very low flowability, slippery, color ranging from dark yellow to black, and with an unpleasant heavy odor. It is light in water and does not dissolve in water. The specific gravity ranges from 0.7 to 0.97 on average. Oil is usually extracted by drilling deep wells under the ground.

Crude oil consists of a mixture of hydrocarbons of different molecular structures. Components of oil: paraffins, naphthenes, and aromatic hydrocarbons. In addition to mixtures of hydrocarbons, it contains a small amount of oxygen, nitrogen, iron, magnesium, calcium, phosphorus, vanadium, sulfur, and a small amount of zinc and cobalt.

Chemical Formation of Oil: There are many theories about the formation of crude oil. The most valid and accepted one of these is that the remains of small organisms (microorganisms) that collapse on the seabed make-up oil (Fig. 8.5). The dead and alive of these small creatures accumulate in the depths of the seas as a result of the drift of the mud layers brought by the rivers. Oil is formed by fermentation by staying under high pressure and temperature for a long time. Hundreds of millions of years ago, the remains of animals and plants living in the seas or being dragged into the seas, in an anaerobic environment, caused by the effects of heat, pressure, and microorganisms, produced similar kerogen to crude oil. Oil is formed

Very small plants and animals living in the seas remain at the bottom of the oceans after they die. Over time, these dead creatures are covered with silt and sand layers. They remain buried in the bottom for millions of years. They are transformed into petroleum and gas with the effect of heat and pressure. Nowadays, sand, silt and rock layers are excavated and petroleum or natural gas deposits are reached and extracted to the Earth.

Fig. 8.5 Oil and gas formation

as a result of heat-receiving reactions in a high temperature and/or pressure environment. Later, during the migration to the upper layers, the kerogen changes gradually and produces crude oil. Despite today's advanced technique, the accumulation of oil underground and in particular its efficiency cannot be easily determined. In this regard, oil exploration and extraction are expensive processes that take a long time.

The chemical structure of oil consists of hydrocarbon chains of different lengths [4]. These chains are separated by distillation process in the treatment process of oil to produce products such as gasoline, jet fuel, and kerosene. The general representation of the alkanes is in the form of C_nH_{2n+2}. For example, the expression of 2,2,4-trimethylpentane, commonly found in gasoline, is in the form of C_8H_{18}, and the reaction of it with oxygen which is exothermic is as follows:

$$C_8H_{18}(s) + 12.5O_2(g) \rightarrow 8CO_2(g) + 9H_2O(g) + \text{Heat}$$

Partial combustion of oil or gasoline causes the release of toxic gases such as carbon monoxide and/or nitric acid.

$$C_8H_{18}(s) + 12.5O_2(g) + N_2(g) \rightarrow 6CO_2(g) + 2CO(g) + 2NO(g) + 9H_2O(g) + \text{Heat}$$

8.1.1.3 Natural Gas

Natural gas is a kind of flammable gas mixture of fossil origin in the earth's crust. It is an oil derivative. It takes second place after crude oil in the ranking of importance as fuel. The majority of natural gas (70–90%) is composed of a hydrocarbon compound

Table 8.5 Properties of natural gas

Components (%)		
Methane (CH_4)	70–90	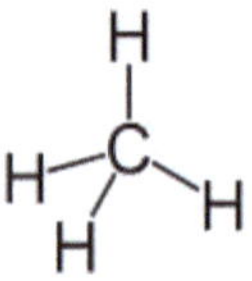
Ethane (C_2H_6)	5–15	
Propane (C_3H_8) and Butane (C_4H_{10})	<5	
CO_2, N_2, H_2S, etc.	The rest	
Features		
Density and phase	0.717 kg/m^3, gas g/cm^3	
Melting point	−182.5 °C (90.6 K) 1 atm 25 °C (298 K) 1.5 GPa °C	
Boiling point	−161.6 °C (111.55 K)	
Hazards		
Main hazards	Highly flammable **(F+)**	
NFPA 704		
Self-ignition point	537 °C	
Flash point	−188 °C	
Explosion limit	5–15%	
Related compounds		
Related alkanes	Ethane, Propane	
Related compounds	Methanol, chloromethane	
If not specified, the data are for the standard form of the material (25 °C, 100 kPa)		

called methane gas (CH_4). Other components are ethane (C_2H_6), propane (C_3H_8), and butane (C_4H_{10}) gases (Table 8.5). It contains very little carbon dioxide (CO_2), nitrogen (N_2), helium (He), and hydrogen sulfide (H_2S).

In the past, natural gas was seen as a useless waste generated during oil production and was incinerated in oil production facilities. Today, it is frequently used in homes and industry as a valuable and important energy source.

Natural Gas in the World: Natural gas is produced in all continents except Antarctica. The largest producer in the world is the Commonwealth of Independent States. The United States, Canada, the Netherlands, and Iran are also major natural gas producers. The most efficient and cheapest transportation method of natural gas is the use of pipelines. Natural gas can also be transported liquefied in pressure tanks. The necessity of keeping liquefied natural gas (LNG) under very high pressure and low temperatures during transport renders this transport method inefficient compared to the pipeline method.

Natural gas, which is the main raw material of various chemical products, meets a significant portion of the world's energy consumption. Today, 24% of energy consumption is met by natural gas. It is estimated that the known natural gas reserves in the world have a lifetime of about 70 years. Known natural gas reserves are equivalent to oil reserves.

All of the pipes, fittings, and equipment used to transport the natural gas from its source to the end is called natural gas installation. Calculations in the installation are determined by taking pressure factors and velocity factors into consideration.

Effects on Human Health: Natural gas is a hydrogen saturated carbon molecule and its multiples. It is found in nature in the free state and in the gas phase, usually underground and in trace amounts of air. It is colorless, odorless. It is blended with an artificial chemical (the smell of skunk) before dispensing to detect leakages by the human nose. Contrary to common knowledge, natural gas is not a toxic gas for humans. Most of the accidents in natural gas use are caused by incorrect chimney applications. Therefore, there is absolutely no poisoning due to natural gas leaks. Carbon monoxide gas, which occurs with the combustion of natural gas, causes poisoning. Shortly after inhalation, dizziness and loss of balance occur and after a while, fainting and death can happen. This is not because of poisoning, but because oxygen breathing stops. As a result of natural gas being lighter than air, it forms a thin film layer in the lungs and covers the alveolar surfaces and stops the contact with air. Although breathing continues, oxygen cannot be absorbed by the lungs, and as a result of a decrease in the amount of oxygen in the brain, dizziness, and fainting occurs and is followed by death. Cases of natural gas poisoning in the media and among the people are actually about drowning. If the person has been exposed to natural gas, it should be taken to fresh air immediately, lying on his back, turning the body 45° to the side and deep breathing should be done to remove the gas inside from the lungs.

Natural gas is not present in nature in the liquid phase, its boiling point is −161.6 °C. Two hundred and fifty-four liters of natural gas can be liquefied with high pressure and compressed up to 22 L. If contact with this liquid phase occurs, severe cold burns will happen on the skin.

8.1.1.4 Liquefied Petroleum Gases

Liquefied petroleum gas (LPG) is a colorless, odorless, heavier than air, and is a flammable gas obtained during the distillation of crude oil in refineries or by separation of natural gas on the oil beds and is liquefied under pressure. In the event of a leak, it is particularly fragrant by the refineries in order to immediately understand the gas leak. The composition of LPG used in many countries is generally 70% butane and 30% propane. When 1 L of LPG becomes gas, it normally reaches a volume of about 250 L of gas. LPG, unlike natural gas, is heavier than air and accumulates in case of leakage and collapses.

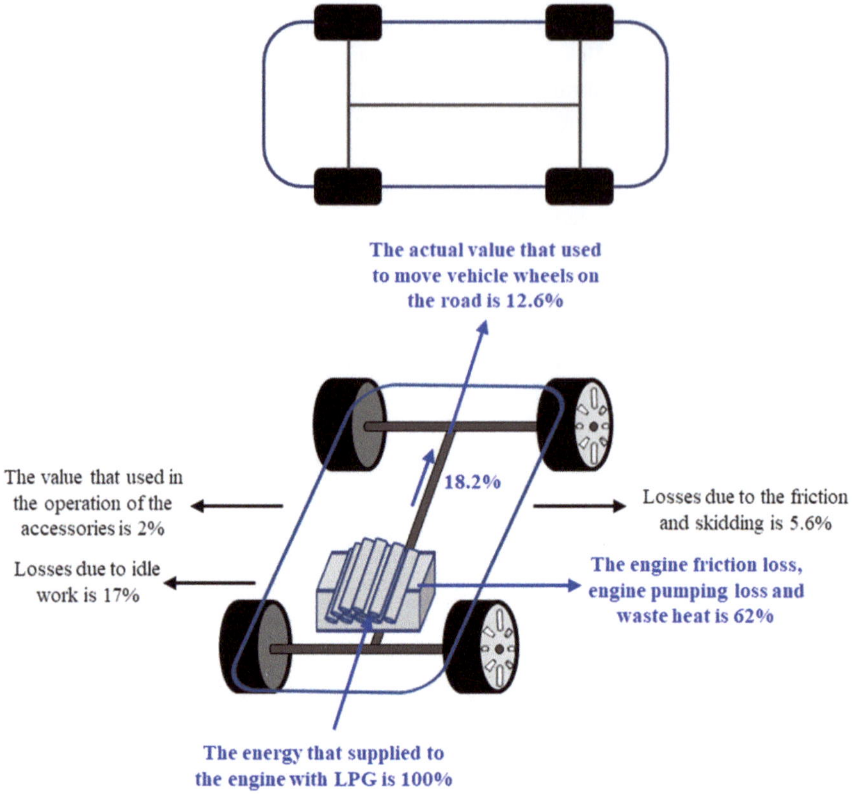

Fig. 8.6 Energy consumption rates in an LPG-powered vehicle in city driving

Energy consumption rates in an LPG-powered vehicle in city driving can be seen in Fig. 8.6.

Autogas is the name given to the form of liquefied petroleum gas obtained from crude oil and natural gas, which consists of a certain combination of butane and propane gases, used as fuel in vehicles. Compared to other petroleum-based fuels such as gasoline and diesel, the emission level is very low and less environmentally hazardous.

Liquefied petroleum gas (LPG) consists of the lightest hydrocarbons in oil. Under normal weather conditions, they become gas and under pressure, they become liquid. They are obtained as by-products when oil is extracted from the wells or oil is processed in refineries. It is also extracted from underground natural gas deposits. When pressure is applied to these gases, their volume is reduced by 230–267 times. For example, when 267 m^3 of gas is liquefied, it fits into a volume of 1 m^3.

LPG Usage Areas: LPG is used in, warming, cooking in the kitchen and as fuel on vehicles. With the widespread use of natural gas in many countries, using it for heating and consumption in the kitchen has decreased, while its use in vehicles

Table 8.6 Benefits of using LPG

General purpose benefits	Benefits of use in engines
• LPG is a fuel that consumers can safely use when the conditions of use are complied with • LPG is a fuel that provides all the benefits of natural gas and does not require any central infrastructure • Due to its simple chemical structure, LPG is a fuel that causes less harm to the environment than other energy sources • LPG is an efficient fuel with a high heat value compared to other types of fuel • LPG is a fuel that can be packaged and transported • LPG can cover almost all energy needs of any living unit	• The amount of energy they give per kilogram of fuel is high (11,858 kcal/kg) • Since it mixes well with the air, combustion efficiency is high • The combustion is too close to full combustion, reducing air pollution • After burning does not leave waste in the cylinders • As the oil film on the cylinder is not torn, the life of the engine parts is prolonged. Reduces repair/maintenance costs • As the oil in the sump is prevented from thinning, the quality of the lubrication does not deteriorate, and the oil change time is prolonged • Since the octane counts are very high, it allows the compression ratio of the motor (10/1–12/1) to increase • Due to the advantages mentioned above, fuel economy of 30% is achieved if LPG is used as engine fuel

became widespread. LPG used in vehicles is also called autogas which is slightly different compared to cylinder gas used in the kitchens. The use of liquefied gases in engines does not require a major change in operation. The different elements in the engine structure consist of a special carburetor, pressure reducing regulators, solenoid valves, and so on. The benefits of LPG for general purpose use in engines are given in Table 8.6. Despite all these advantages, the most important disadvantage of LPG is that the transportation and storage costs of gases are higher compared to liquid fuels.

8.1.1.5 Chemical Structure of Fossil Fuels

Chemical structures of liquid fuels obtained from petroleum consist of carbon (C) and hydrogen (H). These compounds made by carbon are called hydrocarbons and are represented by the closed formula C_nH_m. Liquid fuels can be obtained from all organic materials. However, today the most important natural resource is oil. The chemical structure of hydrocarbons changes the properties of the fuel. The higher the number of hydrogen atoms in the compound compared to the carbon atom, the higher heating (calorific) value of the fuel. Compounds with high molecular weight are less volatile. In hydrocarbons making the same bond, the tear resistance decreases as the number of carbon atoms in the molecule increases [5, 6].

Hydrocarbons, which are the simplest structures of organic compounds, are divided into two main classes:

(1) Aliphatic hydrocarbons,
(2) Aromatic hydrocarbons.

Hydrocarbons are classified according to the way that carbon atoms are bonded to each other. Compounds, where carbon atoms are bonded to each other by a single bond are called saturated, and compounds that are bonded by 2 or 3 bonds are called unsaturated compounds.

Saturated Hydrocarbons: Molecules of hydrocarbons consisting of a single chain bonded to each other in a single bond are called paraffin line or saturated hydrocarbons. The general formula of saturated hydrocarbons is (C_2H_{2n+2}). There are two hydrogens at each carbon on the row, one hydrogen at the beginning and one at the end. Thus, the number of (H) is twice the number of (C) and plus two.

Paraffins: Paraffin class compounds are all present in crude oil and form the major part of engine fuels. The closed formulas are C_nH_{2n+2}. Since the number of hydrogen atoms is high, the heating (calorific) values are high. Their density is low (620–770 kg/m^3). It is divided into two parts according to the way carbon atoms are connected to each other.

Normal Paraffins: Normal paraffins are the ones that make a straight chain bond. Since all four connecting arms of each carbon atom are connected, they are called saturated hydrocarbons. Since the bonds can be easily separated, the ignition tendencies are quite high. The phase state of normal paraffins is as follows according to the number of C atoms: If the number of C atoms is between 1 and 4, the phase of normal paraffin is gas; if it is between 5 and 15, it is a liquid; and if it is greater than 16, it is in solid phase.

The first four of the class up to five carbon atoms are in gaseous form. This group, which becomes liquid under pressure, is used today in kitchen works, heat, and lighting.

$$CH_4 : \text{Methane} \quad C_2H_6 : \text{Ethane} \quad C_3H_8 : \text{Propane} \quad C_4H_{10} : \text{Butane}$$

Compounds with carbon atoms up to 20 are in liquid form. The groups in this section are determined as follows.

$$C_5H_{10} - C_7H_{16} : \text{Gasoline for aircraft} \qquad C_5H_{10} - C_{12}H_{26} : \text{Engine petrol}$$
$$C_9H_{20} - C_{14}H_{30} : \text{Jet fuel (Kerosene)} \qquad C_{12}H_{26} - C_{16}H_{34} : \text{Gas oil}$$
$$C_{15}H_{32} - C_{18}H_{38} : \text{Diesel} \qquad C_{16}H_{34} - C_{20}H_{42} : \text{Mineral oils}$$

Paraffins having more than 20 carbon atoms are in solid phase. Vaseline, paraffin, and so on are members of this section.

Isoparaffins: Some carbon atoms are bifurcated on the general chain structure. Such paraffins which form branched chain bonds are called isoparaffins. They are named according to the number of carbon atoms forming the straight chain part of the group and their branching places. This type of naming is important. Because fuels with the same closed formula, but branched in different ways, may exhibit completely different properties. Ignition gradients, hence, tendency to knock. Iso-paraffins have higher impact resistance than normal paraffins. They are suitable for gasoline engines.

Naphthenes are hydrocarbons of general formula C_nH_{2n}. As their structures are closed in ring form, they are difficult to disintegrate, and their inclination is low. Since the number of hydrogens is low, their heating values are low, and their density is high. Heating values of this group are higher than aromatics and lower than paraffins. They can be used as both gasoline and diesel fuel. Knock resistance is higher than paraffins. Evaporation properties are in accordance with the requirements.

Unsaturated Hydrocarbons

Aromatics: Their closed formulas are as follows $C_nH_{2n\text{-}6}$. They are ring-bound hydrocarbons. They are important as engine fuel because their knock resistance is very high. Due to their ring-shaped structure, a large number of double-bonded carbon atoms, their ignition tendencies are low. They are called aromatics because of their pungent odor. Its main structures are the benzene molecule. They are usually made artificially from coal. They are added to the gasoline to increase the impact resistance. Since they are carcinogenic, the number of additives is limited.

In general, aromatics have high densities (800–850 kg/m^3) and low thermal values. Because the bonds between the carbon atoms are intact, the artifacts that are resistant to knock are high octane numbers. They are suitable to be used as gasoline engine fuel, and because they have low cetane numbers, they are not used as diesel engine fuel. The most important member of the class is benzene (Fig. 8.7). Benzene is used as the starting material of the aromatic compounds. In other words, all aromatic compounds are derived from benzene.

Olefins: Their closed formulas are (C_nH_{2n} or C_nH_{2n-6}). Those with a double bond in the molecule are called mono-olefins (C_nH_{2n}). Those which have two double bonds in the molecule are called diolefins (C_nH_{2n-6}). Some carbon atoms are double bonded unsaturated hydrocarbons. Their thermal (calorific) values are low. Their density is between 620 and 820 kg/m^3. Ignition trends are low. This group can be obtained very little by distillation of crude oil. They are abundant in gasoline, which is obtained by the breakdown of large products. Non-degradable olefins can

Fig. 8.7 Benzene (C_6H_6)

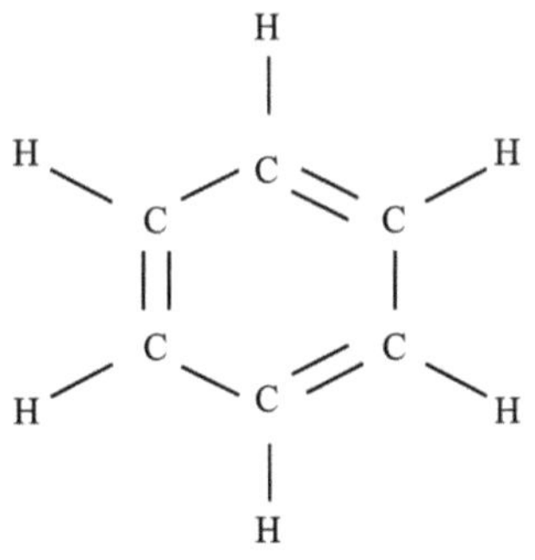

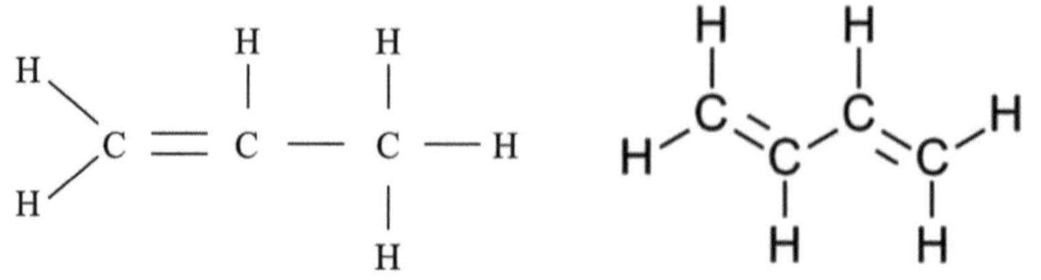

Fig. 8.8 Propylene (C_3H_6) and butadiene (C_4H_6)

be used in gasoline engines. They can be used as diesel fuel if ignition tendencies are increased. Figure 8.8 shows the open and closed formulas of propylene and butadiene hydrocarbons. Olefins have high knock resistance. However, when stored, they are not suitable as motor fuel since they resinify over time.

8.1.1.6 Fossil-Based Liquid Fuel Production

If the oil is used as the form when it is extracted, the hydrocarbons in which it is used are limited and the usage area is expanded. Thus, many products of high commercial value are obtained. The fractionation of the oil is carried out in refineries. This application requires many physical and chemical processes depending on the type of oil and the foreign substances it carries (Fig. 8.9). From oil, petroleum gas, kerosene, gasoline, diesel, fuel oil, lubricating oils, wax, and asphaltic bitumen are obtained. Liquid fuels are generally obtained by distillation of crude oil. However, in order to increase the quantity of certain products which can be obtained in small amounts by distillation, it is possible to crack large molecule hydrocarbons (cracking) or to form new hydrocarbon molecules (polymerization) by combining the essential elements carbon and hydrogen. In addition, some hydrocarbons can be modified to make gasoline.

Cracking: If crude oil is heated above 380 °C, the hydrocarbons will decompose into smaller molecular hydrocarbons. This process is called cracking. The aim of cracking is to break down hydrocarbons with large molecules and high boiling temperatures to obtain small molecules and low boiling hydrocarbons.

Polymerization: It is the opposite of cracking method. At appropriate pressure and temperature conditions, polymerization using appropriate catalysts and molecules joined together is called polymerization. Thus, large-molecule compounds are formed from small molecule compounds. Some molecules polymerize at low temperature. As a result of the polymerization process, the main product is a knot-resistant polymerization gasoline, which can be stored for less time than the distillation gasoline.

Hydrogenation: The process of adding hydrogen chemically to unsaturated heavy hydrocarbons during cracking is called hydrogenation. The purpose of hydrogenation is; to shrink large molecules and to allow hydrogen to accumulate in each molecule.

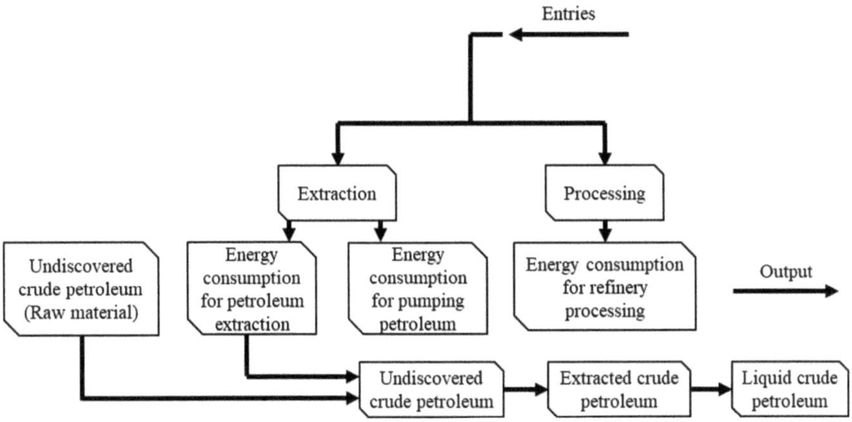

a) Fossil liquid fuel production stages.

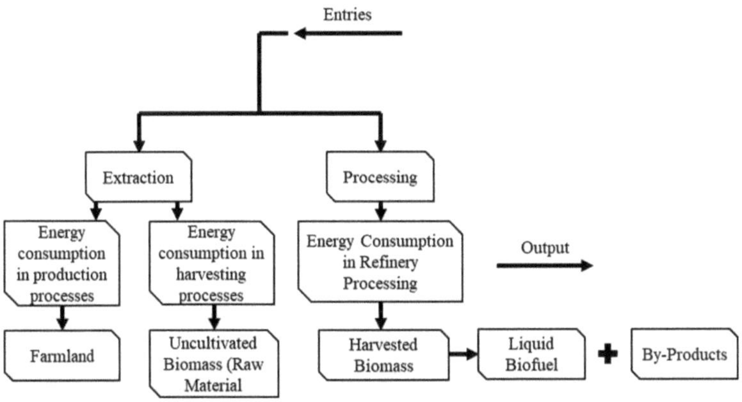

b) Biofuel production processes.

Fig. 8.9 Fossil and biofluid fuel production stages

8.1.2 *Biofuels*

The rapidly increasing world population, industrialization, and the environmental problems caused by the excessive use of fossil resources have gradually changed from a regional and national dimension to a global problem. Countries that do not want to be dependent on external energy and have problems in energy supply are trying to increase and diversify their alternative energy sources due to increasing environmental problems due to fossil resources. It can be said that alternative energy sources, which are unwanted in the past due to high costs, have become more attractive today due to technological developments. New and renewable energy sources are an important opportunity especially for countries that are dependent on foreign sources of energy.

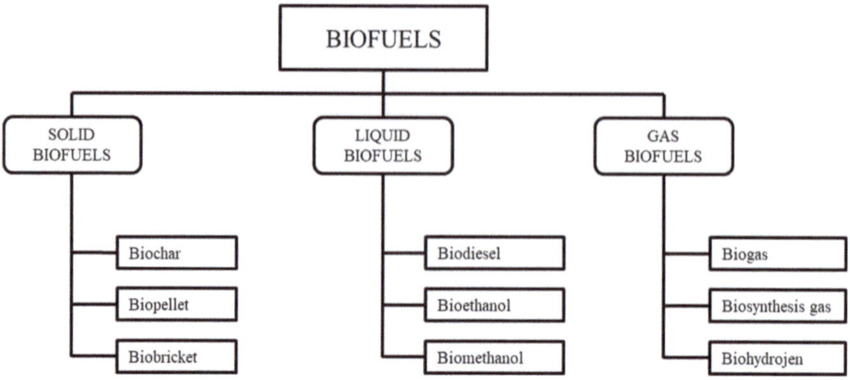

Fig. 8.10 Classification of fuels of biological origin

Biomass can be used as fuel to generate heat energy or electricity. In the evaluation of biomass as energy, various technologies are used to obtain solid, liquid, and gaseous fuels. Biofuels from biomass can also be used in combination with fossil fuels. Fuels of biological origin can be grouped as in Fig. 8.10.

Biofuels produced from biomass energy sources, which have made significant progress in the recent period and directly concern the agricultural sector, are the alternative fuels that attract attention today. Biofuels are derived from organisms that have recently lived or their metabolic output. They are renewable energy sources. Biofuels are defined as any fuel obtained from living organisms that have collected at least 80% by volume of their contents in the last decade.

Solid biofuels: biomass pellets and briquettes can be seen in Fig. 8.11.

Biofuels can be classified in different ways according to their sources and types. Biofuels that can be produced from forests, agricultural and fishery products or municipal wastes, agro-industry, food sector, and food sector products and wastes may be in the form of solid, ethanol, biodiesel, and pyrolysis fuels such as firewood, charcoal, and wood chips, or gas such as biogas. Biofuels: primary (unprocessed) and secondary (processed) biofuels.

(1) Primary biofuels: firewood, wood shavings, and parts thereof are organic materials used in their natural state. Burnt fuels are sources of energy that are often used for energy for cooking, heating, and energy production.

Fig. 8.11 Solid biofuels: biomass pellets and briquettes

(2) Secondary biofuels: solid (charcoal), liquid (ethanol, biodiesel, and other biofuels), and gases (biogas, synthetic gas, and hydrogen).

The most important benefits of biofuels can be summarized as follows:

- Reduces environmental pollution caused by fossil fuels,
- Minimizes the health risks of exhaust emissions,
- Ensures energy security and reduces dependence on energy,
- It helps to realize rural development.

In addition to fuels such as bioethanol, biogas, and biodiesel, also obtained from biomass, it is possible to count many more types of fuel such as fertilizer, hydrogen, methane, and wood briquettes.

The most common biofuels are bioethanol and biodiesel. Bioethanol is a high-octane biofuel originating from agricultural products with starch origin. Bioethanol is obtained as a result of fermentation process after the conversion of starch present in agricultural products to sugar. It can be directly mixed with gasoline. Bioethanol can be obtained from wheat, sugar beet, and corn.

Main features of fossil and biofuel use can be seen in Fig. 8.12.

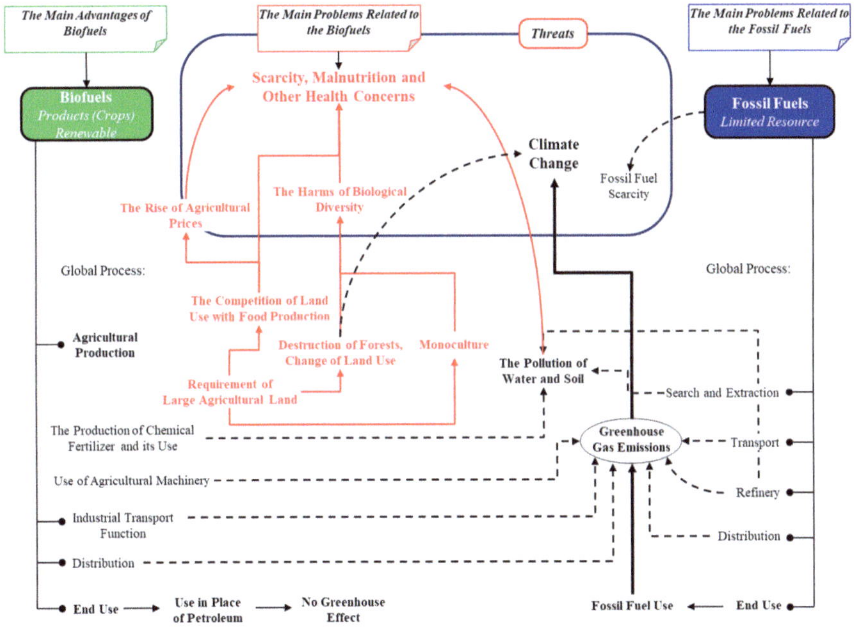

Fig. 8.12 Main features of fossil and biofuel use

8.1.3 Heating (Calorific) Value of Fuels

Organic carbon (C) is the main substance of fuels. The reaction of organic carbon with oxygen (O_2) produces heat. The reaction heat released in this way is utilized in various processes. The energy value per unit mass or volume of fuels is called the heating (calorific) value. The heating value of the fuel is defined as the heat energy produced when the combustion products are brought to ambient temperature as a result of the complete combustion of the fuel in mass or volume. In practice, the heat is divided into upper and lower heating values (LHV).

The higher heating value (HHV) is the amount of heat released when 1 kg or m^3 of fuel at a given temperature is mixed with the minimum air required for full combustion, without any heat loss, and the combustion products are cooled to the initial degree and condensed water vapor in the mixture.

The unit is kJ/m^3 or kJ/kg. The higher heating value is the amount of energy that can be obtained if the heat generated by the combustion of a fuel is utilized until the combustion products have cooled to ambient temperature (20 °C). In this case, the water vapor generated by the combustion of hydrogen in the fuel structure is excreted in the water structure.

The lower heating value is the amount of heat released when 1 kg or m^3 of fuel at a certain temperature is mixed with the minimum air required for full combustion, without any heat loss and when the combustion products are cooled to the initial temperature without condensing the water vapor in the mixture. The unit is kJ/m^3 or kJ/kg. The lower heating value is the energy released by the combustion of the fuel. The lower heating value is approximately 8–9% lower than the higher heating value, as much as the condensation temperature of the water vapor released as a result of combustion. Flue temperature in steam boilers is usually 180–200 °C. If combustion products are far above the ambient temperature; the water vapor contained in the combustion products cannot be condensed and is thrown as steam. It also removes the heat of evaporation. Thus, less heat is obtained from the fuel than the higher heating value, which is called the lower heating value.

All fuels containing hydrogen and moisture contain water vapor in combustion products. The thermal value of the fuel varies depending on whether this water vapor is in the vapor phase or the condensed liquid phase. If water vapor is condensed, the latent heat of evaporation will occur. The thermal value that occurs when this latent heat is taken into consideration is called the higher heating value. If the combustion products are in the gas phase, there is a lower heating value. The lower heating value is lower than the higher heating value as much as the evaporation latent heat of the existing water vapor. However, in practice, the combustion gas temperature is high and the heat of evaporation is not released. Therefore, in the calculations, the considered value is net or lower heating value.

Table 8.7 Lower and higher heating values of some fuels

Fuel type		Unit	Lower heating value		Higher heating value	
			kcal	kWh	kcal	kWh
Wood		kg	2500	2.90	2800	3.25
Coal	Lignite	kg	3000	3.50	3300	3.84
	Export	kg	6000	6.98	6500	6.96
Diesel		kg	10200	11.86	10,800	12.58
Natural gas		m^3	8250	9.59	9155	10.62
Heater fuel		kg	9700	11.28	10,500	12.18
LPG (30% propane, 70% butane)		m^3	26,000	30.16	28,200	32.71
		kg	11,000	12.76	11,900	13.80
Propane		m^3	21,200	23.95	23,000	25.93
		kg	11,100	12.87	12,000	13.98
Electric		kWh	860	1	860	1
Fuel oil		kg	9700	11.28	10,500	12.18

The higher and lower higher heating values are calculated by the following equations:

$$\mathrm{HHV} = 32,796\,\mathrm{C} + 141,886\,(\mathrm{H} - \mathrm{O}/8) + 9300\,\mathrm{S}\quad (\mathrm{kJ/kg}) \tag{8.1}$$

$$\mathrm{LHV} = \mathrm{HHV} - 2440(\mathrm{W} + 9\mathrm{H})\quad (\mathrm{kJ/kg}) \tag{8.2}$$

In these relations, the letters C, H, O, S, and W are percent weights of carbon, hydrogen, oxygen, sulfur, and water vapor in the coal composition, respectively.

The heating (calorific) values of fuels are an important feature. The heating values of some fuels are given in Tables 8.7 and 8.8. Fuels release the energy they contain in two ways: chemical (combustion) or nuclear (fusion).

Lower and higher heating values of some fuels can be seen in Table 8.7.

Heating (calorific) values of some fuels can be seen in Table 8.8.

Lower heating values of energy resources and conversion coefficients to oil equivalent can be seen in Table 8.9.

The temperature that ensures the continuous combustion of fuels is called the combustion point. The flare of the fuels to blazing up themselves by the effect of the heat they receive from the environment is called glare and the temperature at which the glare occurs is called the flash point. The resistance of fluids to flow is called viscosity. In another definition, the viscosity is the internal resistance, the internal friction, of the underlying layer against the movement of a liquid layer. The less fluid a fluid is, the higher its viscosity is, the more fluid it is (the finer it is) the lower its viscosity.

Table 8.8 Heating (calorific) values of some fuels

Fuel	Heating value (MJ/kg)	Fuel	Heating value (MJ/kg)
Animal/plant waste	9.62	Propane	42.70
Asphaltites	18.00	Fuel gas	43.96
Briquette	20.93	Acetylene	59.57
Coal	23.02	Coke gas	16.85
Coke dust (powder)	25.12	Natural gas	34.54
Crude oil	43.96	LPG	45.63
Graphite	33.49	Naphtha	43.54
Industrial lignite	12.56	Fuel oil No: 5	41.97
Lignite plant	8.37	Fuel oil No: 6	41.28
Petro-coke	31.81	Gasoline	43.54
Prina (Olive pomace)	18.00	Gas oil	34.70
Sawdust	12.56	Black liquor	12.56
Wood	12.56	Diesel fuel	42.70

8.2 Combustion

Combustion is a special name for the chemical reaction of fuels with oxygen. During this reaction, chemical energy in the fuel is released. Most of the released energy is spread to the environment in the form of heat (hot gases), and the remaining is in the form of electromagnetic waves (light), electricity (free electrons and ions scattered into the environment), and mechanical energy (sound) [5, 6]. These are generally ignored since very little energy is produced in the form of electrical and mechanical energy. In the same way, the share of electromagnetic energy in the form of light in the combustion process in the total energy is very small. The light emitted during the combustion process now serves to monitor whether or not the combustion continues, for example, in a boiler, by means of light-sensitive resistors.

Combustion Event can be seen in Fig. 8.13.

Combustion is defined as the rapid chemical coupling of flammable elements in the fuel with the oxygen of the air and the release of heat. While the reactants are fuel and air (O_2) during the combustion process, carbon dioxide (CO_2), water vapor (H_2O), and nitrogen (N_2) are also released at the end of combustion (Fig. 8.14). The necessary oxygen source for combustion is air. Air is made up of oxygen, nitrogen, small amounts of water vapor, carbon dioxide, argon, and other elements and is considered as 21% oxygen and 79% nitrogen by volume in the combustion event [7].

Fuels are fossil fuels what we call the primary sources of energy coal, fuel oil, and/or fossil fuels, including diesel and gas. Elemental analysis of each fossil fuel

Table 8.9 Lower heating values of energy resources and conversion coefficients to oil equivalent

Quantity		Energy supply	Density	Lower heating value	Unit	TOE cycle coefficient
1	Ton	Coal	–	6,100,000	kcal	0.610
1	Ton	Coke	–	7,200,000	kcal	0.720
1	Ton	Briquette	–	5,000,000	kcal	0.500
1	Ton	Lignite tarn and industry	–	3,000,000	kcal	0.300
1	Ton	Lignite plant	–	2,000,000	kcal	0.200
1	Ton	Petro-coke	–	7,600,000	kcal	0.760
1	Ton	Prina (Olive pomace)	–	4,300,000	kcal	0.430
1	Ton	Sawdust	–	3,000,000	kcal	0.300
1	Ton	Shell	–	2,250,000	kcal	0.225
1	Ton	Graphite	–	8,000,000	kcal	0.800
1	Ton	Coke dust	–	6,000,000	kcal	0.600
1	Ton	Mine	–	5,500,000	kcal	0.550
1	Ton	Asphaltites	–	4,300,000	kcal	0.430
1	Ton	Wood	–	3,000,000	kcal	0.300
1	Ton	Animal and Plant waste	–	2,300,000	kcal	0.230
1	Ton	Crude oil	–	10,500,000	kcal	1.050
1	Ton	Fuel oil No: 4	–	9,600,000	kcal	0.960
1	Ton	Fuel oil No: 5	0.920 kg/L	10,025,000	kcal	1.003
1	Ton	Fuel oil No: 6	0.940 kg/L	9,860,000	kcal	0.986
1	Ton	Diesel	0.830 kg/L	10,200,000	kcal	1.020
1	Ton	Gasoline	0.735 kg/L	10,400,000	kcal	1.040
1	Ton	Gas oil	0.780 kg/L	8,290,000	kcal	0.829
1	Ton	Black liquor	–	3,000,000	kcal	0.300
1	Ton	Naphtha	–	10,400,000	kcal	1.040
1000	m^3	Natural gas	0.670 kg/m^3	8,250,000	kcal	0.825
1	Ton	Coke gas	–	8,220,000	kcal	0.820
1000	m^3	Coke gas	0.490 kg/m^3	4,028,000	kcal	0.403
1	Ton	Blast furnace gas	–	535,000	kcal	0.054
1000	m^3	Blast furnace gas	1.290 kg/m^3	690,000	kcal	0.069
1000	m^3	Steelworks gas	–	1,500,000	kcal	0,150
1000	m^3	Refinery gas	–	8,783,000	kcal	0.878
1000	m^3	Acetylene	–	14,230,000	kcal	1.423
1000	m^3	Propane	–	10,200,000	kcal	1.020

(continued)

Table 8.9 (continued)

1000	kWh	Electric	–	860,000	kcal	0.086
1000	kWh	Hydraulic	–	860,000	kcal	0.086
1000	kWh	Geothermal	–	860,000	kcal	0.860
1	Ton	LPG	–	10,900,000	kcal	1.090
1000	m^3	LPG	2.477 kg/m^3	27,000,000	kcal	2.700

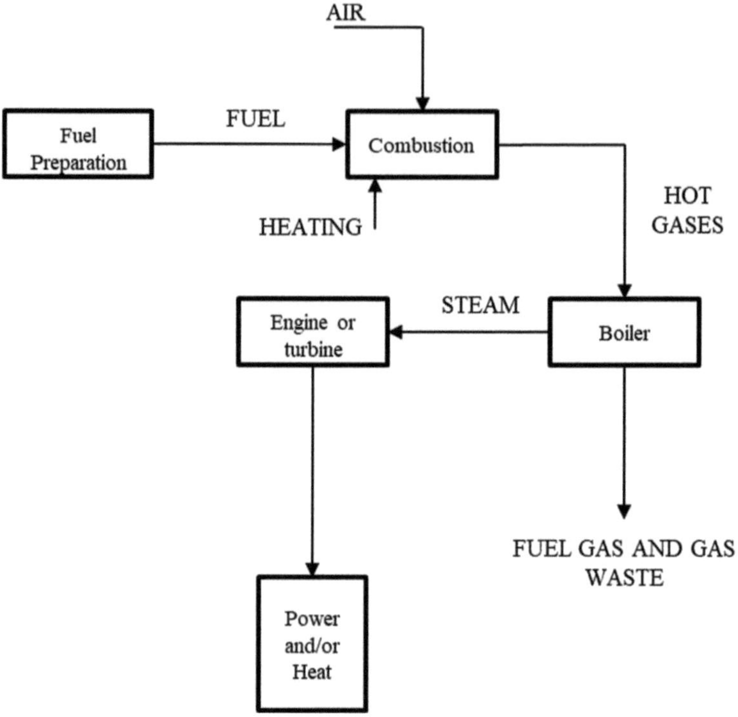

Fig. 8.13 Combustion event

is different. The carbon content of the liquid fuel is richer than natural gas. The carbon content of coal is resource-dependent and can vary within a wide range. The amount of hydrogen in the gas is approximately two times that of liquid fuel and seven times that of coal. Therefore, in the gases, water vapor is produced in large amounts as a combustion product. Liquid fuels and coal are sulfur-rich fuels compared to gas. Gases contain very little sulfur. Although sulfur, as a combustible material, contributes to a certain degree to the heating (calorific) value of the fuel, it is harmful because of its corrosive properties and causes air pollution. The percentage of ash in coal varies over a wide range depending on the source. Particles remaining from combustion cause significant pollution of nature.

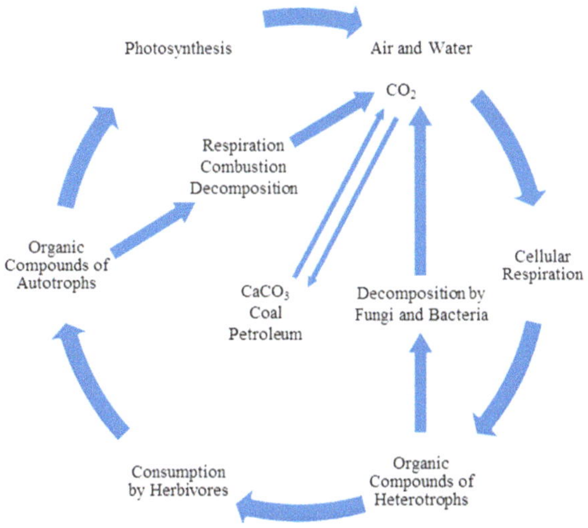

Fig. 8.14 Carbon dioxide cycle

8.2.1 Combustion Reaction

The formation of heat and other combusting products by the reaction of fuels with oxygen is called combustion. Combustion, defined as the conversion of chemical energy to heat energy, is a chemical process. The combustion event depends on the principles applicable to chemical reactions. This process is generally determined by the following reaction:

$$\text{Fuel} + \text{Oxidator} \rightarrow \text{Reaction Products} + \text{Heat}$$

Combustion is defined as the fast-chemical combination of the combustible components in the fuel with the oxygen of the air. This composition should be rapid and heat releasing. The combination of iron with the oxygen of the air must be oxidized and not combusted since it is not fast and liberating. The combination of gunpowder in a bullet with oxygen is an explosion. The main flammable components in fuel, carbon (C), hydrogen (H_2), compounds thereof methane (CH_4), and sulfur. Wastes generated as a result of chemical reaction by combustion are CO_2, water vapor, and some metal oxides. As a result of good combustion, the flue gas should contain 12–13% CO_2 and the CO content should be reduced to the lowest possible level.

$$\text{Carbon} + \text{Oxygen} \rightarrow \text{Carbon Dioxide} + \text{Heat}$$
$$\text{Hydrogen} + \text{Oxygen} \rightarrow \text{Water vapor} + \text{Heat}$$
$$\text{Sulfur} + \text{Oxygen} \rightarrow \text{Sulfur dioxide} + \text{Heat}$$

The carbon content of the fuel is the main component of the combustion heat.

$$1\ \text{kg C} + 2.666\ \text{kg } O_2 \rightarrow 3.666\ \text{kg } CO_2 + 8056\ \text{kcal Heat or}$$
$$C + O_2 \rightarrow CO_2 + 94,013\ \text{kcal/kilo mole Heat}$$

If the carbon is not fully combusted with the oxygen of the air and CO is formed, the thermal value is as follows:

$$1\ \text{kg C} + 1.333\ \text{kg } O_2 \rightarrow 2.333\ \text{kg CO} + 2417\ \text{kcal Heat or}$$
$$C + 0.5\ O_2 \rightarrow CO + 26,384\ \text{kcal/kilo mole Heat}$$

The important thing seen in the above equations is the possibility that carbon can react with oxygen in two ways. In the design of the combustion method, it should ensure that the final product is completely CO_2 in the combustion of carbon in the fuel-oxygen mixture. Otherwise, the combustion efficiency and the amount of heat released are significantly reduced.

8.2.1.1 Ideal Combustion

When the fuel is completely burned, the carbon (C) in it is converted to carbon dioxide (CO_2), hydrogen (H_2) into water vapor (H_2O), sulfur (S) into sulfur dioxide (SO_2).

Full Combustion:

$$C + O_2 \rightarrow CO_2 + 8113\ \text{kcal/kg} - C$$
$$2H_2 + O_2 \rightarrow 2H_2O + 34,650\ \text{kcal/kg} - H$$
$$S + O_2 \rightarrow SO_2 + 2250\ \text{kcal/kg} - S$$

Missing Combustion:

$$2C + O_2 = 2CO + 2467\ \text{kcal/kg} - C$$

As can be seen from here, the amount of energy lost when carbon remains in the form of carbon monoxide as a result of inadequate oxygen is not 70%. With this concern, as a general rule, the air supplied to the fuel is increased to a certain extent to ensure optimum combustion. This is defined by the Excess Air Coefficient (EAC). This coefficient varies depending on the type of fuel, less than necessary and encountered in cases where the negativity is given in Table 8.10. For this reason, flue gas analysis can be easily obtained and evaluated with the help of flue gas analyzers for the optimization of combustion during operation and immediate intervention to burners and boilers.

Ideally, the combustion occurs when the oxygen and fuel elements are mixed in theory at the exact ratios (stoichiometric ratio). However, in a combustion event, more air is always supplied than the theoretical need. Because, depending on the

Table 8.10 Low-high excess air coefficient (EAC)

Excess air coefficient (EAC)	Meaning
High EAC	• Carbon monoxide decreases • The combustion air is heated and discharged from the stack • The combustion process is disturbed • Combustion efficiency decreases
Low EAC	• Carbon Monoxide is formed • The generated heat energy decreases • Soot starts to form • Combustion efficiency decreases

temperature of the air at the determined air flow rate, the O_2 number of moles changes, or the air fan flow rate may change according to the fluctuations in the electricity supply.

When the fuel reacts completely, in other words when the fuel is turned into CO_2 and H_2O, it is called stoichiometric combustion, and the air required for this is called stoichiometric air. Since natural gas contains a great deal of methane (CH_4), the combustion equation requires two moles of O_2 for 1 mol of methane as $CH_4 + 2O_2 \rightarrow CO_2 + 2H_2O$. 10 m^3 of air is required to burn 1 m^3 of methane at 21% oxygen at sea level.

The stoichiometric air flow rate of the actual air flow rate to burn a unit amount of fuel is called the Excess Air Coefficient (EAC, λ). EAC serves to indicate how much the actual air/fuel ratio deviates from the theoretical ideal mass ratio. EAC is the ratio of actually supplied air mass to theoretical air mass required for full (stoichiometric) combustion.

$$\mathrm{EAC}(\lambda) = \frac{\text{Actual air flow rate}}{\text{Stoichiometric Air Flow}} \tag{8.3}$$

EAC is always greater than 1, except in very special cases. The ideal combustion event occurs when the excess air coefficient is close to 1. As EAC increases, combustion process improves. On the other hand, EAC greater than 1.3 slows down the combustion process (Fig. 8.15).

For gasoline engines to operate efficiently, they must have a certain air/fuel ratio. Ideally, the theoretical complete combustion takes place at a mass ratio of 14.7/1. This ratio is also called stoichiometric ratio. In other words, 14.7 kg of air is needed to burn 1 kg of fuel. If defined by volume, approximately 9,500 L of air is needed to completely burn 1 L of gasoline.

Excess air coefficient (EAC) values can be seen in Table 8.11.

The conditions for good combustion are:

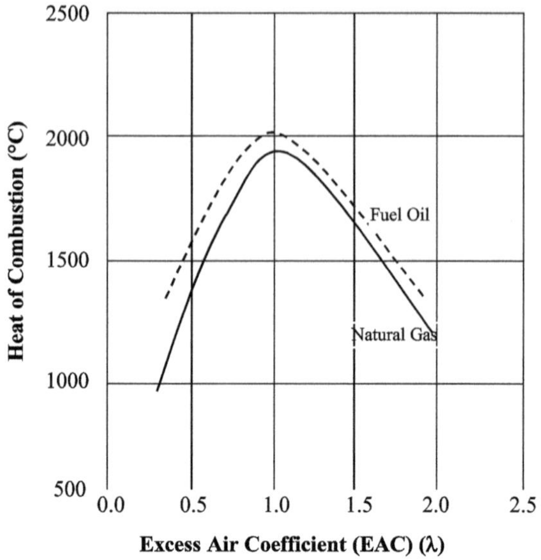

Fig. 8.15 The effect of EAC combustion temperature

Table 8.11 Excess air coefficient (EAC) values

Excess air coefficient (EAC)	Meaning
EAC (λ) = 1	The incoming air mass theoretically conforms to the required mass of air
EAC (λ) > 1	Indicates that the amount of air is excessive. For this reason, it is an indication that the air/fuel mixture is poor
EAC (λ) < 1	Indicates that the amount of air is insufficient. For this reason, it is an indication of the inadequacy of the air/fuel mixture

- Continuity of the selected fuel is important. In gases, natural gas or LPG, it is possible to deliver continuously at the same pressure. In the case of liquid fuels with different viscosity and heating values, the burner also needs to be adjusted to these values.
- It is necessary to maintain and control the optimum fuel-air mixture ratio. The amount of combustion air varies according to the composition of the fuel. For example, 1 Nm^3 of natural gas requires approximately 10 Nm^3 of air, and 1 Nm^3 of LPG requires 26 Nm^3 of air.
- Sufficient combustion time is needed for the fuel to complete the combustion phases. The combustion time is the time after which the fuel burns with a stable flame, releases heat, and turns into combustion products, such as smoke and ash.
- The combustion chamber must be of suitable size and design according to the required flame length and diameter. Inadequate burner sizes prevent the desired

combustion temperature from being reached. Low flame temperatures cause flame ruptures as well as incomplete combustion.
- It is necessary to select the combustion units which will provide combustion with the appropriate type and capacity.

8.2.2 Types of Combustion

There is heterogeneous combustion in different phases at the moment of combustion, homogeneous combustion in the same phase. An example of heterogeneous combustion is coal combustion. The speed of coal combustion is slow, the control is difficult. Full combustion requires more air than the stoichiometric amount. An example of homogeneous combustion is the combustion of natural gas. The combustion process of natural gas is faster than the process of coal and which has full combustion.

Combustion is defined in two ways with respect to the combustion of all the oxidizable components of a fuel:

- Incomplete combustion,
- Complete combustion.

In complete combustion, the heating values of the residues released as a result of combustion are zero. In incomplete combustion, ash, flue, or exhaust gases contain unburned fuel. In the combustion process, it is not clear whether the reacting fuel molecules are fully oxidized or not. Therefore, depending on whether the oxidation is complete or not, the combustion event can be examined in four parts.

(1) Theoretical Full Combustion (TFC),
(2) Full Combustion (FC),
(3) Incomplete Combustion (IC),
(4) Partial Incomplete Combustion (PIC).

If the composition of a fuel is known, the amount of air required to fully and suitably burn that fuel is calculated by means of the combustion equations. The air is rarely dry and often contains water vapor. Water vapor also changes in the same direction depending on the relative humidity and temperature. The composition of the air is by volume: Oxygen (O_2) 20.95%, Nitrogen (N_2) 78.08%, Argon (Ar) 0.93%, and Carbon Dioxide (CO_2) 0.036%.

Considering that the volume of gases to be covered by one kilo mole under normal conditions is 22.4 Nm^3, the combustion equations of these elements can be written as follows:

$$
\begin{aligned}
&\mathrm{C} + \mathrm{O}_2 \rightarrow \mathrm{CO}_2 \\
&(2\ \mathrm{kg})\ \left(22.4\ \mathrm{Nm}^3\right) \rightarrow (2\ \mathrm{kg})\ \left(22.4\ \mathrm{Nm}^3\right) \\
&\mathrm{H}_2 + 1/2\mathrm{O}_2 \rightarrow \mathrm{H}_2\mathrm{O} \\
&(2\ \mathrm{kg})\ \left(1/2 - 22.4\ \mathrm{Nm}^3\right) \rightarrow \left(22.4\ \mathrm{Nm}^3\right)
\end{aligned}
$$

According to these equations, the weight ratios of the elements in the composition are calculated as follows:

$$L_{min} = 100/21(22.4/12 + 11.2/(H - O/8) + 22.4/32S) \qquad (\mathrm{Nm^3/kg\ wet})$$

Since gaseous fuels are measured by volume, the amount of air required for the theoretical complete and ideal combustion of 1 Nm^3 instead of 1 kg is calculated as follows considering the volumetric composition.

$$L_{min} = 100/21(1/2CO + 1/2H_2 + 2CH_4) \qquad (\mathrm{Nm^3/\ Nm^3 wet})$$

Combustion is carried out by mixing the fuel with the oxygen required for combustion in the air. This process determines the properties of combustion, whether the amount of air is less than necessary, more than necessary, or as much as necessary. In other words, it determines the amount of heat released and the type and amount of other combustion products. Therefore, the combustion process is defined according to the amount of air in the fuel-air mixture as follows:

(a) Combustion with incomplete air,
(b) Combustion with excess air,
(c) Full combustion.

8.2.2.1 Theoretical Full Combustion

In the case of Theoretical Full Combustion (TFC), the condition EAC (λ) < 1 applies. The lowest amount of oxygen is used. Reacting fuel molecules as products CO_2, H_2O, SO_2, and N_2. Therefore, it is a form of combustion without O_2 in the exhaust.

8.2.2.2 Full Combustion

In the case of full combustion (FC), the condition EAC (λ) > 1 applies. The amount of air required is greater than in the case of TFC. Exhaust O_2, CO_2, H_2O, SO_2, and N_2 are the form of combustion.

Full combustion is the complete use of the oxidizable components of the fuel in the reactions. In this case, carbon is completely converted to CO_2. In case of insufficient combustion, CO and soot are formed.

At full combustion, the flame is usually light yellow-orange in solid and liquid fuels and blue in gas fuels. Full combustion occurs if no combustible material is present in the combustion products and the remaining parts. There must be no carbon monoxide (CO) in the flue gas. The presence of CO in the flue gas will both harm the environment and increase the amount of fuel. In addition, the O_2 measurement should be investigated to determine whether the combustion is in the direction of excess air. Furthermore, the flue gas temperature plays a direct role in boiler efficiency. Flue gas temperature depends on fuel type and fuel components. In fuels containing sulfur and sulfur compounds, the flue gas temperature should not be reduced below 180 °C. Otherwise, the sulfur combines with oxygen and hydrogen in the air to allow the formation of sulfuric acid. In an ideal combustion, the flue gas should be composed of carbon dioxide and nitrogen gas. However, we usually perform combustion with a little excess air. Since there is no sulfur in the natural gas, the flue gas temperature can be reduced to 56 °C.

8.2.2.3 Incomplete Combustion

In the event of incomplete combustion (IC), the condition EAC (λ) < 1 applies. Due to incomplete oxidation, there are CO, H_2, CO_2, H_2O, SO_2, and N_2 in the exhaust.

8.2.2.4 Partial Incomplete Combustion

In case of partial incomplete combustion (PIC), IC products such as CO and H_2 are seen alongside O_2, although EAC (λ) > 1 condition is valid due to the effects of mixture, temperature change, and residence time insufficiencies. This is not only due to the unevenness of the local fuel/air ratio. At high temperatures, CO_2 and H_2O molecules cause IC products such as CO, H_2, and thermal decomposition. Rapidly bringing these molecules to low temperature does not leave sufficient time for recombination reactions. In other words, freezing occurs at reactions at a certain temperature. Therefore, IC products are seen in exhaust gases.

8.2.2.5 Combustion with Less Air

In combustion with little air, due to insufficient O_2, C molecules form CO instead of CO_2. C particles become soot and dry, reducing heat transfer. This reduces both environmental pollution and boiler efficiency. Over-treatment of smoke pipes and soiling with soot increases the backpressure and prevents the formation of flame.

8.2.2.6 Combustion with Excess Air

Excess air reduces the combustion temperature of the flame and hence the furnace temperature. This reduces the boiler capacity. More fuel is needed to achieve the same capacity. As the flame cools, the hot C molecules turn into soot and ash. Excess air causes an increase in O_2 in the burned gases and therefore a reduction in the volumetric percentage of CO_2.

In this kind of combustion, the flame is light-colored and too bright. In this type of combustion made with excess air, flue gases coming out of the stack will increase in heat, and therefore, the stack loss will increase. In other words, boiler efficiency

Table 8.12 Excess air and emissions required for combustion (carbon dioxide percentages)

Fuel type	Typical excess air quantity (%)	Approximate CO_2 amount in dry flue gas (%)
Coal (grid)	30–50	12–14
Coal (pulverized)	15–35	14.5–16
Natural gas	7–25	10–11.5
Liquid fuel	10–30	12.5–14.5

decreases. In addition, as the flue gas temperatures decrease in both the furnace and other parts of the boiler, the heat transfer balance of the boiler is reduced and the boiler capacity decreases.

Depending on the fuel type, the percentages of carbon dioxide in the excess air and emissions required for combustion are variable (Table 8.12).

8.2.3 Incomplete Combustion Losses

Incomplete oxidation components in IC and PIC cases are CO, H_2, soot, coal or fuel particles, unburned hydrocarbons, and aldehydes. These losses depend on fuel type and burner design. CO, H_2, and soot are generally found in all types of fuels.

8.2.3.1 Liquid Fuel Losses

Lack of air in combustion increases the formation of soot in liquid fuels. The lack of oxygen around the vaporized fuel particle causes the formation of soot molecules of 40–50 carbon and 4–5 hydrogen atoms. These molecules combine and break off soot particles after reaching a certain size and are mixed into the exhaust in the form of soot.

Adding water to the fuel reduces the heat slightly. Also, it is possible to reduce the amount of soot with some surface catalyst such as Cr, Ni, Cu, or some metallic additives to be added to the fuel. For example, 0.1–0.6% by weight of barium/zinc-2 ethyl hexanoate is introduced into the fuel. The mass barium/zinc ratio is 10:1.

The fuel/air ratio cannot be adjusted well when starting the combustion engine or when switching to a new operating condition. This is because the operating principles of the fuel and air equipment cause different inertia. For example, in a continuous burner, the fan is later compatible than the fuel pump. In natural suction systems, the load is controlled by fuel flow. In this case, a rich mixture is formed before the air flow increases. Soot does not occur when gasoline engines work with a rich mixture. H_2 and CO are formed according to the water gas balance in these conditions.

In the case of IC and PIC in liquid fuels, H_2 and CO are lost in relation to the EAC studied. No H_2 is seen in the excess air, and CO is often reduced. In this case, CO is not important in terms of energy but air pollution.

8.2.4 *Calculation of Combustion*

Combustion is the reaction of flammable components in the composition of the fuel with an oxidator (an oxidizer). Combustible components in solid and tight fuels are carbon (C), hydrogen (H_2), and sulfur (S). Air is generally used as an oxidator. However, when high flame temperatures are required, pure oxygen is used.

What must be understood before full combustion is that the carbon in the fuel is completely burned at the CO_2 level. In the real sense, there is no complete combustion, and some of the carbon is present in the flue gas as carbon particles called soot. Furthermore, in insufficient combustion, the flue gas contains CO.

8.2.4.1 Oxygen and Air Amount

The amount of oxygen and air required for combustion is evaluated separately for solid and liquid fuels and for gas fuels. The following reactions occur when a solid or liquid fuel is fully burned:

$$C + O_2 \rightarrow CO_2 \qquad H_2 + \frac{1}{2}O_2 \rightarrow H_2O \qquad S + O_2 \rightarrow SO_2$$

If Y is used as the common symbol for C, H_2, and S, according to stoichiometry, the following equations are obtained:

$$\frac{m_{O_2}}{|v_{O_2}|M_{O_2}} = \frac{m_Y}{|v_Y|M_Y} \text{ or } \frac{m_{O_2}}{m_Y} = \frac{|v_{O_2}|}{v_Y}\frac{M_{O_2}}{M_Y} \tag{8.4}$$

where v_{O2} and v_Y are the stoichiometric coefficients, M_{O2} and M_Y are the mole masses. The minimum amount of oxygen is the amount of oxygen required to burn 1 kg of fuel and is determined as follows:

$$Q_{min}^{Y} = \frac{|v_{O_2}|}{v_Y}\frac{M_{O_2}}{M_Y} \tag{8.5}$$

In this case, the minimum amount of oxygen that required for the combustion of carbon (C), hydrogen (H_2), and sulfur (S) is calculated as follows:

$$Q_{min}^{C} = \frac{1}{1} \times \frac{32}{12.01} = 2.664 \quad kgO_2/kg\ C$$

$$Q_{min}^{H_2} = \frac{0.5}{1} \times \frac{32}{2.016} = 7.937 \quad kgO_2/kg\ H_2$$

$$Q_{min}^{S} = \frac{1}{1} \times \frac{32}{32.06} = 0.998 \quad kgO_2/kg\ S$$

If the concentrations in fuel as a percentage of mass are c, h, s, and o, the minimum amount of oxygen required to burn 1 kg of fuel is determined by one of the following equations.

$$Q_{min} = Q_{min}^{C} \times c + Q_{min}^{H_2} \times h + Q_{min}^{S} \times s - o \quad (8.6)$$

$$Q_{min} = 2.664c + 7.937h + 0.998s - o$$

If the mass percentage of oxygen in the composition of air is considered as 0.232, the minimum amount of air is defined as follows:

$$L_{min} = \frac{Q_{min}}{0.232} \Rightarrow 4.31Q_{min} \quad (8.7)$$

If the minimum air quantity Q_{min} in Eq. (8.7) is replaced by the value defined in Eq. (8.6), the following equation is obtained:

$$L_{min} = 4.31(2.664c + 7.937h + 0.998s - o) \quad (8.8)$$

L_{min} is the minimum amount of air required for full combustion of 1 kg of fuel. The amount of air delivered during combustion is usually greater than the minimum amount of air. For this case, excess air coefficient (EAC, λ) is defined in Eq. (8.9). In this case, for $\lambda > 1$, there is excess air in the combustion and for $\lambda < 1$ there is low air.

$$\lambda = \frac{L}{L_{min}} \quad (8.9)$$

8.2.4.2 Amount of Combustion Products

Knowing the amount of combustion products is of great importance both in the calculation of the heat transfer surfaces of the boilers and in the selection of flue and flue aspirators. The amount of flue gases generated as a result of combustion of 1 kg of fuel can be calculated in the following figure with the aid of combustion equations. In case the minimum air required for full combustion is used:

$$VR_{min} = 8.89C - 21.1(H - O/8) + 3.34S - 1.244(W - 9H) \ \left(Nm^3/kg\right) \quad (8.10)$$

This equation applies to solid and liquid fuels. Equations according to their composition should be used for gas fuels. However, the following experimental equations can be used depending on the heating values for practical quick calculations.

$$\text{For solid fuels: } VR_{min} = 0.227 \times LHV/1000 + 1.375 \ (Nm^3/kg) \tag{8.11}$$

$$\text{For liquid fuels: } VR_{min} = 0.368 \times LHV/1000 + 3.765 \ (Nm^3/kg) \tag{8.12}$$

8.2.5 *Flame*

During combustion, chemical reactions actually occur in a very thin layer. This layer is also called flame. Considering the things going on in the flame, the difficulty of combustion will be better understood. A flame, on the one hand, there is fuel in the gas phase as flammable; on the other hand, there is oxygen in the environment, while at the same time, there are gases produced by the reaction. Both carbon dioxide, nitrogen, and other gases do not participate in combustion, making it difficult for the fuel to meet oxygen. Heat and mass transfer occur simultaneously with the movement of gases in the combustion event or in the flame, even if it is in a continuous regime. In addition, the concentration of gases takes different values at each point. In order to solve this complex structure, heat transfer, mass transfer, and momentum transfer take place simultaneously. Therefore, when examining the combustion event, the basic definitions of gas mixtures and fluid mechanics, heat transfer, and mass transfer should be well known. It should be known how the elements in the fuel composition would interact with each other and with oxygen chemically. Basic chemical knowledge is also required, as combustion is ultimately a chemical reaction and will be necessary in processes such as the equalization of substances entering and leaving the combustion process in the combustion equations.

Flame is a particular combustion event with the formation of light. Homogeneous gas reactions take place in the region where the flame is dominant. Combustion of hydrocarbons is more complex and slower. There are three main regions in hydrocarbon flames. The first zone is the innermost heating zone. The second zone includes hydrocarbon combustion and the third outermost zone is the combustion of CO and other intermediates. In a flaming environment, material conversion and heat transfer occur together.

The increase in the amount of carbon in the fuel causes the color of the flame to turn red, and the increase in the amount of hydrogen results in the increase in the amount of water, which is resulting from combustion.

If the fuel and oxygen molecules coming to the flame zone are mixed beforehand, if the mixture is mixed with the premixed flame or diffusion in the flame zone, the diffusion flame occurs. Bunsen burner, the engine flame of cumene gasoline, and kitchen stoves are examples of pre-mixed flame. Gas lighter, diesel engine, flames in liquid or solid fuel combustion are examples of diffusion flame.

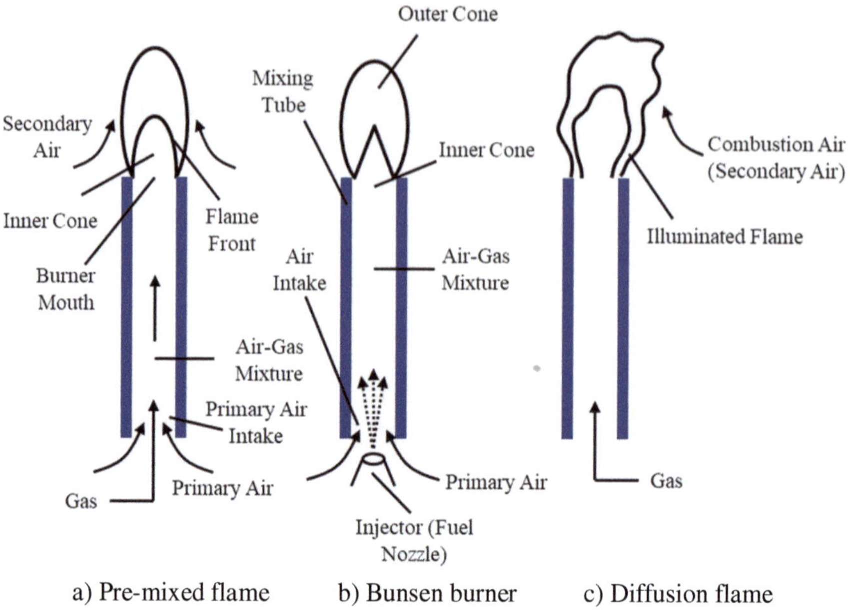

Fig. 8.16 Pre-mixed flame

8.2.5.1 Pre-mixed Flame

Pre-mixed flames are also called pre-ventilated flames (Fig. 8.16). The gas burners used in households generally use a Bunsen burner-type system (Fig. 8.16b). Pre-mix is prepared according to the value of EAC = 0.5. This is called a primary air mixture. Mixing with the secondary air occurs at the burner outlet. As shown in Fig. 8.16a), the primary mixture is the inner flame cone and the secondary mixture is the outer flame cone. The gas under the first cone is a rich mixture and has not yet burned. Its color is green-blue. The outer cone is blue-violet. Here, the fuel injecting pressure from the gas fuel nozzle enters the primary air from the sides into the burner.

8.2.5.2 Diffusion Flame

This type of flame emerges when the fuel from a pipe or duct is ignited while mixing into the air. The flame front occurs between fuel and air. Diffusion flame (Fig. 8.16c), which is not preferred in the use of natural gas, is generally preferred in the use of LPG.

Diffusion flame is abundant. If the gas pressure is low, the flame is amorphous and light. As gas pressure increases, air mixture increases, flame length decreases, and becomes stable. In low flame-fast gases such as natural gas, the flame detaches from the gas jet as the gas pressure increases. For this purpose, the pilot flame is tried to stabilize the main flame.

References

1. Kaya D, Ozturk HH (2012) Air quality management. Umuttepe Publications. Hava Kalitesi Yönetimi, Umuttepe Yayınları
2. https://ourworldindata.org/fossil-fuels
3. Kaya D, Öztürk HH (2014) Sanayide Enerji Yönetimi ve Enerji Verimliliği, Umuttepe Yayınları, Yayın No: 114, Mühendislik Dizisi: 12, Kocaeli/Türkiye
4. Özsarfati R (2020) Combustion basics and burners, Denko heat control technique and service Inc., İstanbul/Turkey. Accessed on 03rd January 2020. In: Özsarfati R (ed), Turkish, Yanma Temelleri ve Brülörler, Denko Isı Kontrol Tekniği ve Servis A.Ş., İstanbul/Turkiye. http://www.pw.com.tr/ss/upload/upload2259.pdf. Accessed 03 Jan 2020
5. McAllister S, Chen J, Fernandez-Pello AC (2011) Fundamentals of combustion processes, 1st (Edn), pp 18–20. Springer, USA
6. Sara McAllister, Jyh-Yuan Chen A (2011) Carlos Fernandez-Pello. Fundamentals of Combustion Processes. Mechanical Engineering Series. Hardcover ISBN 978-1-4419-7942-1. Softcover ISBN 978-1-4614-2865-7. Series ISSN 0941-5122. E-Book ISBN 978-1-4419-7943-8. https://doi.org/10.1007/978-1-4419-7943-8. Springer New York Dordrecht Heidelberg London. Springer
7. http://kisi.deu.edu.tr//serhan.kucuka/YANMA_ve_ALEV.pdf

Chapter 9
Energy Efficiency in Boilers

9.1 Boiler Selection

Boilers are generally defined as closed vessels, which operate under pressure and transmit chemical energy in the form of heat energy to the working fluid (Fig. 9.1). Steam boiler is defined as devices that produce steam at desired pressure, temperature and flow rate.

Boilers are produced in different types according to need; they are very expensive energy generators in terms of initial investment and operating costs. For this reason, the boiler must be selected for its usage purpose, and the necessary care must be taken in its operation and maintenance. In the selection of the boiler, a detailed analysis should be made considering the following factors [1, 2]:

- Intended use of the boiler,
- Amount of steam to be produced,
- Pressure and temperature,
- Feed water inlet temperaturc,
- Water hardness,
- Type of fuel to be used,
- Lower heating value and analysis of fuel,
- Price of fuel.

9.2 Determination of Boiler Efficiency

Boiler thermal efficiency calculation is defined according to the standards. Generally, boiler efficiency is calculated in two ways as direct and indirect methods. In order to calculate the boiler efficiency directly, the following variables must be measured [3, 4]:

D. Kaya et al., *Energy Management and Energy Efficiency in Industry*,
Green Energy and Technology,
https://doi.org/10.1007/978-3-030-25995-2_9

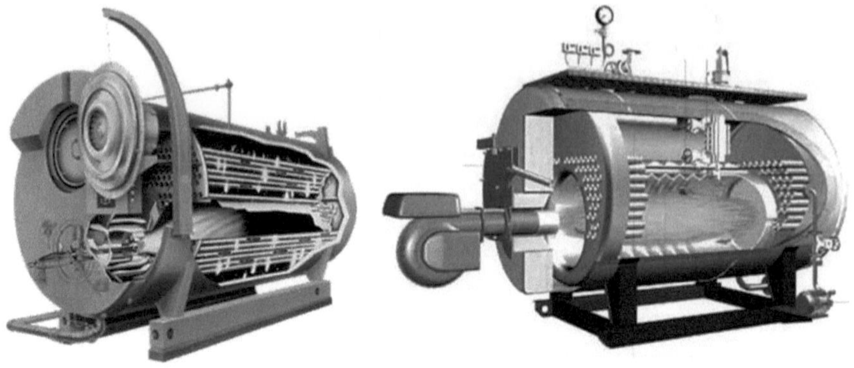

Fig. 9.1 Structure of the boiler

- Feed water and steam quantities,
- Temperature and pressures of feed water and intermediate steam,
- Fuel supply amount,
- Fuel lower heating value.

The boiler efficiency is calculated from the following equation:

$$\eta = \frac{(\dot{m}_s \times h_s) - (\dot{m}_w \times h_w)}{\dot{m}_f \times Hu} \tag{9.1}$$

where

$\dot{m}_s$ = Mass flow rate of steam (kg/h),
$\dot{m}_w$ = Mass flow rate of water (kg/h),
h_s = Steam enthalpy (kcal/kg),
h_w = Enthalpy of feed water (kcal/kg),
$\dot{m}_f$ = The mass flow rate of the fuel (Fuel Consumption) (kg/h),
Hu = LHV = The lower heating value of the fuel (kcal/kg).

Boiler efficiency is calculated by indirect method as follows:

$$\eta = 1 - \sum Z \tag{9.2}$$

where Z (in %) is the various thermal losses.

In order to determine the boiler efficiency by indirect method:

- Flue gas analysis (temperature, velocity, pressure dust measurement in flue gas, gas analysis) is performed.

- The combustion air flow and temperature are measured.
- Boiler outer wall temperatures are measured.
- The amount of blowdown is determined.
- The lower heating value and chemical structure of the fuel are determined.

According to the results of this measurement, specific air quantity and specific flue gas quantities are determined first. Then, the following variables are determined, and the thermal efficiency is calculated accordingly:

- Theoretical specific air volume,
- Theoretical specific amount of smoke,
- Excess Air Coefficient (EAC),
- Actual specific air and smoke content,
- Flue gas heat loss rate,
- Missing combustion loss (soot loss is to be found by benefiting from dust emission) rate,
- Unburned fuel loss rate,
- Blowdown loss rate.

9.3 Factors Affecting Boiler Efficiency

The main factors affecting efficiency in boilers are [5, 6]:

- Incomplete combustion,
- Excess air,
- Heat loss due to water vapor in flue gas,
- Flue gas temperature,
- Fuel type,
- Burners,
- Boiler load,
- Heat losses from the boiler surface,
- Heater surface pollution.

9.3.1 Incomplete Combustion

Incomplete combustion occurs when the combustible substances in the solid and liquid fuels do not burn, remain in the ash, or are discarded as unburned hydrocarbons and carbon monoxide in the flue gas. Whether there is a good combustion in the boilers and sufficient boiler efficiency is achieved or not, it is determined from the flue gas temperature and the O_2, CO_2, and CO values in the flue gases thrown into the stack by leaving the boiler.

In incomplete combustion, it will be necessary to adjust the air/fuel ratio to ensure full combustion in order to achieve full combustion as it causes fuel loss. Therefore,

Table 9.1 Excess air coefficients according to fuel type and combustion system

Fuel type	Combustion system	Excess air coefficients (EAC)
Lump coal	Manually on plane grate	1.6–2
Lump coal	Stoker (ejected feedforward thrust)	1.2–1.6
Lump coal	Lower feed stoker (helical screw)	1.2–1.5
Pulverized coal	Pulverized combustion	1.15–1.2
Natural gas	With forced blown burner	1.07–1.2
Fuel oil	With forced blown burner	1.05–1.10

it is necessary to keep the amount of O_2 in the flue gas at an optimum level. In order for a good combustion to occur in the combustion chamber of a boiler, the following conditions must be met:

- Sufficient amount of fuel and air must be fed into the furnace with sufficient mixing or mixed with the turbulence created in the furnace. (Table 9.1 and Fig. 9.2).
- The ignition temperature of the fuel-air mixture must be at a very high temperature and must burn in the gas phase.
- Sufficient furnace volume (combustion chamber (cell)) must be established to complete combustion.

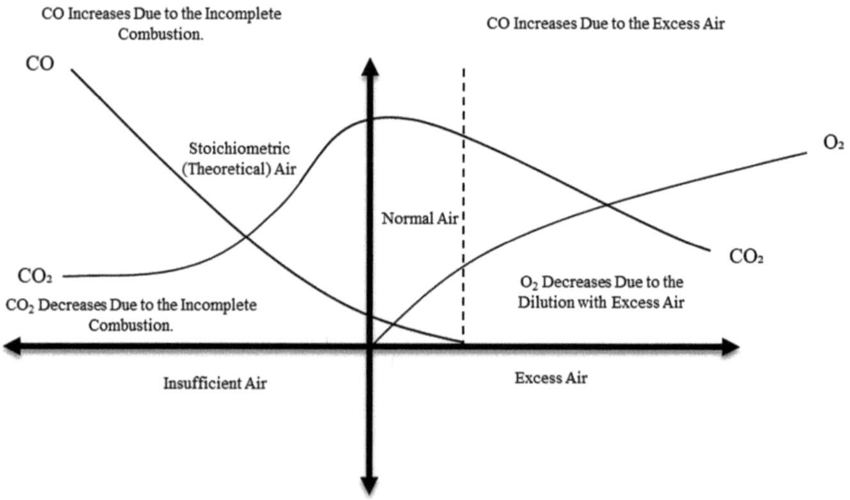

Fig. 9.2 O_2, CO_2, and CO relationship in flue gases leaving the boiler

- In order to gain the required amount of heat energy from the flame and hot smoke gases resulting from combustion, there must be sufficient surface area for heat transfer by thermal radiation and convection.

9.3.2 Air/Fuel Ratio

The combustion system in the boilers should work by providing a minimum air-fuel ratio, without causing combustion problems. Excess air volume increases the amount of flue gas if more than necessary. It causes an increase in flue gas amount and gas flow. Excess air entering the boiler causes the temperature inside the boiler to drop. Depending on fuel type and combustion system, EACs are given in Table 9.1. For these reasons, the excess amount of air should be kept as low as possible. To ensure this, O_2 level in the flue gas should be checked, the oxygen amount should be brought to the lowest possible level by adjusting the air. To this end, the following conditions must be met [7, 8]:

- The air supply to the boiler room must be sufficient and at constant pressure,
- In the case of using liquid fuel, the viscosity of the fuel should be constant and excessive temperature fluctuations should be minimized,
- When using gas fuel, the gas pressure entering the burner must be constant.

9.3.3 Heat Losses from the Flue Gasses

Fuels: Due to their chemical composition, they have moisture in their structure. Moisture in the fuel reveals itself and evaporates during combustion. Moisture, which is released as water vapor, causes some loss of the beneficial energy in the boiler. In terms of energy saving, the free moisture in the fuel must be reduced as much as possible before combustion.

9.3.4 Flue Gas Temperature

One of the important factors affecting boiler efficiency is the flue gas temperature. If the flue gas temperature is above the acceptable values, more energy will be exhausted from the flue into the atmosphere [9]. In this case, the boiler efficiency decreases. There are two main reasons why the energy discharged from the stack is high.

1. Insufficiency of heat transfer surfaces: In this case, air preheater or superheaters are placed in the stack to benefit from the heat of the flue gas.
2. Contamination on heat transfer surfaces: In this case, boiler pipes should be cleaned periodically, and the hardness of the feed water should be checked supplied to the boiler.

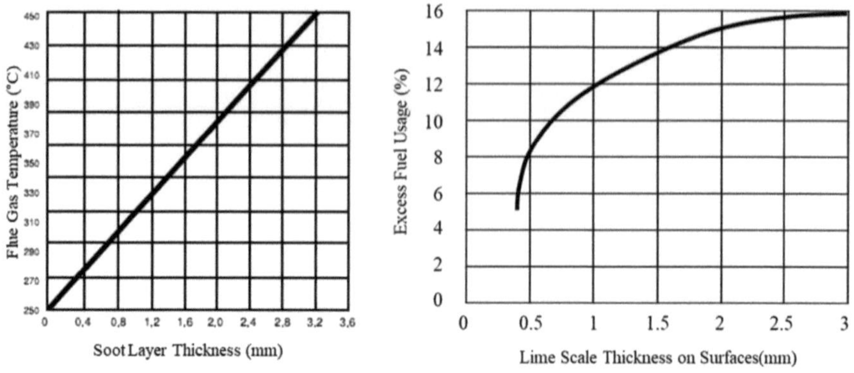

a) Soot layer thickness

b) Lime scale thickness on surfaces

Fig. 9.3 The effects of the soot layer and lime scale thickness on the boiler surface to the flue gas temperature

Every 17 °C increase in the flue gas above the normal temperature causes a decrease in efficiency of approximately 1%. One of the reasons why the flue gas temperature rises above normal values is the accumulation of ash and soot on the heating surfaces.

The effect of soot layer and lime scale thickness on the boiler surfaces on the flue gas temperature is given in Fig. 9.3a, b. In boilers burning fuel oil and coal, even as a good combustion setting, fly ash and soot may accumulate on the heating surfaces. Therefore, cleaning should be done at intervals of 1–2 weeks. Especially in water pipe boilers, manual or automatic soot cleaners are preferred.

Other factors affecting flue gas temperature are:

- If the combustion air is more or deficient than usual,
- Inadequate heat transfer to the water due to the pollution (soot) of gas passageways,
- High flame length in the burner,
- Combustion that may occur in gas passageways (due to soot).

The following precautions should be taken to prevent these problems:

- Combustion must be carried out with sufficient air volume.
- Gas passageways (inside pipe, outside pipe in water pipe boilers, soot holders) should be cleaned periodically according to fuel condition.

In practice, for modern boilers, the flue gas temperatures provided at appropriate O_2 and CO_2 ratios when the boiler is clean are considered to be the most appropriate value. Corrosion problems are encountered in the stack if the flue gas temperature falls below the limit set as the acid condensation temperature. In addition, as a result of excessive lowering of the flue gas temperature, significant decreases occur in the flue draft. These conditions should also be taken into account while reducing the flue gas temperature [10].

9.3.5 Heat Losses from the Stack

In hot boilers, heat losses caused by air circulation are calculated by stack draft as follows. Stack draft effect (ΔP), increases in proportion to the difference between the stack height and the boiler temperature and the outside air temperature.

$$\Delta P = H \times g \times (\rho_2 - \rho_1) \quad [Pa] \tag{9.3}$$

$$\Delta P = H(\rho_2 - \rho_1) \quad [mmSS] \tag{9.4}$$

where

H = Stack height (m),
ρ_1 = Density of air at the boiler temperature (kg/m^3), and
ρ_2 = Density of the air at the ambient air temperature (kg/m^3).

The velocity of hot air in the stack increases with the stack draft.

$$V = \sqrt{2g \times \Delta P/\rho_1} \quad (m/s) \tag{9.5}$$

While stack cross-section (F) increases in directly proportional to the boiler capacity (Q_{bc}) and a stack coefficient (n) which depends on the fuel type, it decreases in inversely proportional to the square root of the chimney height (H).

$$F = n \times \frac{Q_{bc}}{\sqrt{H}} \quad [cm^2] \tag{9.6}$$

Here, the coefficient of n is 0.010–0.012 for natural gas, 0.020 for fuel oil, and 0.030 for solid fuels.

The hot air flow rate ($\dot{V}$) increases in direct proportion to the stack cross section and air velocity.

$$\dot{V} = F \times V \times 3600 \quad \left(m^3/h\right) \tag{9.7}$$

The amount of heat carried by the hot air (Q) increases in direct proportion to the air flow rate, the density of the air, specific heat of the air, and the temperature difference between the boiler and the ambient air temperatures.

$$Q = \dot{V} \times \rho_1 \times C_p \times (T_1 - T_2) \quad (kcal/h) \tag{9.8}$$

where

T_1 = Boiler temperature (°C),
T_2 = Ambient air temperature (°C),
C_p = Specific heat of the air (kcal/kg°C).

9.3.6 Fuel Type

Because different fuels contain different amounts of carbon and hydrogen, their thermal (calorific) values, the amount of moisture in the flue gas as a result of combustion and ash, slag, and soot quantities change.

9.3.7 Burner Type

If in the burners, the fuel pressure and temperature are not at the desired value, this causes the fuel not to be sufficiently atomized and thus incomplete combustion. This has the effect of decreasing efficiency. Today, boiler efficiencies are referred to as annual efficiencies. This value is the average efficiency value of the boilers during an operating season, in terms of total working and waiting times. The efficiency of the burners during the working process is less than the annual average due to the effect of boiler internal cooling losses during the waiting periods. The annual efficiency, the size of the burners stay in operation positively, boiler and burner quality of air leakage caused by the negative effects.

In a hot boiler, which starts to stop, uncontrolled air entering the combustion chamber and smoke pipes with the effect of stack draft, cools the boiler and is thrown out of the stack as heated. Measures to be taken in burner and boiler design are important in reducing internal cooling losses. Single-stage burners usually do not have a suction air damper and remain open directly during stop times. In two-stage and proportionally controlled burners, the existing air damper shuts down during stop times. However, in the case of direct shut-off from the main switch in a portion of the two-stage and proportional burner, the damper may remain open. Therefore, if the burner is to be switched off, it is useful to wait for the thermostat to stop the system. In addition, it is sometimes necessary to check whether the burner air dampers are fully closed or not.

In order to prevent air leaks in the boilers, the front fume covers must be sealed and fully leak-proof, and when closed, the entire cover profile must press properly into the boiler. The burner connection flange must be gasketed and uniform, and the peephole (observation port) must be closed when not in use. Explosion covers should not be bent; they should be sealed and fully closed.

In determining the effect of the internal cooling losses on the annual efficiency of the boilers, it is necessary to make some assumptions for the variables such as the duration of the burners stay, the total operating time of the year, the temperature of the boiler, the change of the outdoor air temperature, and the sealing of the boiler. Therefore, instead of giving some values at this stage regarding the internal cooling losses depending on the type of boiler and fuel, the general results, which do not change from the above-mentioned theoretical expressions, are stated below.

- Burner capacity and flue must not be larger than necessary.
- Double-stage or modulating burners should be used to increase the service life of the burners.
- In automatic control with mixer valves, a boiler water temperature of 5 °C higher should be preferred instead of a constant boiler water temperature of 80–90 °C.
- The burner inlet air damper, connection flange, front fume hoods, explosion cover, sight glass must be sealed and fully sealed.
- In the case of boilers without air inlet dampers, which cannot be sealed with single-stage burners, automatic stack closing damper installation should be considered.
- At certain boiler capacity, as the stack cross section is narrowed depending on the stack height, the hot air flow in the stack remains constant, thus the stack height has no effect on internal cooling losses.

9.3.8 Boiler Load

The low efficiency is obtained when the boilers operate under the low load or excessive load. When the load ratio falls below 50%, the efficiency curve decreases rapidly. As a result of this load drop, the percentage of heat losses from the boiler surface will increase. When operating boilers, these should be observed as much as possible, taking into account the boiler capacity. In case of overload of the boiler, combustion efficiency will decrease, and flue gas temperatures will increase. On the other hand, losses at low loads are mainly due to stop losses. When the boiler stops, it cools both from the external surfaces and from the internal surfaces due to the flue draft.

When the boiler load changes, the amount of fuel burned also changes. The highest efficiency values are generally achieved when the boiler is operating at more than 70% of its full load. Therefore, the boilers must be operated at a load as close to full load as possible.

9.3.9 Heat Losses from Boiler Surface

Heat losses from the boiler surface are caused by radiation and convection. In modern boilers, if the boiler is operating at full load this loss is generally less than 1%. However, in old type boilers, these losses are up to 10%. An insulation made to reduce the boiler surface temperature to a value above 30 °C above the ambient temperature is considered sufficient and appropriate for minimizing such losses.

9.3.10 *Heater Surface Cleaning*

The effect of limestone and soot accumulation on boiler efficiency on heating surfaces is high. The initial thermal efficiency values of boilers are valid for new boilers with clean surfaces where no dirt layer is formed. Contaminants that accumulate on both sides of the heat transfer surfaces (water and smoke sides) significantly inhibit heat transfer. In this case, the hot gases leave the boiler without passing their heat to the water. Thus, boiler efficiency decreases with increasing flue gas temperature. On the smoke side, soot accumulates, especially when fuel oil and coal are burned. This layer is difficult to clean.

It is undesirable that lime deposits form on the boiler heating surfaces (Fig. 9.3b). It is acceptable to build up deposits on surfaces up to a thickness of 1 mm. In this case, the efficiency reduction is 1%. In addition, due to the reduction of heat transfer on the heating surfaces where sediment is formed, heat build-up occurs and the metal temperature increases. This may cause splits and explosions in the boiler. The following measures can be taken to remedy these drawbacks:

- There must be a treatment system suitable for the feed water,
- Hardness of the feed water must be checked continuously,
- Boiler chemicals suitable for the feed water should be used,
- Deposits (scale, residue, etc.) must be removed from the boiler at least once a year.

9.4 Flue Gas Analysis

Flue gas analysis can be easily obtained and evaluated with the help of flue gas analyzers for the optimization of combustion during operation and immediate access to the burners and boilers. Emission limit values of solid, liquid, and gas fuel plants are given in the related regulation on the protection of air quality. The approximate values for the flue gases emitted from the boilers are given in Table 9.2 [11–13].

9.4.1 *Oxygen*

Depending on the type of fuel and the EAC, it is desirable that the oxygen (O_2) ratio in the flue gases is as low as possible so as not to cause carbon monoxide (CO) formation. Two to three percent in natural gas, 3–4% in liquid fuel, and 5–6% in solid fuel are considered ideal values for flue gas analysis.

Table 9.2 Required values for the flue gases emitted from the boiler (as an example)

Boiler type	Fuel type	Flue gas temperature (°C)	O_2 (%)	CO_2 (%)
Fire-tube boiler	Fuel oil	Saturated steam temperature 50–75	3–6	13.7–11.5
	Natural gas		1–3	11.2–10
	Coal		6–9	13–10
Hot water boiler	Fuel oil	180	3–6	13.7–11.5
	Natural gas	80–180	1–3	11.2–10
	Coal	200	6–9	13–10
Hot oil boiler	Fuel oil	Saturated steam temperature 75–100	3–6	13.7–11.5
	Natural gas		1–3	11.2–10
	Coal		6–9	13–10
Condensed hot water boiler	Natural gas	30–60	1–3	11.2–10

9.4.2 *Carbon Dioxide*

Depending on the type of fuel, it is preferable that carbon dioxide (CO_2) is present in the flue gases at a high rate. The values of 11% in natural gas, 14% in liquid fuel, and 14% in solid fuel are stated as appropriate levels for flue gas analysis. As a natural consequence of good combustion, highly desirable CO_2 in flue gases has been accepted in recent years due to the greenhouse effect caused by CO_2. The solution is made possible by the expansion of low-carbon, high-hydrogen fuels, and the limitation of fossil fuel use over time.

9.4.3 *Carbon Monoxide*

Because of the energy loss and pollutant effect, CO is not desirable in the flue gases. By increasing the O_2 given to the fuel, the incomplete combustion must be completed, and the CO must be converted to CO_2. In flue gas analysis, CO content is considered normal up to 100 ppm.

9.4.4 *Sulfur Dioxide*

It is accepted that sulfur dioxide (SO_2) produced by the combustion of sulfur (S) in the fuel, is q dangerous emission for environmental pollution. This gas can only be reduced in flue gases with low sulfur fuels. When considering SO_2 value in the use of natural gas, which is "0" in flue gas, it can be 150–200 ppm in flue gases in the usage of imported coal containing 0.5% sulfur. Sulfur dioxide is known to combine with flue gas at low temperatures with water vapor, resulting in sulfuric acid and cause damage to the boilers.

9.4.5 Nitrogen Oxides

The type of fuel, the EAC, and the amount of nitrogen oxide (NO_x) resulting from the furnace design are considered environmental pollutant emissions. Apart from the ratio allowed by the fuel/air setting, there is no possibility to interfere with nitrogen oxide it is considered as a variable that should be taken into account when choosing the boiler. Today, the flue gas recirculation system and low NO_x burners are accepted as effective methods for controlling nitrogen oxide emissions.

9.4.6 Temperature of the Flue Gas

It is desirable that the flue gases leaving the boiler are as low as possible, depending on the type of fuel and the sulfur content therein. If the fuel flow is too high, the boiler heating surfaces are insufficient and the smoke pipes are dirty, the flue gas temperature will be high. The important factor to be considered here is that the boiler test, in other words, the flue gas analysis is carried out at the fuel flow rate appropriate to the boiler rated power. Because the flue gas temperature is expected to be low at low boiler capacities. High flue gas temperature means loss of efficiency. The lowest values that can be deducted in flue gas temperatures are related to the condensation (dew) temperature of the flue gases. The condensation temperature depends on the sulfur dioxide (SO_2) in the flue gas, i.e., the amount of S in the fuel. Flue gas temperatures of 130–150 °C in natural gas use and 130–175 °C in solid and liquid fuel use can be considered as appropriate values. At high flue gas temperatures, the burner and boiler must be intervened, the flue gas temperature should be reduced by partially reducing the capacity or adding turbulators to the boiler pipes.

9.4.7 Combustion Efficiency

The combustion efficiency can be automatically calculated with the flue gas analyzer by evaluating variables such as oxygen, carbon dioxide, flue gas temperature, and ambient temperature measured in the flue gases. When the operator makes comments on the combustion efficiency, the factors that affect the result can be easily seen. While speaking about boiler efficiency, considering the combustion efficiency, it is necessary to reduce the combustion efficiency by 3–5% due to boiler radiation losses, non-combustible hydrocarbons, ash losses, etc., for non-measurable values depending on the fuel fraction. However, the use of flue gas analyzers in the optimization of the burner is considered the right approach rather than registration of boiler efficiency.

9.5 Case Study on Energy Efficiency in Boilers

In order to determine the energy-saving potential of Steam Boilers Number 1 and 2 belonging to an industrial establishment, gas, velocity, and temperature analysis are measured. Mass and energy balance values were determined using measurement data (temperature, pressure, velocity, gas analysis). Moving from mass and energy balances, the efficiency of each acquisition, the potential areas for energy savings, and the size of savings are calculated. The effects of high EACs and flue gas temperatures on the boiler efficiency of blowdown losses have been determined.

9.5.1 Steam Boiler Number 1

9.5.1.1 Results of the Measurement

According to the data of a facility selected as an example, the boiler efficiency is assumed to produce 40% of the annual average steam requirement of Boiler No. 1. Accordingly, average natural gas and steam flows are calculated using the inlet enthalpy values of steam and water and the results are given in Table 9.3. The oxygen rate of the combustion gas used for boiler operating conditions number 1 is 2.84% and the combustion gas temperature is 175 °C.

Combustion gas analyzes were carried out using the fuel quantity, boiler combustion gas oxygen percentage, and elemental analysis of the fuel, and the results are

Table 9.3 Efficiency calculation for the steam boiler number 1

Variable	Value
Steam flow (m_s, kg/h)	2760
Steam input enthalpy (h_s, kcal/kg)	657
Water intake enthalpy (h_w, kcal/kg)	103
The heat supplied to the water ($Q_s = m_s \times (h_s - h_w)$, kcal/h)	1,528,249
Enthalpy difference ($h_s - h_w$, kcal/kg)	554
Natural gas flow (m, m^3/h)	203
Lower heating value (LHV) (kcal/kg)	8250
The heat of the fuel ($Q_f = m \times LHV$, kcal/h)	1,677,060
General efficiency (Q_s/Q_f, %)	91.13

Table 9.4 Elemental and combustion analysis of fuel for steam boiler number 1

Fuel		Required O_2	Combustion products (m^3/h)					
Analysis	% (ob: original base)		CO_2	SO_2	N_2	Argon	H_2O	O_2
Methane	91.2	370.7	185.9	0.0	1381.7	16.5	403.2	0.0
Ethane	3.2	22.8	13.1	0.0	85.1	1.0	21.6	0.0
Propane	1.2	12.2	7.3	0.0	45.5	0.5	10.8	0.0
Butane	0.2	2.9	1.8	0.0	10.6	0.1	2.4	0.0
Pentane	0.3	5.0	3.1	0.0	18.5	0.2	4.2	0.0
Nitrogen	3.3	0.0	0.0	0.0	3.3	0.0	0.0	0.0
Carbon dioxide	0.6	0.0	0.6	0.0	0.0	0.0	0.0	0.0
Total	100.0	413.6	211.2	0.0	1544.8	18.4	442.2	0.0
Excess air			0.1	0.0	272.2	3.2	6.4	73.0
Gas compound % (ob)* (*ob: original base)			8.2	0.0	70.7	0.8	17.4	2.8

Table 9.5 Boiler flow and excess air ratio for steam boiler number 1

Flow rate	Value
Total stoichiometric combustion gas (m^3/h) (ob)	2217
Theoretical combustion air (m^3/h) (ob)	2010
Excess air (m^3/h)	355
Theoretical total combustion air (m^3/h)	2365
Flue gas flow rate (m^3/h) (ob)	2572
Excess air ratio (%)	17.66

given in Table 9.4. Using the components in Table 9.4, the total stoichiometric combustion gas, theoretical combustion air, excess air, theoretical total combustion air, and boiler exhaust gas emissions and excess air ratio were calculated (Table 9.5). In addition, flow rates are measured at the exit of the stack, and the accuracy of the calculation is ensured. Depending on the measured and calculated values, the energy values for Steam Boiler Number 1 are given in Table 9.6.

9.5.1.2 Potential Energy Saving Areas

Reduction of excess air: It has been determined that the boilers are operated above the EAC (10% for natural gas). In this case, a significant amount of heated air is released into the atmosphere. The boiler should be operated with the appropriate excess air ratio and the flue gas analysis should be carried out at regular intervals. The energy loss caused by excess air is given in Table 9.7.

Table 9.6 Mass and energy values established for steam boiler number 1

Inputs	Flow rate (m^3/h)	Oxygen (%)	Temperature (°C)	Lower heating value (kcal/kg)	C_p (kcal/m^3K)	Q (kcal/h)	Ratio (%)
Natural gas	203	–	10	8250	0.44	1,677,060 (wet) 894 (dry)	84.9 0.01
Air	2.365	20.57	18	–	0.32	13,625	0.7
Water (kg/h)	2.760	–	103	–	1.000	284,280	14.4
Total	–	–	–	–	–	1,975,859	100
Output	Flow rate (m^3/h)	Oxygen (%)	Temperature (°C)	Enthalpy (kcal/kg)	C_p (kcal/m^3K)	Q (kcal/h)	Ratio (%)
Flue gas	2572	2.84	175	–	0.32	144,007	7.3
Steam (kg/h)	2760	–	–	656.7	–	1,812,529	91.7
Losses	–	–	–	–	–	19,323	1.0
Total	–	–	–	–	–	1,975,859	100

Table 9.7 Saving amount of energy by reducing excess air

Variable	Value
Excess air flow rate (m^3/h)	355
Excess air ratio (%)	18
Target excess air ratio (%)	10
Specific heat (kcal/m^3 K)	0.32
Decrease in air (m^3/h)	154
Air inlet temperature (°C)	18
The flue gas temperature (°C)	175
Energy saving (kcal/h)	7734
Annual operating time (hour)	7000
Annual energy saving (kcal)	54,140,039
Financial savings (USD/year)	1642

Reduction of boiler surface losses: In order to determine the surface losses, the entire boiler surface was scanned with a thermal camera and photographs of weakly insulated areas were taken. If the weak insulated areas are identified and strengthened, it will be possible to save energy in these certain areas. Furthermore, the total surface loss was determined from the boiler energy balance. If 25% of this surface loss amount is eliminated, the saving to be provided is given in Table 9.8.

Table 9.8 Savings by reducing surface losses

Variable	Value
Boiler heat losses (kcal/h)	19,323
Target reduction (%)	25
Energy saving (kcal/h)	4831
Annual operating time (hour)	7000
Annual energy saving (kcal)	33,815,582.04
Financial savings (USD/year)	1025

9.5.2 *Steam Boiler Number 2*

9.5.2.1 Results of the Measurement

Assuming that the boiler used in the test produced 60% of the steam. The boiler efficiency is calculated by taking the average enthalpy of steam and water from the average natural gas and steam boiler and the results are given in Table 9.9.

The operating conditions for the Steam Boiler Number 2 were measured as follows: the oxygen ratio of the boiler outlet gas was 4.81 and the combustion gas temperature was 185 °C. The amount of fuel, the gas percentage of oxygen and combustion gas analyzes were carried out according to the elemental analysis of the fuel (Table 9.10).

Using the values given in Table 9.10, the total stoichiometric combustion gas, theoretical combustion air, excess air, theoretical total combustion air and boiler exhaust gas composition and excess air ratio values were calculated (Table 9.11). In addition, flows measured at the exit of the stack, and the accuracy of the calculated is ensured. Depending on the measured and calculated values boiler's input and output values for the Steam Boiler Number 2 are given in Table 9.12.

Table 9.9 Efficiency calculation for the steam boiler number 2

Variable	Value
Steam flow (m_s, kg/h)	4080
Steam input enthalpy (h_s, kcal/kg)	657
Water intake enthalpy (h_w, kcal/kg)	103
Enthalpy difference ($h_s – h_w$, kcal/kg)	554
The heat supplied to the water ($Q_s = m_s \times (h_s – h_w)$, kcal/h)	2,259,151
Natural gas flow (m, m^3/h)	304
Lower heating value (kcal/kg)	8250
The heat of the fuel ($Q_f = m \times LHV$, kcal/h)	2,504,700
General efficiency (Q_s/Q_f, %)	90.20

Table 9.10 Elemental and combustion analysis of fuel for steam boiler number 2

Fuel		Required O_2	Combustion products (m^3/h)					
Analysis	% (ob: original base)		CO_2	SO_2	N_2	Argon	H_2O	O_2
Methane	91.2	553.7	277.6	0.0	2063.6	24.6	602.2	0.0
Ethane	3.2	34.1	19.5	0.0	127.1	1.5	32.2	0.0
Propane	1.2	18.2	11.0	0.0	67.9	0.8	16.2	0.0
Butane	0.2	4.3	2.6	0.0	15.9	0.2	3.7	0.0
Pentane	0.3	7.4	4.7	0.0	27.7	0.3	6.2	0.0
Nitrogen	3.3	0.0	0.0	0.0	3.3	0.0	0.0	0.0
Carbon dioxide	0.6	0.0	0.6	0.0	0.0	0.0	0.0	0.0
Total	100.0	617.7	315.4	0.0	2305.6	27.4	660.4	0.0
Excess air			0.3	0.0	774.2	9.2	18.2	207.7
Gas composition % (ob)			7.3	0.0	71.3	0.8	15.7	4.8

Table 9.11 Flow rate values for steam boiler number 2

Flow	Value
Total stoichiometric combustion gas (m^3/h) (ob: original base)	3309
Theoretical combustion air (m^3/h) (ob)	3003
Excess air (m^3/h)	1010
Theoretical total combustion air (m^3/h)	4012
Flue gas flow rate (m^3/h) (ob)	4318
Excess air ratio (%)	33.63

9.5.2.2 Potential Energy Saving Areas

Excess air reduction: It has been determined that, as just as the Steam Boiler Number 1, the Steam Boiler Number 2 is operated optimum above the EAC (10% for natural gas). In this case, a significant amount of heated air is released to the atmosphere. The boiler should be operated with the appropriate excess air ratio and the flue gas analysis should be carried out at regular intervals. The energy loss caused by excess air is given in Table 9.13.

Reduction of boiler surface losses: In order to determine the surface losses, the entire boiler surface was scanned with a thermal camera and photographs of weakly insulated areas were taken. It has been determined that Steam Boiler Number 2's heat loss is more than Steam Boiler Number 1's. Total surface loss removed from boiler energy balance and if half of this loss amount is eliminated, the saving to be provided is given in Table 9.14.

Table 9.12 Mass and energy values established for steam boiler number 2

Inputs	Flow rate (m^3/h)	Oxygen (%)	Temperature (°C)	Calorific value (kcal/kg)	C_p (kcal/m^3K)	Q (kcal/h)	Ratio (%)
Natural gas	30	–	10	820	0.44	2,504,700(wet) 1336 (dry)	83.9 0.01
Air	4012	20.57	45	–	0.32	57,777	1.9
Water (kg/h)	4080	–	103	–	1.00	420,240	14.1
Total	–	–	–	–	–	2,984,053	100
Output	Flow (m^3/h)	Oxygen (%)	Temperature (°C)	Enthalpy (kcal/kg)	C_p (kcal/m^3K)	Q (kcal/h)	Ratio (%)
Flue gas	4318	4.81	185		0.32	255,655	8.6
Steam (kg/h)	4080	–	–	656.7	–	2,679,391	89.8
Losses	–	–	–	–	–	49,007	1.6
Total	–	–	–	–	–	2,984,053	100

Table 9.13 Saving amount of energy by reducing excess air

Variable	Value
Excess air flow (m^3/h)	1010
Excess air ratio (%)	34
Target excess air ratio (%)	10
Specific heat (C_p, kcal/m^3 K)	0.32
Decrease in air (m^3/h)	709
Air inlet temperature (°C)	45
The flue gas temperature (°C)	185
Energy saving (kcal/h)	31,782
Annual operating time (hour)	7000
Annual energy saving (kcal)	222,471,949
Financial savings (USD/year)	6742

Table 9.14 Saving amount to be provided by reducing boiler surface losses

Variable	Value
Boiler heat losses (kcal/h)	49,007
Target reduction (%)	50
Energy saving (kcal/h)	24,504
Annual operating time (hour)	7000
Annual energy saving (kcal)	171,525,170.8
Financial savings (USD/year)	5198

9.5.3 *Heat Energy Saving*

Heat energy savings will be examined in three sections:

1. Steam production (boiler room, steam boilers, boiler equipment),
2. Steam distribution installation (pipes, control valves, steam traps),
3. Condensate collection system (pipes and tanks).

9.5.3.1 Steam Production

Combustion is the first step of obtaining energy in the devices used in steam production. Optimum combustion provides the following benefits in addition to reducing energy costs:

- Increases the total efficiency,
- Reduces the environmental pollution,
- Increases the lifetime of the devices.

Therefore, the operation of steam boilers with safety and energy saving should be of particular importance for the operation. For this purpose, the following precautions have been taken in order to ensure efficient combustion and efficient use of smoke gases in boilers:

- Automatic adjustment of the fuel-air mixture is required in the burner (proportional control). Proportional control has been made in the existing system.
- Both flue gas energy must be used and the inlet temperature of the boiler feed water to the boiler should be increased by using the economizer. In the present system, economizer is used in boilers.
- The supply air temperature must be increased to the permissible temperature for the burner. In the present system, the air coming from the compressors is used for Boiler No. 2 and the boiler inlet temperature is measured as 33 °C. This value is not sufficient in terms of fuel economy. A "Recuperator" must be used for heating the combustion air.

Use of the recuperator for heating of the combustion air (Fig. 9.4): Boiler efficiency can be increased by heating combustion air with flue gas. Since the amount of energy discharged from the stack will be reduced, heating the combustion air with the flue gases will increase the boiler efficiency. By heating the combustion air to 56 °C, the efficiency can be increased by 2%.

If the combustion air is heated by another inert source, fuel will be saved because of the combustion quality, although there is no change in boiler efficiency.

The amount of savings by using the recuperator: Let us say that the flue gas with a temperature of 185 °C enters the recuperator with a flow rate of 6890 m^3/h and leaves at 110 °C. With this energy, boiler efficiency can be increased by heating the combustion air, which is at 25 °C temperature.

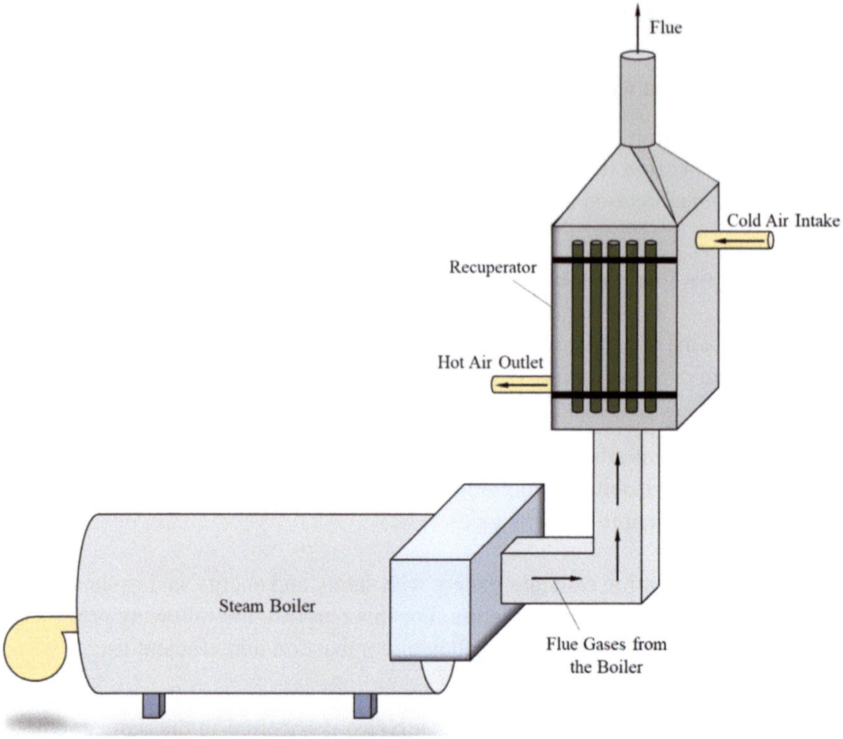

Fig. 9.4 Use of recuperator for combustion air heating in steam boiler

For flue gas with a temperature of 185 °C; $C_p = 0.33$ kcal/m^3 °C
For flue gas with a temperature of 110 °C; $C_p = 0.325$ kcal/m^3°C
For air; C_p= 0.314 kcal/m^3 °C
Recuperator efficiency = 80%
$Q = m \times (C_{p185°C} \times T_i - C_{p110\,°C} \times T_e)$
= 6890 m^3/h (0.33 kcal/m^3 °C×185 °C−0.325 kcal/m^3 °C×110 °C)
= 174,317 kcal/h
= 174,317 kcal/h × 7000 h/year
= 1,220,219,000 kcal/year

If the lower heating value of natural gas is taken as 8250 kcal/m^3 and the price = 0.25 \$/m^3, the amount of natural gas to be saved and the total annual financial savings are calculated as follows.

$$\begin{aligned}\text{Amount of natural gas to be saved} &= (1,220,219,000 \times 0.80)/8250 \\ &= 118,324\ \text{m}^3/\text{year} \\ \text{Total annual financial savings} &= 118,324\ \text{m}^3/\text{year} \times 0.25\,\$/\text{m}^3 \\ &= 29,581\$/\text{year}\end{aligned}$$

The amount of annual financial savings determined here is the cost of the available energy thrown to the environment by the flue gas.

9.5.3.2 Steam Traps

Due to heat losses, the condensed steam becomes water (condensate). The condensate forms a film in the heat exchangers, reducing heat conduction. Inadequate condensate evacuation causes leaks from gaskets and particularly wear on the seating surfaces of valves. In order to operate the plant efficiently and safely, the condensate must be removed as quickly as possible. The condensate remaining inside the installation is drained by the steam at high velocity and hits the elbows and valves. This is called a ram.

The condensate is discharged from the main distribution as well as from the steam lines with steam traps. One of the most important elements of steam installations are steam traps. There may be problems in steam traps for various reasons. If the steam trap remains closed, the steam draw is suddenly reduced. If the steam trap is defective in full or partially open condition, steam consumption will continue. This will cause unnecessary steam or energy loss.

Steam traps have three main functions:

(1) To remove condensate from steam lines,
(2) To remove air and insoluble gases (O_2, CO_2) in steam lines,
(3) To avoid steam leaks.

Since the traps have moving parts, they must be checked in terms of wearing. This control period should be determined as weekly or monthly periods according to the working time of the factory and the size of the trap. In the current (existing) system, the traps are checked every week and control service is taken from another company every three months. No faulty condensate was observed during the observations at the factory. In addition to the factory controls, it is recommended that the numbering of the traps and the mapping of the location help to control effectively.

9.5.3.3 Boiler Water Quality

Various additives are used to remove the hardness of the boiler water and to impart some chemical properties to the water. These additives remain behind as the evaporation takes place as pure water, and the dissolved solid concentration (Total Dissolved Solids (TDS)) in the region near the water surface and consequently the conductivity of the water increases. Foaming starts on the water surface when the conductivity value exceeds the upper limit given by the boiler manufacturers (7000 mS/cm for boilers operating at 7 bar pressure). This formed foam layer makes evaporation difficult, damages the boiler wall, and is dragged with steam, creating unwanted layers in the steam installation pipes and disrupting the armatures.

9.5.3.4 Blowdown

In addition to the problems arising from the negativity of boiler water quality, solid particles coming from the water source or dragged from the condensate line and remaining in the boiler feed water accumulate in the boiler over time to form sludge. These impurities in the boiler affect the efficiency of the boiler and must be removed from the boiler as they damage the boiler. The process of removing these impurities from the boiler is called blowdown. In terms of the efficiency of the steam system, the drawbacks of blowdown are

- Feed water to steam boiler is softened and taken from water purge device by using chemicals. By bluffing, raw water as much as the amount of bluff discarded will be taken back to the boiler by chemical treatment in the purging devices.
- Although the condensate temperature is around 70–80 °C, the feed water temperature is 10–20 °C. This will cause energy loss.

The boiler blowdown is done in two different ways:

There are two types of steam boiler blowdown applications, manual and automatic. Businesses that prefer manual bluffing should take the samples several times according to the program determined and adjust the bluff as necessary. With manual blowdown control, operators can be mistaken in determining the blowdown time and duration. They cannot immediately interfere with the change in feed water and the differences in the steam demand.

Automatic blowdown control constantly monitors the boiler water conductivity and adjusts the blowdown quantity to keep the required water chemistry constant. A sensor measures the conductivity and feeds back to the controller that triggers the blowdown valve. An automatic blowdown control controls the amount of blowdown and minimizes energy losses by keeping the solute concentration close to, but not exceeding, the maximum allowable value.

The cost of purchasing and installing an automatic bluff system is between USD 2,500 and USD 6,000. This investment amortizes itself within a maximum of 1–3 years. The system consists of a low-or-high-pressure type conductivity sensor, temperature compensation, signal evaluation equipment, and a blowdown valve.

Automatically changing the steam boiler blowdown application manually reduces energy consumption by 2%, while providing up to 20% advantages in water costs thrown by bluff.

In order to maintain the TDS level or conductivity of the boiler water within the given limits, the insoluble solids must be removed, i.e., drained by blowdown. This process is done automatically and manually.

In most steam boilers, manual blowdown is carried out, with manual blowdown, 10–12% of the water supplied to the boiler is thrown out. In automatic blowdown, 4–6% of the water taken to the boiler is thrown out. If automatic blowdown is used instead of manual blowdown, the average increase in boiler efficiency will occur. In addition, it will provide significant savings in terms of water and boiler chemicals thanks to automatic blowdown. Automatic blowdown system is installed in the existing system and there is no recovery unit.

The amount of bluffing required is calculated by the following equation:

$$\mathrm{BM} = \mathrm{F}/(\mathrm{B} - \mathrm{F}) \times \mathrm{S} \tag{9.9}$$

where

BM = The amount of bluffing required (kg/h),

F = Feed water TDS (μS /cm),

B = The value of TDS for feed water (upper allowable limit for boiler water)(μS / cm),

S = Boiler capacity (kg/h).

The amount of blowdown required from the boiler was calculated as 712 kg/h, when considering the boiler works 24 h a day, the amount of bluffing required per day will be 17,080 kg/day.

Blowdown heat recovery system can be seen in Fig. 9.5.

The water discharged by continuous blowdown contains useful heat since it is at the boiler water temperature. Some of this energy can be recovered by using plate heat exchangers.

9.5.3.5 Energy Recovery with Flash Steam

The saturated steam gives the heat of evaporation from heat transfer surfaces, it turns into condensate at the same pressure and the enthalpy value is saturated water enthalpy. Condensate containing saturated water enthalpy after heat transfer releases some energy when sent to environments with lower pressures (e.g.,s tanks). The energy released evaporates some of the condensates. This steam is called flash steam. The use of flash steam by separating from condensate means that the same amount of steam is produced less in the boiler, in other words, it saves energy.

The enthalpy of the saturated liquid discharged from the boiler at 7 bar pressure is 721.4 kJ/kg and the enthalpy of saturated liquid at 0.5 bar indicator pressure is 468 kJ/kg. If the blowdown at 7 bar pressure is released to a flash tank at 0.5 bar pressure, an energy of 721.4−468 = 253.4 kJ/kg is released. This energy evaporates a portion of the blowdown boiler water (flash steam).

The vaporization enthalpy of the vapor with a gauge pressure of 0.5 bar is 2226 kJ/kg. In this case, the flash vapor ratio is 253.4/2226 = ~0.114. In other words, 11.4% of the blowdown from 7 bar pressure to 0.5 bar pressure evaporates.

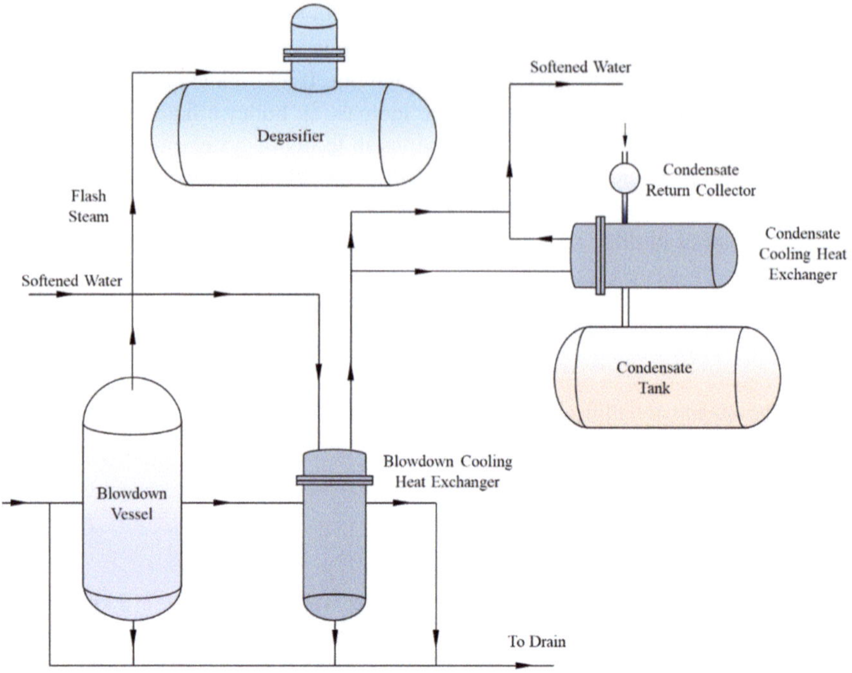

Fig. 9.5 Blowdown heat recovery system

By separating the flash steam from the bluff water, the flash steam obtained from the bluff is saved by 11.4%. In other words, this amount of feed water at 20 °C will enter the boiler less. At the same time, chemicals will not be used for this amount of water.

The amount of blowdown per day = 17,088 kg/day
Flash vapor ratio to be obtained by reducing from 7 bar to 0.5 bar = 11.4%
Amount of flash steam obtained from blowdown = 1947 kg/day
Remaining blowdown = 15,133 kg/day
Evaporation energy of flash steam by liquefying = 500 kcal/kg
Boiler efficiency = 90%
Lower heating value of natural gas = 8250 kcal/m^3
Natural gas price = \$ 0.25/m^3

Saving with Flash Steam Recovery

$$\begin{aligned} Q &= 1947\ \text{kg/day} \times 500\,\text{kcal/kg} \\ &= \sim 974,000\ \text{kcal/day} \times 291\ \text{days} \\ &= 283,434,000\ \text{kcal/year} \end{aligned}$$

The amount of natural gas to be saved is:
= (283,434,000 kcal/year) / (0.90×8250)
= ~38,173 m^3/year
Total annual savings = 9543 $/year

9.5.3.6 Utilization of Blowdown Waste Heat Recovery

Boiler efficiency can be increased by heating the feed water. Feed water can be easily heated by blowdown or flash steam in facilities with an open condensate tank (Preheating Feed Water).

Blowdown temperature = 110 °C (0.5 bar inside the flash tank)
Quantity = 15,133 kg/day

This amount of hot water can be utilized for heating the feed water with the plate heat exchanger.

If the blowdown entering the plate heat exchanger at a flow rate of 15,133 kg/day and a temperature of 110 °C is removed from the heat exchanger at 70 °C, the saving amount to be made is calculated as follows.

$$
\begin{aligned}
Q &= m \times C_p \times (T_i - T_e) \\
&= 15,133\ \mathrm{kg/day} \times (1/24)\ \mathrm{day/h} \times 1\ \mathrm{kcal/kg}\,^\circ\mathrm{C}\ (110 - 70)\,^\circ\mathrm{C} \\
&= \sim 25,222\ \mathrm{kcal/h} \\
&= 25,222\ \mathrm{kcal/h} \times 7000\ \mathrm{h/year} \\
&= 176,554,000\ \mathrm{kcal/year}
\end{aligned}
$$

Considering the lower heating value of natural gas = 8250 kcal/m^3 and its price = 0.25 $/$m^3$, the amount of natural gas to be saved is calculated as follows.

$$
\begin{aligned}
&\text{The amount of natural gas to be saved} \\
&= (176,554,000\ \mathrm{kcal/year}) / (0.90 \times 8250) \\
&= 23,778\ \mathrm{m^3/year} \times 0.25\ \$/\mathrm{m^3} \\
&\text{Total annual savings} = 5945\ \$/\mathrm{year}
\end{aligned}
$$

9.6 An Example of Energy Efficiency in Boiler Fans

9.6.1 Fan Fluid Power Calculation

The useful power of the fan (fluid power) is calculated as follows:

$$P_f = Q \times \Delta p \tag{9.10}$$

$$\Delta p = p_o - p_i = p'_o + p'_i \quad p'_i = p_{fl} - p_i \quad p'_o = p_o - p_{fl}$$

where

P_f = Calculated fan utility power (kW),
Q = Measured air flow (m^3/s),
Δp = Total pressure (kPa),
p'_i = Measured fan inlet pressure (kPa),
p'_o = The measured fan outlet pressure (kPa), and
p_{fl} = Fluid pressure (kPa)

In order to calculate the electric motor power to rotate the fan, it is first necessary to calculate the power transferred to the shaft of the fan, which will give P_f fluid power. The fan power is calculated as follows, taking into account the transmitted power (P_m; kW), fan internal efficiency (η_i), and mechanical efficiency (η_m):

$$P_m = \frac{Q \times \Delta p}{\eta_i \times \eta_m} = \frac{P}{\eta_i \times \eta_m} (\mathrm{kW}) \tag{9.11}$$

Depending on the electric motor efficiency (η_e), the required electric motor power (P_e (kW)) is determined from the following equation:

$$P_e = \frac{P_m}{\eta_e} (\mathrm{kW}) \tag{9.12}$$

The consequence of leakage in the rotary air heater of an industrial entity and the operation of the boiler with a high EAC is the load of FD and ID fans. The amount of energy savings to be provided when the flow is controlled by means of the outlet closure or by the frequency converter is given in Table 9.15.

Table 9.15 Saving amount by reducing leakage air and fresh air consumption

Left rotary air heater	Number of fresh air fans (Piece)	1
	Fresh air fan power (Label value, kW)	90
	Fresh air fan power (Measured value, kW)	69.77
	Power to be saved by reducing excess air in an existing damper system (kW)	6.96
	Power to be saved by reducing excess air with frequency converter (kW)	34.9
	Number of forced draft fans (Pieces)	1
	Forced drafting fan power (Label rating, kW)	320
	Forced draft fan power (Measured value, kW)	184.2
	Power to be saved by reducing excess air in an existing damper system (kW)	12.76
	Power to be saved by reducing excess air with frequency converter (kW)	81.04
Right rotary air heater	Number of fresh air fans (Piece)	1
	Fresh air fan power (Label value, kW)	90
	Fresh air fan power (Measured value, kW)	61.82
	Power to be saved by reducing excess air in the existing damper system (kW)	7.98
	Power to be saved by reducing excess air with a frequency converter (kW)	34
	Number of forced draft fans (Pieces)	1
	Forced drafting fan power (Label value, kW)	320
	Forced draft fan power (Measured value, kW)	178.5
	Power to be saved by reducing excess air in an existing damper system (kW)	16.32
	Power to be saved by reducing excess air by frequency converter (kW)	85.66

9.6.2 Investments and Payback Periods

9.6.2.1 Examination of FD Fans

(a) **Left rotary air heater FD fan**: The average power output of the existing electric motor was measured as 69.77 kW at a flow rate of 18.23 m^3/h. The rated power of the motor is given as 90 kW. In the present embodiment, the high fan flow due to leakage and excess air in the rotary air heater will be reduced to the required flow levels (12.70 m^3/s) with dampers. Since the power of the electric motor is high, the effect of the idle power of the motor remains high at this rate which will lead to waste electricity and waste energy.

Reduction of excess air with existing damper (see Table 9.15):

$$\begin{aligned}\text{Annual Savings} &= (6.96\ \text{kW} \times 5840\ \text{hours/year} \times 0.07\ \text{USD/kWh}) \\ &= 2845\ \text{USD/year}\end{aligned}$$

Solution 1: By keeping the existing fan and electric motor the same and reducing the speed of the electric motor with the frequency converter, the fan capacity can be adjusted, and thus, energy saving can be realized. For the existing fan motor operating at low voltage (380 V), the cost of the frequency converter for speed reduction/adjustment is 5000 € + VAT (Value Added Tax).

Annual Savings = (34.89 kW×5840 hours/year×0.07 USD/kWh)

= 14,263 USD / year

Frequency converter investment cost = 5000 € + (5000 € x 18% VAT)

= 5900 € or 8024 USD

Payback Period = 8024 USD / (14,263 USD / year) ≈ 6.8 months (0.562 year)

(b) **Right rotary air heater FD fan**: The average power output of the existing electric motor was measured as 61.82 kW at a flow rate of 19.61 m^3/h. The rated power of the motor is given as 90 kW. In the present embodiment, the high fan flow due to leakage and excess air in the rotary air heater will be reduced to the required flow levels (12.77 m^3/s) with dampers. Since the power of the electric motor is quite high, the effect of the idle power of the motor will be wasted.

Reduction of excess air with the damper:

$$\begin{aligned}\text{Annual Saving Amount} &= (7.98\ \text{kW} \times 5840\ \text{h/year} \times 0.07\ \text{USD/kWh}) \\ &= 3,262\ \text{USD/year}\end{aligned}$$

Solution 1: By keeping the existing fan and electric motor the same and reducing the speed of the electric motor with the frequency converter, the fan capacity can be adjusted, and thus, energy saving can be realized. For the existing fan motor operating at low voltage (380 V), the cost of the frequency converter for speed reduction/adjustment is 5000 € + VAT.

Annual Saving Amount = (34 kW×5840 h/year×0.07 USD/kWh)

= 13,900 USD/year

Frequency converter investment cost = 5 000 € + (5000 € x 18% VAT)

= 5900 € or 8024 USD

Payback Period = 8024 USD / (13,900 USD / year) ≈ 7 months (0.577 year)

9.6.2.2 Examination of ID Fans

(a) **Left rotary air heater ID fan**: The average power output of the existing electric motor was measured as 184.2 kW at a flow rate of 22.27 m^3/h. The rated power of the motor is given as 320 kW. In the present embodiment, due to leakage and excess air in the rotary air heater, the high fan flow rate will be reduced with dampers to the required flow levels (16.74 m^3/h). Reduction of excess air with the damper:

$$\begin{aligned}\text{Annual Savings} &= (12.76\ \text{kW} \times 5840\ \text{h/year} \times 0.07\ \text{USD/kWh}) \\ &= 5216\ \text{USD/year}\end{aligned}$$

Solution 1: By keeping the existing fan and electric motor the same and reducing the speed of the electric motor with the frequency converter, the fan capacity can be adjusted, and thus, energy saving can be achieved. The cost of the frequency converter for reducing/adjusting the speed of the existing fan motor operating at medium voltage (3300 V) is 120,000 € + VAT.

Annual Saving Amount
= (81.04 kW × 5840 h/year × 0.07 USD/kWh)
= 33,129 USD / year
Frequency converter investment cost is:
= 120,000 € + (120,000 € × 18% VAT)
= 141,600 € or 192,576 USD
Payback Period = 192,576 USD / (33,129 USD/year) ≈ 70 months (5.812 year)

Solution 2: The existing fan may remain the same, but the costs are as follows when replacing the existing medium voltage electric motor with the new low-voltage electric motor and applying a low-voltage (400 VAC) frequency converter.

New electric motor (320 kW) + frequency converter
= 40,000 € + VAT
Savings Amount = 33,129 USD / year
Investment Cost = 40,000 € + (40,000 × 18% VAT) = 47,200 € or 64,192 USD
Payback Period = 64,192 USD / (33,129 USD/year) ≈ 23.3 months (1.937 year)

(b) **Right rotary air heater ID fan**: The average power output of the existing electric motor was measured as 178.47 kW at a flow rate of 23.65 m^3/h. The rated power of the motor is given as 320 kW. In the present embodiment, the fan flow rate, which is high due to leakage and excess air in the rotary air heater, will be reduced to the required flow levels (16.81 m^3/s) with dampers. Since the electric motor power is high, there will be a waste of energy due to the idle power of the motor.

Reduction of excess air with the existing valve:
Annual Savings = (16.32 kW × 5840 h / year × 0.07 USD/kWh)
= 6671 USD/year

Solution 1: By keeping the existing fan and electric motor the same and reducing the speed of the electric motor with the frequency converter, the fan capacity can be adjusted, and thus energy can be saved. The cost of the frequency converter for reducing/adjusting the speed of the existing fan motor operating at medium voltage (3300 V) is 120,000 € + VAT.

Annual saving amount
= (85.66 kW × 5840 h / year × 0.07 USD/kWh)
= 35,018 USD/year
Frequency converter cost = 120,000 € + (120,000 × 18% VAT)
= 141,600 € or 192,576 USD
Payback Period = 192,576 USD / (35,018 USD/year) ≈ 66 months (5.5 years)

Solution 2: The existing fan may remain the same, but the costs are as follows when replacing the existing medium voltage electric motor with the new low-voltage electric motor and applying a frequency converter to operate at low-voltage (400 VAC).

New electric motor (320 kW) + frequency converter = 40,000 € + VAT
Savings Amount = 35,018 USD/year
Investment Cost = 40,000 € + (40,000 € × 18% VAT) = 47,200 € or 64,192 USD
Payback Period = 64,192 USD / (35,018 USD/year) ≈ 22 months (1.833 year)

9.6.3 Energy Saving in Fans

The high EAC of the boiler and excessive leakage air losses in the rotary air heaters cause overloading of ID and FD fans. Decreasing the excess air and leakage losses to the boiler design values will reduce the electrical load of the fans and thus save electrical energy.

It is assumed that while calculating the resulting electric load loosed the ID and FD fans operate at 100% flow with increasing airflow and increasing flow rate with leakage. The target fan flow rate (TFFR) is calculated based on the target flow rate value.

TFFR = Target fan flow rate × 100/Flowrate in the existing situation

Using this value, the fan power percentage (FPP) and annual energy saving potential (AESP) are calculated from the following equation.

$FPP = -0.0088 \times TFFR^{2} + 1.8299 \times TFFR + 5.2433$
Excess electricity consumed = Existing electric load − Existing electric load × *FPP*
AESP = Consumed excess electrical load × Annual operating hours

Example 1: The amount of power to be saved is calculated as follows if the excess air is reduced by the damper on the left-hand rotary air heaters fresh air fan:

Existing fan flow rate = Theoretical combustion air + Excess air flow rate + Leakage air flow rate

Theoretical combustion air = 34,220 m^3/h
Excess air flow rate = 12,322 m^3/h
Leakage air flow rate = 19,088 m^3/h
Fan flow rate in the existing situation = 34,220 + 12,322 + 19,088 = 65,630 m^3/h
High excess air flow rate = 19,913 m^3/h
Target fan flow rate = Existing fan flow rate—high excess air flow rate
Target fan flow rate = 65,630 m^3/h − 19,913 m^3/h = 45,717 m^3/h
$TFFR$ = (Target fan flow rate × 100) / Existing flow rate
$TFFR = (45{,}717 \times 100) / 65{,}631 = 69.657$ m^3/h
$FPP = (-0.0088 \times TFFR^{2} + 1.8299 \times TFFR + 5.2433$
$FPP = (-0.0088 \times (69.557)^2) + (1.8299 \times 69.557) + 5.2433 = 89.949$
Excess electrical load consumed = Existing electrical load—(Existing electrical load × FPP)
= 69.77 −(69.77× 0.9001) = 6.97 kW
AESP = Consumed excess electrical charge × Annual operating hours
AESP = 6.97 kW × 5840 h = 40,704.8 kWh
Financial savings = 40,704.8 kWh × 0.07 USD/kWh = 2,849 USD

Example 2: The amount of power to be saved by reducing the excess air with the frequency converter in the fresh air fan of the boiler left-hand rotary air heaters is calculated as follows:

$TFFR = (45{,}718 \times 100) / 65{,}630 = 69.66$
In the absence of excess air, the percentage of fan power is 69.66% of the prepared graphs.
Excess electric charge consumed due to excess air (kW)
= Existing fan power drawn by excess air—Fan power drawn by FC (Frequency Converter) = 69.77 −(0.50 × 69.77) = 34.885 kW
AESP = Consumed excess electrical charge × Annual operating hours
AESP = 34.885 kW × 5840 h = 203,728.4 kWh
Financial savings = 203,728.4 kWh × 0.07 USD/kWh = ~14,261 USD

9.7 Better Operation of Boilers

9.7.1 Heat Recovery from Flue Gases

Waste gases coming out of the stack are generally discharged from the stack at a temperature higher than 40–80 °C than the temperature of the steam produced. Recovering some of the waste heat will increase boiler efficiency and save fuel. Heat recovery can be accomplished using either an economizer to preheat the boiler feed water or a recuperator to preheat the combustion air.

9.7.1.1 Determination of Economizer Usage Opportunities

There is no point in placing a very expensive heat recovery system in a boiler which can be improved by some simple and inexpensive modifications. The lowest level of flue gas temperature depends on the type of fuel used. In fuels containing very little sulfur such as natural gas and wood, the flue gas temperature can be reduced to 120 °C. The flue gas temperature specified here is the value taken when the water and gas side heat transfer surfaces of the boiler are clean.

Example: A flue gas temperature of a cleaned and adjusted fuel oil fired boiler is assumed as 350 °C. The oxygen content in the gas is generally around 3%. Consider the fact that the cost of fuel oil is 1100 USD/ton and that an economizer is installed in such a boiler. Considering the boiler efficiency of 83% and the steam cost factor of about 0.077 and the steam cost is calculated as follows:

$$\text{Cost of steam} = 0.077 \times 1100\ \text{USD/ton} = 84.7\ \text{USD/ton}$$

If the flue temperature is 350 °C, the expected saving rate is 9.8%. Thus, savings in steam costs

$$= 0.098 \times 84.7\ \text{USD/ton} = 8.3\ \text{USD/ton}.$$

If a boiler with a capacity of 20 tons/hour operates 8000 h per year, the total savings are calculated as follows:

$$(8.3\ \text{USD/ton}) \times (20\ \text{tons/hour}) \times (8000\ \text{hours/year}) = 1,328,000\ \text{USD/year}$$

In cases where the flue gas temperature is too high, the heat recovery process is limited by the amount of heat that the feed water can absorb before evaporation. This limitation varies depending on the system pressure. If the boiler flue gas temperature shows a greater recovery than the system pressure limitations, a specially designed economizer such as an evaporator economizer must be installed. In this regard, the following factors should be considered:

- The flue gas temperature should be tested after the boiler setting has been made. Setting the boiler means cleaning the heat transfer surfaces and keeping the excess air rate at the optimum level during operation.
- If this temperature is higher than the specified limit values for condensation and corrosion, the savings potential must be determined.
- Using the cost of steam, the savings corresponding to the annual steam production should be calculated.
- The investment required for the economizer must be compared with the amount of savings to be achieved.

9.7.2 Improvement of Liquid Fueled Boiler Efficiency

In many industrial plants, boilers consume a large part of the energy. Therefore, a significant amount of energy can be saved by monitoring the operation of the boilers.

9.7.2.1 Flue Gas Control

The composition and temperature of the flue gas are indicators used to determine boiler efficiency. Large boilers are fitted with devices that show the composition and temperature of the flue gas. Small boilers, on the other hand, should be checked regularly, usually using low-cost portable devices. Analyzes should be performed at least once a day. The flue gas analyzer consists mainly of the following units:

- Oxygen and/or CO_2 analyzer
- Thermometer (0–500 °C)
- Smoke tester

These equipments are supplied from many manufacturers and their prices depend on preferences and the control sensitivity of the microprocessor.

Flue Gas Measurement Locations: Gas analysis is required to ensure efficient operation of boilers and furnaces. Therefore, it is important to determine the appropriate measuring locations. The gas sample should be taken as close as possible to the combustion chamber outlet. This is because the temperature of the gas drops at distant points and inaccurate results are obtained. Otherwise, there is a risk that the air will constantly leak into the stack if the measuring points are after the connection chambers in the stack. Therefore, the analysis will not give accurate results.

If there is an economizer or an air heater in the boiler, the gas sample should be taken after the equipment. Air may leak from the equipment. Therefore, samples taken before the equipment are checked by comparing them with each other. Thus, it is determined whether or not air leakage occurs. If there is any air leak, it should be prevented as soon as possible.

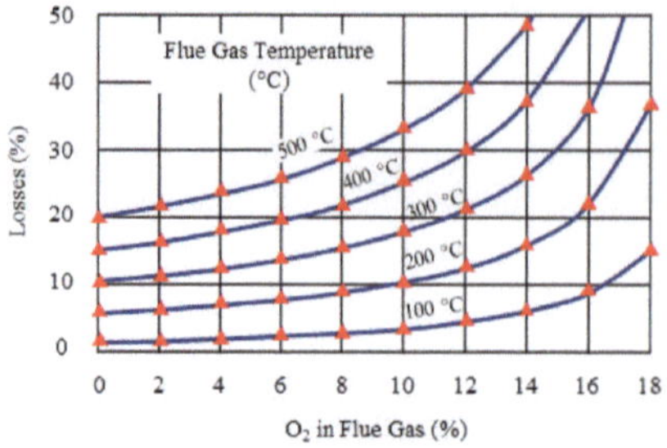

a) For heavy fuel oil

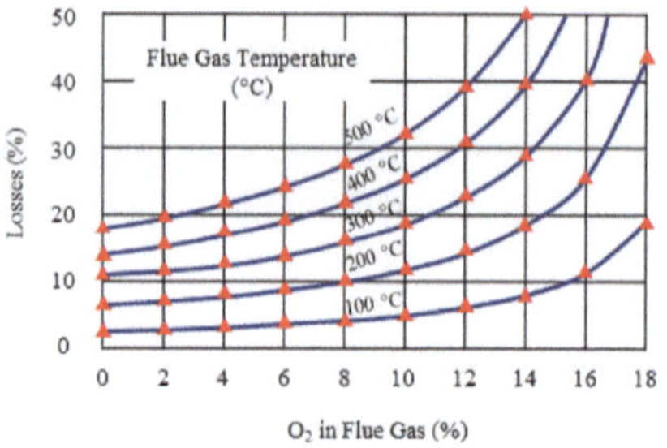

b) For diesel

Fig. 9.6 Flue gas losses in boilers

During the measurement, care must be taken to prevent air from entering the measuring point. It is better to drill a small hole on the stack. These holes must correspond to the dimensions of the measuring probes of the devices. The perimeter of these holes must be sealed with insulating material to ensure the sealing during measurement.

9.7.2.2 Calculation of Flue Gas Losses

In order to calculate flue gas losses; gas compositions (O_2 percentage-volumetric) and gas temperature (°C) measurements must be done. Although there are correlations to calculate flue losses from these measurements, it is more convenient to use ready-made graphics for quick calculations. Graphs for heavy fuel oil and diesel fuel boilers are given in Fig. 9.6. In these graphs, the amount of O_2 in the flue gas is defined in the x-axis and the flue gas losses in the flue gas temperatures between 100 and 500 °C are defined in the y-axis.

These graphs in Fig. 9.6 have been developed for an ambient temperature of 15 °C. If the ambient temperature is different from 15 °C, there may be a small error.

Example: Four percent of oxygen by volume and in the case the temperature is 210 °C, the loss rate is determined as 9.5% from Fig. 9.6a. In a similar manner, 5.5% of oxygen by volume and, in the case the temperature is 300 °C, the loss rate is determined as 14% from Fig. 9.6b.

9.7.2.3 Efficiency

Flue gas loss in boilers is not the only source of loss. There are also blowdown losses and heat losses from the boiler surfaces. If the outside temperature of the boiler is known, surface losses can be calculated. In well-insulated cylindrical boilers, standard values can be used for quick calculations. Standard values for surface losses in cylindrical boilers using fuel oil and gas are given in Table 9.16.

Table 9.16 Surface losses in cylindrical boilers

Rated capacity (ton/h)	Losses in full load (%)
11	1.9
2	1.5
4	1.1
8	0.7
10	0.6
15	0.5
20	0.3

Example: If it is assumed that the boiler which burns heavy fuel oil given in the above example is a 10 tons/hour capacity boiler, the efficiency of the boiler is found to be about 100−9.5−0.6 = 89.9%.

9.7.2.4 Smoke

The presence of smoke can be determined by checking the flue gas. Although the stack is often seen clean at first glance, careful and close inspection may reveal that there are quite a few soot pieces. The smoke pump is an inexpensive device for extracting the flue gas sample from the filter paper. The fine soot particles are held by the paper and are easily visible on a white background. The amount of soot depends on the smoke density. Thus, the degree of smoke can be decided. The operating instructions for the smoke pump should include a test method and a graph showing the smoke level. A high amount of soot indicates a bad air-fuel ratio (insufficient air). However, it indicates that the burner needs maintenance.

It is also useful to control the amount of CO in the flue gas for combustion control. If high CO (400 ppm and above) is present with oxygen above 4%, this indicates further problems in thc fuel/air mixture. The following measures should be taken to eliminate common problems:

- Check the boiler operating conditions regularly.
- If the amount of O_2 in the flue gas exceeds 3%, the air entering the burner must be reduced.
- If the boiler emits smoke due to low air, the air/fuel mixture may be faulty. In this case, if there is a problem with the burner, the burner must be serviced or replaced.
- There might be poor atomization problem due to improper fuel temperature or fuel pressure. For heavy fuels, the temperature of the fuel in the burner should be around 120 °C.
- In addition, there may be problems with air intake. If the air inlet valves are worn, the air will combine with the fuel at an inappropriate rate and improper combustion will occur. This causes the formation of smoke.

9.7.2.5 Efficient Combustion

It is possible to operate a boiler with 2–3% O_2 in the flue gas with good maintenance and a good combustion setting. For simple units without special heat recovery equipment, the flue gas temperature should be between 170 and 190 °C. The smoke number (according to the Bacharach Code) should be less than 5. With these values, the flue gas loss is around 8% and therefore the efficiency is 91–92%. The amount of soot in the gas is so small that it does not quickly contaminate the heat conduction surfaces of the boiler.

9.7.3 Improvement of Gas-Fired Boiler Efficiencies

Boilers with natural gas combustion do not need as much control as boilers using liquid and solid fuel. Natural gas is a clean gas and ensures that the burners stay clean for a long time. However, it is necessary to check the flue gas composition at least once a week. In changing load situations, the amount of control must be increased. Because, in such cases, it is not always easy to improve combustion conditions.

9.7.3.1 Required Equipment

As with liquid fuel boilers, a gas analyzer is needed to measure the amount of O_2 or CO_2 in the flue gas and a thermometer is required to measure the flue gas temperature. Smoke testing is not required in gas-fired boilers, as there is a low likelihood of soot particles in the flue gas. However, it may be useful to perform a smoke test from time to time. The presence of soot particles in the flue gas indicates a problem with the fuel/air mixture in the burner.

By controlling the amount of CO in the flue gas and gases such as methane in the unburned state, an idea of the operation of the burner can be obtained. The amount of methane (CH_4) can be determined by commonly known Draeger tube equipment. In flue gas, unburned gases may form even in case of excess air. Modern electronic gas analyzers with microprocessor circuitry, which can calculate the combustion efficiency instantly, are available from the market.

9.7.3.2 Flue Gas Losses

The graph has been developed for flue gas losses that occur when the flue gas does not contain CH_4 and CO (Fig. 9.7).

Example: If the oxygen content is 3% and the flue gas temperature is 130 °C, it can be seen from the graph that the higher heating value of the burned gas is lost from the flue gases by 15.5%.

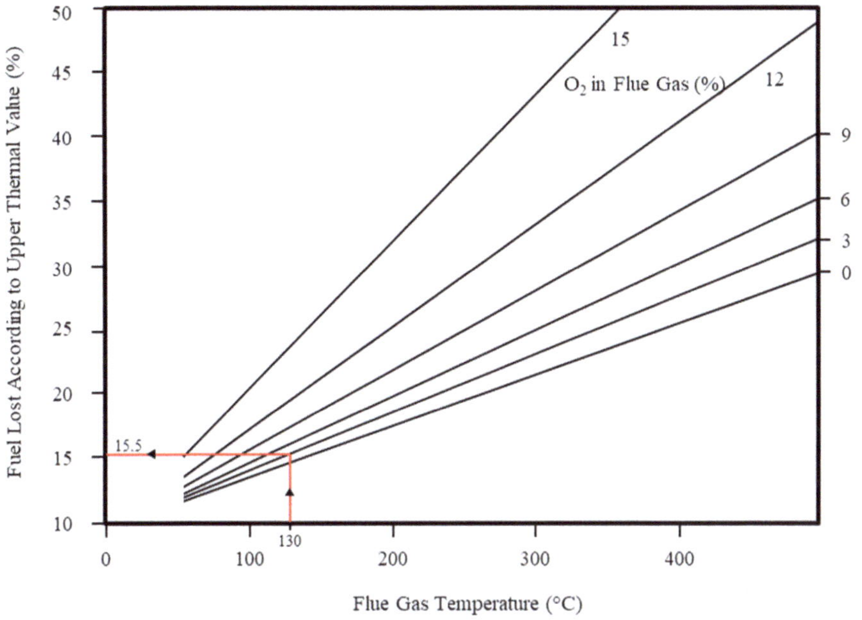

Fig. 9.7 Flue gas losses in boilers burning natural gas

Modern gas burners are capable of operating with very little air. They can work even when the O_2 level in the flue gas is 1.5% or less. The CO content should also be less than 100 ppm. Since natural gas contains very little sulfur, it is possible to work with low flue gas temperatures (such as 120 °C) without corrosion problem in the stack. This means that flue gas losses are reduced to 13–14% (according to the higher heating value). In space heating applications, there are boilers designed to condense the water vapor in the flue gas and recover the latent heat. These boilers operate at very high efficiencies (over 90%).

The following factors should be considered to improve the efficiency of gas-fired boilers:

- Check the combustion efficiency regularly.
- The air ratio should be adjusted to keep the O_2 content in the flue gas at 1–2%. CO content should be kept below 100 ppm. Methane content should be around 100 ppm.
- If the O_2 level in the flue gas is lowered and the CO amount increases, it should be checked for a problem in the fuel/air mixture.
- When the flue gas temperature is above 130 °C, heat recovery should be considered by installing devices such as air preheater and economizer. In such cases, economic assessments should be made.
- If the combustion device of a previously fired oil boiler is turned into a natural gas combustion device, this results in changes in the heat transfer properties of

the flame and consequently the increase in the flue gas temperature. Replacing such a boiler with a modern gas boiler is economically more efficient in the long run.

- For certain applications, such as space heating or hot water supply, it may be economical to add a special section to the stack to condense the water vapor in the flue gas and provide heat recovery. Furthermore, cleaning the flue gas with cold water can be an economical method of producing hot water. Both methods can assure the saving amount of 3–5%.

9.7.4 *Improvement of Coal-Fired Boiler Efficiencies*

Coal-fired boilers are more difficult to control than liquid-fired boilers. The most important differences from the other boilers are the unburned carbon in the ash formed in the boiler and the CO, which can contain significant amounts of flue gases. Another factor to consider is the moisture content of coal. In order to accurately calculate the efficiency of coal-fired boilers, coal and ash analysis should be carried out on a regular basis, as well as regular control of the flue gas temperature and composition.

There is a need for data such as the chemical composition of coal, the C content in ash, and the thermal (calorific) value of coal. Coal and ash samples should be analyzed regularly. Periodic analysis of flue gas is important in terms of controlling the operating conditions of the boiler and improving its efficiency.

The losses in coal-fired boilers are not only caused by flue gas losses. Losses from unburned fuel, such as CO in flue gas and C in unburned state, are also very important. Moisture content in coal can vary. Since the heat will disappear from the flue gas by evaporating the water, the actual thermal value obtained from the coal changes.

In boilers, the targeted O_2 amount and gas temperature in the flue gas can vary significantly depending on the boiler design, the coal composition, and the physical (size, moisture content) quality of the coal. Boiler manufacturers should be consulted for this. With a rough approach, for most boilers, the amount of oxygen in the flue gas should generally be less than 7%. In modernly designed pulverized coal-fired boilers, this ratio should be around 5%.

The following factors should be considered in order to improve the efficiency of coal-burning boilers:

- The flue gas composition and temperature must be continuously monitored.
- Values higher than normal temperature indicate that the pipes need to be cleaned.
- In manual loading furnaces, it is more convenient to load during certain periods. Such an application prevents large quantities of coal from having to be burned with very low air. If random loading is carried out, there are large losses initially due to CO, and then increases of the flue gas losses as a result of burning a small amount of coal with a high rate of combustion air.
- In order to prevent the coal thickness to be too high in one place (CO formation increases as a result of insufficient air) and too low in another place (high amount

of air and increase the amount of O_2 in the flue gas), the coal should be disposed as smoothly as possible on the grid. If the coal is not properly distributed, the O_2 and CO ratios in the flue gas will be high.

- Air dampers must be used to properly control the air flow rate. The amount of air should be adjusted according to the flue gas analysis, not simply by the combustion appearance of the coal.
- The amount of dust in the feed coal must be kept to a minimum level. The high dust level affects the combustion efficiency. Large coal grains also reduce efficiencies. For this reason, sieving should be applied to keep coal sizes in proper condition.
- The thermal (calorific) values of coal quite vary. Especially in lignite coal, this difference is more pronounced. Furthermore, when the coal is stored in the open air, its thermal value may be significantly reduced due to oxidation and moisture. Therefore, proper storage of coal provides economic savings.
- When the air intake to the boilers is low, a certain amount of fuel will remain unburned. This will increase operating costs. In this case, the control unit mounted on the feeder can provide significant benefits. For this purpose, manufacturers should be consulted to find the most suitable equipment.
- In order to understand that the fuel/air mixture is at an appropriate level, the method generally used is flue gas analysis. By determining the temperature value and composition of the flue gas, the losses from the flue gas are calculated.
- The optimum boiler efficiency depends on the fuel used, the type of the boiler, the method of operation, the maintenance status of the boiler, the skill of the operator, and the combustion equipment.
- The air/fuel control units in modern boilers can provide the desired excess of air ratio throughout the operation by the feeders that are well maintained and adjusted to the fuel used and operated according to the manufacturer's instruction.
- Flue gas losses are quite high due to the excessive adjustment of the feed controls in the application and the excess air is surplus. Therefore, it is useful to act in accordance with the manufacturer's instructions. Boiler efficiency can be improved by installing air/fuel control elements in large capacity boilers.

9.7.5 *Better Operation of Boilers*

9.7.5.1 Testing of Flue Gas

In mechanical feed boilers with an excess air rate of 30–80%, the CO_2 content of the flue gas is in the range of 10–14% or less. It is recommended to reach more CO_2 levels to reduce heat losses. Continuous monitoring of CO and O_2 amounts in the flue gas is important to ensure that the excess air is properly controlled. During flue gas sampling, it is necessary to avoid air leaks.

9.7.5.2 Keeping Clean of Heat Transfer Surfaces

In water pipe boilers, the formation of boiler stones prevents heat conduction and increases the temperature difference between water and combustion gases due to the accumulation of ash and soot particles on the outer surfaces of the pipe. In this case, the boiler efficiency decreases. Therefore, it is necessary to keep the heat transfer surfaces clean. This problem is not usually encountered in gas-fired boilers. However, this is particularly important in boilers burning fuel oil or coal.

9.7.5.3 Efficiency Increasing Applications

As a first step, the flue gas temperature should be monitored carefully. Temperature indicators should be available even in very small boilers. These are not expensive devices. The temperature should be measured immediately after the heat transfer surfaces are cleaned and the rise in temperature should be monitored as the contamination increases gradually.

The flue gas temperature also varies depending on the steam load of the boiler. For this reason, it should be checked whether the readings increase in parallel with the steam production.

If the flue gas temperature is 30 °C above the required temperature when the surfaces are clean, the boiler needs to be cleaned. In this case, extra losses for a well-tuned boiler are around 1–2%.

The best method to maintain the cleanliness of a boiler operating with fuel oil is to make appropriate burner settings. Adjustments can be made using the smoke tester. If dark black smoke comes out of the chimney, this is a situation that creates an undesired amount of soot. The appearance of the smoke may affect the brightness of the air and the natural state of the sky. During the operation of a well-maintained and well-operated burner, the measured soot amount of the flue gas is usually 4–5 from the Bacharach Scale. At the values below 7, there is no serious soot formation. If around the value of 5 is read, it is sufficient to clean the soot once a week.

Ash, which is usually formed in boilers using heavy fuel oil, can cause deposits and corrosion in the hottest parts of the boiler, such as serpentine pipes, albeit in very small quantities. To avoid this, fuel additives can be used. In fact, some additives contain high levels of ash. This may result in high ash build-up rather than less ash.

In coal-fired boilers, soot and ash accumulated in the pipes should be cleaned at least once a day. Soot blowing equipment is very useful even in small smoke pipe boilers. Simple stationary devices are cheap and suitable for small boilers. In contrast, large boilers should be equipped with automatic systems operated by the operator or operating at fixed intervals according to the flue gas temperature.

Steam blast cleaning is not expensive, but sometimes a compressed air system is preferred. This is because the water formed by the condensation of steam may agglomerate the puddles, making further cleaning difficult. The amount of steam used for blowing should be measured. Discrete blown systems using intermittent high-pressure air are designed to keep air flow to a minimum.

Special care must be taken when purchasing the economizer. Of these, the large surface (finned tube) may easily accumulate between the wings. This causes serious operating problems. In cases where the system temperature is low, periodic washing with a water spraying device may be considered.

In order to improve boiler efficiencies, the following factors should generally be considered:

- Flue gas temperature should be monitored. The maximum permissible temperature value of the flue gas must be determined, and the boiler operating personnel should be advised of cleaning or soot blowing.
- The burner settings should be made by applying the smoke test in systems operating with fuel oil. If proper values cannot be achieved, fuel pressure and temperature should be checked, burner parts inspected, air valves checked, and all other necessary maintenance performed.
- In coal-fired boilers, the amount of fly ash and coal dust amount should be reduced with proper operation. In systems where loading is done manually, the coal layers should be of equal thickness and sudden changes in the rate of steam production should be avoided.
- Heat transfer surfaces should be cleaned frequently. It is very difficult to stop and clean the boiler manually, so soot blowers that work with the boiler during operation should be considered.
- The institution should contact the sellers for the prices of the soot blower elements. The economic status of the soot blowers should be investigated, especially in the post-mounted period.

References

1. Onat K, Genceli O, Arısoy A (1988) Steam boilers thermal calculations, denklem printing house, İstanbul/Turkey 1988. In: Onat K, Genceli O, Arısoy A (eds), Turkish, Buhar Kazanları Isıl Hesapları, Denklem Matbaası, İstanbul/Türkiye
2. Bilgiç M For efficient use of energy in industrial boiler rooms; matters to be considered from fuel to flue gas. J Turkish Plumb Eng Associat. In: Bilgiç M (ed), Turkish, Endüstri Kazan Dairelerinde Enerjinin Etkin Kullanılması için; Yakıttan Baca Gazına Kadar Dikkate Alınması Gereken Hususlar, Türk Tesisat Mühendisleri Derneği Dergisi
3. Kaya D, Eyidogan M (2010) Energy conservation opportunities in an industrial boiler system. J Energy Eng 136(1):18–25
4. Kaya D, Eyidogan M (2009) Energy conservation opportunity in boiler systems. J Energy Res Technol 131(3):032401
5. Çanka Kiliç F (2017) Energy efficiency and emission reduction opportunities in industrial boilers, Gazi University. J Sci Part C Des Technol. e-ISSN: 2147–9526, GU J Sci, Part C, 5(2): 147–158, Ankara/Turkey, 2017. In: Çanka Kiliç F (ed), Turkish, Endüstriyel Kazanlarda Enerji Verimliliği ve Emisyon Azalımı Fırsatları, Gazi Üniversitesi, Fen Bilimleri Dergisi, Part C: Tasarım ve Teknoloji, e-ISSN: 2147-9526, GU J Sci, Part C, 5(2): 147–158, Ankara/Türkiye
6. Çanka Kiliç F, Eyidoğan M, Sapmaz S (2018) Investigation of energy efficiency increasing solutions in an automobile assembly plant, Gazi University. J Sci Part C Des Technol GU J Sci, Part C, 6(1), e-ISSN 2147-9526, 149–162. https://dx.doi.org/10.29109/http-gujsc-gazi-edu-tr.331104, Ankara/Turkey. In: Çanka Kiliç F, Eyidoğan M, Sapmaz S (eds), Turkish, Bir Otomobil

Montaj Işletmesinde Enerji Verimliliği Artırıcı Çözümlerin Irdelenmesi., Gazi Üniversitesi Fen Bilimleri Dergisi Part C: Tasarim Ve Teknoloji, GU J Sci, Part C, 6(1), e-ISSN 2147-9526, 149–162. https://dx.doi.org/10.29109/http-gujsc-gazi-edu-tr.331104, Ankara/Türkiye
7. Kılıçaslan I, Özdemir E (2005) Energy economy with a variable speed drive in an oxygen trim controlled boiler house. Int J Energy Res 127:59–65
8. Siitonen S, Tuomaala M, Ahtila P (2010) Variables affecting energy efficiency and CO_2 emissions in the steel industry. Energy Policy 38:2477–2485
9. Çanka Kılıç F, Kaya D, Eyidoğan M, Sapmaz S Çoban V, Çağman S (2014) Energy conservation and efficiency increasing studies in an industrial boiler, 23–24 September 2014 International Energy and Security Congress, 23–24 September 2014. (In Turkish: 23–24 Eylül 2014 Uluslararası Enerji ve Güvenlik Kongresi, 23–24 Eylül 2014.), New Technologies in Energy Sector, pp 997–1022
10. Enerkon (1987) Improving steam boiler operating efficiency
11. Kaya D, Eyidoğan M, Çanka Kiliç F, Çay Y, Çağman S Çoban V (2014) Energy saving and emission reduction opportunities in mixed-fueled industrial boilers. Environm Progr Sust Energy 33(4):1350-1356. ISSN:1944-7450. https://doi.org/10.1002/ep.11925
12. Nikhil D, Samsher SSK, Rajesh A (2014) GTA modeling of combined cycle power plant efficiency analysis. Ain Shams Eng J 6(1)
13. Kaya D, Öztürk HH (2014) Sanayide Enerji Yönetimi ve Enerji Verimliliği, Umuttepe Yayınları, Yayın No: 114, Mühendislik Dizisi: 12, Kocaeli/Türkiye

Chapter 10
Energy Efficiency in Furnaces

Technical units, which increase the process temperature by heating the materials placed or charged continuously in an economical way and if necessary, keep this temperature for the required time, are called furnaces. Furnaces, especially annealing furnaces that operate at high temperatures, are the systems that must be operated as efficiently as possible in terms of fuel consumption and pollution caused by the waste gases it gives to the environment.

Annealing furnaces are continuous rolling furnaces that are used to heat steels at a temperature range of 1050–1300 °C, placed on a rolling band of steel. An industrial annealing furnace used in the industry is shown in Fig. 10.1.

Furnaces are the leading units that use energy most intensively in industrial plants. Industrial furnaces perform processes such as melting, heat treatment, baking, drying, tempering, etc. The furnaces must be supplied with the necessary raw materials and energy in order to carry out the processes designed for them. When designing a new furnace to be used for heating the parts for hot forming, the following factors must be taken into account in order to improve the quality [1–3]:

- Parts must be able to be removed from the furnace outlet at the same temperature,
- There must be no temperature difference in the length or width of the parts,
- There must be no temperature difference between the wall and core of the parts and the upper surface and the lower surface,
- Heating costs should be reduced with better use of flue gases,
- The water used to cool the furnace (hot water, boiling water, steam) should be used in boilers or other units,
- Reduction of scaling on parts should be attempted,
- Parts of different sizes can be heated in the furnace and the feed rate should be changed if desired,
- No mechanical distortion, such as bending or surface damage, should be in progress. Parts must not stick together due to overheating,
- Burners that can adapt to the fuel change in the furnace should be used,

D. Kaya et al., *Energy Management and Energy Efficiency in Industry*,
Green Energy and Technology,
https://doi.org/10.1007/978-3-030-25995-2_10

Fig. 10.1 An example of the industrial annealing furnace

- The utilization rate of the base surface in the furnace should be high,
- The furnace plant should be automatically adjustable and work in coordination with other furnace related units,
- The furnace should not break down easily, maintenance and repair costs should be low,
- In case of operational faults in the plant, it should be possible to intervene immediately,
- Safety measures must be taken against explosions and other accidents. In particular, storage and transportation of fuel should be carried out with caution,
- The furnace and the plant must operate with minimum noise.

10.1 Thermal Efficiency in Furnaces and the Factors that Affecting Efficiency

10.1.1 Thermal Efficiency in Furnaces

The thermal efficiency of the furnaces is calculated by two methods: direct and indirect.

10.1.1.1 Direct Method

In the direct method, the thermal efficiency is calculated by dividing the heat of the material by the heat of the fuel used. In this method, the following measurements should be performed:

- Amount of material entering and leaving the furnace,
- Temperature of the material during entering and leaving the furnace,
- Fuel supply amount, and
- Fuel lower heating value.

Based on these measurements, the thermal efficiency (η; %) is generally calculated with the following equation:

$$\eta = \frac{\dot{m}_m \times C_{p_2} \times T_2 - \dot{m}_m \times C_{p_1} \times T_1}{\dot{m}_f \times LHV} \tag{10.1}$$

where

$\dot{m}_m$ = Mass flow rate of the material supplied to the furnace (kg/h),
C_{p_1} = Specific heat of the material before furnace (kcal/kg °C),
C_{p_2} = Specific heat of the material (kcal/kg °C),
T_1 = Temperature of the material before furnace (°C),
T_2 = Temperature of material (°C),
$\dot{m}_f$ = The flow rate of the fuel (m^3/h), and
LHV = The lower heating value of the fuel (kcal/m^3)

10.1.1.2 Indirect Method

In the indirect method, the thermal efficiency is calculated by subtracting the furnace losses:

$$\eta = 1 - \sum Z \tag{10.2}$$

where Z = heat losses from the furnace [%]. These losses consist of; flue gas losses, span (gap) losses, scale/material losses, wall losses, cooling water losses, and other uncalculated losses.

The following measurements should be taken, and the result data are taken into account when calculating the thermal efficiency by indirect method:

- The amounts and temperatures of the materials entering and leaving the furnace,
- Fuel supply amount,
- Mixed fuel ratios,
- Lower heating value and elemental analysis of fuel,
- Air-fuel ratio,

- Flue gas analysis (temperature, velocity, pressure, and gas measurements in flue gas),
- Combustion air and furnace temperature,
- Outer wall temperatures of the furnace,
- Exhaust gas temperature before and after recuperator,
- Oxygen content before and after recuperator,
- Cooling water flow, inlet and outlet temperatures.

Based on these measurement results, the following data are obtained, and the thermal efficiency of the furnace is calculated using all these data:

- Theoretical specific air volume,
- Theoretical specific flue gas quantity,
- Combustion air flow,
- Excess air coefficient,
- Actual specific air volume,
- Flue gas heat loss rate,
- Missing combustion loss rate,
- Unburned fuel loss rate.

10.1.2 Factors Affecting the Efficiency in Furnaces

The main factors affecting the efficiency of the furnaces are [4–8]:

- Incomplete combustion,
- Fuel type and air/fuel ratio,
- Flue gas temperature,
- Recuperators,
- Wall losses,
- Scale/material and cooling water losses,
- Loss of openness,
- Characteristics of the combustion unit and material filling temperature.

10.1.2.1 Incomplete Combustion

Incomplete combustion occurs when the combustible substances in the gas fuel do not burn and are discarded as unburned hydrocarbons and carbon monoxide in the flue gas. Since incomplete combustion causes loss of fuel, it is important to adjust the air-fuel ratio smoothly for complete combustion. Therefore, it is necessary to keep O_2 quantity of flue gas at an optimum level [9].

10.1.2.2 Air/Fuel Ratio

In the furnaces, the combustion system should be optimized by adjusting the minimum air-fuel ratio that will not cause combustion problems. If the excess air quantity is higher than necessary, the amount of flue gas increases. The increasing amount of air heats up to the flue gas temperature and receives energy. This causes energy to be thrown out of the stack. In addition, the increase in the amount of flue gas causes an increase in gas flow rate; in other words, velocity, and consequently, this causes decrease in heat transfer ratio. Excess air (O_2) in the furnace also causes an increase in scale/material loss. Therefore, the excess air ratio must be kept at a minimum level. To provide this, the oxygen (O_2) level in flue gas must be controlled strictly, the amount of oxygen (O_2) should be adjusted to the lowest possible level by adjusting the air. For this purpose, the following actions should be taken [10]:

- The air supply to the furnace must be sufficient amount and at constant pressure,
- The gas pressure entering the combustion unit must be constant.

10.1.2.3 Temperature of Flue Gas

One of the important factors affecting the furnace efficiency is the temperature of flue gas. If the flue gas temperature is above the acceptable values, more energy will be exhausted from the flue into the atmosphere. As a result of this, the furnace efficiency decreases. Factors affecting flue gas temperature are

- The high energy output from the flue is caused by insufficient heat transfer rate in the recuperator,
- If the combustion air is more or less than usual,
- Very high flame length in the furnace,
- Combustion in gas passageways.

To prevent such drawbacks stated

- Combustion must be carried out with sufficient air,
- Gas passageways (especially recuperators) should be periodically cleaned according to the condition of the fuel.

In practice, the flue gas temperatures provided in O_2 and CO_2 ratios suitable for modern furnaces are considered to be the optimum value. In case the flue gas temperature falls below the limit determined as *acid condensation temperature*, corrosion problems are encountered in the flue. In addition, excessive reductions in flue gas temperature may result in significant reductions in flue traction. These conditions need to be taken into account when lowering the flue gas temperature.

10.1.2.4 Recuperators

Recuperators are heat exchangers that transfer waste heat from flue gas to combustion air. In recuperators, while the combustion air passes through the pipes, the flue gas passes through these pipes; in the same direction, opposite direction, or cross direction. The pipes in the recuperator are made of heat resistant steel material. However, over time, due to the corrosive effect of the flue gas, these pipes are damaged and the amount of leakage from the high-pressure medium to the low-pressure medium also increases Since there is forced draft in the fresh air line, discharge fan (positive pressure) and flue gas line, the direction of leakage is from the high-pressure fresh air line to the low-pressure flue gas line. In order to determine the amount of air leakage in recuperators, gas composition analysis should be performed before and after the recuperator.

10.1.2.5 Wall Losses

Wall losses occur as a result of the loss of heat from the wall, ceiling, and floor by convection and radiation. When heat reaches the outer surface of the furnace, it spreads to the environment or is lost by air currents. Although modern furnaces are very well insulated, heat losses through convection and radiation have a significant effect on furnace efficiency. An insulation made to reduce the furnace surface temperature to a value above about 30 °C above the ambient temperature is sufficient and suitable for minimizing such losses.

10.1.2.6 Scale (Oxide Layer)/Material Losses

This is a problem encountered in annealing furnaces. When the steel is heated to the rolling temperature, it is known that scaling occurs when oxygen reacts with the metal surface. The amount of scaling depends on the heating time, the quality of the heated steel, the atmosphere of the furnace, the type, and use of the heating equipment. In general annealing practice, the oxidation property of burned gases is consistently very high. As the surface temperature of the steel increases, the amount of scale is increased, and this increase is due to the increasing percentages of O_2, CO_2, and water vapor during annealing. Oxidation of the log (billet) can be minimized by controlling the air/fuel ratio to reduce scale loss. The effects of air/fuel ratio (a), temperature (b), gas velocity (c), and annealing time (d) for scaling formation are given in Fig. 10.2.

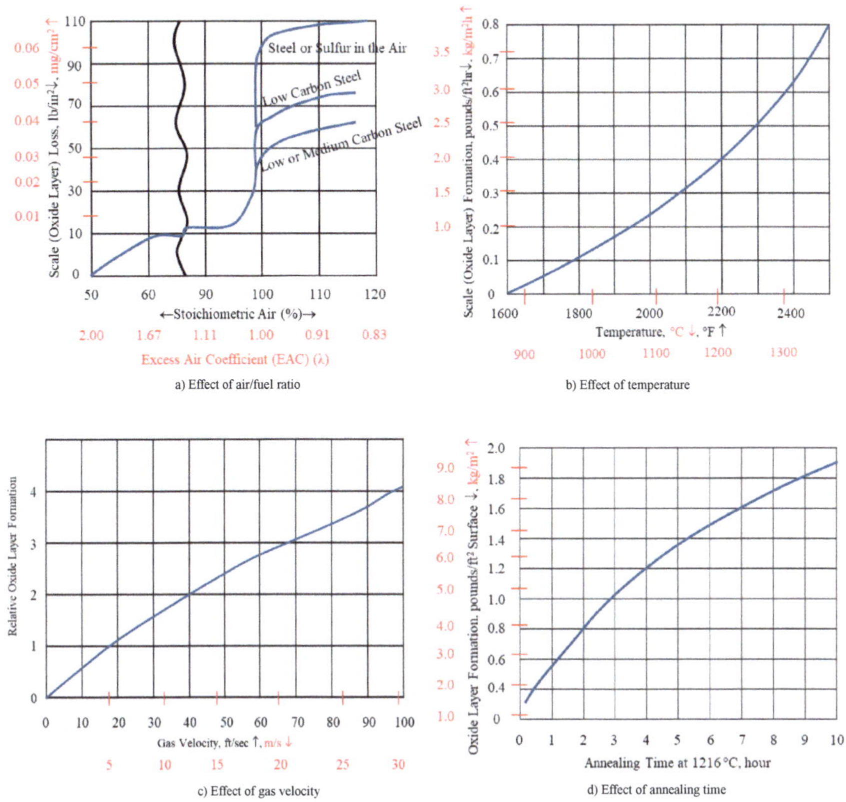

Fig. 10.2 The effect of different factors on scale formation

10.1.2.7 Cooling Water and Clearance Losses

Water cooling systems, which are necessary for the protection of skid pipes, conveyor rollers, door frames from excessive heat, reduce heat efficiency by taking heat. Radiation heat loss occurs through observation holes, open door openings, firing holes, and similar ranges in the furnaces. In addition, a significant amount of heat is lost during the opening and closing of the filling and draining lids.

10.1.2.8 Fuel Types and Burners

Since different fuels contain different amounts of carbon and hydrogen, their thermal (calorific) values are different, and the moisture content of the flue gas changes after combustion. In the burner, if the fuel pressure and temperature are not at the desired value, the fuel cannot sufficiently atomize, and this causes incomplete combustion. This has the effect of reducing efficiency.

10.2 Combustion in Furnaces

10.2.1 Theoretical Principles

The oxidation reaction of a fuel giving a large amount of energy is called *combustion*. The most commonly used oxidizer in the combustion process is air. Pure oxygen is used only in certain special applications where air is not used, such as cutting and welding.

During combustion, nitrogen acts as an inert gas, and a small amount of nitrogen does not react with other elements other than the formation of oxide. However, even the presence of nitrogen significantly affects the result of combustion. Because, generally, nitrogen enters the combustion chamber at low temperatures and exits at high temperatures by taking up most of the chemical energy generated during combustion. In most combustion processes, moisture and H_2O formed during combustion can be assumed as inert gas like nitrogen. When the combustion gases cool below the dew point of the water vapor, some moisture condenses. It is very important to know *the dew point* in advance as the water droplets combine with the SO_2 present in the combustion gases to form sulfuric acid which causes corrosion.

During combustion, the components found before the reaction are called *inputs*, and the components found after the reaction are called *products*. For example, in the combustion of 1 kmol of carbon with 1 kmol of pure oxygen by forming carbon dioxide, according to the following reaction, C and O_2 are inputs that existed prior to combustion. The CO_2 released as a result of the reaction is the product because it is in the environment after combustion.

$$C + O_2 \rightarrow CO_2 \tag{10.3}$$

If carbon burns with air instead of pure oxygen, N_2 will be present on both sides of the combustion equation. Chemical reactions are compensated based on the principle of mass conservation. The principle of mass conservation can be expressed as *the total mass of each element remains constant during the chemical reaction.* However, the total number of moles of inputs may not be equal to the total number of moles of products. In other words, the total number of moles in a chemical reaction may not remain constant.

Air combustion stoichiometry for general hydrocarbon fuels, $C_\alpha H_\beta O_\gamma$, can be defined as follows:

$$C_\alpha H_\beta O_\gamma + \left(\alpha + \frac{\beta}{4} - \frac{\gamma}{2}\right)(O_2 + 3.76N_2) \rightarrow \alpha CO_2 + \frac{\beta}{2}H_2O + 3.76\left(\alpha + \frac{\beta}{4} - \frac{\gamma}{2}\right)N_2 \tag{10.4}$$

Here, α, β, and γ are the atomic numbers of the elements C, H, and O, respectively. For the complete combustion of a fuel, the minimum amount of air required is called *stoichiometric* or *theoretical air*. In actual combustion processes, it is common practice to use more air than the stoichiometric amount to ensure complete combustion and to control the temperature of the combustion chamber. The amount of air used more than stoichiometric amount is called *excess air*. The amount of excess air (EA; %) is defined as the percentage of excess air depending on the stoichiometric air

$$EA\ (\%) = 100 \times \left(\frac{m_a - m_{sa}}{m_{sa}} \right) \tag{10.5}$$

where

m_{sa} = stoichiometric amount of air and
m_a = amount of air.

In the analysis of the combustion process, the size that is used to define the amount of fuel and air numerically is the air/fuel ratio (A/F) (or Air-Fuel Ratio (AFR)). It is defined as the ratio of mass of air to mass of fuel in combustion processes.

$$AFR = \frac{m_a}{m_f} \tag{10.6}$$

where

m is the mass of matter and is given by $m = N \times M$
N is based on the number of moles, M is the molar mass.

The actual air/fuel ratio divided by the stoichiometric air/fuel ratio gives the *equivalence ratio* (φ).

$$\varphi = \frac{AFR}{AFR_s} \tag{10.7}$$

Here, If $\varphi < 1$, it is incomplete combustion; if $\varphi = 1$, it is stoichiometric combustion; if $\varphi > 1$, it is rich combustion.

Another term used in connection with the combustion of fuels is the heating value of the fuel. *The heating value of the fuel* is defined as the amount of heat generated when the fuel is completely burned and turned into inputs/products. The heating value depends on the phase of the water in the products. When the water in the products is in the liquid form, the thermal (calorific) value is called the *higher heating value* (HHV); when the water in the products is vapor form, the thermal value is called the *lower heating value* (LHV).

10.2.2 Energy Equivalence

During the period of operation of many kinds of machines used in engineering, such as turbines, compressors, nozzles, and so on, and their input, output, and other operating conditions do not change. Thermodynamic analysis of these machines is made with a realistic model called *continuous flow open system*. In a *continuous flow open system*, the flow of fluid through the control volume is continuous.

In *continuous flow open systems*, the total energy in the control volume is constant (ECV = constant). Thus, the total energy change in the control volume is zero (ΔECV = 0). Therefore, in all forms, the amount of energy entering the control volume must be equal to the amount of energy exiting the control volume. The energy conservation for continuous flow open systems can be written as follows:

$$\dot{E}_i = \dot{E}_o \tag{10.8}$$

The energy balance for rolling mill annealing furnaces can be written as follows:

$$\dot{Q}_f + \dot{Q}_{ca} + \dot{Q}_{li} + \dot{Q}_{cwi} = \dot{Q}_{lo} + \dot{Q}_{fg} + \dot{Q}_{cwo} + \dot{Q}_{loss} \tag{10.9}$$

where

$\dot{Q}_f$ = Energy of fuel,
$\dot{Q}_{ca}$ = Energy of combustion air,
$\dot{Q}_{li}$ = The energy of the log (billet) at the furnace inlet,
$\dot{Q}_{cwi}$ = The energy of the cooling water at the inlet of the furnace,
$\dot{Q}_{lo}$ = The energy of the log (billet) at the furnace outlet,
$\dot{Q}_{fg}$ = The energy of flue gas,
$\dot{Q}_{cwo}$ = The energy of the cooling water at the furnace outlet, and
$\dot{Q}_{loss}$ = Lost energies.

The energy entering the system by burning the fuel is calculated as follows:

$$\dot{Q}_f = \dot{m}_f \times LHV \tag{10.10}$$

where

$\dot{m}_f$ = The amount of fuel
LHV = The lower heating value of the fuel

Combustion air energy $(\dot{Q}_{ca})$ is calculated based on the amount of combustion air ($\dot{m}_{ca}$), specific heat $(C_{p(ca)})$, and temperature (T_{ca}):

$$\dot{Q}_{ca} = \dot{m}_{ca} \times C_{p(ca)} \times T_{ca} \tag{10.11}$$

The energy transferred to the log in the furnace is calculated from the following equation:

$$\dot{Q}_{log} = \dot{Q}_{lo} - \dot{Q}_{li} = \left(\dot{m}_{lo} \times C_{p(lo)} \times T_{lo}\right) - \left(\dot{m}_{li} \times C_{p(li)} \times T_{li}\right) \tag{10.12}$$

where

$\dot{m}_{lo}$ = Flow rate of the log at the furnace outlet,

$\dot{m}_{li}$ = Flow rate of the log at the furnace inlet,
$C_{p(lo)}$ = The specific heat of the log at the furnace outlet temperature,
$C_{p(li)}$ = The specific heat of the log at the furnace inlet temperature,
T_{lo} = The temperature of the log at the furnace outlet, and
T_{li} = The temperature of the log at the furnace inlet.

The energy transferred to the cooling water in the furnace is calculated as follows:

$$\dot{Q}_{cw} = \dot{Q}_{cwo} - \dot{Q}_{cwi} = \left(\dot{m}_{cwo} \times C_{p(cwo)} \times T_{cwo}\right) - \left(\dot{m}_{cwi} \times C_{p(cwi)} \times T_{cwi}\right) \quad (10.13)$$

where

$\dot{m}_{cwo}$ = Flow rate of the cooling water at the furnace outlet,
$\dot{m}_{cwi}$ = Flow rate of the cooling water at the furnace inlet,
$C_{p(cwo)}$ = Specific heat of the cooling water at the outlet of the furnace,
$C_{p(cwi)}$ = Specific heat of the cooling water at the inlet of the furnace,
T_{cwo} = The furnace outlet temperature of the cooling water, and
T_{cwi} = The furnace inlet temperature of the cooling water.

The energy transferred to the flue gas in the furnaces is calculated as follows:

$$\dot{Q}_{fg} = \dot{m}_{fg} \times C_{p(fg)} \times T_{fg} \quad (10.14)$$

where

$\dot{m}_{fg}$ = Flue gas flow rate from the furnace,
$C_{p(fg)}$ = The specific heat of the flue gas exiting from the furnace, and
T_{fg} = The temperature of the flue gas coming out of the furnace.

10.3 Energy Saving in Metal Melting Process

In metal main industry and soil industry which constitutes approximately 65% of all industrial consumption, the share of energy in total costs varies between 20 and 60%. Therefore, in most energy-intensive countries, there is a potential of 20% energy savings, corresponding to approximately \$1 billion per year.

Furnaces with different properties are used in the melting process (Table 10.1). Natural gas and electricity are generally used as energy sources in the furnaces. Measures to be taken to save energy by reducing energy consumption in furnaces and possible savings rates are given in Table 10.2. The measures given herein are applicable to small capacity single melting technologies. However, in the entire iron and steel sector, preheating and oxygen enrichment of the charge (scrap) is applied in all melting processes [11–14]. In gas-fired melting furnaces, preheating the combustion air is a very common method. Thermal efficiency in melting furnaces can be seen in Table 10.1.

Table 10.1 Thermal efficiency in melting furnaces

Melting furnace type	Thermal efficiency (%)
Gas-fueled crucible furnace	7–9
Cupola	40–50
Arc furnace	35–45
Induction	50–76
Electric reflecting furnace	50–76
Gas reflecting furnace	30–45
Rotary furnace	35
Flue type furnace	40–45

Table 10.2 Energy saving values in new melting technologies

Melting technology	Thermal efficiency (%)
Preheating of filler	5–10
Cooling	5–10
Air preheating	10–20
Mixing molten metal	5–30
Workplace improvement	0–30
Oxygen enrichment technology	0–40

Metal melting time depends on the structure of the furnace, the type of energy, the quality of the refractory used, the size of the raw material, the auxiliary equipment, and the amount of additional fuel. Energy management of the melting device theoretically depends on the amount of energy, melting temperature, purity of the metal, alloying, loading, and making [15–17].

10.4 Case Study for Energy Survey in Furnaces

10.4.1 Measurement Methods and Measuring Instruments

In the furnace, to establish the balance of mass and energy between the furnace inlet and outlet and the recuperator inlet and outlet; velocity, pressure, temperature, and combustion gas measurements of the flows were made, in addition, values were read from the existing meters on the system. The measured values and the values obtained from the current meters were used to establish the mass and energy balances [18]. Schematic representation of the annealing furnace operating and measuring system is shown in Fig. 10.3.

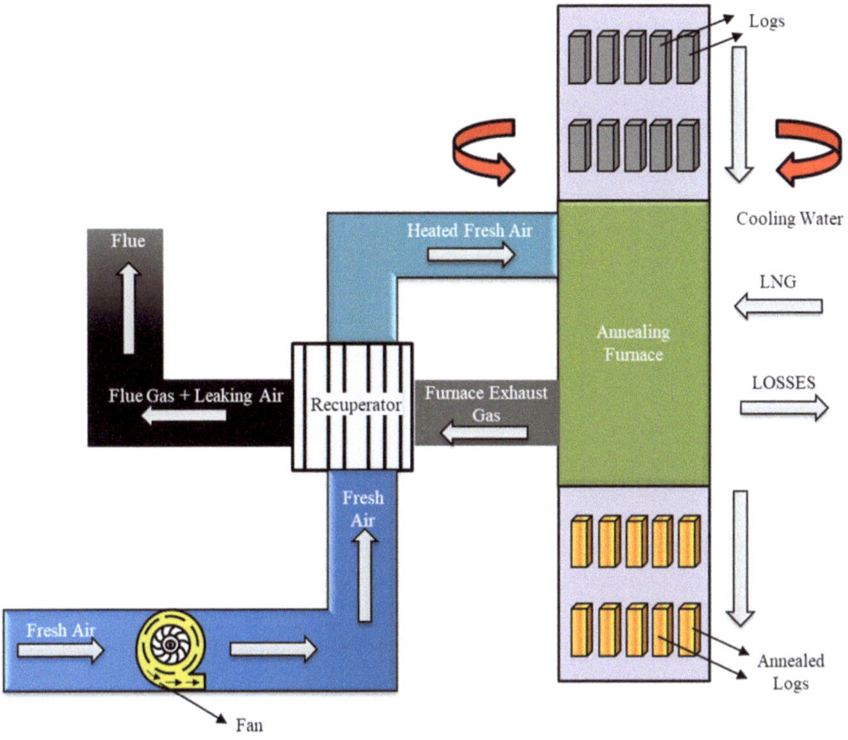

Fig. 10.3 Schematic representation of the annealing furnace operating and measuring system

LNG (liquefied natural gas) flow rates used as fuel in the furnace were taken from the control room. The components and flow rates of the combustion products were calculated using the content of LNG. The calculated theoretical values were also compared with the gas analysis, pressure, velocity, and temperature measurements at the furnace outlet and flue outlet and their accuracy was checked. A model of a brand (X1) and (X2) gas analyzers working with electrochemical detector method were used for flue gas, exhaust gas analysis of output of furnace and exhaust gas analysis of output of recuperator.

Flue gas velocity and pressure values were measured with a device (X3). Flue gas flow rates were calculated based on these values. After measuring the flue gas flow rates, the furnace combustion gas flow rate was calculated using the flue outlet O_2 percentage and the furnace outlet O_2 percentages. The information about the amount of log supplied to the system and information about the amount of scaling in the system was obtained from the enterprise. In addition, their input and output temperature values were measured with temperature measuring devices.

10.4.2 Evaluation of Measurement and Calculation Results

In order to determine the energy-saving potential of annealing furnaces in iron and steel rolling mill of an industrial enterprise; combustion gas analysis, velocity, pressure, and temperature measurements were performed. Mass and energy balances were established using the measurement data. Based on the mass and energy balances, efficiency, potential saving areas, and amounts of savings were calculated for each furnace. The following conditions which are effective in efficiency losses have been evaluated:

- Operation of furnace with high excess air coefficients,
- Low efficiency of the current recuperator,
- Flue gas losses.

10.4.2.1 Rolling Mill Annealing Furnace Measurement Results

Measurement Results: Iron and steel rolling mill annealing furnace before and after recuperator oxygen content and exhaust gas temperature measurement values and fuel flow rates are given in Table 10.3. LNG analysis results fed to the furnace are given in Table 10.4. The amount of fuel, the combustion gas oxygen percentage of furnace, and the combustion gas analysis using the elemental analysis of fuel were made and the results are given in Table 10.5.

Using the values given in Table 10.5, total stoichiometric combustion gas, theoretical combustion air, excess air, theoretical total combustion air, furnace outlet gas flow rates, and excess air ratio was calculated and given in Table 10.6. Accuracy of the calculated flow rates has been ensured by measuring at the flue outlet.

Annealing Furnace Efficiency: The mass flow rate of the log is calculated based on the amount of heat given to the log and fuel heat (Table 10.7).

Table 10.3 Annealing furnace measurement values and fuel flow rates

Rolling mill annealing furnace		
	O_2 (%)	Temperature (°C)
Before recuperator	8.64	587
After recuperator	8.64	458
LNG (m^3/h)	506	

Table 10.4 LNG analysis values

Fuel composition (Mass, %)	CH_4	C_2H_6	C_3H_8	C_4H_{10}	CO_2	N_2	Lower heating value (kcal/kg)
Original base	94.9	3.1	1.0	0.3	0.5	0.2	8.414

Table 10.5 Annealing furnace LNG analysis, combustion products, and flow rates

Fuel			Combustion products (m^3/h)					
Fuel analysis	% (ob)	Stock. O_2	CO_2	N_2	Argon	H_2O	O_2	Other
CH_4	94.9	960	482	3579	43	1044	0	0
C_2H_6	3.1	55	31	205	2	52	0	0
C_3H_8	1.0	25	15	94	1	22	0	0
C_4H_{10}	0.3	10	6	37	0	8	0	0
CO_2	0.5	0	0	0	0	0	0	0
N_2	0.2	0	3	0	0	0	0	0
Total	**100%**	0	0	0	0	0	0	0
Excess air (m^3/h)			1	3123	37	73	838	0
Gas composition % (ob)			5.55	72.57	0.86	12.38	8.64	0.00

Table 10.6 Annealing furnace combustion air and furnace combustion gas flow rate

Variable	Value
Theoretical combustion air (m^3/h)	5106
Theoretical furnace output gas flow rate (m^3/h)	5626
Excess air volume (m^3/h)	4073
Total furnace output gas flow rate (m^3/h)	9699
Air entering the furnace (m^3/h)	9179
Excess air rate (%)	79.77

Table 10.7 Rolling mill annealing furnace efficiency

Variable	Value
The mass flow rate of the log (m_{log}, kg/h)	13,900
Heat transfer amount to the log, Q_{log} (kcal/h)	2,276,820
Total fuel heat energy Q_f (kcal/h)	4,257,484
General efficiency (Q_{log}/Q_f, %)	**52.76**

Energy and Mass Balance of the System: Using the measured and calculated values, energy, and mass balances were established for the furnace and recuperator and the results are given in Tables 10.8, 10.9 and 10.10.

Total energy balance of annealing furnace Sankey diagram can be seen in Fig. 10.4.

Table 10.8 Rolling mill annealing furnace energy and mass balance

Input	Flow rate (m^3/h)	Temperature (°C)	C_p ($kcal/m^3K$)	Q (kcal/h)	%
LNG (Burning heat)	506		8414	4,257,484	90.54
LNG (Sensible heat)	506	18	0.43	3916	0.08
Combustion air entering the furnace	9179	142	0.32	412,284	8.77
Log	13,900 *kg/h	18	0.11 **kcal/kg°C	28,523	0.61
Total				**4,702,207**	
Output	Flow rate (m^3/h)	Temperature (°C)	C_p ($kcal/m^3K$)	Q (kcal/h)	%
Log	13,553*	1050	0.16**	2,276,820	48.42
Furnace outlet gas	9699	587	0.38	2,163,449	46.01
Scale loss (2.5%)	348*	1050	0.16**	58,380	1.24
Other losses				203,559	4.33
Total				**4,702,207**	

*kg/h
**kcal/kg °C

Table 10.9 Annealing furnace recuperator energy and mass balance

Input	Flow rate (m^3/h)	Temperature (°C)	C_p ($kcal/m^3K$)	Q (kcal/h)	%
Furnace outlet gas	9699	587	0.38	2,163,449	97.69
Combustion air entering the furnace	9179		0.31	51,253	2.31
Recuperator leaking air	0	18	0.31	0	0.00
Total				**2,214,702**	
Output	Flow rate (m^3/h)	Temperature (°C)	C_p ($kcal/m^3K$)	Q (kcal/h)	%
Flue gas	9699	458	0.37	1,643,585	74.21
Furnace combustion air	9179	142	0.32	412,284	18.62
Other losses				158,833	7.17
Total				**2,214,702**	

*kg/h
**kcal/kg °C

Table 10.10 Total energy and mass balance of annealing furnace

Input	Flow rate (m^3/h)	Oxygen (%)	Temperature (°C)	C_p (kcal/m^3K)	Q (kcal/h)	%
LNG (Burning heat)	506			8414	4,257,484	98.07
LNG (Sensible heat)	506		18	0.43	3916	0.09
Air entering the furnace	9179	21	18	0.31	51,253	1.18
Log	13,900*		18	0.11**	28,523	0.66
Recuperator leaking air	0		18	0.31	0	0.00
Cooling water	0		0	1	0	0.00
Total					**4,341,177**	
Output	Flow rate (m^3/h)	Oxygen (%)	Temperature (°C)	C_p (kcal/m^3K)	Q (kcal/h)	%
Log	13,553*		1050	0.16**	2,276,820	52.45
Flue gas	9699	8.64	458	0.37	1,643,585	37.86
Recuperator leaking air	0		458	0.37	0	0.00
Scale loss (2.5%)	348*		1050	0.16**	58,380	1.34
Cooling water	0		18	1	0	0.00
Total					**362,392**	**100**

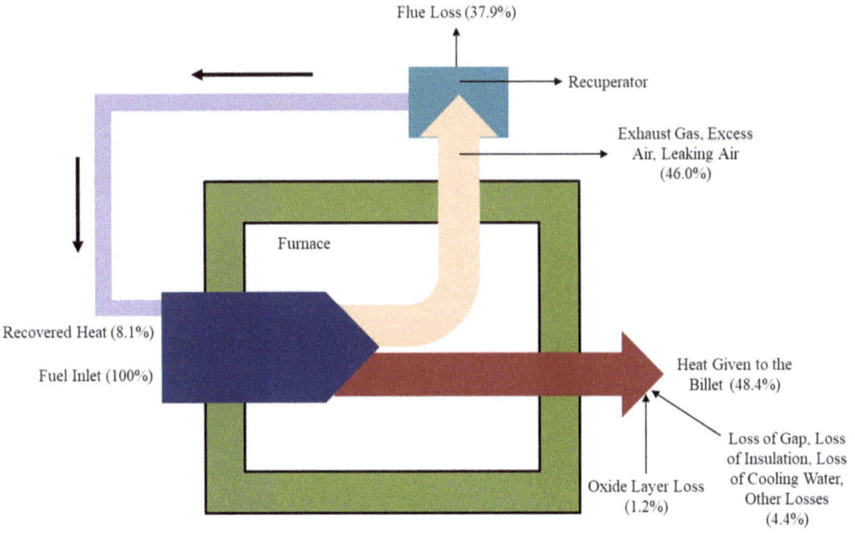

Fig. 10.4 Total energy balance of annealing furnace Sankey diagram

Table 10.11 Energy saving by reducing excess air in an annealing furnace

Variable	Value
Excess air flow rate (m^3/h)	4073
Excess air rate (%)	79.77
Target over-air rate (%)	10
C_p ($kcal/m^3K$)	0.37
Reduction in air (Nm^3/h)	3563
Air inlet temperature (°C)	18
Flue gas temperature (°C)	458
Energy saving (kcal/h)	583,837
Furnace annual working time (h)	2300
Annual energy saving (kcal)	1,342,825,400
LNG equivalent of savings (m^3/h)	76.3

10.4.3 Potential Saving Areas

10.4.3.1 Reduction of Excess Air

According to measurements in annealing furnace, it was determined that the furnace operates over optimum (10%) excess air coefficient. According to gas analysis carried out in the furnace outlet (pre-recuperator), although the furnace works over the optimum excess air coefficient, it is seen that a high amount of CO is produced in the analysis results. This is due to the fact that the air delivered to each burner is not sufficient to provide full combustion in that burner. In other words, while not enough air is delivered to one burner, more than the required amount of air is delivered to the other burner. In this case, on the one hand, there is incomplete combustion, on the other hand, a significant amount of air is heated and released into the atmosphere. When each burner in the furnace is operated with the optimum excess air coefficient, it will be possible to achieve significant energy savings. The amount of energy saving to be made in case the excess air coefficient is drawn to the optimum value is given in Table 10.11.

10.4.3.2 High-Capacity New Recuperator Manufacturing

Since the efficiency of the present recuperator is low, a very large proportion of the flue gas heat cannot be transferred to the furnace combustion air (fresh air). This energy, which is discharged from the flue without lowering the flue gas temperature below the limit specified as the *acid condensate temperature* (110–120 °C), can be recycled to the furnace with a high-efficient recuperator. The energy-saving amount to be made with the production of a new high-capacity recuperator is given in Table 10.12.

Table 10.12 Saving in case of replacing the existing annealing furnace recuperator with a high-capacity recuperator

Variable	Value
Current recuperator output flue gas temperature (°C)	458
New recuperator outlet flue gas temperature (°C)	180
Recuperator outlet gas flow rate (m^3/h)	6136
Energy saving (kcal/h)	872,402
Furnace annual working time (h)	2300
LNG equivalent of savings (m^3/h)	114.1

10.4.3.3 Prevention of Scale (Oxide Layer) Losses

When the steel is heated to the rolling temperature, it is known that scaling occurs when oxygen reacts with the metal surface. The amount of scaling that occurs, depends on the excess coefficient of the air in the furnace, the heating time, the quality of the steel being heated, the atmosphere of the furnace, the type, and use of the heating apparatus.

In the gas analysis conducted at the annealing furnace outlet (before the recuperator), it was determined that the amount of combustion air sent to the furnace was more than necessary. When the amount of combustion air is too much, the temperature of the furnace decreases and scaling occurs in the material to be annealed due to the excess air and material loss occurs. According to the information received from the plant engineers, the average scaling loss in the annealing furnace is 2.5%. In the researches, it has been understood that this value can be reduced to 1.8–2% on average.

10.4.3.4 Total Saving Amount

The total energy savings to be made in case of the above-mentioned saving potentials and the financial equivalent of this saving are given in Table 10.13.

Table 10.13 Total saving amount of annealing furnace

Savings subject	Energy (kcal/h)	LNG provision of savings (m^3/h)	Annual financial value (USD)
Reduction of excess air	583,837	69.4	95,151
Replacing the existing recuperator with a high-efficient (with high capacity) recuperator	872,402	114.1	142,180
Total	**1,456,239**	**183.5**	**237,331**

1 m^3 LNG = 8414 kcal/Nm^3

Table 10.14 Investments and repayment periods for annealing furnace

Savings subject	Investment to be made	Investment cost (USD)	Amount of savings (USD)	Repayment period (Month)
Reduction of excess air	Annealing furnace output oxygen measuring system	10,000	95,151	1.3
Replacing the existing recuperator with a high-efficient recuperator	Manufacturing a high-efficient recuperator	55,000	142,180	4.6

10.4.3.5 Investments and Repayment Periods

Investment amounts and repayment periods for savings amounts to be made by "reducing excess air" and "manufacturing a high-efficient recuperator" from the potential savings areas calculated above are given in Table 10.14.

References

1. Kilinç E, Kaya D, Çanka Kiliç F, Eyidoğan M, Özkaymak M, Taylan O, Pedrycz W (2014) An energy efficiency analysis of an industrial reheating furnace and an implementation of efficiency enhancements methods. Energy Explor Exploit 32(6): 989–1003. https://doi.org/10.1260/0144-5987.32.6.989, ISSN:0144-5987
2. Çanka Kiliç F, Sert MÖ, Eyidoğan M, Kaya D, Özdemir NC (2017) Energy saving in industrial annealing furnaces through an ORC system application. In: IIIth International Iron and Steel Symposium (UDCS'17) Iron and Steel Institute, Karabuk University, Karabuk, Turkey, 3–5th April 2017, III. International Iron and Steel Symposium (UDCS'17) Iron and Steel Institute, Karabuk University, Karabuk, Turkey, 3-5th April 2017, 3rd Iron and Steel Symposium (UDCS'17), 3–5th April 2017 Karabuk-Turkey, pp 83–88
3. Chen WH, Chung YC, Liu JL (2005) Analysis on energy consumption and performance of reheating furnaces in a hot strip mill. Int Commun Heat Mass Transfer 32:695–706
4. Trinks W, Mawhinney MH, Shannon RA, Reed RJ, Garvey JR (2004) Industrial Furnaces, 6th edn. Wiley, USA
5. Tütünoğlu Y, Güven A, Öztürk İT (2011) Energy analysis in glass tempering furnace, III. Energy Efficiency Congress, Kocaeli/Turkey, pp 153–166. (In Turkish: Tütünoğlu Y, Güven A, Öztürk İT, Cam Temperleme Fırınında Enerji Analizi, III. Enerji Verimliliği Kongresi, Kocaeli/Türkiye, pp 153–166, 2011)
6. Topbaş MA (1991) Industrial Furnaces, (Volume 1), Kurtiş Printing House, İstanbul/Turkey. (In Turkish: Topbaş, M. A., Endüstri Fırınları, (Cilt 1), Kurtiş Matbaası, İstanbul/Türkiye, 1991)
7. Topbaş MA (1992) Industrial Furnaces, (Volume 2), Damla Printing House, İstanbul/Turkey. (In Turkish: Topbaş, M. A., Endüstri Fırınları, (Cilt 2), Damla Matbaacılık, İstanbul/Türkiye, 1992)
8. Xu C, Cang D (2010) A brief overview of low CO_2 emission technologies for iron and steel making. Int J Iran Steel Res 17(3):1–7

9. Manatura K, Tangtrakul M (2010) A study of specific energy consumption in reheating furnace using regenerative burners combined with recuperator. Silpakorn U Sci Tech J 4(2):7–13
10. Karimi HJ, Saidi MH (2010) Heat transfer and energy analysis of a pusher type reheating furnace using oxygen ehanced air for combustion. Int J Iron Steel Res 17(4):12–17
11. Kirschen M, Badr K, Pfeifer H (2011) Influence of direct reduced iron on the energy balance of the electric arc furnace in steel industry. Energy 36:6146–6155
12. Kaya D, Çanka Kiliç F, Uzun E, Eyidoğan M, Taylan O (2018) Energy efficiency in glass industry, science and social sciences international Marmara Congress 2018. In: Proceedings Book Imascon the 2018 International Marmara Science and Social Sciences Congress, 23–25 November 2018, Kocaeli/Turkey, ISBN: 978-605-245-598-2, 899-905, 2018. (In Turkish: Kaya D., Çanka Kiliç F., Uzun E., Eyidoğan M., Taylan O. Cam Endüstrisinde Enerji Verimliliği, Uluslararası Marmara Fen ve Sosyal Bilimler Kongresi 2018, Bildiriler Kitabı Imascon 2018, Uluslararası Marmara Fen ve Sosyal Bilimler Kongresi, 23–25 Kasım 2018, Kocaeli/Türkiye, ISBN: 978-605-245-598-2, 899-905, 2018)
13. Trejo E, Martell F, Micheloud O, Teng L, Llamas A, Montesinos-Castellanos A (2012) A novel estimation of electrical and cooling losses in electric arc frnaces. Energy, 1–11
14. Worrell E, Martin N, Price L (1999) Energy Efficiency Opportunities in Electric Arc Steelmaking, LBNL-42775, California
15. Mullinger P, Jenkins B (2008) Industrial and process furnaces, 1st Edition, Butterworth-Heinemann, Hungary, 304, 10
16. Mullinger P, Jenkins B (2008) Industrial and process furnaces. Butterworth-Heinemann, UK
17. Turkish: Ertem G, Çelik B, Yeşilyurt S (2008) Heat Equivalence Calculations and Determination of Energy Efficiency in Industrial Annealing Furnaces, IV. Aegean Energy Symposium, Izmir/Turkey, 1–8th. (In Turkish: Ertem G, Çelik B, Yeşilyurt S, Endüstriyel Tav Fırınlarında Isı Denkliği Hesaplamaları ve Enerji Verimliliğinin Belirlenmesi, IV. Ege Enerji Sempozyumu, İzmir/Türkiye, 1-8th, 2008)
18. Çanka Kiliç F, Sert MÖ, Eyidoğan M, Kaya D, Özdemir NC (2018) Energy saving with the application of an ORC system in industrial annealing Furnaces. 6 (1), 33–43, e-ISSN 2147-9526. (In Turkish: Çanka Kiliç F., Sert M. Ö., Eyidoğan M., Kaya D., Özdemir N. C., Endüstriyel Tav Fırınlarında Bir ORC Sistemi Uygulaması ile Enerji Tasarrufu, 6 (1), 33-43, e-ISSN 2147-9526, 2018)

Chapter 11
Energy Efficiency in Pumps

One of the areas where the energy-saving potential is high is the pumping systems in energy conservation work done [1]. In a study conducted by the American Hydraulic Institute, 20% of the energy consumed in developed countries is consumed by pumps. A good system design and selection of suitable pumps have been announced to save 30–50% of this energy. This has led to a search for more efficient production and operation of the system in pump manufacturers and users. In addition, legal regulations have been introduced in some countries. For example, in the European Union, the natural labeling of circulation pumps ($P < 2.5$ kW) is its final stage. In circulation pumps produced in Germany, it was also obligatory to put energy-efficient letters on pump labels. In addition, in the European Union, diagrams have been published showing how the efficiency of the pump, known as the overpressure, discharge head, and number of revolutions, should be determined so that the pump efficiency can be controlled when centrifugal pumps are purchased [2, 3].

Using high-efficient pumps alone is not enough or a pump system to operate at maximum efficiency. The efficiency of the pump systems to operate efficiently depends only on the design of the pump, the design of the complex system, and the operating conditions. Otherwise, it is inevitable that the most efficient pump will become inefficient in a system that is improperly designed and improperly assembled [4–6].

When purchasing the pump and the motor, you should make a choice according to the Life Cycle Cost calculation (LCC), not just the first purchase cost [7]. Industrial and urban pumping systems usually have service periods of 15 years or more. This is a valid measure of the total cost of the project, taking into account total energy costs, maintenance, and other factors. LCC analysis is an effective method for determining the total costs of the projects. All cost elements such as purchasing, maintenance, energy cost, loss of production loss in case of failure, cost of dismantling are included in the life cycle cost [8]. The distribution of the LCC components in a pump system

D. Kaya et al., *Energy Management and Energy Efficiency in Industry*,
Green Energy and Technology,
https://doi.org/10.1007/978-3-030-25995-2_11

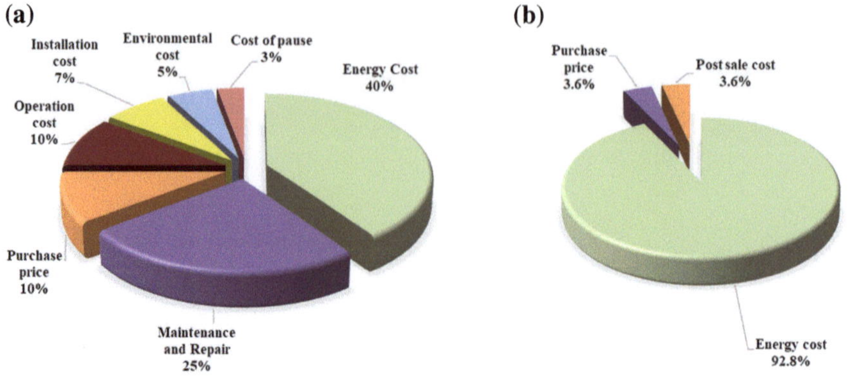

Fig. 11.1 **a** Total cost of the components **b** Cost of a 65 kW pump [1, 8]

is given in Fig. 11.1a. In a general pump system, the purchase cost of the pump is 10%, the energy cost is 40% and the maintenance cost is 25%. After the purchase of the pump, the total cost reaches 90%.

As a result of the LCC analysis conducted by Europump and Hydraulic Institute for a 65 kW deep well pump, the cost distribution is given in Fig. 11.1b. Here, the annual running time for the pump in question is 4000 h, the electricity cost is 0.075 €/kWh, and the service life is taken as 10 years. As seen in Fig. 11.1b, the share of the first purchase cost of the pump in the lifetime cost is only 3.6%, while the energy cost is 92.8% and the maintenance cost is 3.6%. As a result of this cost analysis, we can conclude that the use of high-efficient and durable pumps will be more useful.

11.1 Types of Pump

11.1.1 Centrifugal Pump

Centrifugal pumps are the most common type of pump used in plumbing systems. Figure 11.2 illustrates a cross section of a typical centrifugal pump. A pump does not create pressure, it only provides flow. Pressure is a measure of the amount of resistance to that flow. A centrifugal pump has two main components, one moving and one stationary. The moving component consists of an impeller and a shaft [9, 10].

Fluid enters the inlet port at the center of the rotating impeller, or the suction eye (Fig. 11.3). As the impeller rotates, the fluid is drawn into the blade passage at the impeller eye, the center of the impeller. The inlet pipe is axial and therefore fluid enters the impeller with a very little whirl or tangential component of velocity and flows outwards in the direction of the blades. The fluid receives energy from the impeller while flowing through it and is discharged with increased pressure and

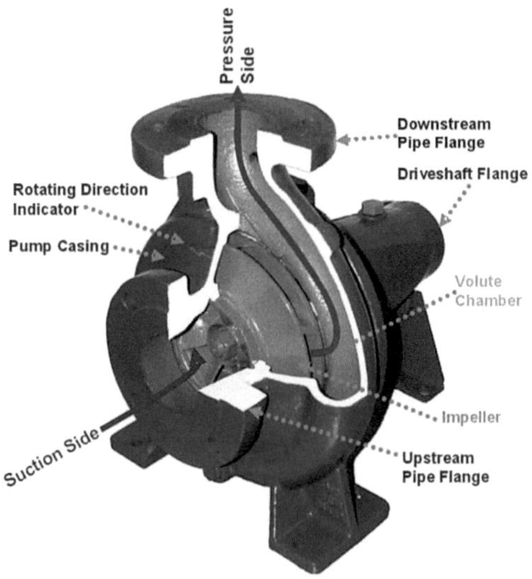

Fig. 11.2 Main components of centrifugal pump

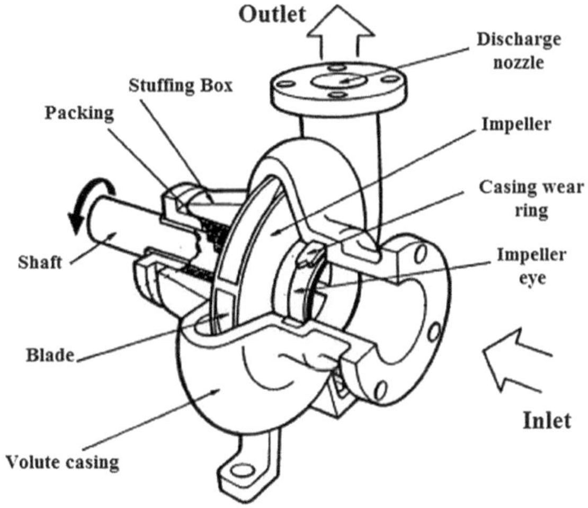

Fig. 11.3 Working principle of centrifugal pump [11]

velocity into the casing. To convert the kinetic energy or fluid at the impeller outlet gradually into pressure energy, diffuser blades mounted on a diffuser ring are used.

The stationary blade passages so formed have an increasing cross-sectional area which reduces the flow velocity and hence increases the static pressure of the fluid. Finally, the fluid moves from the diffuser blades into the volute casing which is a passage of gradually increasing cross section and also serves to reduce the velocity of fluid and to convert some of the velocity head into static head. Sometimes pumps have only volute casing without any diffuser.

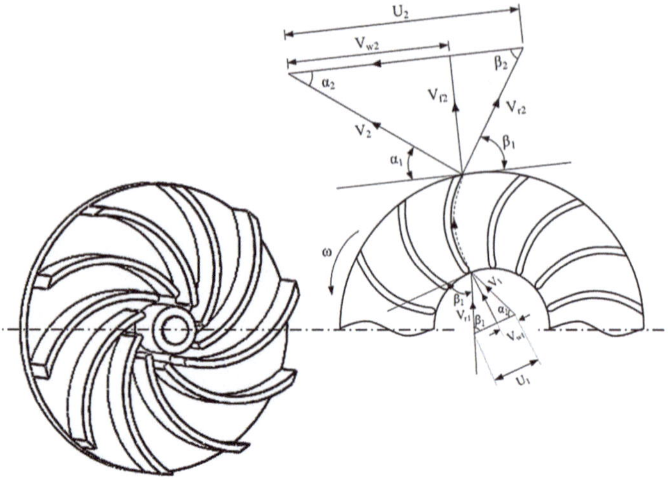

Fig. 11.4 Velocity triangles for centrifugal pump impeller

Figure 11.4 shows an impeller of a centrifugal pump with the velocity triangles drawn at inlet and outlet. The blades are curved between the inlet and outlet radius. A particle of fluid moves along the broken curve shown in Fig. 11.4.

As the impeller spins in a counter-clockwise direction, it thrusts the fluid outward radially, causing centrifugal acceleration (Fig. 11.4). As it does this, it creates a vacuum in its wake, drawing even more fluid into the inlet. Centrifugal acceleration creates energy proportional to the speed of the impeller. The faster the impeller rotates, the faster the fluid movement and the stronger its force. This energy is harnessed by introducing resistance.

Fluid is moved by a centrifugal pump through the use of centrifugal force. Fluid is taken into the center of the impeller through the inlet connection called suction eye (fluid entrance) (Fig. 11.5). Most centrifugal pumps prefer a positive inlet pressure to prevent cavitation (lack of enough positive inlet pressure to prevent liquid vaporization). This fluid is then caught by the vanes of the impeller as it spins. This

Fig. 11.5 Fluid moving into centrifugal pump

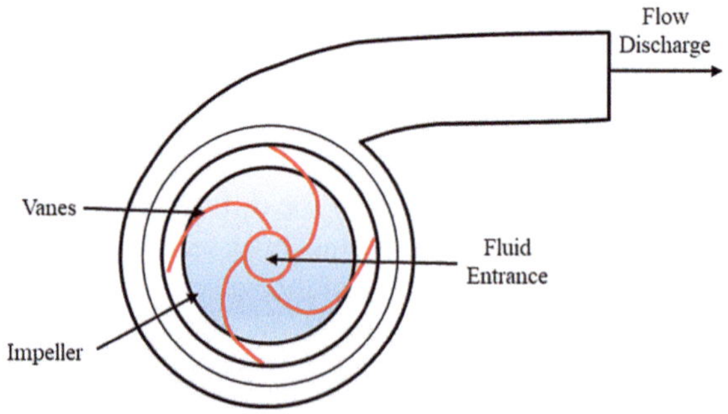

rotation of the fluid mechanically by the vanes "throws" the fluid to the outside of the impeller and toward the discharge port of the liquid end of the pump. This mechanical movement of the fluid creates the discharge pressure of the pump (Fig. 11.5). Variables like inlet fluid supply pressure, impeller diameter, motor horsepower, and closed face versus open face all affect the flow and pressure of the pump. Each of these variables can be manipulated to achieve a desired flow and/or pressure.

Impellers are the rotating blades that actually move the fluid. They are connected to the drive shaft that rotates within the pump casing. The impeller is designed to impart a whirling or motion to the liquid in the pump. Impellers are classified in several different ways (Table 11.1).

Open and semi-open impellers are less prone to clogging but require manual adjustment to the volute or back-plate to prevent internal re-circulation. Closed impellers require to wear rings, which must be replaced periodically, presenting a maintenance problem. Figure 11.6 shows types of centrifugal pumps according to the type of casing [9].

Vortex impellers are effective for solids and fibrous materials, but they are less efficient than other designs. Figure 11.7 provides a visual example of how fluid might flow through these different types of pumps [9].

Table 11.1 Classification of impellers for centrifugal pump

Direction of flow relative to the axis of the shaft	Type of suction	Mechanical construction
• *Radial* flow • *Axial* flow • *Mixed* flow	• *Single-suction*: Fluid inlet on one side • *Double-suction*: Fluid inlet on both sides	• *Open*: No shrouds or wall to enclose the vanes • *Closed*: Shrouds or sidewall enclosing the vanes • *Semi-open* or *vortex* type

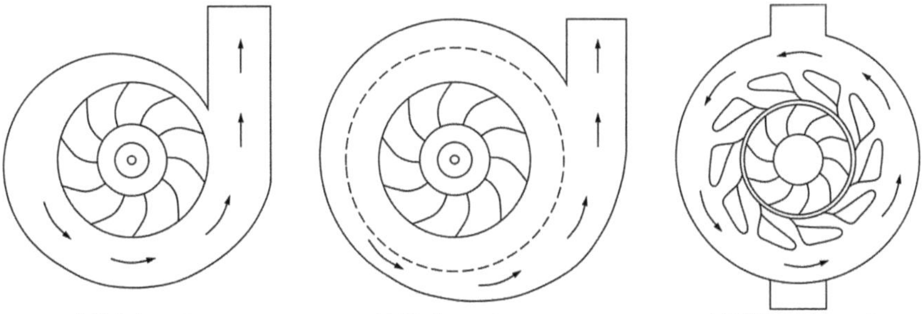

a) Volute casing b) Vortex casing c) Diffuser ring casing

Fig. 11.6 Types of centrifugal pumps according to the casing types [9]. *Source:* https://www.educationdiscussion.com/centrifugal-pump/

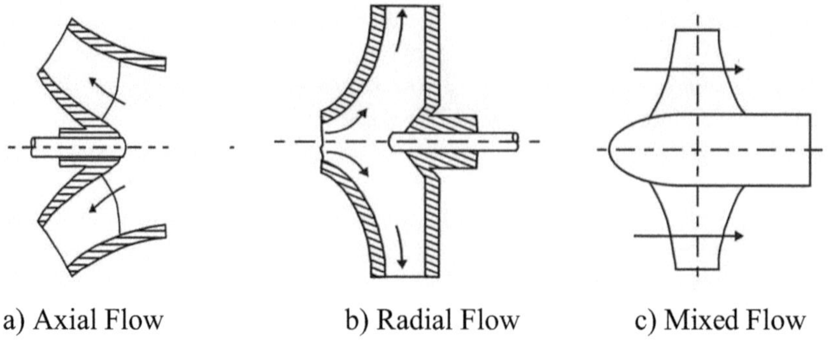

Fig. 11.7 Flow directions through different types of pumps

11.1.2 Axial Pump

The axial flow or propeller pump is the converse of axial flow turbine and is very similar to it in appearance. The impeller consists of a central boss with a number of blades mounted on it. The impeller rotates within a cylindrical casing with fine clearance between the blade tips and the casing walls (Fig. 11.8) [12].

Fluid particles, in course of their flow through the pump, do not change their radial locations. The inlet guide vanes are provided to properly direct the fluid to the rotor. The outlet guide vanes are provided to eliminate the whirling component of velocity at discharge. The usual number of impeller blades lies between 2 and 8, with a hub diameter to impeller diameter ratio of 0.3–0.6.

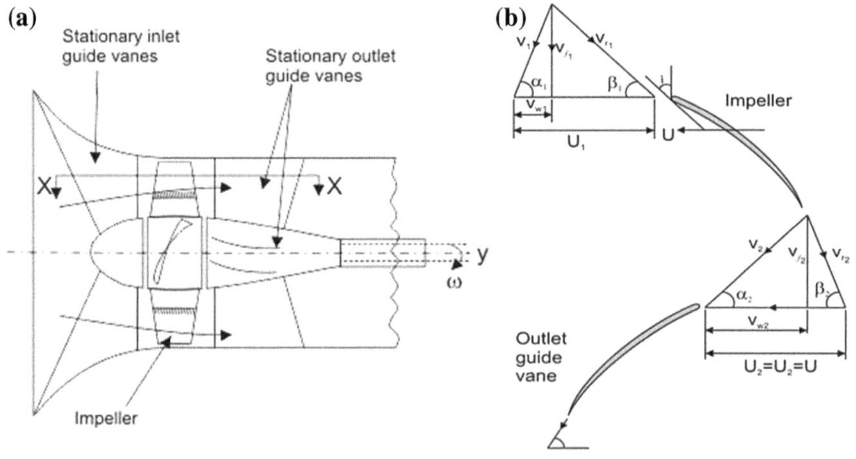

Fig. 11.8 **a** A propeller of an axial flow pump. **b** Velocity triangles *Source:* https://nptel.ac.in/content/storage2/courses/112104117/chapter_8/8_8.html [12]

Figure 11.8 shows an axial flow pump. The flow is the same at inlet and outlet. An axial flow pump develops a low head but has high capacity. The maximum head for such pump is in the order of 20 m. The section through the blade at X–X (Fig. 11.8a) is shown with inlet and outlet velocity triangles in Fig. 11.8b.

11.2 Energy Efficiency in Pump Usage

Effective use of energy in pumps can be considered in two stages, during the design and the operation (Table 11.2) [13–15].

Pump installation and the operation point of the pump can be seen in Fig. 11.9a [16] and b.

In situations where a high, constant pressure is required, consider adding speed control to the final pump in a series (Fig. 11.10). This configuration achieves the high pressure that is needed, while keeping a low flow, because the fixed-speed pump feeds into the speed-controlled pump, which adjusts its output with a pressure transmitter to add only enough head to maintain a constant pressure.

The combined curve for parallel pumps is created from the addition of the flow capacities of each pump (Fig. 11.11). Two of the same pumps will result in double the flow while two different pumps will result in the addition of the flows.

Figure 11.12 shows equal-sized fixed-speed pump and speed-controlled pump connected in series. A pressure transmitter PT together with a speed controller is making certain that the pressure is constant at the outlet of P_2.

Figure 11.13a shows two speed-controlled pumps connected in parallel (same size). The orange curve shows the performance at reduced speed. In Fig. 11.13b, one pump at full speed is compared to two pumps at reduced speed. In this case, the two pumps have the highest total efficiency.

Table 11.2 Efficient use of energy in pumps

Design of pump	Use of pump
• Proper capacity and type of pump selection and piping design should be done • An electric motor of adequate strength must be selected • Highly efficient electric motor should be preferred • Appropriate auxiliary equipment (packing, bed, etc.) should be selected • An economical system (frequency converter etc.) should be selected for variable-flow systems • Increasing the number of pumps and switching the pumps in parallel according to their needs can save energy especially in variable flow systems	• The obstructions that may occur in valves, pipelines, and pumps are to be removed • Piping circuit should be sealed • Filters should be maintained regularly • Thermal insulation should be done in heating circuits • Maintenance of belts, pulleys and bearings should be done • Vibration should be avoided • Frequency control application should be examined in the existing pumps • Surface coating should be done on worn pump fans

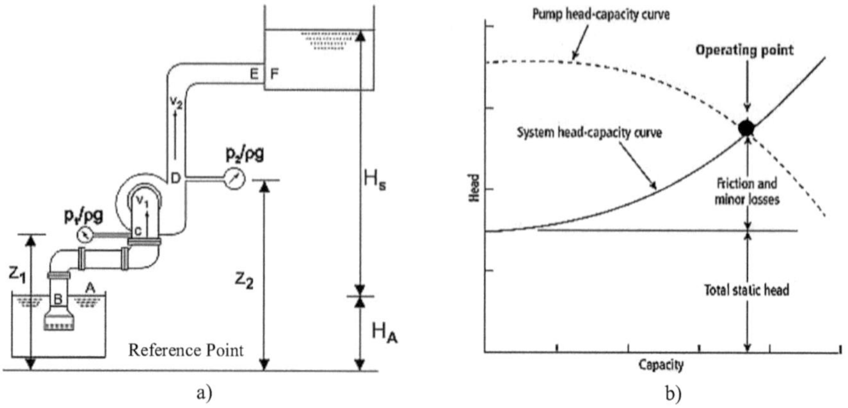

Fig. 11.9 **a** Pump installation and **b** Operation point of the pump [16]. *Source:* https://www.hkdivedi.com/2019/08/total-head-developed-by-centrifugal-pump.html [16]

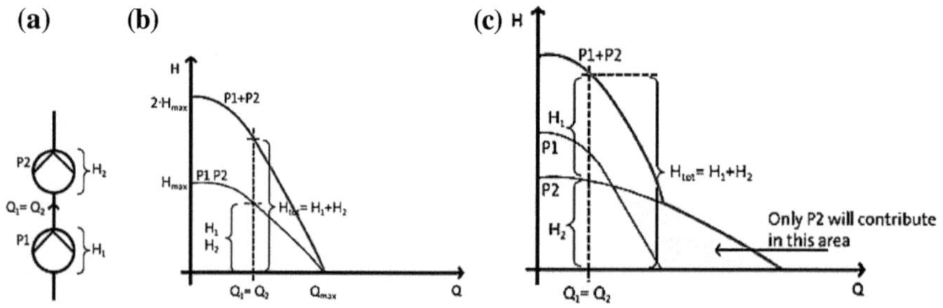

Fig. 11.10 **a** Pumps connected in series **b** Two equal-sized pumps connected in series **c** Two different-sized pumps connected in series

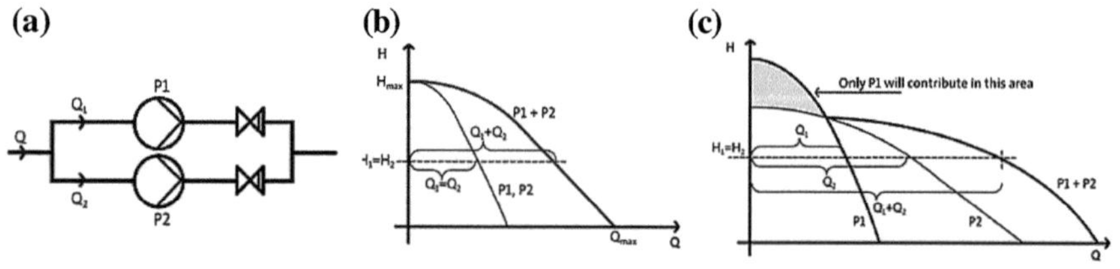

Fig. 11.11 **a** Pumps connected in parallel **b** Two pumps connected in parallel with similar performance curve **c** Two pumps connected in parallel with unequal performance curve

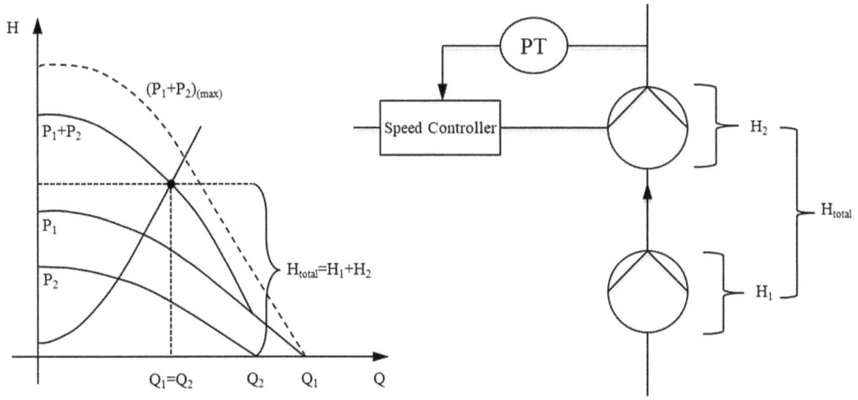

Fig. 11.12 Equal-sized fixed-speed pump and speed-controlled pump connected in series

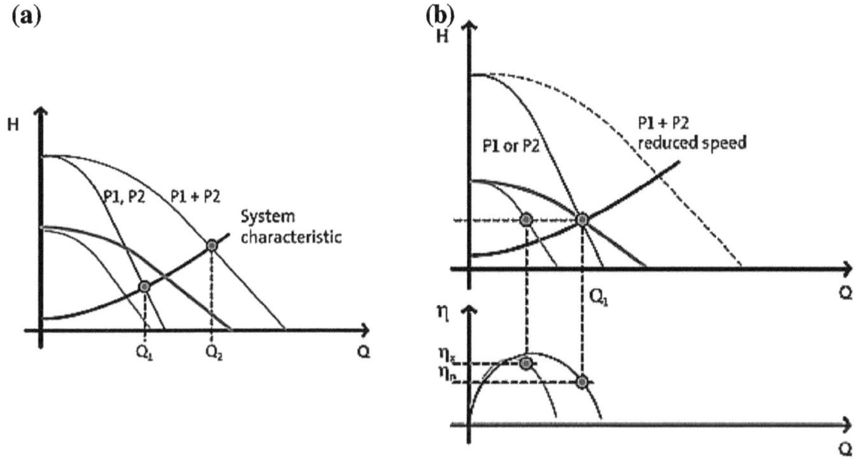

Fig. 11.13 **a** Efficiency at reduced speed. **b** The highest efficiency

11.2.1 *Efficiency in Pump Design*

11.2.1.1 Appropriate Pump Selection and Piping Design

When designing a pump selection, it is necessary to know the exact need of the process in order to select the most effective and efficient system [17]. The system should be well known throughout the year for the flow-time interval and the head pressure.

The system should be chosen to meet the highest capacity, but economically, it should be known at what capacity the system will work in most cases. Once they are known, the piping system can be designed. If maximum capacity is needed for a very short time, large-diameter pipes are not needed. If the system is operating at maximum capacity in the majority of the work, this must be taken into account when determining the pipe diameter.

While designing the pipe system, the system curve must also be drawn. It is very important to select the pump with the highest efficiency and the optimum working range. In the lifetime costs of the pumps, the initial purchase costs are only between 3 and 5%, requiring operators to make more careful choices when buying the pump [7].

11.2.1.2 Selection of Electric Motor

The selection of motors at the appropriate power is crucial for efficient operation. In general, motor capacities are chosen so that they can meet extra loads. This situation causes the engines to operate at low load and thus inefficient. Normally, the motors work more efficiently at 75% of the rated load and above. Engines operating at loads less than 50% of the rated load are chosen to be large in terms of capacity, operate inefficiently, and due to the increase in reactive current, power factors also decrease. These types of motors do not consume energy efficiently because they are chosen very strong according to their needs. The electricity consumed at low loads is converted to heat instead of mechanical power and increases the risk of malfunction resulting from overheating in the motors and shortens the life of the motor [18]. These engines should be replaced as soon as possible with engines of suitable capacity, and energy-efficient engines should be preferred when purchasing new engines. Factors affecting engine efficiency are given in Fig. 11.14.

The efficiency value indicated on the motor label indicates the efficiency when the motor is at full load. The efficiency value at different loads is different from the value stated on the label. Figure 11.15 shows the change in motor efficiency depending on loading. The efficiency of the engine is determined by looking at the efficiency load curve. The output value equals the maximum value when the engine is running at a

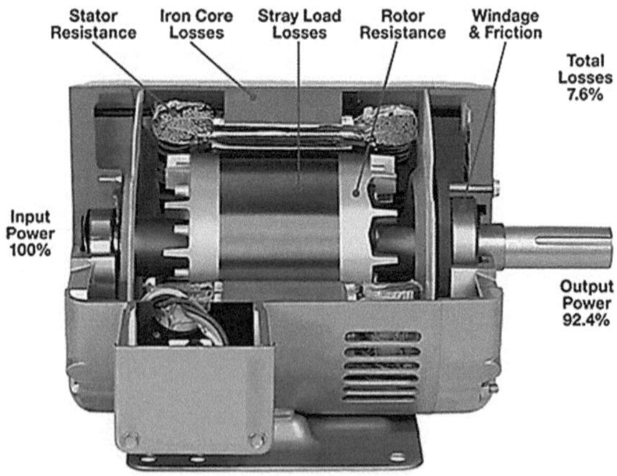

Fig. 11.14 Factors affecting motor efficiency [18]

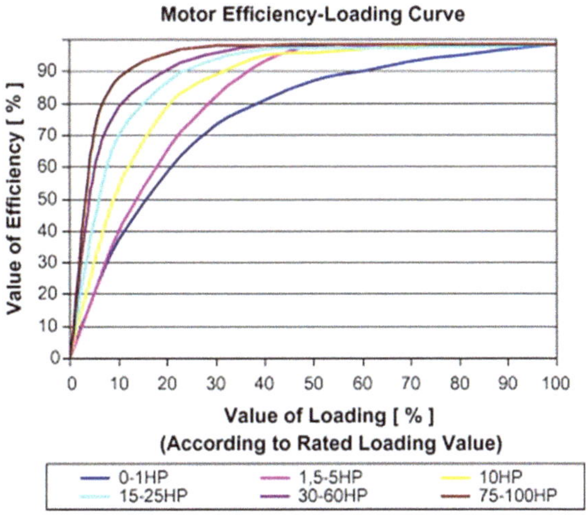

Fig. 11.15 Change of motor efficiency depending on loading

load value of 75% or greater. The preferred working area for the motors is 60–90% of the nominal load and the ideal situation is to operate at full load [19].

11.2.1.3 Selection of the High-Efficient Electric Motor

The energy consumed by electric motors constitutes about 65% of total energy consumption in plants. The cost of purchasing a typical motor is even less than 2% of the total cost of that motor [20]. The cost of energy can reach 98% of the total cost. For this reason, it is important to choose "high-efficient" motors in the business. Decision makers of some multinational companies have fully grasped this fact and have taken the decision to replace the failed standard engines with the most efficient ones and put this decision into practice [20].

The cost of high-efficient motors developed in recent years is 10–20% more expensive than standard motors, but this difference is recovered in a short time due to the fact that operating costs are often low. By increasing the cross section of the copper conductor used in the windings of these motors, the primary can reduce $I^2 \times R$ losses. Iron core losses can be limited by the increase in the size of the stator core, usually with a reduction in flux density. However, these losses can also be reduced by reducing the plate thickness and using high-quality alloys. Also, due to the reduced losses in high-efficient motors, the need for the heat energy to be released is reduced.

Like all other types of motors, electric motors cannot turn all the energy they use into mechanical energy. The ratio of the mechanical power output of the motor to the electric power drawn is called *motor efficiency*. Motor efficiency varies between 70 and 96%, depending on motor type and size. In addition, the efficiency of partial-duty motors is also low. These efficiencies also vary from motor to motor. For example, when a motor is 90% full load, 87% for semi-load, and 80% for 1/4 load, another

Table 11.3 Electric motor efficiency comparison

Nominal motor power (HP)	Nominal motor power (kW)	Average efficiency for standard motors	Average efficiency for high-efficient motors
1	0.746	0.825	0.865
1.5	1.119	0.840	0.894
2	1.492	0.840	0.888
2.5	1.865	0.812	0.870
3	2.238	0.875	0.895
4	2.984	0.827	0.889
5	3.73	0.875	0.902
7.5	5.595	0.895	0.917
10	7.46	0.895	0.917
15	11.19	0.910	0.930
18	13.428	0.878	0.924
20	14.92	0.910	0.936
25	18.65	0.924	0.941
30	22.38	0.924	0.941
40	29.84	0.930	0.945
50	37.3	0.930	0.950
60	44.76	0.936	0.954
75	55.95	0.941	0.954
100	74.6	0.945	0.958
125	93.25	0.945	0.954
150	111.9	0.950	0.958
200	149.2	0.950	0.958
250	186.5	0.954	0.962
300	223.8	0.954	0.962

Note Values are average values of eight companies and they are valid for when the engine is at full load

motor of the same specification can operate at 75% efficiency at 1/4 load, while full-load efficiency is 91%. Comparisons between standard and high-efficient motors are given in Table 11.3 and Fig. 11.16.

11.2.1.4 Selection of Variable Flow System

Different methods of obtaining a variable-flow pump system are

- Running the pump as needed (intermittent operation),
- Send a part of the fluid back to the tank by running the pump continuously (bypass system),

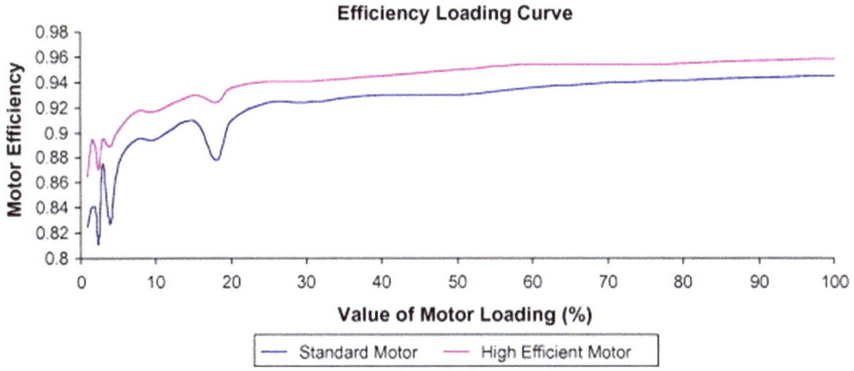

Fig. 11.16 Motor efficiency for standard and high-efficient motors

- Run the pump at the depot level by feeding the system to a depot,
- Adjust the flow rate by changing the system curve with the flow control valve at the pump outlet,
- Install a parallel working pump system,
- Changing the pump cycle with the belt pulley system,
- Set the pump speed according to the flow rate or pressure requirement by placing a hydraulic or electric clutch between the fixed-speed electric motor and the pump,
- Use the frequency converter.

Among the above methods, the frequency converters are the most commonly used and increasingly broadcast systems.

Frequency Converter Systems As a result of recent developments in electronics, the frequency converters (FC) have become cheaper and become more economical to use in pumps. The extent to which the frequency converter (FC) actually save is dependent on the fact that the output is variable. For systems operating on variable loads, the use of FC instead of valve reduction will reduce energy consumption. If the flow rate is not variable, the best solution is a constant-speed pump running at the best efficiency point (BEP). Generally, FC devices spend between 2 and 6% power. These losses increase as the engine speed (frequency) decreases. However, it can be seen that these losses can be neglected when compared to the energy saved by operating the pump at low speeds. If an existing motor FC is to be adjusted, care must be taken to ensure that the electrical characteristics of the motor and the frequency converter match. Otherwise, it is inevitable that the system will encounter problems [21].

The following factors must be considered in order to examine the energy savings provided by the FC economically:

- As a first step, the percentage distribution of the system within the study period should be determined.
- Depending on the percent change in the flow, the change in power should be determined comparatively.

- Weighted power distribution should be determined.
- The power value should be determined at full speed (in current conditions).
- Weighted power saving should be determined.
- The cost of the frequency converter should be determined.
- The monetary value of the energy to be saved according to the annual working time should be determined.

The variation of the pump efficiency depending on the speed can be explained by the similarity rules. However, affinity rules cannot be used alone without evaluating the system. System and system curves should be considered.

Similarity Rules:

- Change of flow rate, pressure, and power depending on the number of revolutions:

$$Q_2 = Q_1(n_2/n_1) \quad \Delta P_2 = \Delta P_1(n_2/n_1)^2 \quad N_2 = N_1(n_2/n_1)^3 \quad \rho \text{ and } D = \text{constant}$$

- Change of flow rate, pressure, and power depending on the diameter of rotor:

$$Q_2 = Q_1(D_2/D_1)^3 \; \Delta P_2 = \Delta P_1(D_2/D_1)^2 \; N_2 = N_1(D_2/D_1)^5 \; \rho \text{ and } D = \text{constant}$$

- Change of pressure and power depending on the density:

$$\Delta P_2 = \Delta P_1(\rho_2/\rho_1) \qquad N_2 = N_1(\rho_2/\rho_1) \qquad n, \text{ D and Q} = \text{constant}$$

According to these equations, the following comments can be made:

- The flow rate is directly proportional to the number of revolutions.
- The pressure is directly proportional to the square of the number of revolutions.
- The power changes proportionally to the number of cycles.

Theoretically, if the number of cycles is reduced to 80%, the flow rate is reduced to 80%, the pressure is reduced to 64% (=0.8 × 0.8), and the power is reduced to 51.2% (= 0.8 × 0.8 × 0.8).

Similarity rules give good results, especially for the flow rate and high pressure. As the speed decreases too much, the efficiency curves change, as the friction forces change in proportion to the hydraulic forces. In variable speed pumps, the flow rate of the pumps must not be reduced below about 60% of the nominal flow (the most productive flow) of that period. If the flow rate at the operating point of pump (Fig. 11.17) drops below 60% of the rated flow, the flow in the impeller changes and the return flows begin on the suction and/or discharge sides. In this case, in the pump vibrations and excessive noise occur and the efficiency of the pump falls excessively. In systems with a high static head, these problems will be more frequent. If the static discharge head is too high, the pump can operate in areas that should not be operated continuously, even if the speed is very low. A further decrease in speed may cause the pump to operate at zero flow. In this case, the pump needs less energy to operate at full speed. On the other hand, when the discharge pressure of

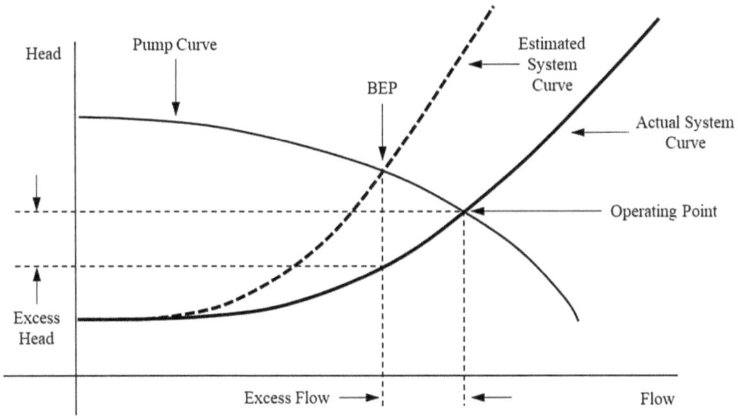

Fig. 11.17 The operating point of pump

the pump is zero, the energy delivered to the pump is transferred to the fluid and causes it to warm up. In systems where there is little or no static discharge head and variable, pump systems have FC that can save energy considerably. In systems where the static discharge head is too high, pump systems have the FC that may not benefit from energy consumption or even increase energy consumption [22].

11.2.1.5 Auxiliary Equipment

Pump auxiliary equipment helps the pump to work successfully, it is necessary for durable and reliable operation of the pumps. Valves, cooling, heating, and lubrication equipment are some of the pieces of pump auxiliary equipment. Seal problems constitute 70% of the problems experienced in pumps. Appropriate and reliable seal selection is very important for the long life of the pumps [7].

Thanks to newly developed sealing systems, mechanical unshielded seals are used in very high temperatures.

11.2.2 Efficiency in Pump Use

The most significant efficiency loss in the operation phase of the pumps is due to the pumps being operated at partial load. The highest efficiency is obtained if the pumps are operated at their nominal capacity. Moreover, if the nominal flow rate of the centrifugal pumps is assumed to be 100%, if they are working at about 40%

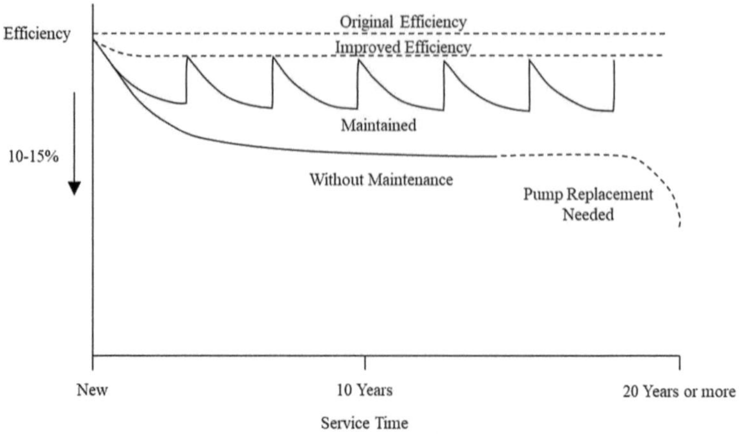

Fig. 11.18 Effect of pump periodic maintenance on pump efficiency

flow rate, generally vibration occurs, radial loads increase, excessive noise occurs, and efficiency decreases. For this reason, care should be taken that the pumps are operated close to their nominal capacity [23].

Elimination of blockages that may occur in valves, pipelines, pumps; ensuring pipe circuit sealing; regular maintenance of belts, pulleys, bearings, and filters; insulation in heating circuits and operations such as vibration prevention in the system provide significant energy and financial savings pumps, like any machine, wear out over time, and their flow rate and head decrease. The change in the efficiency of the pump is shown in Fig. 11.18. The change in pump efficiency is shown in Fig. 11.18 when the pump in this case is repaired and not repaired [24].

Despite the extra cost, coating of the pump surface and polishing to remove surface roughness can increase the efficiency of the pump. This situation is much more effective, especially in small power pumps. Depending on the operating conditions, the pumps complete their economic life within a certain lifetime. Pumps in this situation should be renewed in a specific investment plan [25].

11.3 Case Study on Energy Efficiency of Pumps

11.3.1 Introduction to Measured Pumps and Systems

Centrifugal Pumps (Unit 1) There are seven centrifugal pumps in the pump room belonging to unit 1 of an industrial establishment. These pumps are operated so that three of them are in operation at the same time. Pumps are powered by electric motors. The labeling information for the pump and motion systems are given in Table 11.4.

Bottom Pumps (Unit 2) There are seven bottom pumps in the pump bay (pump house). These pumps are operated at the same time as six of them are in operation. Pumps are powered by electric motors. The pumps are connected in parallel and their outputs converge to one line (Table 11.5).

Bottom and Centrifugal Pumps (Unit 3) There are two bottom pumps and one centrifugal pump in the industrial facility (Table 11.6). One of the pumps is in operation and the other two are backup.

11.3.2 Measurement Methods and Measurement Results

Within the scope of the energy efficiency project in the pumps, the measurements made in the factory consist of two parts, electrical and mechanical. Electrical measurements include measurements from electric motors that give motion to the pump. In mechanical measurements, the flow, pressure, and temperature values of the pumps are measured.

11.3.2.1 Electrical Measurements

In electrical measurements, the power supply voltage of the motor, the current drawn from the mains, and the power factor of the motor are measured in the motors of the pumps. Using the measurement data, the apparent power, the active power, the reactive power, the load, the working efficiency, and the power value of the electric motors are calculated, and the results are evaluated. The following equations are used in the calculation of apparent power, active power, and reactive power:

$$\text{Visible Power (VP)} = \frac{\sqrt{3} \times \text{V} \times \text{A}}{1000} \quad \text{(kVA)} \tag{11.1}$$

$$\text{Active Power (AP)} = \frac{\sqrt{3} \times \text{V} \times \text{A} \times \text{PF}}{1000} \quad \text{(kW)} \tag{11.2}$$

$$\text{Reactive Power (RP)} = \sqrt{(\text{VP})^2 - (\text{AP})^2} \quad \text{(kVAr)} \tag{11.3}$$

where

V = The voltage of the electric motor draws from the network,
A = The current of the electric motor draws from the network,
PF = Power factor.

Table 11.4 Label information of pumps and motion systems (Unit 1)

Specifications		Number of the Pump						
		1	2	3	4	5	6	7
Pump Specifications	Type of pump	Centrifugal	Centrifugal	Centrifugal	Centrifugal	Centrifugal	Centrifugal	Centrifugal
	Revolutions Per Minute (RPM)	725	725	725	970	730	730	650
	Flow rate (ton/h)	2300	2300	2300	2300	2300	2300	3000
	Water temperature (°C)	25	25	25	25	25	25	25
Electric motor	Power (kW)	210	210	210	115	120	210	132
	Voltage (V)	3300	3300	3300	3300	3300	3300	400
	Current (A)	48	48	48	27	28	48	261
	Revolutions Per Minute (RPM)	725	725	725	970	730	730	650

Table 11.5 Label information of pumps and motion systems (Unit 2)

Specifications		Number of the pump						
		1	2	3	4	5	6	7
Pump specifications	Type of pump	Bottom	Bottom	Bottom	Bottom	Bottom	Bottom	Bottom
	Revolutions Per Minute (RPM)	960	960	960	960	960	960	960
	Flow rate (ton/h)	3600	3600	3600	3600	3600	3600	3600
	Water temperature (°C)	25	25	25	25	25	25	25
Electric motor	Power (kW)	400	400	400	400	400	400	400
	Voltage (V)	3300	3300	3300	3300	3300	3300	3300
	Current (A)	89	89	89	89	89	89	89
	Revolutions Per Minute (RPM)	960	960	960	960	960	960	960

Table 11.6 Label information of pumps and motion systems (Unit 3)

Specifications		Number of the Pump		
		1	2	3
Pump Specifications	Type of pump	Bottom	Bottom	Centrifugal
	Revolutions Per Minute (RPM)	970	970	1470
	Flow rate (ton/h)	1600	1600	1135
	Water temperature (°C)	25	25	25
Electric motor	Power (kW)	75	75	75
	Voltage (V)	380	380	380
	Current (A)	147	147	147
	Revolutions Per Minute (RPM)	975	975	975

Assumptions: During measurements made on all pumps, it is assumed that there are no large sudden load changes which will change the behavior of the system. Short time values were taken on the motors.

11.3.2.2 Measuring Methods

In the measurements, a model of a brand electric energy analyzer device was used. The measurements were made in 3-phase 1-wire form. The picture of the device is given in Fig. 11.19a [26] and b. In the measurements, three voltage probes and 200 A current probe belonging to the analyzer were used.

Three voltage probes and 200 A current probes of the analyzer were used for the measurements. Measurements, for medium-voltage level (3300 V) motors, current, and voltage transformers are located in the secondary part of the supply point on the main panel of the motor. In the measurement, three voltage probes of the energy

(a)

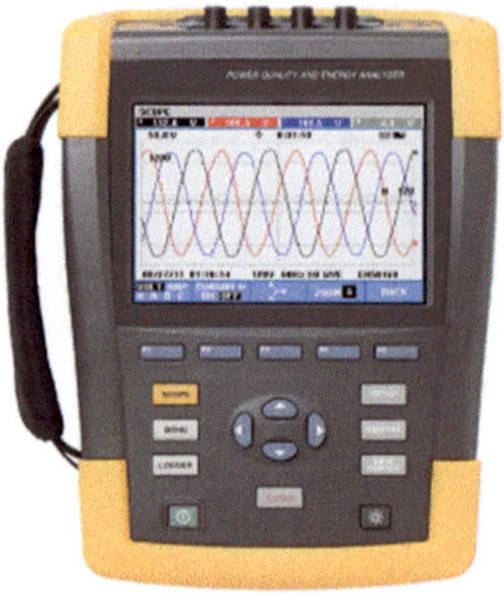

(b)

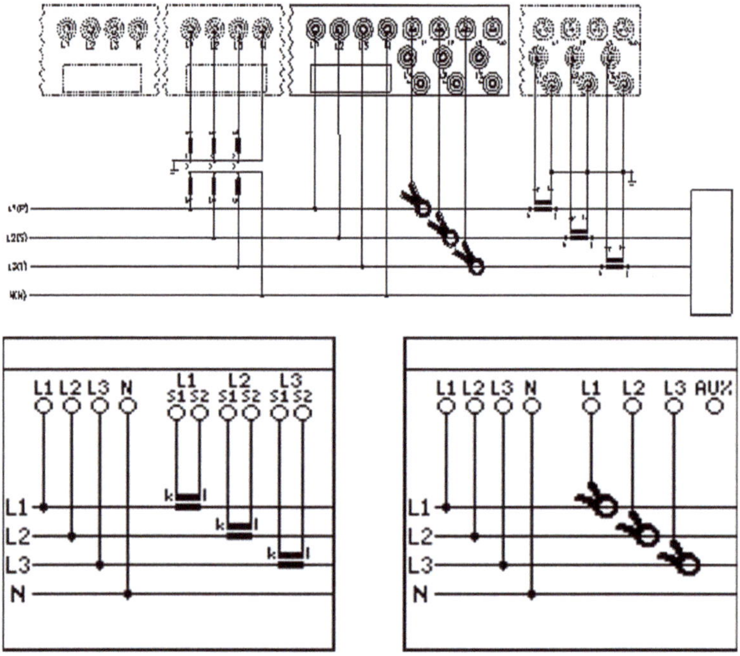

Fig. 11.19 **a** Energy analyzer [26] *Source:* https://www.fluke.com/ **b** Schematic diagram of electrical measurements

analyzer are connected to the secondary part of the voltage transformers and a 200 A current probe is connected to the secondary part of the current transformer. The measurement method is shown schematically in Fig. 11.19b Low-voltage (400 V) level of the supply voltage motors; voltage probes are connected directly to the supply point on the main pylon of the motor, and motor current is measured via a current transformer using a 200 A probe. All measurements were made while operating the existing pump of the motors during normal operation.

11.3.2.3 Measurement Points

Measurements have been made for the electric motor of 16 pumps in the factory. The names of the measuring points and the nominal values of the measured electric motors are given in Table 11.7. The measurement results for the electric motors of the pumps are given in Table 11.8.

Table 11.7 Measured electric motors and label values

Unit	Electric motors	Nominal values					
		Power		Voltage [V]	Current [A]	Revolutions per minute [rpm]	Power factor [-]
		[HP]	[kW]*				
1	1 central pump	282	210	3300	48	725	0.85
	2 central pump	282	210	3300	48	725	0.85
	3 central pump	282	210	3300	48	725	0.85
	4 central pump	154	115	3300	27	970	0.86
	5 central pump	161	120	3300	28	730	0.84
	6 central pump	282	210	3300	48	730	0.85
	7 central pump	177	132	400	261	650	0.78
2	2 tower pump	537	400	3300	89	–	0.84
	3 tower pump	537	400	3300	89	–	0.84
	4 tower pump	537	400	3300	89	–	0.84
	5 tower pump	537	400	3300	89	–	0.84
	6 tower pump	537	400	3300	89	–	0.84
	7 tower pump	537	400	3300	89	–	0.84
3	1 point pump	101	75	380	147	975	0.86
	2 point pump	101	75	380	147	975	0.86
	3 point pump	101	75	380	147	970	0.86

[a] 1 HP = 0.745 kW

Table 11.8 Power measurement in electric motors

Unit	Electric motors	Voltage [V]	Current [A]	Visible power [kVA]	Active power [kW]	Reactive power [kVAr]	Power factor [-]
1	1 central pump	3280	25.9	146.97	124.9	77.46	0.85
	2 central pump	3280	28	158.88	135.1	83.61	0.85
	3 central pump	3280	45	255.35	217.05	134.50	0.85
	4 central pump	3368	27	157.32	135.29	80.28	0.86
	5 central pump	3280	25.9	146.97	123.45	79.75	0.84
	6 central pump	3280	26.4	149.80	127.3	78.96	0.85
	7 central pump	389	252	169.59	132.28	106.12	0.78
2	2 tower pump	3280	58.6	332.52	279.32	180.41	0.84
	3 tower pump	3280	56.6	321.17	269.78	174.26	0.84
	4 tower pump	3280	56.8	322.31	270.34	175.55	0.84
	5 tower pump	3280	56.6	321.17	269.78	174.26	0.84
	6 tower pump	3280	56.6	321.17	269.78	174.26	0.84
	7 tower pump	3280	55	312.09	262.16	169.32	0.84
3	1 point pump	397	130	89.29	76.79	45.56	0.86
	2 point pump	397	134	92.03	79.15	46.95	0.86
	3 point pump	397	112	76.92	66.15	39.25	0.86

11.3.3 Mechanical Measurements

Within the mechanical measurements in the pumps, flow rate, inlet and outlet temperatures of fluid, and pressure values are measured. The flow rate of pumps was measured using the ultrasonic flowmeter X4 (a model of a brand). Two transducer pipes belonging to the flow meter are connected externally. The first transducer

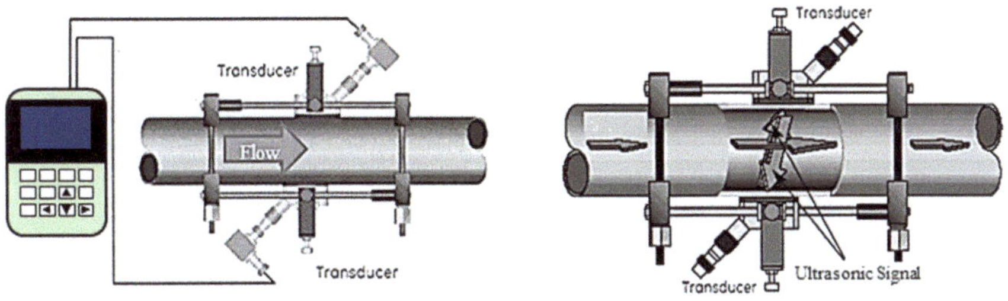

Fig. 11.20 Schematic representation of the flow measurement system

parallel to the flow is the signal generator and the second is the signal receiver. The signal arrival duration was measured, and the difference between the sound velocity and the sound velocity was determined as the fluid velocity. Since the device also measures the pipe diameter, the flow rate is measured online. The measurement system is schematically shown in Fig. 11.20.

Fluid temperatures are measured with a thermal imager at the pump inlets and outlets. A surface temperature loss value of +2 °C is added to these measured values.

11.3.4 Loading and Efficiency of Electric Motors

The mechanical power (P_{mec}) transmitted by the motor to the pump connected to the shaft is calculated as follows, depending on the active power ($P_{network}$) of the electric motor drawn from the network and efficiency value (η_m) [27]:

$$P_{mec} = P_{network} \times \eta_m \tag{11.4}$$

Power values ($P_{network}$) of the motors of the pumps taken from the network are measured in the plant. The efficiency values of these motors and the efficiency curves showing the variation of motor efficiency according to load are not available. For this reason, it is calculated by calculating which load and efficiency the motors are operating [28, 29].

Schematic representation of the motor-pump system can be seen in Fig. 11.21.

Field measurements and motor label values are used in these calculations. The loading value of the electric motors is calculated according to the current measurement technique. In calculating the efficiency of the motor operating at this loading value, the calculated loading value, the power of the motor from the network, and the label (nominal) power values are used. The motor loading value is calculated by the following formula:

$$\text{Loading}\ (\%) = \frac{I_{network}}{I_{label}} \times \frac{V_{network}}{V_{label}} \times 100 \tag{11.5}$$

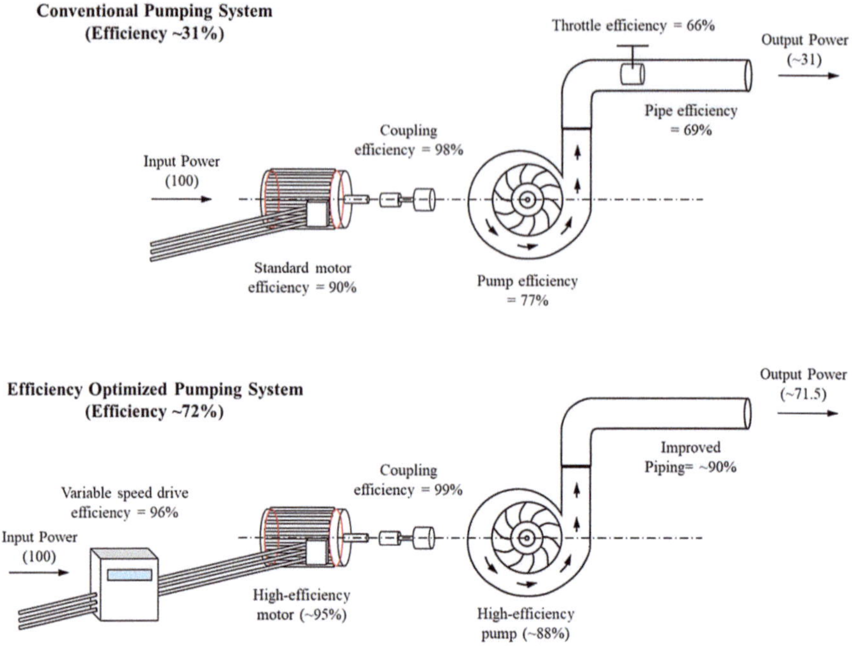

Fig. 11.21 Schematic representation of the motor-pump system

I_{label} and $I_{network}$ = Nominal current and current drawn from network of electric motor (A)

V_{label} and $V_{network}$ = Nominal voltage and the voltage measured at its terminals of motor (V)

The efficiency of the electric motor is calculated by proportioning the effective power of motor to the power drawn from the network.

$$\eta(\%) = \frac{\text{Loading} \times P_{label}}{P_{network}} \tag{11.6}$$

The loading and efficiency values of the electric motors are given in Table 11.9. The mechanical power value transferred from the motor to the pump connected to the shaft is calculated by Eq. (11.4). The loading values of pumps 1, 2, and 6 of unit 1 on electric motors are determined to be lower than 60% of the rated load. Pumps 3, 4, 5, and 7 on unit 1 and pumps 1 and 2 on unit 3 were found to be higher than 90% of the nominal load of the electric motors. The operating efficiency of the motors is in the range of 86–93%, which is in the proper range for the electric motor.

As a result of the examinations, it has been determined that electric motors 1, 2, and 6 with 210 kW power rating on unit 1 are operating at low loads. The loadings of the motors are 53%, 57%, and 54%, respectively, and the powers transmitted to the pumps are 112 kW, 121 kW, and 114 kW, respectively. If these values are lower

Table 11.9 The values of electric motors

Unit	Electric motors	Power obtained from network ($P_{Network}$) [kW]	Loading[a] [%]	Efficiency[b] [%]	Power transmitted to the pump (P_{mec}) [kW]
1	Pump 1	124.92	53.63	90.16	112.63
	Pump 2	135.05	57.98	90.16	121.76
	Pump 3	217.05	93.18	90.16	195.68
	Pump 4	135.29	102.06	86.75	117.37
	Pump 5	123.45	91.94	89.37	110.33
	Pump 6	127.33	54.67	90.16	114.80
	Pump 7	132.28	93.90	93.70	123.94
2	Pump 2	279.32	65.44	93.72	261.77
	Pump 3	269.78	63.21	93.72	252.84
	Pump 4	270.74	63.43	93.72	253.73
	Pump 5	268.78	63.21	93.72	252.84
	Pump 7	262.16	61.42	93.72	245.69
3	Pump 1	76.79	92.39	90.24	69.29
	Pump 2	79.15	95.23	90.24	71.43
	Pump 3	66.15	79.60	90.24	59.70

[a]The current measurement is based on the calculation of the load value of the electric motor
[b]The load value of the motor is also taken into account in the calculation of the working efficiency

than the motor's rated power (210 kW) and the pump efficiency is low, it will be suggested to replace it with a lower power electric motor.

The loadings of electric motors with pumping motors 1, 3, 4, 5, and 7 were determined as 93%, 102%, 91%, and 93%, respectively. The loadings for electric motors of pumps 1 and 2 on unit 3 are 92% and 95%, respectively. It is seen that the loading values in the pumps of unit 2 are between 60% and 65%. The powers transmitted to the fluid range from 240 to 265 kW. These values will be lower than the label power of the motors (compared to 400 kW) and will be recommended to replace with a lower power motor if the efficiency of the pump being operated is low. The efficiencies of the pumps at the pump stations are calculated using the flow, inlet and outlet pressures, and the electrical power transferred to the pump (Table 11.10).

11.3.5 Potential Savings and Suggestions

Potential areas of potential savings when working on industrial pump systems are

- Pumps that have low efficiency should be replaced with new ones.
- The efficiency of the pumps should be maintained in certain proportions.

Table 11.10 The values of pump efficiency

Unit	Electric Motors	Q ton/h	P_1 bar	P_2 bar	$P_2 - P_1$ bar	P_{fl} kW	P_{elec} kW	η (%)
1	Center pump No 1					112.63		
	Center pump No 2					121.76		
	Center pump No 3	2285	−0.5	0.8	1.3	82.51	195.68	42.17
	Center pump No 4	1805	−0.35	0.9	1.25	62.67	117.37	53.40
	Center pump No 5	2011	−0.5	0.8	1.3	72.62	110.33	65.82
	Center pump No 6	2280	−0.5	0.8	1.3	82.33	114.80	71.72
	Center pump No 7	2265	−0.5	1	1.5	94.38	123.94	76.14
2	Tower pump No 2	3147	−0.5	1.4	1.9	166.09	261.77	63.45
	Tower pump No 3	3240	−0.5	1	1.5	135	252.84	53.39
	Tower pump No 4	3607	−0.5	1.25	1.75	175.34	253.73	69.10
	Tower pump No 5	3416	−0.5	1.7	2.2	208.76	252.84	82.56
	Tower pump No 6	No measurement was taken due to malfunction						
	Tower pump No 7	3148	−0.5	1.5	2	174.89	245.69	71.18
3	Point pump No 1	1055	−0.5	1.15	1.65	48.35	69.29	69.78
	Point pump No 2	1145	−0.5	1.15	1.65	52.48	71.43	73.47
	Point pump No 3	890	−0.5	0.65	1.15	28.43	59.70	47.62

Q = Fluid flow rate (ton/h), P_1 = Inlet pressure of the fluid (bar), P_2 = Outlet pressure of the fluid (bar), $P_2 - P_1$ = Pressure difference (bar), P_{fl} = Power transmitted to the fluid = ($Q \times (P_2 - P_1)/36$, kW) as accepted, P_{elec} = Electrical power (kW), η = Total efficiency (P_{fl}/P_{elec}, %)

- Electric motors with high power selected should be replaced with those with appropriate power.
- Electric motors that have high efficiency should be used.
- Attention should be paid to the efficient operation of the system.

11.3.5.1 Replacement of Low Efficient Pumps

Pumps of Unit 1 As a result of the pump efficiency calculations, it was determined that the efficiencies of pumps 3 and 4 of unit 1 are lower than 60% and the efficiency of pump 5 is 65% in operating conditions. Producers have received new pump proposals with the same pressure and flow capacities as the mentioned pumps. In order to determine the flow and pressure values under the measurement conditions, the electric motor power and the pump efficiency value are calculated by using the efficiency, power, pressure, and flow diagram of the proposed pump (Table 11.11).

When existing pumps in operation are replaced with new ones, the efficiencies will increase to the values of 20–43%. In the case of replacing existing pumps and electric

Table 11.11 The existing and recommended pump values

Name of the pump		Power transmitted to the pump [kW]	Power transmitted to the fluid [kW]	Pump efficiency [%]
Pump no 3	Existing pump	195.68	82.51	42.17
	Recommended pump	99	84.6	85.5
Pump no 4	Existing pump	117.37	62.67	53.40
	Recommended pump	99	84.6	85.5
Pump no 5	Existing pump	110.33	72.62	65.82
	Recommended pump	99	84.6	85.5

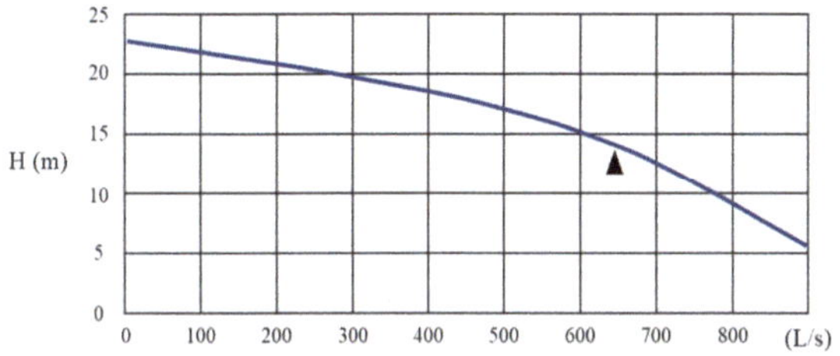

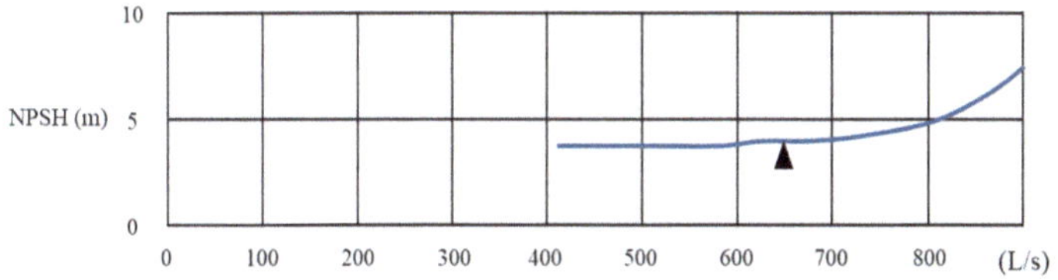

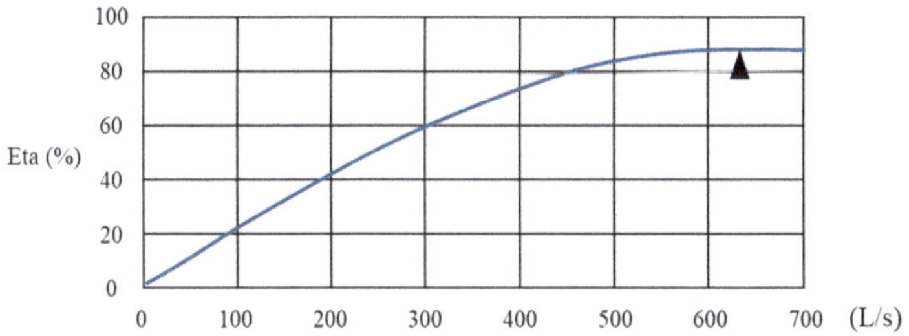

(continued)

motors with new ones, the annual monetary savings, the amount of investment, and the repayment period of the investment are given in Table 11.12.

A total of eight pumps are alternately operated in a way with three pumps in continuous operation on unit 1. Taking efficiency and energy consumption into consideration, when pumps and electric motors 3, 4, and 5 are renewed and operated continuously the payback periods of the pumps are given in Table 11.13.

Table 11.11 (continued)

Name of the pump	Power transmitted to the pump [kW]	Power transmitted to the fluid [kW]	Pump efficiency [%]

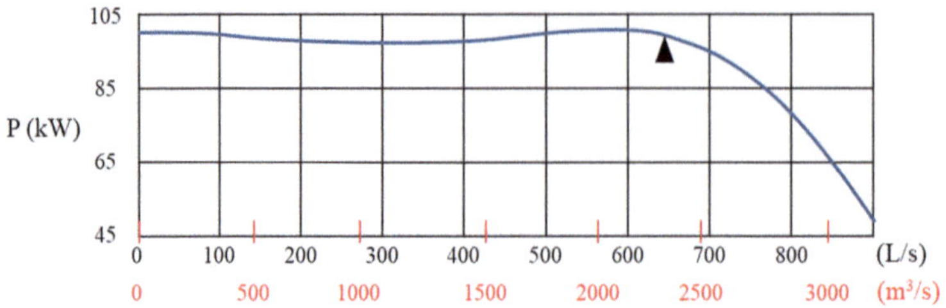

Table 11.12 Annual monetary savings, cost, and repayment period of investment in the case of replacement of existing pumps and electric motors

Pump no	Network power of the existing motor [kW]	Network power of the suggested motor [kW]	Energy saving [kWh]	Operating time [hour]	Annual financial saving[a] [€]	Investment amount[b] (Pump + Motor) [€]	Payback period [Month]
3	217.05	104.5	112,55	4128	32,522.45	47,195	17.4
4	135.29	104.5	30.79	5664	12,207.62	47,195	46.4
5	123.45	104.5	18.95	3264	4329.70	47,195	130.8

[a]Cost of 1 kWh is accepted as 7 cent EURO
[b]It is a budget offer and can be discounted during the procurement phase

Table 11.13 Annual monetary savings, investment cost, and payback periods in the case of replacement and continuous operation of the lowest efficient of three pumps and the electric motors

Pump no	Network power of the existing motor [kW]	Network power of the suggested motor [kW]	Energy saving [kWh]	Operating time [hour]	Annual financial saving[a] [€]	Investment amount (Pump + Motor) [€]	Payback period [Month]
3	217.05	104.5	112.55	8760	69,015.66	47,195	8.2
4	135.29	104.5	30.79	8760	18,880.43	47,195	30.0
5	123.45	104.5	18.95	8760	11,620.14	47,195	48.7

[a]Cost of 1 kWh is accepted as 7 cent EURO

Table 11.14 Annual monetary savings, amount of the investment, and payback periods of the investment in the case of the replacement of three pumps that have the lowest efficiency and running continuously

Pump no	Network power of the existing motor [kW]	Network power of the suggested motor [kW]	Energy saving [kWh]	Operating time [hour]	Annual financial saving[a] [€]	Investment amount (Pump) [€]	Payback period [Month]
3	217.05	110	107.05	8760	65,643.06	38,163	7.0
4	135.29	114.35	20.94	8760	12,840.41	38,163	35.7
5	123.45	111	12.45	8760	7634.34	38,163	60.0

[a]Cost of 1 kWh is accepted as 7 cent EURO

For the option given in Table 11.13, the repayment periods that will occur by replacing only inefficient pumps without renewing the electric motors are given in Table 11.14. As it can be seen in Table 11.14, from the pumps 3, 4, 5, 6, and 7, by replacing continuous operation of the pumps that are 3, 4, and 5, the repayment of the investment in case the other pumps (6 and 7) are left unchanged and hold in reserve and repayment periods vary between 7 and 60 months. In this case, it is more convenient to replace the pumps with low efficiency and to keep them running continuously and spare other pumps.

Compared to Tables 11.13 and 11.14, it is seen that the replacement of the pump only for 3 pumps + motor system and both the pump and the electric motor in the 4 and 5 pumps + motor system would be more appropriate.

In the above Tables, the modified pumps and electric motors are given only their own payback periods. On the other side, if the unit 1 pump system is taken as a whole, as a result of investments and changes made, the payback period will be even lower when annual energy consumption is taken into consideration. Table 11.15 shows the payback period of the unit 1 pump system.

Taking into account the total investment and total savings, pump system in unit 1 will start to make monetary payments of €85,045 per year after 16 months, which is the annual financial gain shown in Table 11.15.

Table 11.15 The payback period of the system for the lowest cost case

System	Energy consumption before the investment [kWh]	Energy consumption after the investment [kWh]	Yearly energy saving[a] [kWh]	Annual financial income [€]	Total investment amount [€]	Payback period [Month]
Unit 1	4,009,381	2,794,440	1,214,941	85,045.92	114,489	16.15

[a]Cost of 1 kWh is accepted as 7 cent EURO

Pumps of Unit 2 As a result of the pump efficiency calculations, it was determined that the efficiency of pump 3 on unit 2 is 53.39% and the efficiency of pumps 2, 4, and 7 is 63–72%. Since the efficiency of electric motors is about 93%, it does not need to be changed. There is a difference of about 1% between the efficiency values of the mentioned electric motors and the proposed new motors. When compared with a 1% efficiency increase (5 kWh) and an investment to be made (€22,000), the payback period is 7.1 years.

Electric motor power and pump efficiency values were determined using the efficiency, power, pressure, and flow diagram of the proposed pump (Fig. 11.22) to ensure the flow and pressure values under the measurement conditions. Calculation of the existing and proposed pump is given in Table 11.16.

As it can be seen in Table 11.16, when the pumps 2, 3, and 4 are replaced by new ones, an increase of 20–30% will be achieved. In the case of replacement of the

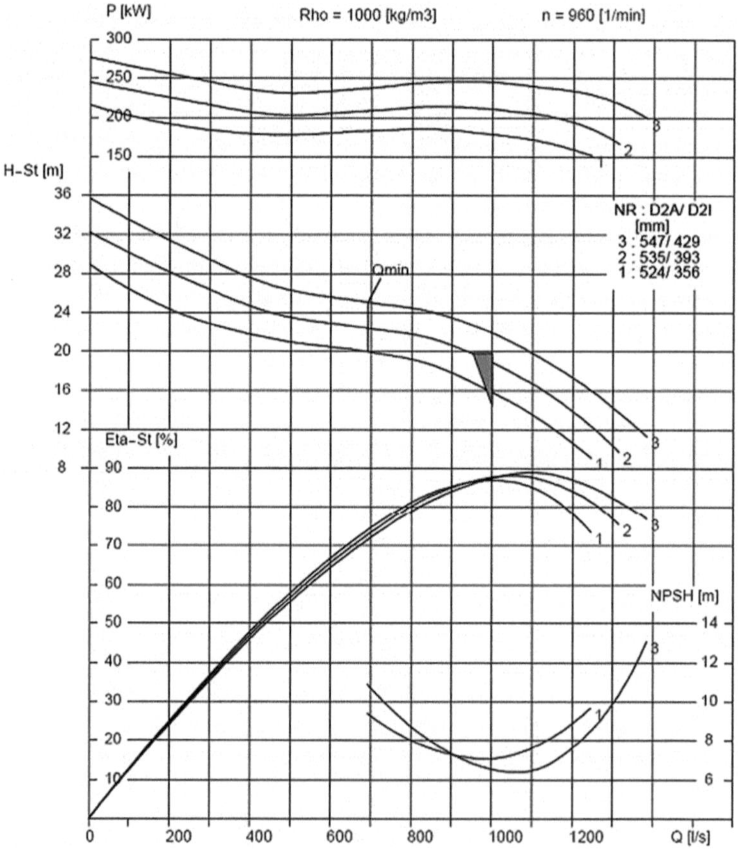

Fig. 11.22 Characteristic curves for the recommended pump

Table 11.16 Electric motor ratings of the existing and recommended pumps

Pump no		Power transmitted to the pump [kW]	Power transmitted to the fluid [kW]	Pump efficiency [%]
Pump no 2	Existing pump	261.77	166.09	63.45
	Recommended pump	231.6	191.3	82.6
Pump no 3	Existing pump	252.84	135	53.39
	Recommended pump	231.6	191.3	82.6
Pump no 4	Existing pump	253.73	175.34	69.10
	Recommended pump	231.6	191.3	82.6
Pump no 5	Existing pump	252.84	208.76	82.56
	Recommended pump	231.6	191.3	82.6
Pump no 7	Existing pump	245.69	174.89	71.18
	Recommended pump	231.6	191.3	82.6

pumps and electric motors with new ones, the annual monetary savings, the amount of investment, and the repayment period of the investment are given in Table 11.17.

As given in Table 11.17, the amount of saving that occurs when the entire pump and electric motors of unit 2 are replaced, and the payback period is between 5 and 7 years. A total of seven pumps are alternately operated in unit 2 with six pumps operating continuously. When the efficiency and energy consumption are taken into account, the pumps and electric motors 2, 3, and 4 are renewed and operated continuously, the payback period of the pumps is given in Table 11.18. The payback periods arising

Table 11.17 Annual monetary savings, amount of the investment, and repayment periods of the investment in the case of the replacement of existing pumps and electric motors

Pump no	Network power of the existing motor [kW]	Network power of the suggested motor [kW]	Energy saving per hour [kWh]	Operating time [hour]	Annual financial saving[a] [€]	Investment amount (Pump + Motor) [€]	Payback period [Month]
2	279.32	242.7	36.62	4632	11,873.67	81,000	81.9
3	269.78	242.7	27.08	8136	15,422.6	81,000	63.0
4	270.74	242.7	28.04	6120	12,012.34	81,000	80.9

[a]Cost of 1 kWh is accepted as 7 cent EURO

Table 11.18 Replacement and continuous operation of three pumps and electric motors that have the lowest efficiency

Pump no	Network power of the existing motor [kW]	Network power of the suggested motor [kW]	Energy saving per hour [kWh]	Operating time [hour]	Annual financial saving[a] [€]	Investment amount (Pump + Motor) [€]	Payback period [Month]
2	279.32	242.7	36.62	8760	22,455.38	81,000	43.3
3	269.78	242.7	27.08	8760	16,605.46	81,000	58.5
4	270.74	242.7	28.04	8760	17,194.13	81,000	56.5

[a]Cost of 1 kWh is accepted as 7 cent EURO

Table 11.19 Annual monetary savings, amount of the investment, and repayment periods of the investment in the case of replacement of three pumps that have the lowest efficiency

Pump no	Network power of the existing motor [kW]	Network power of the suggested motor [kW]	Energy saving per hour [kWh]	Operating time [hour]	Annual financial saving[a] [€]	Investment amount (Pump + Motor) [€]	Payback period [Month]
2	279.32	247.1	32.22	8760	19,757.3	59,720	36.3
3	269.78	247.1	22.68	8760	13,907.38	59,720	51.5
4	270.74	247.1	23.64	8760	14,496.05	59,720	49.4

[a]Cost of 1 kWh is accepted as 7 cent EURO

from the replacement of only inefficient pumps without renewal of the electric motors are given in Table 11.19.

As shown in Table 11.19, the duration of the repayment of the investment varies between 36 and 51.5 months as a result of continuous operation of the pumps by changing the pumps 2, 3, and 4 and reducing the working times of the other pumps. This result indicates that it is more appropriate to renew and continuously operate pumps with low efficiency. Since the efficiency of existing electric motors of the specified pump is about 93% and the difference between the recommended new motors is about 1%, it will not be economical application for changing.

In the above Tables, the modified pumps and electric motors are given only their own payback periods. However, as a result of the investments and changes made, the pump system on unit 2 is considered as a whole and when the annual energy consumption is considered, the repayment period will be further reduced. The payback period of the pump system on unit 2 is calculated and given in Table 11.20.

Considering the total investment and total savings, the company will start to make monetary savings in operating costs of around €42,529.37, which is the annual financial gain shown in Table 11.20 after 50.55 months of the unit pumps system.

Table 11.20 The payback period of system for the lowest cost

System	Energy consumption before the investment [kWh]	Energy consumption after the investment [kWh]	Annual energy saving[a] [kWh]	Annual financial gain [€]	Total investment amount [€]	Payback period [Month]
Unit 2	10,698,852	1,009,128	607,562.4	42,529.37	179,160	50.55

[a]Cost of 1 kWh is accepted as 7 cent EURO

Pumps of Unit 3 As a result of the pump efficiency calculations, it has been determined that the efficiency of pump 3 is lower than 50% and the efficiency of pump 1 is close to 70% in unit 3. Producers have been offered two new pumps with the same pressure and flow capacities as the pumps, horizontal and vertical.

In order to obtain flow and pressure values under the measurement conditions, the electric motor power and pump efficiency values are determined using the efficiency, power, pressure, and flow rate of the proposed pump (Fig. 11.23). Calculation of the existing and proposed pump is given in Table 11.21. When replacing pumps 1, 2, and 3, the efficiency will be improved by 9–34%. In the case of replacing existing pumps with new ones, annual monetary savings, investment amount, and payback period of investment are calculated and are given in Table 11.22.

Horizontal Placement of Pumps A total of three pumps are alternately operated with one pump continuously running in unit 3. When the efficiency and energy consumption are taken into consideration, the pump's payback period is given in Table 11.23 in case the pump and electric motor with 2 or 3 are renewed and operated continuously and the remaining two pumps are left in reserve. The reimbursement periods that will occur by replacing only inefficient pumps without renewing the electric motors are given in Table 11.24.

If pump 2 is continuously operated by changing and pumps 1 and 3 are not changed and left in reserve, the payback period of investment is 14.8 months (Table 11.24). In this case, the pump with a low supply and high electricity consumption should be replaced, it must be continuously operated, and the other pumps should be left in reserve.

Compared to Tables 11.23 and 11.24, it is seen that only the pump replacement is more suitable in the pump + motor system 2. In the above Tables, only their own payback periods are given for the modified pumps and electric motors. However, as a result of the investments and changes made, when the unit 3 pump system is considered as a whole and the annual energy consumption is taken into consideration, the payback period will decrease further. Considering the total investment and savings, unit 3 will start to make an annual monetary savings of € 20,921.51 after 14.5 months (Table 11.25).

Vertical Placement of Pumps As seen in Table 11.26, when pumps 1 and 3 are replaced by pump alternatives, their efficiency will increase by 3–25%. In the case of

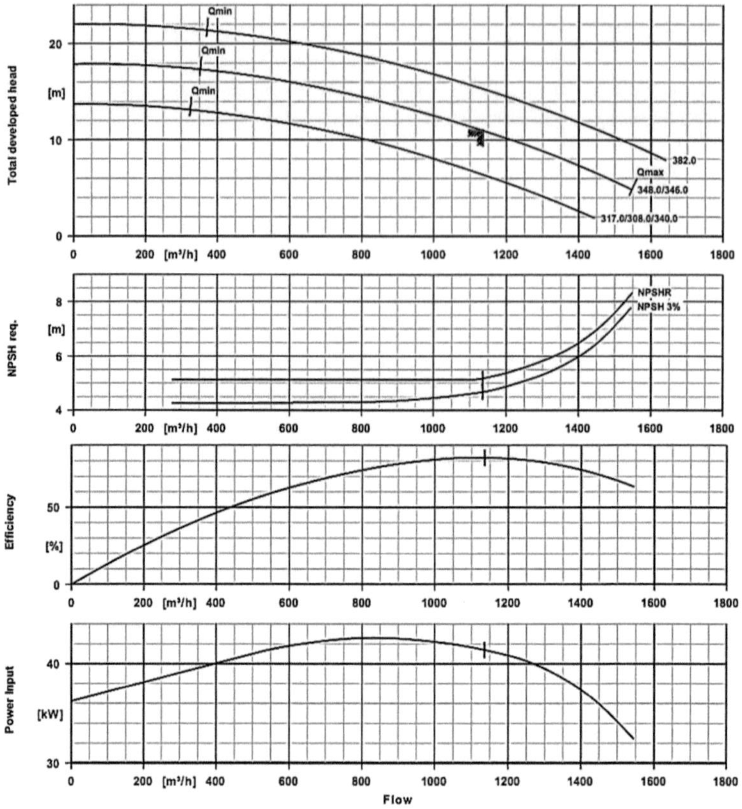

Fig. 11.23 Characteristic curves for recommended horizontal pump

Table 11.21 The existing and recommended pump values

Name of the pump		Power transmitted to the pump [kW]	Power transmitted to the fluid [kW]	Pump efficiency [%]
Pump no 1	Existing pump	69.29	48.35	69.78
	Recommended pump	41.45	34	82.1
Pump no 2	Existing pump	71.43	52.48	73.47
	Recommended pump	41.45	34	82.1
Pump no 3	Existing pump	59.70	28.43	47.62
	Recommended pump	41.45	34	82.1

Table 11.22 Annual monetary savings, the investment amount, and investment payback periods in the case of replacing existing pumps and electric motors with new ones

Pump no	Network power of the existing motor [kW]	Network power of the suggested motor [kW]	Energy saving [kWh]	Operating time [hour]	Annual financial saving[a] [€]	Investment amount (Pump + Motor) [€]	Payback period [Month]
1	76.79	44.42	32.37	1728	3915.48	30,152	92.4
2	79.15	44.42	34.73	4776	11,610.93	30,152	31.2
3	66.15	44.42	21.73	2880	4380.77	30,152	82.6

[a]Cost of 1 kWh is accepted as 7 cent EURO

Table 11.23 Annual monetary savings, the investment amount, and investment payback periods in the case of replacements and continuous operation of the lowest efficient pump and electric motor

Pump no	Network power of the existing motor [kW]	Network power of the suggested motor [kW]	Energy saving [kWh]	Operating time [hour]	Annual financial saving[a] [€]	Investment amount (Pump + Motor) [€]	Payback period [Month]
2	79.15	44.42	34.73	8760	21,296.44	30,152	17.0
3	66.15	44.42	21.73	8760	13,324.84	30,152	27.2

[a]Cost of 1 kWh is accepted as 7 cent EURO

Table 11.24 Annual monetary savings, amount of investment, and repayment of investment in the case of the replacement of pumps have the lowest efficiency and continuously running

Pump no	Network power of the existing motor [kW]	Network power of the suggested motor [kW]	Energy saving [kWh]	Operating time [hour]	Annual financial saving[a] [€]	Investment amount (Pump + Motor) [€]	Payback period [Month]
2	79.15	45.93	33.22	8760	20,370.5	25,175	14.8
3	66.15	45.93	20.22	8760	12,398.9	25,175	24.4

[a]Cost of 1 kWh is accepted as 7 cent EURO

Table 11.25 The payback period of the system for the lowest cost

System	Energy consumption before the investment [kWh]	Energy consumption after the investment [kWh]	Annual energy saving[a] [kWh]	Annual financial gain [€]	Total investment amount [€]	Payback period [Month]
Unit 3	701,225.52	402,346.8	298,878.72	20,921.51	25,175	14.5

[a]Cost of 1 kWh is accepted as 7 cent EURO

Table 11.26 The existing and recommended pumps and motor values

Name of the pump		Power transmitted to the pump [kW]	Power transmitted to the fluid [kW]	Pump efficiency [%]
Pump no 1	Existing pump	69.29	48.35	69.78
	Recommended pump	48.6	35.186	72.4
Pump no 2	Existing pump	71.43	52.48	73.47
	Recommended pump	48.6	35.186	72.4
Pump no 3	Existing pump	59.70	28.43	47.62
	Recommended pump	48.6	35.186	72.4

Table 11.27 Annual monetary savings, amount of the investment, and payback periods of the investment in the case of replacement of existing pumps and electric motors

Pump no	Network power of the existing motor [kW]	Network power of the suggested motor [kW]	Energy saving [kWh]	Operating time [hour]	Annual financial saving[a] [€]	Investment amount (Pump + Motor) [€]	Payback period [Month]
1	76.79	52.1	24.69	1728	2986.5	45,134	181.4
2	79.15	52.1	27.05	4776	9043.36	45,134	59.9
3	66.15	52.1	14.05	2880	2832.48	45,134	191.2

[a]Cost of 1 kWh is accepted as 7 cent EURO

replacing existing pumps with new ones, the annual monetary savings, the amount of investment, and the repayment period of the investment are calculated and presented in Table 11.27.

A total of three pumps are alternately operated with one pump continuously running in unit 3. When the efficiency and energy consumption are taken into consideration, the re-payment period is given in Table 11.28 in case the renewal and continuous operation of the pump 2 and electric motor and the leaving of the pumps 1 and 3 in the stand by. The payback periods due to the replacement of only inefficient pumps without replacement of the electric motors are given in Table 11.29.

As seen in Table 11.29, in case pump 2 is continuously operated by changing and pumps 1 and 3 are left in reserve without being changed, the payback period of the investment is 23.3 months. This result shows that it is more appropriate to replace and continuously running of pump has low efficiency and high electricity consumption and to keep the other pumps in reserve.

Compared to Tables 11.28 and 11.29, only the replacement of the pump seems to be more appropriate in the pump and motor system 2. In the above tables, the modified pumps and electric motors are given only their own payback periods. However, as

Table 11.28 Annual monetary savings, amount of the investment, and payback periods of the investment in the case of replacement of pumps that having the lowest efficiency and running continuously

Pump no	Network power of the existing motor [kW]	Network power of the suggested motor [kW]	Energy saving [kWh]	Operating time [hour]	Annual financial saving[a] [€]	Investment amount (Pump + Motor) [€]	Payback period [Month]
2	79.15	52.1	27.05	8760	16,587.06	30,152	21.8
3	66.15	52.1	14.05	8760	8615.46	45,734	63.7

[a]Cost of 1 kWh is accepted as 7 cent EURO

Table 11.29 Annual monetary savings, amount of the investment, and payback periods of the investment in the case of replacement of the pumps that having the lowest efficiency and running continuously

Pump no	Network power of the existing motor [kW]	Network power of the suggested motor [kW]	Energy saving [kWh]	Operating time [hour]	Annual financial saving[a] [€]	Investment amount (Pump + Motor) [€]	Payback period [Month]
2	79.15	53.85	25.3	8760	15,513.96	30,152	23.3
3	66.15	53.85	12.3	8760	7542.36	40,850	65.0

[a]Cost of 1 kWh is accepted as 7 cent EURO

a result of the investments and changes made, the pump system 3 is considered as a whole, and when the annual energy consumption is considered, the payback period will be further reduced (Table 11.30). When total investment and savings are taken into account, pump system 3 will start to make annual monetary payments of €16,064.97 per year after 22.5 months.

Table 11.30 The Payback period of the system for the lowest cost

System	Energy consumption before the investment [kWh]	Energy consumption after the investment [kWh]	Annual energy saving[a] [kWh]	Annual financial gain [€]	Total investment amount [€]	Payback period [Month]
Unit 3	701,225.52	471,726	229,499.52	16,064.97	30,152	22.5

[a]Cost of 1 kWh is accepted as 7 cent EURO

11.3.5.2 Improvement of Efficiency of Existing Pumps

Pumps 6 and 7 of Unit 1 It has been determined that the efficiency of pump 6 is about 71% and the efficiency of pump 7 is about 76% in measurements made under operating conditions. Producers have received new pump proposals with the same pressure and flow capacities as the mentioned pumps. To obtain the flow and pressure values under the measurement conditions, the electric motor power and pump efficiency values were determined using the new pump's efficiency, power, pressure, and flow rate diagram. Values of power transmitted to the pump, the power transmitted to the fluid, and the efficiency for the existing and the recommended pump are all given in Table 11.31.

As it can be seen in Table 11.31, existing pumps are about 9–13% less efficient than the new pump. Existing pump is taken into revision, wing coating, polishing, and bed care, etc., the efficiency can be increased by 5%. In this case, the annual monetary savings, the amount of investment, and the payback period of the investment are calculated and are given in Table 11.32.

Pump 7 of Unit 2 It has been determined that the pump efficiency is about 71% in measurements made under operating conditions. Producers have received new pump proposals with the same pressure and flow capacities as the pump. In order to

Table 11.31 Values of power transmitted to the pump, the power transmitted to the fluid, and the efficiency for the existing and the recommended pump

Name of the pump		Power transmitted to the pump [kW]	Power transmitted to the fluid [kW]	Pump efficiency [%]
Pump no 6	Existing pump	114.80	82.33	71.77
	Recommended pump	99	84.6	85.5
Pump no 7	Existing pump	123.94	94.38	76.14
	Recommended pump	99	84.6	85.5

Table 11.32 Annual monetary savings, amount of the investment, and payback periods of the investment in the case of revision of existing pumps

Pump no	Network power of the existing motor [kW]	Network power of the suggested motor after revision [kW]	Energy saving [kWh]	Operating time [hour]	Annual financial saving[a] [€]	Investment amount (Pump + Motor) [€]	Payback period [Month]
6	127.33	120.9	6.36	2136	951.9	3000	37.8
7	132.28	125.6	6.61	4392	2033	3000	17.7

[a]Cost of 1 kWh is accepted as 7 cent EURO

Table 11.33 Values of power transmitted to the pump, the power transmitted to the fluid, and the efficiency for the existing and the recommended pump

Name of the pump		Power transmitted to the pump [kW]	Power transmitted to the fluid [kW]	Pump efficiency [%]
Pump no 7	Existing pump	245.69	174.89	71.18
	Recommended pump	231.6	191.3	82.6

provide the flow and pressure values under the measurement conditions, the electric motor power and the pump efficiency value were determined using the efficiency, power, pressure, and flow of the new pump. Values of power transmitted to the pump, the power transmitted to the fluid, and the efficiency for the existing and the recommended pump are all given in Table 11.33.

As it can be seen in Table 11.33, the existing pump works approximately 11% less efficiently than the new pump. When the existing pump is revised by realizing the blade coating, polishing, and bed care, etc., and such necessary maintenance is performed, its efficiency can be increased by 5%. In this case, annual monetary savings, investment amount, and the payback period of the investment are calculated and given in Table 11.34.

Pumps 1 and 2 of Unit 3 Pump efficiencies in unit 3 were found to be 69.78% and 73.47%, respectively, when measured in three operating conditions. Producers have received new pump proposals with the same pressure and flow capacities as the pump. In order to obtain the flow and pressure values under the measurement conditions, the efficiency, power, pressure, and flow rate of the new pump were used to determine the electric motor power and efficiency values. Values of power transmitted to the pump, the power transmitted to the fluid, and the efficiency for the existing and the recommended pump are all given in Table 11.35.

As can be seen in Table 11.35, existing pumps are about 8–13% less efficient than the new pump. Existing pump is taken into revision, wing coating, polishing, and bed care, etc., the efficiency can be increased by 5%. In this case, the annual

Table 11.34 Annual monetary savings, amount of the investment, and payback period of the investment in the case of revision of existing pumps

Pump no	Network power of the existing motor [kW]	Network power of the suggested motor after revision [kW]	Energy saving [kWh]	Operating time [hour]	Annual financial saving[a] [€]	Investment amount (Pump + Motor) [€]	Payback period [Month]
7	262.16	249	13	7344	6738	3000	5.3

[a]Cost of 1 kWh is accepted as 7 cent EURO

Table 11.35 Values of power transmitted to the pump, the power transmitted to the fluid, and the efficiency for the existing and the recommended pump

Name of the pump		Power transmitted to the pump [kW]	Power transmitted to the fluid [kW]	Pump efficiency [%]
Pump no 1	Existing pump	69.29	48.38	69.78
	Recommended pump	41.45	34	82.1
Pump no 2	Existing pump	71.43	52.48	73.47
	Recommended pump	41.45	34	82.1

Table 11.36 Annual monetary savings, amount of the investment, and payback period of the investment in the case of revision of existing pumps

Pump no	Network power of the existing motor [kW]	Network power of the suggested motor after revision [kW]	Energy saving [kWh]	Operating time [hour]	Annual financial saving[a] [€]	Investment amount (Pump + Motor) [€]	Payback period [Month]
1	76.79	72.95	3.84	1728	464	3000	77.5
2	79.15	75.19	3.96	4776	1323.9	3000	27.2

[a]Cost of 1 kWh is accepted 7 cent EURO

monetary saving, the amount of investment, and the payback period of the investment are calculated and given in Table 11.36.

11.3.5.3 High-Efficient Electric Motor Use

If the pump motors operated by the electric motor are replaced by high-efficient motors, the energy savings to be provided are calculated. It has been investigated how much energy can be saved if the pump and current operating conditions are the same and only the motor is replaced by a high-efficient motor. High-efficient electric motors are preferred instead of existing standard electric motors, resulting in more efficient use of energy and therefore energy savings when new compressors, HVAC, and pump systems are purchased as a result of major revisions that complete the economic lifetime.

If a standard motor is replaced by a high-efficient motor, the amount of energy to be saved can be calculated as follows:

$$\mathrm{ES} = \mathrm{NM} \times \mathrm{NP} \times \mathrm{OT} \times \mathrm{LC} \times \mathrm{UF} \times \left(1/\eta_{\mathrm{standard}} - 1/\eta_{\mathrm{high-efficient}}\right). \quad (11.7)$$

where ES—Energy saving, NM—The number of motors that have the same power, NP—Nominal power, OT—Operating time, LC—Load coefficient, UF—Usage factor (for continuously running motors: UF = 1), $\eta_{standard}$—The efficiency of the standard motor and $\eta_{high\text{-}efficient}$—The efficiency of the high-efficient motor.

The efficiency comparison of standard and high-efficient motors is given in Table 11.37. The efficiency rating of high-efficient motors with a label power greater than 224 kW is unknown. The values given in Table 11.37 are the average values of the eight companies and are valid for the case where the engine is at full load.

If high-efficient motor is used, the monthly demand power saving (*DPS*) and, the monthly kilowatt-hour energy usage saving (US) is calculated as follows:

$$\text{DPS} = \text{NM} \times \text{NP} \times \text{LC} \times \left(1/\eta_{\text{standard}} - 1/\eta_{\text{high-efficient}}\right) \tag{11.8}$$

Table 11.37 Comparison of motor efficiencies

Nominal motor power (HP)	Nominal motor power (kW)	Standard type motor for the average efficiency	High-efficient motor for the average efficiency
1	0.75	0.825	0.865
2	1.12	0.84	0.894
2	1.49	0.84	0.888
3	1.87	0.812	0.87
3	2.24	0.875	0.895
4	2.98	0.827	0.889
5	3.73	0.875	0.902
8	5.60	0.895	0.917
10	7.46	0.895	0.917
15	11.19	0.91	0.93
18	13.43	0.878	0.924
20	14.92	0.91	0.936
25	18.65	0.924	0.941
30	22.38	0.924	0.941
40	29.84	0.93	0.945
50	37.30	0.93	0.95
60	44.76	0.936	0.954
75	55.95	0.941	0.954
100	74.60	0.945	0.958
125	93.25	0.945	0.954
150	111.90	0.95	0.958
200	149.20	0.95	0.958
250	186.50	0.954	0.962
300	223.80	0.954	0.962

$$\text{US} = \text{DPS} \times \text{OT} \times \text{UF} \tag{11.9}$$

Example In a facility where the unit price of electricity is 0.07 €/kWh, energy saving is calculated as follows in the case of replacing 36 motors with a nominal power of 45 kW working 7000 h/year at full load and with a continuous power:

$DPS = (45\ \text{kW} \times 36 \times 1) \times [(1.0/0.936) - (1.0/0.954)] = 32.656\ \text{kW/month}$
$US = (32.656\ \text{kW/month}) \times (7000\ \text{h/year}) = 228{,}592\ \text{kWh/year}$
Monetary equivalent of annual savings
$MAS = US \times (\textit{average electricity price})$
$MAS = 228{,}592\ \text{kWh/year} \times 0.07\ €/\text{kWh} = 16{,}001.4\ €/\text{year}$

The Nominal Power (*NP*), Annual Operating Times (A*OT*), Loading Coefficients (*LC*), and Utilization Factor (*UF*) of the electrically operated pump motors are given in Table 11.38. When the label values of the pumps belonging to the pump are examined, there are ten pump motors with a power less than 224 kW. Since the efficiency values of powerful high-efficient motors larger than this value are not

Table 11.38 Annual operating time of electric motors

Unit	Motor no	Number of the Motor (*NM*)	Power (kW)	Power drawn by the motor (kW)	Annual operating time (A*OT*) (hour)	Load coefficient (*LC*)	Usage factor (*UF*)
1	1	1	210	124.92	4584	53.63	1
	2	1	210	135.05	3840	57.98	1
	3	1	210	217.05	4128	93.18	1
	4	1	115	135.29	5664	102.06	1
	5	1	120	123.45	3264	91.94	1
	6	1	210	127.33	3136	54.67	1
	7	1	132	132.28	4392	93.90	1
2	1	1	400	262.16	6432	61.42	1
	2	1	400	279.32	4632	65.44	1
	3	1	400	269.78	8136	63.21	1
	4	1	400	270.74	6120	63.43	1
	5	1	400	268.78	7224	63.21	1
	6	1	400	262.16	1392	61.42	1
	7	1	400	76.79	7344	92.39	1
3	1	1	75	79.15	1728	95.23	1
	2	1	75	66.15	4776	79.60	1
	3	1	75	124.92	2880	53.63	1

[a]Because all motors are running continuously $UF = 1$

Table 11.39 Energy saving with energy-efficient electric motors

Unit	Pump	*DPS* (kW/Month)	*US* (kWh)	*UCS* [€]	Investment cost [€]
1	1	6.9	31,814.96	2227.05	500
	2	7.5	28,789.35	2015.25	500
	3	12.1	49,763.55	3483.45	500
	4	11.8	66,950.93	4686.57	500
	5	7.3	23,948.18	1676.37	500
	6	7.0	21,927.64	1534.93	500
	7	1.8	7874.07	55,119	500
3	1	3.6	6210.96	43,477	500
	2	3.0	14,275.22	99,927	500
	3	2.0	5775.10	40,426	500
Total		63	257,330	18,013	5000

known, calculations are only made for ten electric motors with power less than 224 kW.

In case of replacement of these motors with high-efficient motors, Monthly Demand Power Savings (DPS), Usage Savings (US), and Monetary Equivalent of Annual Savings (MAS) are given in Table 11.39. As given in the table, monthly total demand energy saving is 63 kW and annual energy usage saving (total energy saving) is 257,330 kWh. When the unit cost of electricity is taken as 0.07 €/kWh and when ten examined motors are replaced with high-efficient motors, the monetary value of the total savings to be earned each year is 18,013 €/year.

When a high-efficient motor is purchased instead of a standard motor, the surcharge period of the extra paid money, can be calculated by considering the price difference between the standard motor and the high-efficient motor. For an engine at 110 kW power, this price difference was an average of €500.

Payback period = (Investment cost)/(Annual financial saving)

Payback period = [(5000 €)/(18,013 €/year)] × 12 ay/year

Payback period = 3.33 months

Once the period of repayment has been completed, annual savings of 257,330 kWh/year or 18,013 €/year will begin to be achieved.

11.3.5.4 Cavitation Control

Cavitation can be defined as the formation of essential vapor-filled gas spaces in a fluid and their sudden collapse. Gas cavities are formed in places where the fluid pressure falls close to the evaporation pressure in relation to temperature. They are carried by flow and collapse where the fluid pressure rises. These gaps are bubbles, swirls, sacs, clusters, and so on. After the air or steam bubble forms, it moves in the flow direction and then disappears there by going to the wall where the static pressure is the lowest, and this destruction process destroys the material. As a result, this undesirable condition in pumps causes damage to the pump impeller and casing. It shortens the life of the pump performance in a short time and causes the pump to become unusable after a while [27].

Regarding the operating conditions and fluid temperatures of the pumps in the enterprise, cavitation calculation of each pump has been made and the related results are given in Table 11.40. While calculating cavitation, the following equations are used:

$$h_{s_{max}} = \frac{P_{amb}}{\rho \times g} - \frac{P_{sat}}{\rho \times g} - NPSH - h_{loss} \quad (11.10)$$

$$P_s = \rho \times g \times h_{s_{max}} \quad (11.11)$$

where

P_{amb} = The ambient pressure in the area where the pump is installed (N/m^2)

P_{sat} = Saturation pressure dependent on the water temperature at the pump inlet (N/m^2)

P_s = Suction line inlet pressure (N/m^2)

Table 11.40 Cavitation calculation of the pumps

Unit No	Pump No	P_{amb} [Pa]	P_{sat} [Pa]	Q [ton/h]	n [d/min]	Δy [m]	h_s [m]	P_s [bar]	P_{inlet} (P_1) [bar]	Result
1	1	97,000	3169	–	725	–	–	–	–	–
	2	97,000	3169	–	725	–	–	–	–	–
	3	97,000	3169	2285	725	6.05	2.51	−0.25	0.50	No
	4	97,000	3169	1805	970	7.63	0.94	−0.09	0.35	No
	5	97,000	3169	2011	730	5.61	2.95	−0.29	0.50	No
	6	97,000	3169	2280	730	6.10	2.46	−0.24	0.50	No
	7	97,000	3169	2265	650	5.20	3.36	−0.33	0.50	No
2	1	97,000	3169	4025	960	12.84	−4.27	0.42	0.50	No
	2	97,000	3169	3147	960	10.90	−2.33	0.23	0.50	No
	3	97,000	3169	3240	960	11.11	−2.54	0.25	0.50	No
	4	97,000	3169	3607	960	11.93	−3.37	0.33	0.50	No
	5	97,000	3169	3416	960	11.51	−2.94	0.29	0.50	No
	6	97,000	3169	–	960	–	–	–	–	–
	7	97,000	3169	3148	960	10.90	−2.33	0.23	0.50	No
3	1	97,000	3169	1055	970	5.33	3.23	−0.32	0.50	No
	2	97,000	3169	1145	970	5.63	2.93	−0.29	0.50	No
	3	97,000	3169	890	1470	8.29	0.28	−0.03	0.50	No

NPSH = According to the catalog or calculated values Net Positive Suction Head (m)

h_{loss} = Pressure loss in pipes of suction lines and regional elements (m)

$h_{s_{max}}$ = The maximum height of the suction line (m)

In order for the pump to operate without cavitation, the value of P_s must be smaller than the value of P_{inlet} (P_1). Otherwise, the pump will operate with cavitation. As seen in Table 11.40, no cavitation was found in the pumps being measured.

References

1. Ertöz AÖ (2003) Energy efficiency in pumps. In: 6th National plumbing engineering and congress, Tesisat Journal, TES-22, Istanbul/Turkey (In Turkish: Ertöz A. Ö., Pompalarda Enerji Verimliliği, 6. Ulusal Tesisat Mühendisliği ve Kongresi, Tesisat Dergisi, TES-22, İstanbul/Türkiye, 2003)
2. ISO 9905 (2000) Radial (Centrifugal) Pump Technical Specifications, Class I, POMSAD Publications, No: 7, 2000 (In Turkish: ISO 9905, Radyal (Santrifüj) Pompa Teknik Özellikleri, Sınıf I, POMSAD Yayınları, No: 7, 2000)
3. ISO 9908 (2000) Radial (Centrifugal) Pump Technical Specifications, Class III, POMSAD Publications, No:8, 2000 (In Turkish: ISO 9908, Radyal (Santrifüj) Pompa Teknik Özellikleri, Sınıf III, POMSAD Yayınları, No: 8, 2000)
4. Kaya D (2001) The effect of the impeller blade construction on the pump overall efficiency in axial pumps, PhD Thesis, Sakarya University Institute of Science, Sakarya/Turkey, 2001) (In Turkish: Kaya D., Eksenel Pompalarda Çark Kanat Konstrüksiyonunun Pompa Genel Verimine Etkisi, Doktora Tezi, Sakarya Üniversitesi Fen Bilimleri Enstitüsü, Sakarya/Türkiye, 2001)
5. ISO 3555 (1998) Radial, Mixed Flow and Axial Pumps Acceptance Test Principles, Class B, POMSAD Publications No: 3, 1998. (In Turkish: ISO 3555, Radyal, Karışık Akımlı ve Eksenel Pompalar Kabul Deneyi Esasları, B Sınıfı, POMSAD Yayınları, No: 3, 1998)
6. ISO 2548 (1998) Radial, Mixed Flow and Axial Pumps Acceptance Test Principles, Class C, POMSAD Publications No: 4, 1998. (In Turkish: ISO 2548, Radyal, Karışık Akımlı ve Eksenel Pompalar Kabul Deneyi Esasları, C Sınıfı, POMSAD Yayınları, No: 4, 1998)
7. Lifetime Cost in Pumps: Lifetime Cost Calculation for Pump Plants, POMSAD Publications, No: 12, 2001.) (In Turkish: Pompalarda Ömür Boyu Maliyet: Pompalı Tesisler İçin Ömür Boyu Maliyet Hesabı, POMSAD Yayınları, No: 12, 2001)
8. Pump Life Cycle Costs: A Guide to LCC Analysis for Pumping Systems, Hydraulic Institute
9. https://www.educationdiscussion.com/centrifugal-pump/
10. https://theconstructor.org/practical-guide/centrifugal-pump-working-types/2917/
11. Hassan A (2015) Experimental and computational study of semi-open centrifugal pump. https://doi.org/10.13140/rg.2.1.5190.3766
12. https://nptel.ac.in/content/storage2/courses/112104117/chapter_8/8_8.html
13. Kaya D, Saraç HI, Olgun H, Tırıs M (2000) A study of axial flow pump impellers-effects of pump design parameters on its performance. In: Proceedings of the twelfth international symposium, on transport phenomena, Istanbul, pp 701–705
14. Pump Terms Guide, POMSAD Publications, No: 2, 1997. (In Turkish: Pompa Terimleri Kılavuzu, POMSAD Yayınları, No: 2, 1997)
15. Achievable Efficiencies for Worm Pumps, POMSAD Publications, No: 10, 2001 (In Turkish: Salyangozlu Pompalar İçin Ulaşabilir Verimler, POMSAD Yayınları, No: 10, 2001)
16. https://www.hkdivedi.com/2019/08/total-head-developed-by-centrifugal-pump.html

17. Toklu E, Kılıçaslan İ, Yiğit KS (1996) Selection of optimum characteristics in pump design, 2. Pump Congress, Istanbul/ Turkey, April 3–5, 1996) (In Turkish: Toklu E., Kılıçaslan İ., Yiğit K. S., Pompa Dizaynında Optimum Karakteristiklerin Seçimi, 2. Pompa Kongresi, İstanbul/Türkiye, Nisan 3-5, 1996)
18. Energy Efficiency in Electric Motor Systems, General Directorate of Electrical Works Survey Administration. (In Turkish: Elektrik Motor Sistemlerinde Enerji Verimliliği, Elektrik İşleri Etüt İdaresi Genel Müdürlüğü)
19. Kaya D An experimental study on regaining the tangential velocity energy of axial flow pump. Energy Conv Manag 44:1817–1829
20. http://marineinfobox.blogspot.com/2017/06/what-are-centrifugal-pump-its-operating.html
21. Theoretical and Experimental Investigation of Optimum Cycle Number in Centrifugal Pumps, 8. Engineering Week, Istanbul/ Turkey, May 26–28, 1994. (In Turkish: Çallı İ., Yiğit K. S., Korkmaz Y., Toklu E., Santrifüj Pompalarda Optimum Devir Sayısının Teorik ve Deneysel İncelenmesi, 8. Mühendislik Haftası, Isparta/Türkiye, Mayıs 26–28, 1994)
22. Yiğit KS, Çallı İ, Sözbir N, Saraç Hİ, Brawn DM (1995) Experimental investigation of optimum centrifugal pumps speeds. In: Second international conference on pumps and fans, vol – II, Beijing, Chine, October 1995
23. Operation of Rotodynamic Pumps Beyond Design Values, POMSAD Publications, No: 11, 2001. (In Turkish: Rotodinamik Pompaların Tasarım Değerlerinin Uzağında Çalıştırılması, POMSAD Yayınları, No: 11, 2001)
24. Yiğit KS (1994) Theoretical and experimental investigation of gap loss of open vane pumps used in vehicle engines, PhD Thesis, Kocaeli University, Institute of Science, Kocaeli/Turkey, 1994. (In Turkish: Yiğit K.S., Taşıt Motorlarında Kullanılan Açık Kanatlı Pompaların Aralık Kayıplarının Teorik ve Deneysel İncelenmesi, Doktora Tezi, Kocaeli Üniversitesi, Fen Bilimleri Enstitüsü, Kocaeli/Türkiye, 1994)
25. Kaya D (2002) Experimental study of the effect of open and closed axial flow pump impellers on pump performance. Eng Mach 508: 47–53. (In Turkish: Kaya D., Açık ve Kapalı Eksenel Akışlı Pompa Çarklarının Pompa Performansına Etkisinin Deneysel Etüdü, Mühendis ve Makina, 508:47–53, 2002)
26. https://www.fluke.com/
27. Kaya D, Yagmur EA, Yigit KS, Canka Kilic F, Eren AS, Celik C (2008) Energy efficiency in pumps. Energy Convers Manag 49(6):1662–1673
28. Yiğit KS, Yağmur A, Kaya D, Eren AS, Çelik C (2009) Energy efficiency in high capacity water pumps. Tesisat J 157: 104–110. (In Turkish: Yiğit K. S., Yağmur A., Kaya D., Eren A. S., Çelik C., Yüksek Kapasiteli Su Pompalarında Enerji Verimliliği, Tesisat Dergisi, 157: 104–110, 2009)
29. https://www.engineersedge.com/pumps/centrifugal_pump.html

Chapter 12
Energy Efficiency in Electric Motors

12.1 Asynchronous Motors

Asynchronous motors (ASM) (Fig. 12.1) are widely used in industrial, commercial, agricultural, transport and home applications due to their simple and robust, torque-speed characteristics being smooth, easy to use, and operate directly from the network. A significant part of these applications also constitutes speed monitoring and non-adjusting applications. Three-phase ASMs used to drive such loads; the efficiency is greatly reduced because of the reasons such as the selection of large powers, the necessity of small loads or idling in a significant part of the daily working periods. As a result, ASMs lead to extra energy consumption, reduced usable capacities of transmission and distribution networks, and economic losses [1].

When ASMs are operated at loads below 50% of rated value, such as in all electric machines, their efficiency is significantly reduced. As a result, energy consumption increases. There are two ways to avoid extra energy consumption. The first is the use of high-quality materials and improvements in design to improve engine efficiency. With this method, the efficiency increase is possible for 75% and larger loads, whereas for smaller loads it is not possible in today's technology. The second method is to reduce energy losses by performing optimal energy control and ensure that the motor runs at maximum efficiency at every load [1].

Nowadays, inverter-based speed-controlled drives are used for energy-saving purposes for loads such as heating, ventilation, air-conditioning and pumps and other loads that require speed monitoring. Because in such loads since the input power varies depending on the square or cube of the speed, the use of transducer-based speed-controlled drives is mandatory in order to maximize energy savings. In applications that do not require speed adjustment and control, ASM's use of soft starters in energy-saving supervision is preferred over transducer-based speed drives for cost and ease of control.

D. Kaya et al., *Energy Management and Energy Efficiency in Industry*,
Green Energy and Technology,
https://doi.org/10.1007/978-3-030-25995-2_12

Fig. 12.1 Asynchronous motors

12.2 Energy Saving in Electric Motors

Electric motors are divided into 3 classes depending on their efficiency:

(1) EFF3: Standard motors,
(2) EFF2: Electric motors with increased efficiency, and
(3) EFF1: Highest efficiency motors.

Despite the fact that energy prices in the industrial sector are only slightly lower, high-efficient engines amortize their cost differences, usually in as little as 12 months. In countries with high energy conservation consciousness, attention is paid to the use of highly efficient engines.

Structural components of the asynchronous motor can be seen in Fig. 12.2.

For example, the efficiency of a standard motor with a power of 15 kW is approximately 88%. The efficiency of a high-efficient motor of the same power increases to 91%, and in the highest efficiency engines this value reaches 93%. In other words, while the standard motors convert 12% of the electricity they consume to heat and throw it to the environment as waste heat, the highest efficiency motors convert 5%

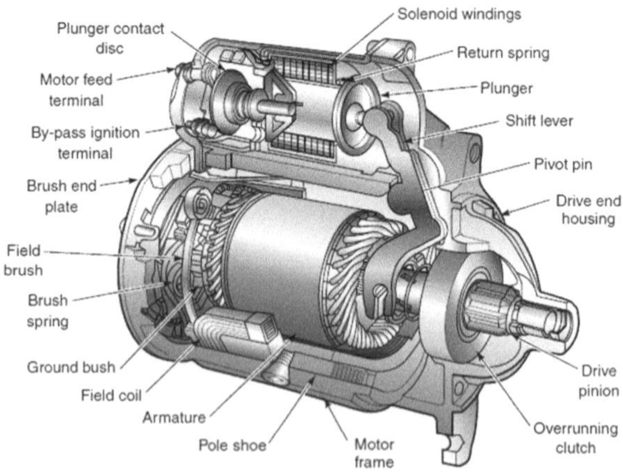

Fig. 12.2 Structural components of the asynchronous motor

of this 12% to usable mechanical power and only 7% convert it to waste heat. A low waste heat rate allows the engine to run cooler, prolongs its life and makes the engine more resistant to overloading and abnormal operating conditions. It makes the poor quality of the electric current and voltage more tolerant [2].

Selecting a highly efficient engine with an efficiency of 93% would save 4102 kWh of electricity a year instead of a standard engine with an efficiency of 88.3%, with 15 kW power an average 75% load factor in operating at 6000 h per year. Given that the average 0.65 kg of CO_2 released in every kWh of electricity production in an example country, the high-efficient motor will prevent 3 tons of greenhouse gases that entering the atmosphere. In addition to energy savings and environmental contribution, high-efficient motors reduce operating costs with higher reliability, in other words lesser breakdowns, less cause for lost production, and lower maintenance costs [2–4].

Electric Motor Efficiency Classes

In 2000 the European Committee of Manufacturers of Electrical Machines and Power Systems (CEMEP) issued a voluntary agreement of motor manufactures on efficiency classification with three efficiency classes: EFF1 for highest efficiency, EFF2 for standard efficiency, and EFF3 for lowest efficiency. From 2004, the electric motors with EFF3 class are banned in Europe. International Standard IEC 60034-30 defined new energy efficiency classes for electric motor in 2008: IE1, IE2, IE3, and IE4 [5].

In accordance with this new identification:

IE1: Standard Efficiency (According to CEMEP: EFF2),
IE2: High Efficiency (According to CEMEP: EFF1),
IE3: Premium Efficiency, and
IE4: Super Premium Efficiency.

The error tolerance limits for motor efficiency are $[(1 - \text{Efficiency }\%) \times 0.15]$. If the actual electric motor efficiency is tolerated with $\pm$ this calculated value, it is in the specified class. EFF and IE efficiency values are roughly equal. Small differences between EFF and IE values are due to additional losses and different calculation methods. EFF additional losses were always allocated as 0.5% in the old standards, it changes between 0.5 and 2.5% in the new IE standards depending on frame size.

Example: An old electric motor with a power of 75 kW and a speed of 1500 rpm will be replaced with a standard (EFF2) motor or a high-efficient (EFF1) motor operating at the same speed and upgraded in the new case. The current electric motor is operating at 6000 h per year with a 75% load factor. The efficiency of the standard motor is 93.6% and efficiency of the high-efficient motor is 94.8%. The prices of standard and high-efficient motors including VAT are 2405 USD and 2994 USD, respectively, and the unit price of electricity is 0.091 USD/kWh. In this case, calculate the annual energy and cost savings that the high-efficient motor will provide and the payback period for the additional cost.

Solution:

$$\begin{aligned} \textit{Energy saving} &= \textit{Power}\,(kW) \times \textit{Load factor} \times \textit{Operation time}\,(\textit{hours}) \times (1/\eta_s - 1/\eta_h) \\ &= 75 \times 0.75 \times 6000 \times [(1/0.936) - (1/0.948)] = 4564\ \text{kWh/year} \\ \textit{Cost saving} &= \textit{Energy saving} \times \textit{Energy price} \\ &= (4564\ \text{kWh/year}) \times (0.091\ \text{USD/kWh}) = 415.32\ \text{USD/year} \\ \textit{Payback Period} &= (\textit{Installation cost})/(\textit{Cost saving}) \\ &= (2994\ \text{USD} - 2405\ \text{USD})/415.32\ \text{USD/year} = 1.41\ \text{year} \end{aligned}$$

where

η_s *Standard Efficiency Engine,*
η_h *High-Efficiency Engine*

The high-efficiency motor with only a 1.2% efficiency difference will pay the cost difference of 589 USD, the energy saved in one and a half years. This motor will continue to save 415.32 USD per year to the operation. As energy prices increase, the annual savings will also increase. In addition, the high-efficiency motor will reduce the greenhouse effect by preventing the release of about 3 tons of CO_2 per year to the atmosphere.

The total cost of the standard motor energy consumed every year over a 20-year lifetime;

$$\begin{aligned} \text{Total energy cost} &= [(\text{kW} \times \text{Load factor} \times \text{Operation time} \times (1/\eta_\text{s})] \times \text{Energy price} \\ &= [75 \times 0.75 \times 6000 \times (1/0.936)] \times 0.091 = 32,812.5\ \text{USD} \end{aligned}$$

As you can see, this motor, which has a purchase price of 2405 USD, uses about 14 times more energy than the purchase cost for one year. The average life span of a motor in a 75 kW power group is approximately 25 years. The cost of energy consumed by this motor during its working life is close to 820,312 USD. Even small differences in engine efficiency can lead to energy and cost savings and can make the investment return quickly. If the old motor is running at 90% efficiency and this motor is considered to be running for 5000 h a year, then if this old motor is replaced by a high-efficiency motor, it will pay back its initial cost of 2994 USD in about 2 years with the energy it will save.

Since standard ASMs are designed to operate at maximum efficiency at full load, with loads less than 60% their efficiency and PF are greatly reduced. In other words, ASM applies two energy-saving methods to increase efficiency and PF at light loads, to ensure high efficiency at all loads:

(1) Energy-saving made by special arrangement in motor design (high-efficiency ASM).
(2) Energy saving by means of driver usage in standard motors.

In ASM running on a starter with a light load, the energy control is done by reducing the voltage value to the lowest value that can meet the load moment. If the load is reduced, the motor stator voltage is reduced to the minimum value that the magnetic flux value in the air gap can meet the load torque.

There is more loss at all working points outside of the optimum working point. In this case, the efficiency decreases. For this reason, the motor voltage should not be lowered below the minimum value that will produce the required current for the load torque. Otherwise, if the generated moment is proportional to the magnitude of the voltage, the slip will increase, and the current drawn by the motor and the active input power will start to increase again. If the voltage drops too much, the engine will stop.

When using electric motors, the following factors must be considered in order to save energy:

- A breakdown of all electric motors in operation must be removed. A list of the motor and plate information (nominal power, speed, efficiency, etc.) and annual working hours should be prepared. Attention should be paid to standard efficient motors with a power greater than 18 kW and used more than 2000 h per year. The voltages and currents of the servomotors must be measured.
- In order to achieve economic and energy efficiency enhancements, a motor repair/change policy should be prepared. The motors should be labeled for the most suitable application. For example, when the motor fails replace it immediately with a high-efficient motor, or send it for the repairing process to have a proper winding when it malfunctions.
- Motors must be selected in accordance with the load. The motor must not be selected as large as necessary. Thus, it is possible to prevent the motors from operating at low power and hence low efficiency compared to rated power written on the labels. As the load increases, the efficiency increases. Motor output generally reaches a maximum level of 75% overload. Electricity consumed at low loads is converted into heat instead of mechanical power. In this case, due to overheating of the motor, the risk of failure increases, and the life of the motor shortens.
- Variable speed drive systems are also known as inverter or variable frequency drive systems. These systems change the frequency of the alternating current, and therefore the speed of rotation of the motor, preventing the motor from overloading unnecessarily. Thus, the same work can be done by consuming much less energy. By adding an inverter system to the motors, up to 50% energy savings can be achieved. In other words, the amount of electricity the engine consumes for the same job can be reduced by half. The cost of the motors equipped with the inverter is higher. However, in well-selected applications such as pumps and compressors, variable speed drive systems pay for the energy they cost, usually within two years or less. According to some analyses, only about 10% of the energy-saving potential in motor systems can be achieved with increased efficiency. The remaining 90% can only be achieved by equipping the engines with inverter systems.
- In systems where the motor power is indirectly transmitted by straight belt or standard V belts instead of direct connection, losses occur between 2% and 8% due to belt slip and friction. This loss and belt warming problem can be avoided by replacing standard belts with knurled high-efficiency V belts.

12.3 Motor Load Characteristics

It is possible to divide the ASM loads into two as loads with variable and constant torque-speed characteristics according to their torque-speed characteristics, and to divide them into two as passive and active loads in terms of the effects on the system.

In passive loads, the operating point of the system is directly connected to the motor and when the motor torque is zero, the load torque becomes zero. These are the loads that work against air or liquid friction like air conditioner, pump, fan, etc., and have variable moment velocity characteristics. In active loads, the operating point of the system is completely dependent on the load and when the motor torque is set to zero, the load torque is not zero. Elevator, crane, conveyor belt, escalator, and so on loads with constant torque-speed character fall into this group. These characteristics are taken into account when selecting motor drives. The following describes the torque-speed characteristics of these loads, the selection of the motor drive, the motor drives, and the energy-saving methods used in these drives.

12.3.1 Variable Torque-Speed Characteristics Load

The loads in this group are generally known as HVAC; which creates passive loads such as ventilators, pumps, air compressors, and refrigeration appliances. Since the torque at these loads changes in proportion to the speed of the crank ($M \sim \omega^2$), the speed of the power is proportional to the cube ($P \sim \omega^3$), reducing the motor speed at a very low value causes a large decrease in the drawn force. The common feature of the loads in this group is that they are the loads that act against air or liquid friction. In the ASM driver that feeds these loads, the amount of air or liquid exiting is provided by adjusting the engine speed. Energy saving in ASM drives is achieved by reducing the speed in order to reduce the amount of air or liquid being vented. When the speed is reduced for this purpose, momentarily, indirectly, the input power is greatly reduced. For example, when the motor speed is reduced by 10%, the power taken by the motor is reduced by 27.1%. An example of this practice is air-conditioning. In sunny times when the weather is not too cold in winter, the speed of the engine is reduced to save energy in order to reduce the heat that the air conditioner gives to the environment [6, 7].

Energy saving with the inverter and converter is done by speed adjustment, keeping the voltage/frequency ratio (V/f) constant. Figure 12.3 shows the variable torque characteristic curve and the energy-saving method applied to loads with these characteristic curves. As shown in Fig. 12.3, the motor first operates at point 1 with the frequency f_1 and at speed of ω_{m1}. At this operating point, the motor moment is M_{e1}, the power is P_{g1}. When the frequency is reduced to f_2 in order to save energy, the motor operates at point 2 with the speed of ω_{m2} but at the same slip value. At this new operating point, the moment drops to M_{e2} and the power drops to P_{g2}. Since power is proportional to the speed of the cube, the reduction in power drawn is much

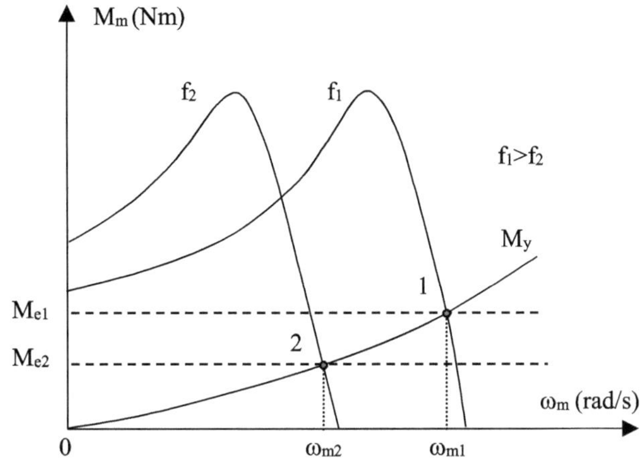

Fig. 12.3 High-energy-saving method with variable torque-speed characteristics

greater. When the engine is running at this point, it is additionally possible to save some energy by lowering the voltage while keeping the frequency constant.

12.3.2 Constant Torque-Speed Characteristics Load

For loads with a constant torque-speed characteristic, the torque is independent of speed. If all speeds are the same, energy saving cannot be done with speed adjustment. These loads, as mentioned earlier, fall into the active load class. Figure 12.4 shows the constant torque characteristic curve and the energy-saving method applied to loads with these characteristic curves.

As shown in the graphic in Fig. 12.4, at point 1 the motor runs at full load (M_{yN}), nominal speed (ω_{mN}), and slip (s_N) with moment torque characteristic at nominal voltage V_{mN}. When the load torque is reduced to M_y value and the load is reduced, the motor accelerates to work at point 1 at ω_{m1} speed. At this operating point, the motor starts to work at a low efficiency and over-excitation at PF (Power Factor), as the slip (s_1) is smaller than the normal value. If the motor voltage is reduced to bring a smaller value which is sufficient for the flux value, the torque characteristic curve (inner curve) with lower starting and tipping moment values are obtained. The motor starts to operate at point 2 with the speed (ω_{m2}) and slip (s_2) value close to the speed and slip value at point 1. As this is very close to the full-load slip given in the working point label, the motor duty increases to near normal value. With V_{mN} this method, for each load moment value, there is an optimal slip value close to the full-load value.

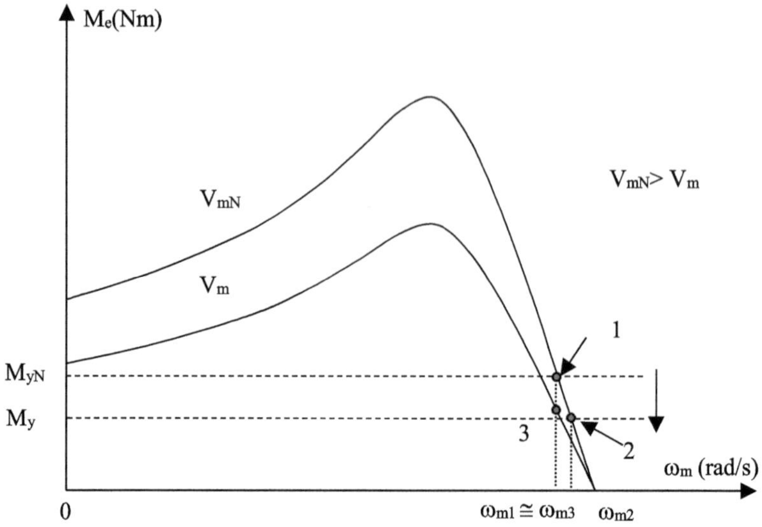

Fig. 12.4 High-energy-saving method with constant torque speed

12.4 Driver Selection for Asynchronous Motor

In selecting the motor drive for energy saving, the load torque, the speed characteristic setting, and whether the speed control are necessary to play an important role. The first criterion in driver selection for the load is whether speed adjustment or speed control is required. In applications where speed control or speed adjustment is required, absolutely inverter drives are used. Soft starters are more economical for speed setting and non-control applications.

The second criterion in motor drive selection is the load moment torque characteristic. For loads with constant torque-speed characteristics, the energy control is made at each load case by adjusting the voltage according to the load torque. For this purpose, soft road markers are used which are cheap and easier to control. With variable speed characteristics, torque loads with passive loads (passive loads and HVAC systems), variable speed drives are superior and economical in terms of energy saving. In these loads, energy-saving control with soft starters is not economical as it will be limited only by minimizing the losses.

12.5 High-Efficient Motor Use

Like all motors, electric motors cannot convert all the energy they use into mechanical energy. The ratio of the mechanical power output of the motor to the electric power drawn is called *motor efficiency*. Motor output varies between 70% and 96%, depending on motor type and size [5–7]. In addition, the efficiency of partial-duty motors is also low. These efficiencies also vary from motor to motor. For example, when a motor is 90% full load, 87% for semi-load, and 80% for 1/4 load, another motor of the same specification can operate at 1/4 load with 75% efficiency while full-load efficiency is 91%. For example, if EFF3 > 91% motor is used instead of EFF3 < 88.4% motor at 11 kW power, an energy saving of 2288 kWh/year is considered as 8000 h usage per year. Standard and high-efficiency motors are compared in Table 12.1.

The cost of high-efficiency motors is 15–25% higher than that of standard motors, but this difference is recovered in a short time due to the fact that operating costs are often low. Primer I^2R losses can be reduced by increasing the cross section of the copper conductor used in the windings of these motors. The iron core losses can be limited by the reduction of the flux density and usually by increasing the stator core's length. However, these losses can also be reduced by reducing the plate thickness

Table 12.1 Comparison of motor efficiency

Nominal motor power (HP)	Nominal motor power (kW)	Average efficiency for standard motor	Average efficiency for high-efficiency motor
1	0.746	0.825	0.865
1.5	1.119	0.840	0.894
2	1.492	0.840	0.888
2.5	1.865	0.812	0.870
3	2.238	0.875	0.895
4	2.984	0.827	0.889
5	3.73	0.875	0.902
7.5	5.595	0.895	0.917
10	7.46	0.895	0.917
15	11.19	0.910	0.930
18	13.428	0.878	0.924
20	14.92	0.910	0.936
25	18.65	0.924	0.941
30	22.38	0.924	0.941
40	29.84	0.930	0.945
50	37.3	0.930	0.950
60	44.76	0.936	0.954
75	55.95	0.941	0.954
100	74.6	0.945	0.958
125	93.25	0.945	0.954
150	111.9	0.950	0.958
200	149.2	0.950	0.958
250	186.5	0.954	0.962
300	223.8	0.954	0.962

and using high-quality alloys. Also, due to the diminished losses in high-efficiency motors, the need for heat release is reduced. In case of replacing a standard motor with a high-efficiency motor, the energy to be saved can be calculated with help of the following formula.

$$\text{Energy saving} = \text{NMS} \times \text{Nominal power} \times \text{OT} \times \text{LC} \times \text{UF} \times \left(1/\eta_{\text{standard}} - 1/\eta_{\text{high-efficient}}\right) \tag{12.1}$$

where NMS—Number of motors that have the same power, OT—Operating time, LC—Load coefficient (the ratio of the real load to the full load), UF—Usage factor, η_{standard}—The efficiency of the standard motor and $\eta_{\text{high-efficient}}$—The efficiency of the high-efficient motor.

As a result of the development of power electronics with increasing technology every year, the spread of starter or driver usage in electric motors has made it possible to use the energy taken from the network more efficiently. It is clear that driver applications will have a more pronounced benefit in load situations where the load change is high. Apart from driver applications, one of the energy-saving possibilities for electric motors is the use of high-efficiency electric motors. In the case of an electric motor that malfunctions or is supposed to be replaced in the plant, it is possible to use high-efficiency motors instead of the existing standard motors to use the energy more efficiently and thus, save energy.

The P_E power drawn from the network is converted to P_M mechanical power, which defines the motor's function, by compensating for the P_L lost power, and by the efficiency of η. Where η is the ratio of the mechanical output power to the input stator (network) power.

$$P_E = P_M + P_L \tag{12.2}$$

$$\eta = (P_M/P_E) \times 100 \tag{12.3}$$

The mechanical force is given by the following equation, which describes the active work the motor has done.

$$P_M = (\eta \times P_E)/100 \tag{12.4}$$

If you need to write power and efficiency equations for two electric motors with different efficiencies doing the same active job, the following power and efficiency equations can be written:

$$P_{E_1} = P_M + P_{L_1} \tag{12.5}$$

$$P_{E_2} = P_M + P_{L_2} \tag{12.6}$$

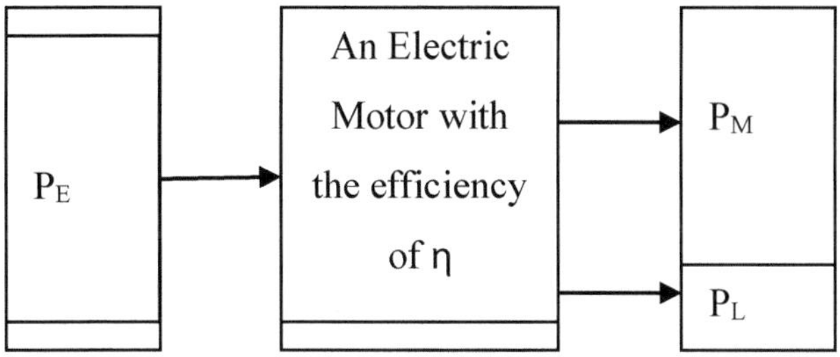

Fig. 12.5 Power flow diagram for the electric motor

$$\eta_1 = \left(P_M/P_{E_1}\right) \times 100 \tag{12.7}$$

$$\eta_2 = \left(P_M/P_{E_2}\right) \times 100 \tag{12.8}$$

When a more efficient electric motor is used, the electric power drawn from the network which is lower than the old one is determined as follows:

$$P_{E_2} = P_{E_1} \times (\eta_1/\eta_2) \tag{12.9}$$

P_L is the loss power, a power that decreases as the motor efficiency increases. This power consumption represents electrical, magnetic, and mechanical losses. With today's technology, it is impossible to completely eliminate these losses. However, the engines with the least losses are high-efficiency motors. Due to the fact that the loss of power is related to the physical properties and material technology, the efficiency difference of high-efficiency motors compared to standard efficiency motors is especially noticeable in motors with power up to 75 kW. Power flow in the electric motor can be summarized simply as shown in Fig. 12.5.

12.5.1 An Example of High-Efficiency Motor Application

The energy-saving values to be obtained in the case of replacing the motors found in an industrial facility (Table 12.2) with the high-efficiency motors are calculated and given in Table 12.3. The efficiency reduction of the motor due to the obsolescence share is assumed to be 3%, and the higher efficiency motor's efficiency value is added to it. Thus, the total efficiency gain obtained for the relevant motor is given in the column "efficiency difference value (%)".

The investment to be made on a department basis, the total savings and repayment periods are examined in detail. When all the motors are replaced, the total savings is $13,206.14/year. The payback period for these motors (engines) can be determined by the price difference between the high-efficiency engine and the standard engine.

Table 12.2 Electrical motors measurement values of engineering and twisting department

Motors	Motor Number (NM)	Power (kW)	Power drawn (kW)	Phase difference cos φ	Operating Time (OT) (hour/year)	Load Coefficient (LC)	Usage Factor (UF)
Main motor 1 for compressor	1	75	69.00	0.85	7000	1.00	1.00
Main motor 2 for compressor	1	75	75.00	0.85	7000	1.00	1.00
Burner motor for boiler 4	1	13.5	8.50	0.86	7000	1.00	1.00
Burner motor for boiler 1	1	9	4.20	0.71	7000	0.60	1.00
Twisting Machine (1–24)	24	45	13–35	0.5–0.85	7000	0.50	0.90
Twisting Machine (25–36)	12	45	25.00	0.70	7000	0.70	0.90
Twisting or Folding Machine(1–2)	2	30	18.00	0.75	7000	0.70	0.80
Folding Machine (1–7)	6	11	3–6.8	0.4–0.7	7000	0.70	0.90
Twisting or Folding Machine 1	1	30	12.00	0.80	7000	0.70	0.70
Twisting or Folding Machine 2	1	30	Not working		7000	0.70	0.70
Twisting or Folding Machine 3	1	37	12.00	0.80	7000	0.70	0.70
Twisting Machine (1–6) sub-motor	6	7.5	4.4–5.6	0.6–0.8	7000	0.70	0.80
Twisting Machine (1–6) top motor	6	11	4.1–7.3	0.5–0.8	7000	0.70	0.80
Twisting Machine (7–8) sub-motor	2	22	Not working		7000	0.70	0.80
Twisting Machine (7–8) top motor	2	22	Not working		7000	0.70	0.80

(continued)

Table 12.2 (continued)

Motors	Motor Number (NM)	Power (kW)	Power drawn (kW)	Phase difference cos φ	Operating Time (OT) (hour/year)	Load Coefficient (LC)	Usage Factor (UF)
Twisting Machine (9–10) sub-motor	2	10	2.40	0.80	7000	0.70	0.80
Twisting Machine (9–10) top motor	2	10	3.00	0.85	7000	0.70	0.80
Main motor 3 for compressor	1	75	backup				
Main motor 4 for compressor	1	55	backup				
Related Machine	1	15	Not working				

Electrical Motors Measurement Values of Engineering and Twisting Department can be seen in Table 12.3.

Payback period = (Investment cost)/(Annual Financial Savings) × 12 months
Payback period = (Total *RPD (Engine Replacement Price Difference)*/Total *AFS*) × 12 months
Payback period = [(20,320)/(13,206.14)] × 12 months/year
Payback period = 18.46 months

12.6 Using Frequency Converters in Asynchronous Motors

Electric motors use a very important part of the total electric charge. An electric motor can be driven in a variety of ways. These methods are used to provide sufficient current to the motor to drive the motor to the nominal load speed, creating an initial torque. Each method has its own advantages and disadvantages (Table 12.4). Among them, the frequency converters provide the most energy-saving. Generally, three methods are used to drive electric motors:

(1) Full voltage starting,
(2) Deceleration with reduced voltage, and
(3) Starting with a frequency converter.

The simplest starting method, full voltage starting, uses a motor contactor. The full mains voltage is applied to the mains motor terminal. Reduced voltage starting can be applied by several methods, such as autotransformer or star-triangle.

Table 12.3 Energy saving in engineering and twisting departments

Motor name	DPS (kW/month)	US (kWh)	MAS = AFS ($/year)	RPD ($)	TRPD ($) (*)	Efficiency difference
Main motor 1 for compressor	1.07	7507.50	563.06	580.00	580.00	0.0143
Main motor 2 for compressor	1.07	7507.50	563.06	580.00	580.00	0.0143
Burner motor for boiler 4	0.77	5358.15	401.86	220.00	220.00	0.0567
Burner motor for boiler 1	0.14	1013.04	75.98	140.00	140.00	0.0268
Twisting Machine (1–24)	10.91	68,720.40	5154.03	360.00	8640.00	0.0202
Twisting Machine (25–36)	7.64	48,104.28	3607.82	360.00	4320.00	0.0202
Twisting or Folding Machine (1–2)	0.71	3998.40	299.88	220.00	440.00	0.0170
Folding Machine (1–7)	1.09	6869.02	515.18	190.00	1140.00	0.0236
Twisting or Folding Machine 1	0.36	1749.30	131.20	240.00	240.00	0.0170
Twisting or Folding Machine 2	0.36	1749.30	131.20	240.00	240.00	0.0170
Twisting or Folding Machine 3	0.44	2157.47	161.81	260.00	260.00	0.0170
Twisting Machine (1–6) sub-motor	0.71	3986.64	299.00	120.00	720.00	0.0226
Twisting Machine (1–6) top motor	1.24	6933.70	520.03	190.00	1140.00	0.0268
Twisting Machine (7–8) sub-motor	0.60	3363.36	252.25	220.00	440.00	0.0195
Twisting Machine (7–8) top motor	0.60	3363.36	252.25	220.00	440.00	0.0195
Twisting Machine (9–10 sub-motor	0.33	1850.24	138.77	195.00	390.00	0.0236
Twisting Machine (9–10) top motor	0.33	1850.24	138.77	195.00	390.00	0.0236
	Total savings		13,206.14		20,320	

(*) TRPD (Total Engine Replacement Price Difference)

Table 12.4 Current-torque relation according to starting methods

Method		Voltage	Initial current [A]	Provided torque
Direct		Full	7 × 100 = 700	Good
Soft	Auto transformer	Reduced	4 × 100 = 400	Good/medium
	Star-delta	Reduced	3 × 100 = 300	Weak
	Soft driver	Reduced	2 × 100 = 200	Low on departure, then good
Frequency converter		Reduced	1 × 100 = 100	Good [*]

[*] It can be regarded as excellent when evaluated with the starting current

For each starter, the full-load currents for a motor with a rated current of 100 A are compared in Table 12.4. When using a frequency converter, the starting torque is the lowest and the torque provided is excellent.

12.6.1 Frequency Converter

The speed of an induction motor is directly proportional to the frequency of the AC grid. If it was possible to adjust the frequency of the network, it would be possible to control the speed of the motor. The frequency converter is a starting method used to control the speed of the motor. Frequency converters are electronic devices that convert the input of constant frequency AC power into a variable frequency output. It is used to control the AC rotational speed, in other words, the speed of the motor, by controlling the frequency of the electrical power supplied to the motor.

12.6.1.1 Benefits of Frequency Converter

Due to decreasing costs, energy awareness and improvements in processes, the use of frequency converters in motor systems has been increasing since the 1990s. Together with technological improvements, the frequency converter dimensions and costs are greatly reduced. Thanks to the energy savings they provide, repayment times have changed between two and three years, so they are widely used. The benefits of using frequency converters can be summarized as follows:

- It can save energy up to 50% by controlling capacity requirements.
- It has the lowest start-up current compared to other road users.
- It reduces the thermal and mechanical stresses in the motor and load system. Thus, the life of the system extends.
- Installation and network connection are simple.
- The external power factor removes the correction capacitors from the center. Thus, it provides a high-power factor.

- It reduces voltage drops and power interruptions, resulting in lower total power (kVA). Thus, other machines in the system are less affected.
- It provides high-speed capability, limited by the mechanical properties of the motor.
- It removes the need for expensive mechanisms, such as control valves, which are expensive and cause energy waste.

Frequency converters are the most energy-efficient systems in pump and fan applications. Many installations, centrifugal pumps, and fans operate at a constant speed. Traditionally, the fluid velocity is changed by a mechanical system. The motor speed can be changed electronically with the frequency converter. By adjusting the speed of a pump or fan to the desired value, energy costs can be reduced. Frequency converters, which remove most of the useless/unreliable mechanical systems that change speed or pressure, reduce maintenance costs.

The motors in small powers are designed to work at two times the nominal speed. As the motor's power increases, the motor's ability to lift high-speed drops to zero. Above the nominal speed, the frequency converter and the motor system operate in constant power mode. Torque decreases linearly at high speeds up to 150%, usually at the speeds above it, in the square of the V/Hz ratio. At speeds lower than the nominal speed, the frequency converter and motor system operate in constant torque mode.

12.6.1.2 Effects of Frequency Converter on Electric Motors

Along with the increased use of frequency converters, new electric motor malfunctions are also emerging. These faults come from the PWM output form; reflective waves, harmonics, engine cooling, shaft currents, etc., these kinds of defects. Frequency converters using PWM technology directly affect the electrical, mechanical, and thermal capacity of an electric motor in the following ways:

- Motor overheating due to harmonic distortion,
- Forced fan usage requirement,
- Voltage reflection,
- Cable length, and
- Carrier frequencies.

12.6.1.3 Characteristics of Electric Motors

A very important issue to consider when applying frequency converters to low-voltage motors is stresses in motor winding insulation. The frequency converter generates the AC output waveform from the DC source. The DC/AC transition is usually made by semiconductor switching devices, some of which can have very fast dV/dt rise times. High dV/dt, stress loads on motor insulation.

Motors suitable for use with a frequency converter have to be specially designed to be able to withstand the voltages of the output wave of the frequency converter. The features of this design are

- In order to increase the dielectric strength capacity, it is necessary to use a coil type suitable for driving with a converter.
- For higher dielectric strength, the ability to insulate among phases and phases should be developed.
- The motor windings have to be changed from concentrate (half thickness) to lap winding (full mold).
- Multiple immersion and drying or vacuum pressure application of varnish should be done to the windings.
- To reduce the dV/dt effect, the first few coils of the windings must be tightly sealed by wrapping the head.

In motors driven by a frequency converter, a high-quality coiled stator insulation system with a good quality should have the following characteristics:

- Winding insulation,
- Gutter inner surface insulation,
- Grooving insulation,
- In-groove interphase isolation,
- Stator package end gutter reinforcement insulation,
- Coil head phase separation isolation,
- Tightening of coil heads, and
- Varnish insulation.

12.7 Replacement of Low Load Motors

The efficiencies of asynchronous motors vary greatly depending on the generated power. If the motor is operated with a value close to the label value, for example, 85–90%, the best efficiency is achieved. When the power drawn by the motor drops, the power factor will decrease accordingly. Excessively large selected motor causes direct loss and indirect reactive loss as well as indirect loss. This is the main cause of the low power factor for many installations.

The main engines in the plants should be inspected and the actual power consumption of each one measured. If the actual consumption is below 60% of the nominal design value, the reason why these motors are chosen as large as necessary should be investigated. If there is no significant reason, it should be researched whether it is economical to replace the engine with lower strength. High-efficiency motors should be used in places where a large power requirement is required during operation in order to reduce the loss of efficiency at low loads.

Example: A fan is operated by an asynchronous motor with medium efficiency at 75 kW. In fact, 22.5 kW power is drawn from the system measured. How much savings can be achieved by replacing the motor?

$$\begin{aligned} \text{Real power} &= 0.3 \times \text{Nominal power} \\ Required\ mechanical\ power &= Power\ drawn \times Motor\ efficiency \\ &= 22.5 \times 0.72 = 16.2\ \text{kW} \end{aligned}$$

At full load, a 20 kW motor has 89% duty (power factor 0.9).

$$\begin{aligned} \text{Power drawn} &= 16.2/0.89 = 18.2\ \text{kW} \\ \text{Power saving} &= 22.5 - 18.2 = 4.3\ \text{kW} \end{aligned}$$

If the annual working time is taken as 8400 h and the unit price of electricity is taken as 0.091 USD/kWh:

$$\text{Yearly saving} = 4.3 \times 8400 \times 0.091 = 3286.92\ \text{USD/year}$$

12.8 Correction of Power Factor in Electric Motors

Depending on the tariffs, the low power factor causes high electricity bills for the factories. The placement of capacitors for reactive power compensation is the most common method. This raises the power factor of the installation. Generally, the capacitor groups are placed near the measuring devices in the main transformer substation. It should not be forgotten that the low power factor will further increase the network losses of the installation. For this reason, placing a capacitor close to the main user may also be a justifiable economic reason.

Generally, the resistance losses of the distribution lines are designed to be 3% of the nominal power. Voltage drop in long lines is a limiting factor. Power losses can be calculated from the voltage value:

$$\begin{aligned} Power\ losses\ in\ \%\ of\ nominal\ power &= Maximum\ voltage\ drop/Normal\ voltage \times 100 \\ Actual\ loss\ in\ \%\ of\ transmitted\ power &= Design\ loss\ in\ \%\ of\ nominal\ power/(Power\ factor)^2 \end{aligned}$$

12.8.1 Reduction of Idle Running Time in Electric Motors

In many facilities, some energy-consuming equipment are operated when not in load. If possible, operating such equipment at full load and turning it off when not in use will save a great deal of energy and money. The saving value to be made by preventing idle operation is determined as follows.

The energy drawn by the motor at full-load energy $= MP \times OT_f \times LF/\eta_f$
The energy drawn by the motor at partial load $= MP \times OT_p \times LF/\eta_p$
The energy drawn by the motor at idle running $= MP \times OT_i \times LF/\eta_i$
Energy saving = The energy that the motor draws at partial load
+ The energy that the motor draws at idle running
− The energy that the motor draws at full load
Cost saving = *Energy saving* × *Unit cost of electricity*

where

MP = Motor power (kW),
OT_f = Operating time at full load (h),
OT_i= Operating time at idle running (h),
η_f= Motor efficiency at full load (%),
η_p= Motor efficiency at partial load (%),
η_i= Motor efficiency at idle running (%), and
LF = The loading factor of the motor.

References

1. Öztürk HH, Kaya D (2012) Secondary energy generation and electric motors. Umuttepe Publications, Publication No: 73. ISBN: 978-605-5936. (In Turkish: Öztürk, H.H., Kaya, D., İkincil Enerji Üretimi ve Elektrik Motorları, Umuttepe Yayınları, Yayın No: 73, ISBN: 978-605-5936, 2012)
2. Nadel S, Shepard M, Grenberg S, Katz G, Almeida A (1991) Energy efficient motor systems, a handbook on technology, program, and policy opportunities. American Council for Energy-Efficient Economy, Washington D.C
3. General Directorate of Electrical Affairs Study Administration, Energy Efficiency in Electric Motor Systems (In Turkish: Elektrik İşleri Etüt İdaresi Genel Müdürlüğü, Elektrik Motor Sistemlerinde Enerji Verimliliği
4. Değişken Frekanslı Sürücü (VFD) Teknolojisi İşletme ve Uygulamaları,OnxControl, 2020
5. ELK Motor. Electric Motor Efficiency Classes, http://www.elkmotor.com.tr/elkmotor-enerji-verimliligi.aspx
6. http://www.acdrive.org/ac-drives-basics.html
7. Kaya D, Çanka Kılıç F (2004) Energy conservation opportunity in VSD systems—a case study. the world energy engineering congress in Austin, Texas, USA. In: The world energy engineering congress in Austin, Contents No: 22, CD Code: 0541, Texas/USA

Chapter 13
Energy Efficiency in Compressed Air Systems

Compressed air is the fourth basic raw material in the industry after fuel, electricity, and water. The compressed air system is the third highest power user in many industrial plants and it gives the most important opportunity to reduce energy costs. Despite the high cost of compressed air, it has been determined that many industrial plants have losses up to 30% due to leaks in compressed air systems, misuse, inadequate maintenance, and poor control system. Compressed air is used in the industry for pneumatic equipment and other necessary places.

When the costs are analyzed in terms of the compressor, it is seen as a distribution as in Fig. 13.1. Generally, it is known that the one-year operating expense of compressors is equal to the approximate sales cost. A 1% improvement in energy cost would correspond to a 4% improvement in sales cost [1]. Therefore, in addition to short-term costs in compressor investments, long-term operating expenses should be particularly evaluated. In the graphic given in Fig. 13.1, the costs of the compressor were evaluated during the 10-year operation period. Although the compressor sales cost is 18%, the energy cost constitutes 73% of the cost.

13.1 Basic Equipment of Compressed Air Systems

The compressed air system starts with the compressor unit (Fig. 13.2). Air passes through the inlet filters, comes to the compressor, and is compressed to increase its pressure. During this process, the temperature of the air rises to 80–170 °C. At the same time, oil is smeared from the mechanical equipment of the compressor. Compressed air is passed through oil filters and oil and dirt inside are cleaned. At this stage, the temperature of the air must be reduced. Air is cooled to condense the water inside. The condensed water is discharged through steam traps. The air is then sent to the air tanks ready for use. Air tanks are tanks where compressed air is stored. Thanks to these tanks, short-term and variable loads can be met in a balanced way.

D. Kaya et al., *Energy Management and Energy Efficiency in Industry*, Green Energy and Technology,
https://doi.org/10.1007/978-3-030-25995-2_13

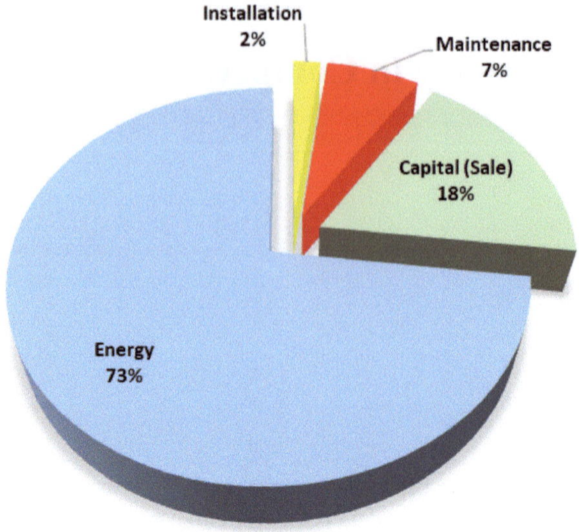

Fig. 13.1 The cost ratios of a compressor at the end of ten years [2]

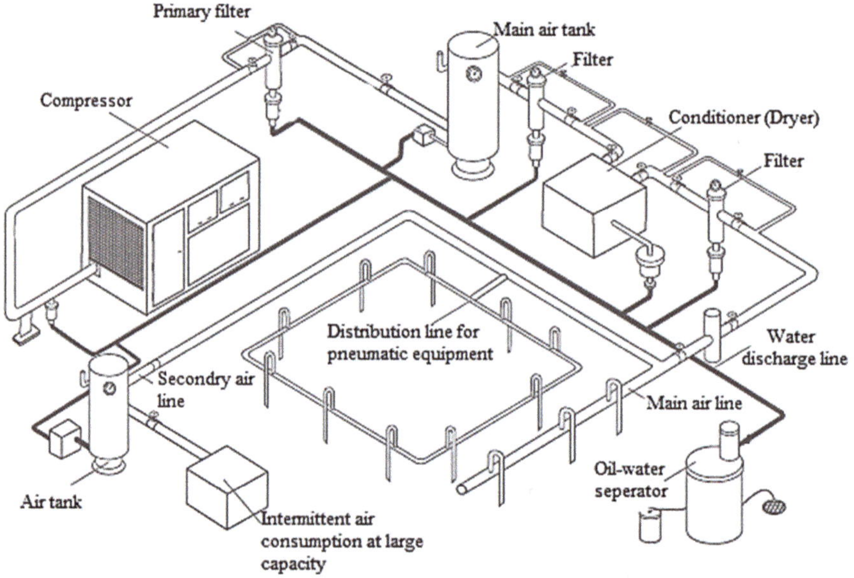

Fig. 13.2 Main components of the compressed air system

Air tanks are connected to the airline to be delivered to end-use points. The airline must be partially inclined to drain the water that may condense over time and the condensate water must be drained with steam traps.

Industrial systems that provide pressurized air consist of many sub-stages. These sub-stages comprise general compressor, drive system, control unit, maintenance unit, distribution system, and accessories. Drive unit is an electric motor or an internal combustion engine that activates the compressor. The control unit adjusts the amount of pressurized gas. The maintenance unit removes unwanted substances from the gas. The distribution system also provides the gas within the system and delivers it to the places to be used.

In energy-saving studies, compressed air system is found to be one of the highest energy-saving potential areas. Compressed air is an indispensable input that is widely used in the industry and its use is becoming increasingly widespread. A malfunction in the compressed air system causes many fabrication processes to stop. Today, it is almost impossible to think about a factory without a compressor. Almost all of the compressors work with electricity, which is expensive energy, as well as the types driven by the four-stroke engine.

Significant savings can be achieved if simple and regular checks and maintenance are made in a compressed air system, which is quite expensive to produce. The main energy-saving opportunities identified in compressed air systems in energy-saving activities at different industrial establishments are

- Selecting the correct type and size of compressors,
- Providing to the compressor cold, clean, and dry air intake,
- Use of compressor cooling air,
- Regular maintenance of compressed air equipment,
- Provision of necessary instruments for efficiency improvement,
- Elimination of air leaks,
- Operating the system at the lowest possible pressure,
- If economical, recovery of cooling energy, and
- Appropriate lubrication of the equipment.

13.1.1 Compressors

Compressors are used to raise air pressure by compressing. Compressors provide compressed air by compressing the fluid they receive at the inlet, in different shapes, according to their working principle. As the pressure of the air compressed by the compressors increases, most of the energy transferred to the air becomes heat energy. From this point of view, heated compressed air is obtained at the compressor outlet. However, in order to prevent damage to the equipment as a result of the condensation of the inside air, the air needs to be cooled. The energy content of the warming air is a very small fraction of the energy transmitted by the shaft. For this reason, compressors are low efficient machines.

Considering the cost, the necessity or alternatives of compressed air should be seriously considered and evaluated. Compressed air applications are a necessity for the use of pneumatic equipment where the use of electricity is dangerous, especially since the risk of explosion is high. On the other hand, it is more economical to use partially low-pressure blowers instead of compressed air produced by compressors in simple drying and cleaning operations.

Compressors consume more energy than most equipment used in the industry. For this reason, improvements to compressors and compressed air systems can cause significant reductions in cost items. The basic improvements to be made in terms of energy saving in compressed air applications are

- The efficiency of the compressors should be evaluated,
- The type of compressor and quality of the air should be suitable for processing,
- Leaks must be identified and eliminated,
- Compressed air should not be used unnecessarily,
- In the installation, pressure drop and leakage should be minimum,
- Pressure regulators should be installed at the connection points for equipment that can operate at a lower pressure than the mainline pressure, and
- At the outlet of the compressor waste heat in the air should be recovered.

Compressors are devices used to increase pressure of various compressible fluids, or gases, the most common example of these is air. The pressure levels of the fluid (usually gas) at the compressor inlet can vary from vacuum pressure to values above the atmospheric pressure. Likewise, the pressure value obtained at the compressor outlet may vary from very low atmospheric pressures to very high atmospheric values. The diagram related to energy recovery in compressors is given in Fig. 13.3 [3].

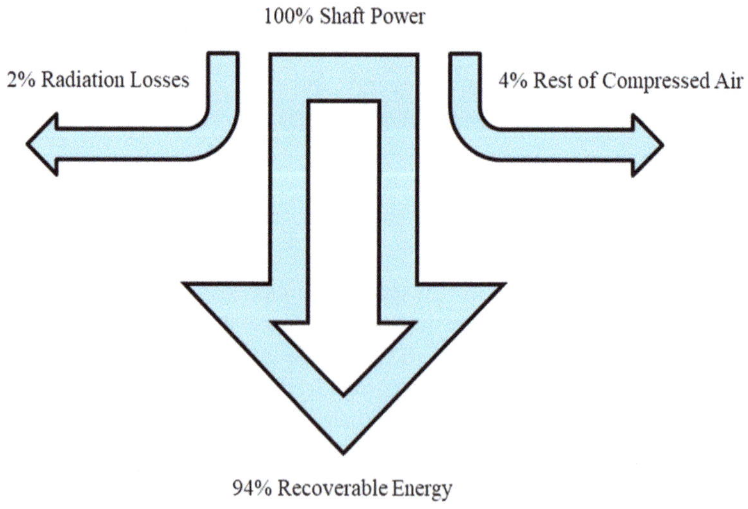

Fig. 13.3 Recoverable energy ratio in compressors [3]

Compressors consume more electricity than any other type of equipment in industrial activities. This means that energy savings can be achieved at very high rates thanks to improvements in compressor units. Improvements to these systems also reduce maintenance costs.

The compressors are produced in capacities ranging from several kW to 10,000 kW and are among the most energy-consuming equipment in many industrial plants [4]. Therefore, if a system can be used instead of compressed air in the system, this should be preferred. Energy loss due to inadequate installation and maintenance in compressed air systems can reach 50% of the energy consumed by the compressor. It is practically possible to avoid half of these energy losses with simple operating measures [5].

The cost of electricity spent running a compressor for one year usually exceeds the purchase price of the compressor. For example, assuming that a compressor with an electric motor of 100 kW and 90% efficiency operates for 6000 h per year, the annual energy expenditure for the energy unit price of 0.07 USD/kWh is 46,600 USD. This example illustrates the numerical size of the savings in the compressed air system. However, a significant amount of energy and financial savings can be achieved through some practical measures in these systems [6]. During the economic life of a compressor operating in a compressed air system, 86% of the cost distribution consists of energy expenses, 9% maintenance expenses, and 5% investment expenses.

13.1.2 Types of Compressors

Compressors can be classified as positive displacement compressors and dynamic compressors (Table 13.1). Positive displacement compressors operate according to the principle of reducing the volume of a certain amount of air by taking it into the system, in other words, increasing the pressure. Dynamic compressors, on the other hand, are based on the principle of increasing the pressure while decreasing the speed of the diffuser after increasing the flow rate of the air [7].

Specific power consumption in compressors is the power consumed by the compressor to compress 1 L/s compressed air (Table 13.2). Its unit is kW/(L/s). Its value depends on the compressor size and design. The power consumed by the compressor when idle running is called unloaded power consumption. In screw compressors, the power drawn depends on the system pressure of approximately

Table 13.1 Classification of compressors

Positive displacement compressors	Dynamic compressors
• Piston compressor • Rotary channel compressor • Diaphragm compressor • Screw compressor	• Centrifugal turbocompressor • Axial turbocompressor

Table 13.2 Specific power consumption in compressors [8]

Compressor type	Specific power consumption (kW/(L/s)) (Working pressure 7 bar)
Screw compressor: small capacity, oil-injected	0.36–0.43
Screw compressor: large capacity, oil-free	0.34–0.40
Rotary channel compressor: oil-injected	0.40–0.43
Small piston compressor	0.36–0.54
Large piston compressor	0.29–0.36

70%. In reciprocating compressors, the power drawn at idle running is independent of system pressure.

13.1.2.1 Piston Compressors (Reciprocating Compressors)

Piston compressors are positive displacement compressors. In the air, the intake valve is opened at the intake time, and the piston is filled. After the air is filled, the intake valve is closed, and the piston is pushed and the air inside is compressed to increase the pressure. Maintenance is costly and detailed because the friction is high in the system. There are oil-injected piston and dry piston types. In the oil-injected system, oil filters are used to clean the oil mixed into the air. These filters cause a partial pressure drop. Teflon gaskets are used in dry piston types. However, Teflon particles that are broken due to wear are mixed into the air and sent to the system. In addition, the wear-out of the Teflon equipment requires a higher maintenance cost of the system [7].

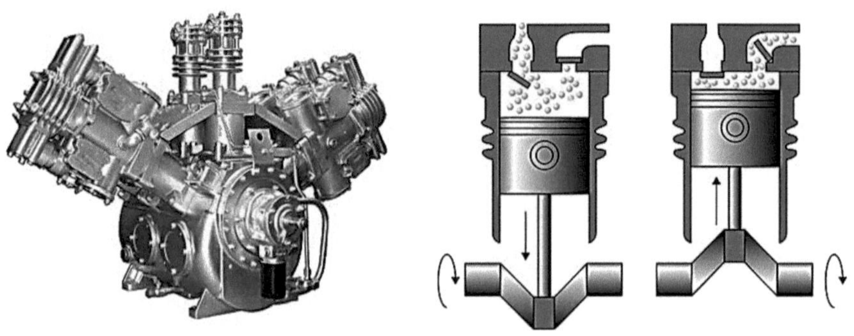

Fig. 13.4 Piston compressor

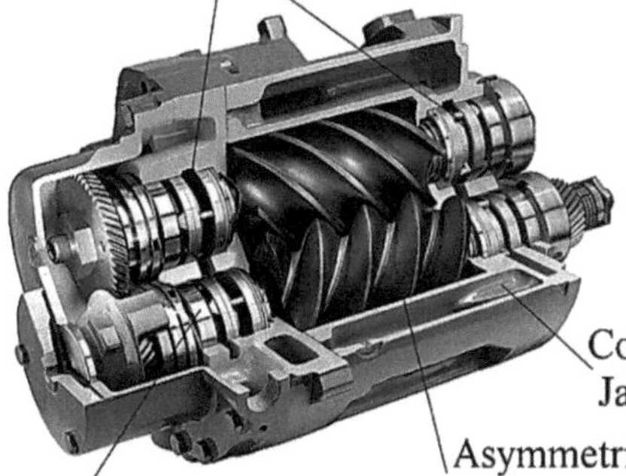

Fig. 13.5 Screw compressors

Piston compressors (Fig. 13.4) are machines with high installation, sales, and maintenance costs. They need a wide area for installation and accessory equipment. They work noisily. However, reciprocating compressors are successful in high-pressure applications up to 400 bar and compression of special gases.

13.1.2.2 Screw Compressors

Screw compressors are positive displacement compressor type (Fig. 13.5). Screw compressors have rotors in the compression chamber. The air is sucked in between the rotating moving screws and compressed and pushed down through the rotor. The screw compressor is the compressor type that has the lowest installation and sales costs. These compressors can operate for long periods of time without any maintenance other than scheduled maintenance. For this reason, maintenance costs are also partly lower. In addition to its low cost, it is also easy to install and does not need much space. This is the most common type of compressor used due to its superiority.

Oil-injected and oil-free screw compressors can be supplied in single stage from 2.2 to 400 kW. More efficient double-stage screw compressors can be supplied between 75 and 900 kW. The most important drawback is that the efficiency of partial loads is low.

13.1.2.3 Centrifugal Compressors

Centrifugal compressors are dynamic displacement compressors. The air is absorbed by the propellers rotating at high speed from the outside. Air is accelerated through propellers and sent to the diffuser. The pressure is increased by decreasing the air velocity at the diffuser. Although installation costs are not high, centrifugal compressors are expensive because they are precision machines. It is the most efficient solution system for applications that require high capacity air. Although centrifugal compressors are produced at 200 L/s capacity, screwed compressors up to 1000 L/s are more efficient than centrifugal compressors.

13.1.3 Compressor Control Systems

As far as compressor efficiency is concerned, capacity utilization control of the compressor is also important. Capacity utilization of compressors should be managed by determining capacity utilization patterns, which indicate in which range and in what amount of air used in the compressed air line. A compressor with a full capacity for constant load and a smaller compressor with VSD control for the variable load can be selected.

On/Off Control: When there is a demand for compressed air, it is transmitted to the compressor by the pressure sensor in the air tank and the compressor starts operating. This type of control system is inefficient for systems with variable air demand. Constant stop-and-start of the compressor not only increases the maintenance cost but also reduces the life of the compressor.

Load On/Load Off Control: In this system, when there is no air demand, the compressor motor will continue to operate. However, by closing the air inlet valve completely, air intake is blocked. Thus, the power consumption of the compressor is reduced. In the case of compressors running on a zero-load principle, when the air is not needed, the compressor runs idle and consumes about 25% of the energy.

Throttle Pressure: The air inlet valves are controlled by opening or closing the valve according to the demand.

VSD Control: In variable speed drive systems, capacity control is achieved by adjusting compressor speed and frequency.

13.2 Low-Pressure Use

The lowest pressure at which the equipment can operate must be used. If higher pressures are needed in a single application, replacement of the equipment or installation of a small, new compressor for that equipment should be made. The power reduction ratios for different set pressures in compressors are given in Table 13.3.

Table 13.3 Power decrease rates in compressors at set pressures (%)

Available Compressor Set Pressure (bar)	New Compressor Set Pressure (bar)								
	4.0	4.5	5.0	5.5	6.0	6.5	7.0	7.5	8.0
4.0	0								
4.5	7	0							
5.0	13	6	0						
5.5	17	11	5	0					
6.0	22	16	10	5	0				
6.5	25	19	14	9	4	0			
7.0	28	23	18	13	8	4	0		
7.5	31	26	21	16	12	8	4	0	
8.0	33	28	24	19	15	11	7	3	0

13.3 Prevention of Air Leaks

Air leaks are the leading cause of energy losses in the compressed air system. A compressor needs to work longer to prevent the pressure drop caused by air leaks. According to various studies, about 25% of the produced compressed air is lost due to leaks [9]. The complete prevention of these losses is non-practical and the reduction to 10% is accepted as an acceptable limit [10, 11].

The monetary value of pressurized air leaks equals the cost of the energy required to compress the atmospheric pressure air into the compressor output set (Table 13.4). The amount of air leakage depends on the line pressure, the temperature of the leaking point of the compressed air, the air temperature at the compressor suction, and the diameter of the hole through which the air escapes. Therefore, by determining the lowest pressure required at the point of use, lowering with pressure regulators will

Table 13.4 Power loss in different diameter and line pressure (kW)

Compressor set pressure (bar)	Line pressure (bar)	Hole diameter (mm)						
		0.25	0.5	1	2	3	4	5
4.5	4.0	0.01	0.04	0.17	0.68	1.52	2.71	4.23
5.0	4.5	0.01	0.05	0.20	0.79	1.78	3.17	4.95
5.5	5.0	0.01	0.06	0.23	0.91	2.06	3.66	5.72
6.0	5.5	0.02	0.07	0.26	1.04	2.34	4.17	6.51
6.5	6.0	0.02	0.07	0.29	1.17	2.64	4.70	7.34
7.0	6.5	0.02	0.08	0.33	1.31	2.95	5.25	8.20
7.5	7.0	0.02	0.09	0.36	1.45	3.27	5.81	9.08
8.0	7.5	0.03	0.10	0.40	1.60	3.60	6.40	10.00

Note It has been accepted as the system has a screw compressor and 0.5 bar pressure difference between the compressor outlet and leakage point, escaping air temperature 20 °C, atmospheric pressure 1.01 bar, Cd = 0.8

also reduce the cost of leaks. In addition, the pressure of the air used for cleaning must also be reduced. Using a cheaper method instead of cleaning with compressed air will save energy in large quantities. Connecting a hose to the water discharge valves in compressed air lines is a very bad practice. Instead of this, the pressure must be reduced with a separate connection and an air gun must be fitted to the hose outlet. Cleaning is done with an air gun with low air. In addition, the trigger on the pistol can close the air outlet when it is not needed.

Usually, air leaks occur at fittings of pipes, flashes, reductions in couplings and elbows, valve bodies, filters, hoses, check valves, extensions, and devices using compressed air. Temperature changes and vibration are the main causes of loosening of the connections and thus of leaks. For this reason, the periodic maintenance of the joints of the pipes is the first thing to be done in this regard. Leaks are generally at the end-use location or where the compressed air line is connected to the equipment. In such places where the compressed air intake is frequently turned on and off, the seals rapidly deteriorate. For this reason, it is necessary to replace the old ones by periodically maintaining the gaskets. In addition, the compressed air must be dwelled on and the users must be educated. Due to the increase in air leaks, it is impossible to reduce the compressor outlet pressure. This creates an additional cost [12].

13.3.1 Detection of Air Leaks

Pressurized air leaks can be determined in a number of ways. At the beginning of these methods is *listening by ear*. Large holes create a sound that can be heard comfortably with the ear. However, in noisy environments, it is very difficult to find leaks with this method. Controlling the joints with *soap foam* is much more effective than listening with the ear.

Another practical method of finding the total amount of leaks on the plant is *monitoring the pressure drop* all units operating with compressed air are stopped to determine the amount of leakage by monitoring the total pressure drop. The operating time of the compressor at idle and overload is measured and recorded at least 5 times and the amount of leakage is determined by the following equation:

$$\text{Total leak quantity (L/s)} = \text{Capacity of compressor (L/s)} \times \mathrm{t_{load}}/(\mathrm{t_{idle}} + \mathrm{t_{load}}) \quad (13.1)$$

Here, $\mathrm{t_{load}}$ and $\mathrm{t_{idle}}$ are the on-load running (operation) and idle running times per seconds, respectively. If the power is wasted, the energy taken by the compressor can be found by measuring it with a power analyzer.

The most effective method to identify leaks is to use an *ultrasonic sound detector*. These devices operate by recognizing, through a microphone, the voices at the level that the human ear cannot hear, due to air leaks, to the level that the ear can hear. In the calculations of losses of compressed air system, the compressor efficiency is used depending on compressor type (Table 13.5).

Table 13.5 Efficiency of compressors

Type of compressors	η_{comp}
Single-stage piston	0.88
Multi-stage piston	0.75
Screw piston	0.82
Rotary channel	0.72
Single-stage centrifuge	0.80
Multi-stage centrifuge	0.70
Turbo fan	0.70

As the hole diameter increases, pressure losses increase exponentially. The change in energy loss due to the diameter of the hole in the compressed air system is given in Fig. 13.6. The curve given in Fig. 13.6 is based on the values of screw compressor, engine efficiency is 0.90, compressor outlet pressure is 700 kPa, line pressure is 650 kPa, escaping air temperature is 20 °C, atmospheric pressure is 101 kPa bar, and $C_d = 0.8$.

In the compressed air system, the ratio of atmospheric pressure to line pressure is checked. If this ratio is less than 52%, the flow (V_f, m^3/s) of the air escaping from the hole can be calculated as follows, assuming that the flow is a leak.

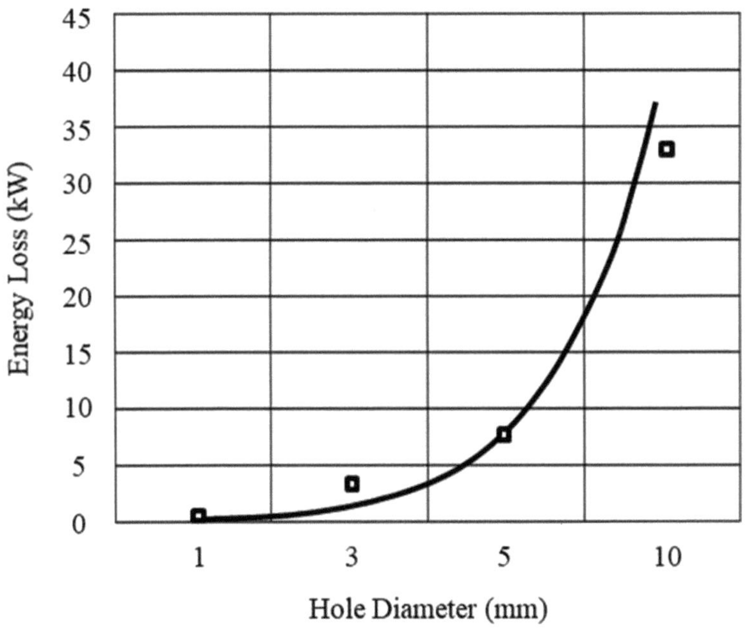

Fig. 13.6 Change of energy loss in the compressed air system [10]

$$V_f = \frac{N_L \times (T_i) \times \frac{P_l}{P_{amb}} \times C_1 \times C_d \times \frac{\pi D^2}{4}}{\sqrt{T_l}} \tag{13.2}$$

where

N_L = Number of air leaks,
T_i = Air (suction air) temperature in the compressor (K),
P_l = Line pressure where there is a hole (kPa),
P_{amb} = Atmospheric ambient pressure (kPa),
C_1 = Volumetric sonic flow constant (13.29),
C_d = Square section orifice coefficient (0.8),
D = The diameter of the hole (m), and
T_l = The average temperature of the line (K).

13.3.2 Energy Losses Due to Air Leaks

Power loss due to air leaks (L, kW) is calculated by Eq. (13.3). The compressor adiabatic efficiency in Eq. (13.3) is $E_a = 0.88$ for the single-stroke piston compressor, $E_a = 0.75$ for the multi-stroke piston compressor and $E_a = 0.82$ for the screw type compressor.

$$L = \frac{P_i \times V_f \times \frac{k}{k-1} \times N \times \left[\left(\frac{P_0}{P_i} \right)^{\left(\frac{k-1}{k \times N} \right)} - 1 \right]}{E_a \times E_m} \tag{13.3}$$

where

V_f = The flow rate of escape air (air leakage) (m^3/s),
k = Specific heat ratio of the air (1.4),
N = Stage number,
P_o = Compressor working pressure (kPa),
E_a = Compressor adiabatic efficiency, and
E_m = Compressor motor efficiency.

Annual energy savings are calculated as follows:

$$\text{Energy saving} = L \times H \tag{13.4}$$

where

H = Time during which the air line is under pressure annually (hours).

The annual economic earnings can be calculated as follows:

$$\text{Economic Gain} = \text{Energy gain} \times \text{Unit cost of the use} \tag{13.5}$$

Table 13.6 Compressed air system variables

Variables	Value
Compressor suction air temperature (°C)	36
Air temperature at the leak point (°C)	34
Compressor working pressure (kPa)	720
Compressor working pressure (Absolute Pressure) (kPa)	821
Line pressure at the leak point (kPa)	670
Total compressor engine power (kW)	110
Compressor engine efficiency (%)	91
Compressor type	Screw
Compressor working hours (yearly)	8640
Electricity unit price ($/kWh)	0.07

Example: In the context of the energy audit studies conducted in a factory, 10 leaks in compressed air system, with a diameter of 0.5 mm, 12 leaks with a diameter of 0.75 mm, and 8 leaks with a diameter of 1 mm were identified. Other variables required to calculate the cost of air leaks were measured and are given in Table 13.6. Due to the length of the compressed air lines, it is assumed that the compressed air temperature is equal to the ambient temperature. Calculate the total energy loss, financial loss, investment amount, and repayment period of the system?

As an example, the following values are calculated for 10 holes, actually 0.5 mm. The air that escapes through the hole (V_f):

$$V_f = \frac{10 \times (36+273) \times \frac{670}{101} \times 13.29 \times 0.8 \times \frac{\pi \times 0.0005^2}{4}}{\sqrt{34+273}} = 0.00243\,\mathrm{m^3/s}$$

Power loss due to escapes (L):

$$L = \frac{101 \times 0.00243 \times \frac{1.4}{1.4-1} \times 1 \times [\left(\frac{821}{101}\right)^{\left(\frac{1.4-1}{1.4 \times 1}\right)} - 1]}{0.82 \times 0.91} = 0.9468\,\mathrm{kW}$$

$$\text{Annual energy savings} = 0.9468\,\mathrm{kW} \times 8640\mathrm{h} = \sim 8180\,\mathrm{kWh}$$

$$\text{Cost Savings} = 8180\,\mathrm{kWh} \times 0.07 = \sim 573\$$$

The same procedures were also applied for leaks of 0.75 and 1 mm and the energy loss and its financial cost were calculated, depending on the hole diameter and the count (how many), and the results are given in Table 13.7.

In general, pressurized air leaks can be repaired with the process like replacement of hoses or couplings, replacement of gaskets, the shutdown of compressed air in an hour when the factory does not produce, repair of defects in the lines, etc. (Table 13.8).

Table 13.7 Cost of pressurized air leaks

Hole diameter (mm)	Count of the leaks	Air leak (m^3/s)	Power loss (kW)	Energy loss (kWh/year)	Cost ($/year)
0.5	10	0.00243	0.9468	8180	573
0.75	12	0.006573	2.556	22,200	1555
1	8	0.00974	3.029	26,311	1842
Total	30	0.0187	6.5318	56,691	3970

Table 13.8 Approximate expenditures to prevent air leaks

Quantity	Unit	Explanation	Unit cost ($)	Total cost ($)
30	Piece	Air leaks (Materials and Workmanship)	15	450
Total				450

Repayment period can be calculated by dividing the realization cost by the calculated annual savings amount.

$$\begin{aligned}\text{Repayment period} &= \text{Cost of implementation} \div \text{Annual saving amount} \\ &= (\$450)/(3970\$/\text{year}) = 0.11 \text{ year } (1.36 \text{ month}).\end{aligned}$$

13.4 Reduction of Compressor Outlet Pressure

In pressurized air systems, another issue where energy is wasted too much is the compression of air to higher pressures than equipment requires. As known, the amount of energy consumed to compress as the pressure rises increases. Therefore, by examining equipment using compressed air, the minimum amount of pressure required must be determined and the compressor outlet pressure must be adjusted accordingly. If necessary, equipment requiring pressure at different levels should be fed from separate compressors and lines. Just separating the line and lowering the pressure with the regulator at the beginning of the line will also reduce the leakage losses on that line.

There are pressure drops, especially in case of sudden air use, due to leaks in compressed air lines and inadequate selection of pipe sections. In order for the machines to operate without being affected by pressure drops, the commonly applied method is to raise the compressor outlet pressure. As a result, the amount of energy consumed by the compressor is also increasing. In sudden air shots, air tanks close to the end-use locations should be installed so that the line pressure does not fall. Another reason for the fall of the line pressure is the use of narrow cross-sectional pipes at the installation.

As can be seen from the following equations, the amount of work required to compress the air is also increasing, depending on the increase of the outlet pressure (P_h). By reducing the compressor pressure to P_2, the power consumption ratio (PCR) is calculated by the following equation.

$$PCR = \frac{[(P_2 + P_{atm})/P_{atm}]^{(k-1)/k} - 1}{[(P_h + P_{atm})/P_{atm}]^{(k-1)/k} - 1} \tag{13.6}$$

where P_h-compressor existing output pressure (set value), P_2-compressor recommended outlet pressure, P_{atm}- compressor suction pressure, and k-specific heat ratio of air (1.4).

The amount of energy that can be saved in a year by lowering the pressure in a compressor and its economic benefit can be estimated with the help of the following equations.

$$\text{Energy Saving} = CEP \times CU \times AOT \times (1 - PCR) \div \eta \tag{13.7}$$

$$\text{Economic Gain} = (\text{Energy Gain} \times \text{Unit Cost of Use})$$

where CEP—Compressor engine power, CU—Capacity utilization rate of compressor, AOT—Annual operating time of the compressor, PCR—Power consumption ratio, and η—Compressor engine efficiency.

Example: In the energy audit studies conducted in a factory; It has been determined that the majority of the machines in the facility operate with 4 bars pressure and some machines need 6 bars pressure. The cost of electricity in this facility is 0.07 $/kWh. For the smooth operation of the machines, the output pressure of the compressors, which are given in Table 13.9, is set to 7.5 bar. In the case of compressor pressure reducing from 7.5 bar to 7.3, 7.0, and 6.7 bar, respectively; calculate energy savings, financial savings, investment amount, and payback period?

For Compressor 1, PCR is calculated by Eq. (13.6). Similar operations are performed for Compressors 2 and 3; annual energy-saving values are given in Table 13.10.

Table 13.9 Compressed air system variables

Variables	Compressor 1	Compressor 2	Compressor 3
Compressor working pressure, bar	7.5	7.5	7.5
Targeted work press, bar	7.3	7.0	6.7
Compressor engine power, kW	90	37	30
Compressor engine efficiency, (%)	91	91	91
Compressor type	Screw	Screw	Screw
Annual working time (hours)	8640	8640	8640

Table 13.10 Energy and cost saving for compressors

Compressor	Engine power (kW)	Power decay coefficient	Annual operating time (h/year)	Capacity usage ratio	Engine efficiency	Energy saving (kWh/year)	Saved money ($/year)
1	90	0.99	8640	0.9	0.91	11,535.82	807.5
2	37	0.96	8640	0.6	0.91	7913	554
3	30	0.94	8640	0.3	0.91	5202	364
					Total	24,650.82	1725.5

$$\text{PCR} = \frac{[(730+100)/100]^{(1.4-1)/1.4}-1}{[(750+100)/100]^{(1.4-1)/1.4}-1} = 0.985$$

The PCR value is determined to be 0.962 and 0.939 for compressors 2 and 3 by similar calculation, respectively.

$$\text{Saving Energy} = 90 \times 0.9 \times 8640 \times (1-0.985)/0.91 = 11{,}535.82\ \text{kWh/year}$$

$$\text{Saving Cost Equivalent} = 11{,}535.82\ \text{kWh/year} \times 0.07\$/\text{kWh} = 807.5\$/\text{year}.$$

13.5 Taking Compressor Suction Air from Outside

In order for the compressors to operate efficiently, the absorbed air must be cool, clean, and dry. The effect of inlet air temperature on energy consumption in compressors is given in Table 13.11, the power drop ratios for different outdoor temperatures are given in Table 13.12.

As density increases with decreasing temperature, more air can be compressed with less power. A reduction of the intake air temperature by every °C, energy consumption is reduced by 1%. Compressors are usually operated in indoor areas and as suction is made through the indoor area, they work with warmer air compared to the external atmosphere. Instead, as shown in Fig. 13.7, it may be helpful to extract air from an external environment with an air duct.

The amount of energy to be saved by decreasing the temperature can be calculated by the following equation.

$$\text{Saving Energy} = \text{PFR} \times \text{Nominal Power} \times \text{OT} \times \text{LF}/\eta_{\text{engine}} \tag{13.8}$$

where OT—Operating time, LF—Load factor (actual load to full-load ratio), η_{engine}—Engine efficiency, and PFR—Power failure ratio.

Table 13.11 The effect of inlet air temperature on energy saving in compressors

Temperature of incoming air (°C)	Air volume required for 1000 m^3 flow at 21 °C (m^3)	Saving or overfitting by 21 °C temperature (%)
−1	925	7.5 saving
5	943	5.7 saving
10	962	3.8 saving
16	981	1.9 saving
21	1000	No saving
27	1020	1.9 excess consumption
32	1040	3.8 excess consumption
37	1060	5.7 excess consumption
43	1080	7.5 excess consumption
49	1100	9.5 excess consumption

Table 13.12 Power dissipation rates at different outdoor temperatures

Current temperature (°C)	Outside temperature (°C)								
	0	5	10	15	20	25	30	35	40
5	0.02	0.00							
10	0.04	0.02	0.00						
15	0.05	0.03	0.02	0.00					
20	0.07	0.05	0.03	0.02	0.00				
25	0.08	0.07	0.05	0.03	0.02	0.00			
30	0.10	0.08	0.07	0.05	0.03	0.02	0.00		
35	0.11	0.10	0.08	0.06	0.05	0.03	0.02	0.00	
40	0.13	0.11	0.10	0.08	0.06	0.05	0.03	0.02	0.00
45	0.14	0.13	0.11	0.09	0.08	0.06	0.05	0.03	0.02

The power failure ratio (PFR) can be calculated from the following equation depending on the temperature.

$$\mathrm{PFR} = 1 - \frac{\mathrm{T_o} + 273}{\mathrm{T_i} + 273} \tag{13.9}$$

where $\mathrm{T_i}$—Indoor temperature (°C) and $\mathrm{T_o}$—Outdoor temperature (°C).

Example: In energy audit studies conducted in an industrial enterprise; it has been determined that the air intake of all compressors used in the test is from a closed environment. The compressor is operating 8640 h per year, compressor motor duty is 0.91, and compressor usage factor is 1. The outdoor temperature ($\mathrm{T_o}$) was measured as 10 °C and the compressor suction temperature ($\mathrm{T_i}$) was 16 °C. Electricity cost is

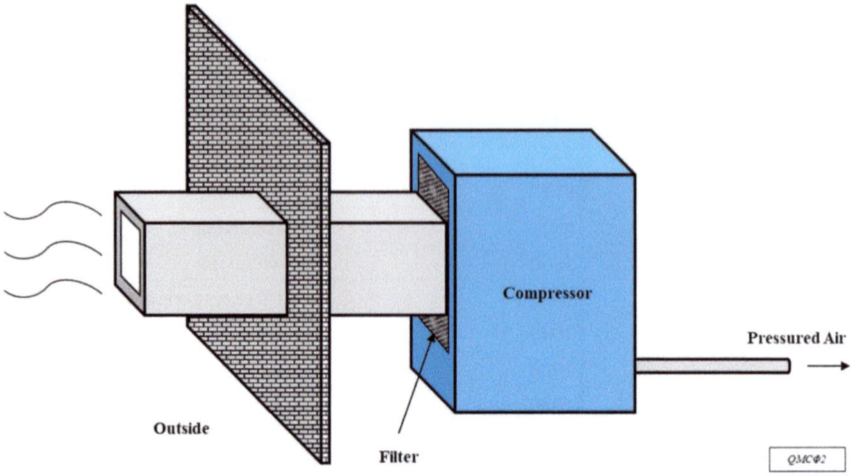

Fig. 13.7 Use of external air to save energy

0.07 $/kWh. In the case where the compressor suction air is taken from the outside by a channel, what are the annual savings and the repayment period?

The power drop ratio is calculated from Eq. (13.9) as follows.

$$\text{PFR} = 1 - \frac{10 + 273}{16 + 273} = 0.020$$

The amount of energy to be saved is calculated from Eq. (13.8).

$$\begin{aligned}
&\text{Energy Saving} = 0.020 \times 75 \times 8640 \times 1/0.91 = 14,241.758 \ \text{kWh}\\
&\text{Monetary compensation for annual savings from electricity use}\\
&\text{(MAS)}:\\
&\text{MAS} = \text{ES} \times (\text{electric unit price})\\
&\text{MAS} = 14,241.758 \ \text{kWh/year} \times 0.07 \ \$/\text{kWh} = {\sim}997 \ \$/\text{year}\\
&\text{Total savings for 6 compressors with 75 kW power will be}\\
&= 6 \times 997 \ \$/\text{year} = 5982 \ \$/\text{year}.
\end{aligned}$$

Investment Amount: Approximately 8 m of the metal channel will be used for the investment, the total price of which is about $250. The labor cost required for this job is estimated at $250. The total investment amount is $500.

Payback Period: The actual cost can be calculated by dividing the amount of annual savings calculated.

$$\text{Payback period} = (\$500)/(5982\$/\text{year}) = 0.083\text{year (about 1 month)}.$$

13.6 Use of Compressor Cooling Air

Due to rising energy costs and increased environmental awareness, most compressor users have begun to realize the potential of the heat in the compressors and the heat that is thrown out. Compressor manufacturers remove the heat that comes out during the compression by using a fan or water-cooled heat exchangers.

To obtain compressed air, 90% or more of the electricity consumed in compressors can be recovered as heat energy. If electricity, gas, or liquid fuel is used for production purposes during the production or processing phase, it is possible that one of these methods will be partially or completely replaced by the heat energy to be obtained from the compressor. Heat energy drawn from various refrigerators by oil, water, or hot air can be used for space heating, boiler combustion air or feed water preheating, process heating, and other purposes. While determining the recovered heat energy gain, the temperature levels to be determined determine the possible usage areas. As it is mentioned above, the diagram related to energy recovery in compressors is given in Fig. 13.3. Here, the types of energy that are in the 94% segment; heat recovery from low-pressure stage, heat recovery from intercooler, heat recovery from high-pressure stage, and heat recovery from after cooler.

The investment cost of the additional equipment that will be used to recover the waste heat repays itself in a short time with the savings to be made. If the heat recovery system is designed on the basis that the compressor will operate on the load, the targeted saving levels will not be achieved in case of a decrease in the use of air capacity. With a simple assumption, we can say that the amount of heat that can be obtained is directly proportional to the use of air.

Example: In a rolling mill of an enterprise, the annealing furnace combustion air is provided from the outside environment. Instead, two compressors' cooling air with a power of 166 kW and 138 kW can easily be used as furnace combustion air. The compressors are on–off controlled and cooled by air, label information is given in Table 13.13. For this, air ducts and an air fan are required at a distance of about 150 m from the compressor outlets.

Saving by using the cooling air in the furnace:

$$P_{aeg} = N \times P_{comp} \times f \times k \times t \tag{13.10}$$

Table 13.13 Label information of rolling mill compressors

Brand of compressor	A_1	A_2
Type of compressor	Screw	Screw
Engine power	166 kW	138 kW
Max. operating pressure	7.5 bar	8.5 bar
Unit of compressor (Piece)	1 piece	1 piece

Table 13.14 Investment cost

Investment type	Investment cost
Channel cost	13,142.85 USD
Fan cost (15,000 Nm^3/h)	2,571.15 USD
Total	**15,714 USD**

where P_{aeg}—Annual energy gain (kW), N—Unit of compressors (piece), P_{comp}—Power of compressor (kW), f—Heat energy conservation factor (80%), k—Safety factor (5% loss-leakage) and t—Operation time (8500 h/year).

$$P_{aeg} = 2 \times ((138 + 166)/2)\ \text{kW} \times 0.80 \times 0.95 \times 8500\ \text{h/year}$$
$$P_{aeg} = 1{,}963{,}840\ \text{kWh/year and } 1\ \text{kWh} = 860\ \text{kcal;}$$
$$P_{aeg} = 1{,}688{,}902{,}400\ \text{kcal/year}$$
$$\text{Annual fuel economy} = 1,688,902,400\text{kcal}/5847\ \text{kcal/kg}$$
$$\text{Annual fuel savings (coal equivalent)} = 288.849\ \text{ton}$$
$$\text{Annual monetary saving} = 288.849 \times 200\ \text{USD/ton} = 57,760\ \text{USD/year.}$$

Investment and Operation Cost: In order for the cooling air to be used in the annealing furnace, an air duct of approximately 150 m length should be made from the compressor room to the furnace recuperator entrance. In order to avoid losing the energy of hot air in this channel or for the least losses, the duct circumference must be insulated. In addition, it will be useful to install a reinforcing fan for the convenience of flow in the duct. In order to do all these works, a labor cost is required. As a result of the offers received from the companies in the market, the investment planned for all these is given in Table 13.14. In addition, the section calculation of the channel planned to be made is given below.

$$\dot{Q}_{\text{flowrate}} = A_{\text{channel}} \times V \tag{13.11}$$

where, $\dot{Q}_{\text{flowrate}}$ =—Compressor cooling air flow (10,960 m^3/h = 3.04 m^3/s) A_{channel}—Channel cross-sectional area (m^2), and V—Hot air velocity in the channel (5 m/s).

$$3.04 = A \times 5\,\text{m/s} \quad \rightarrow \quad A = 0.608\,\text{m}^2$$

A square channel design with dimensions of approximately 75 × 75 cm has been found suitable. The electric motor used for the added fan is 5 kW. Here the operating cost is calculated as follows.

$$5\ \text{kW}\ 8500\ \text{h/year}\ 0.070\ \text{USD/kWh} = 2975\ \text{USD}$$

Table 13.15 Economic analysis

Annual energy gain (kcal)	Annual coal equivalent fuel savings (ton)	Annual monetary savings (USD)	Operating cost (USD)	Annual net savings (USD)	Investment cost (USD)	Repayment period (months)
1,688,902,400	288.8	**57,769**	2975	54,794	15,714	**4.1**

Payback Period: The data obtained as a result of this investment which are operating cost, investment cost, monetary saving, and repayment period are given in Table 13.15.

13.7 Compressed Air Flow Control and Energy Economy

In accordance with the conditions required for operation, the air flow provided by the compressors varies with time. The correct selection of the flow control system, which enables the system to operate at maximum efficiency as a whole, is of great importance, especially in terms of energy savings.

On/Off Control: In this method, depending on the need of air, the compressor works either at full load or unloaded at all. In the case of no-load operation, the compressor continues to rotate, and when the compressor is running at full load, approximately 30% of the energy consumed is consumed in no-load operation.

Load Control with Compressor Air Inlet Damper: The compressor is running continuously. The capacity of the compressor is controlled by the valve in the inlet air duct. This method, according to the "on/off control" operating method, keeps the compressor output more stable, but starts to run inefficiently when the compressor load falls below 90%. In this method, even when no air is needed, it continues to consume about 50% of the energy consumed when working at full load.

Variable Speed Control (VSD): It is known that changing the engine speed is the most convenient way to adjust the compressor output air flow. It is possible to use electric systems with frequency converters to change the speed in the compressor motor. In systems with frequency converters, the compressor output air flow and the energy consumption in the compressor are directly proportional. It is also possible to keep the pressure of the air at the compressor outlet at a stable value. For the three methods described above, the relationship between compressor outlet airflow and energy consumption is shown in Fig. 13.8.

Frequency Converter Application in Compressors: When measuring the compressors of an operation, compressor number 1 operates at a load close to full load. Compressor number 2 is activated and deactivated according to the amount of air needed. Since one of the compressors operates at a load close to full load, only one

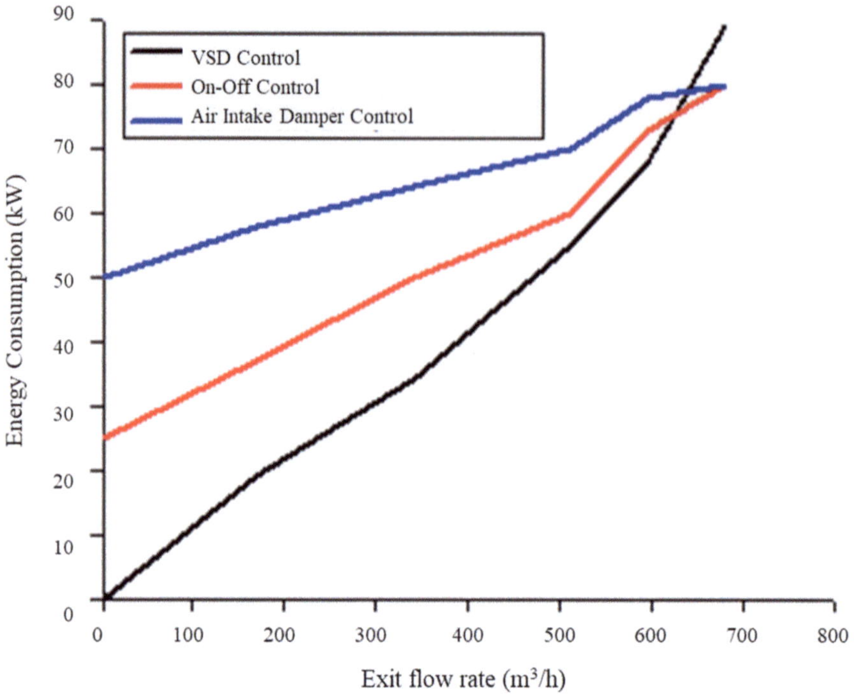

Fig. 13.8 Relationship between compressor outlet air flow and energy consumption for three methods used for compressed air flow control

compressor frequency converter application in this compressor room may be sufficient. Two of the compressors can be run alternately at full load and a compressor frequency converter can be applied, so variable flow demand can be financially affordable.

Compressor capacity can be adjusted by reducing the frequency of the electric motor and the frequency converter without changing the existing compressor and motor. Thus, energy saving can be achieved. The cost of the frequency converter for reducing/adjusting the current compressor motor operating at low voltage (380–400 V) is 7000 € + VAT.

As shown in Fig. 13.8, while frequency-controlled compressors operate at low loads, they provide energy savings between 15 and 25% of the energy consumed when operating at full load. When the proposed compressor is considered to provide 15% energy saving while operating at low load;

Annual Saving Amount = (138 kW × 0.15 × 8000 h/year × 0.07 USD/kWh)

= 11,592 USD/year

Investment Cost (Frequencyconverter) = 7000 € + 18%VAT* = 8260 € or 13,216 USD

Repayment Period = 13,216 USD/(11,592 USD/year) = 1.14 year ≈ 13.68 month

(**Note*: The percentage of VAT varies by the country, where this calculation is made.)

13.8 Closing of Compressors and Main Valves

Screw compressors draw power up to 85% of nominal load even when they are not loaded. For this reason, it will make sense to stop the compressors completely when there is no production on the lines the compressors supply. In addition, since the pressure drops with the leakage of air, the compressor will have to work on the load to compensate for the leakage. Piston compressors will also cause energy wastage if they leak in and out of the circuit. Also, the loss due to leakage can be reduced by closing the valve immediately at the outlet of the air tank. Compressors can be switched off manually or automatically in conjunction with the compressor's air-fed unit. Since people tend to forget, automatic controls always work better.

13.9 Recommendations for the Operation of Compressors

- Air leaks should be prevented.
- The lowest possible set pressure should be selected.
- Cold, clean, and dry air intake should be provided.
- Filters on the suction side should be cleaned regularly, filters that reduce pressure less should be used. By placing a pressure gauge at the filter inlet and outlet, filter replacement times should be adjusted accordingly.
- Compressed air should not be used for cleaning purposes, the pressure should be reduced with the pressure regulator where necessary and the nozzle should be installed at the end of the hose.
- Water drain valves should be kept shut off and the tightness check should be done frequently.
- Synthetic oils should be used for lubrication.
- Geared (knurled) V-belt should be used in motors, pulleys and belt tension should be checked continuously.
- Unused equipment and lines should be disabled.
- Measurement and control instruments should be examined regularly.
- Unnecessary hose connections should be canceled.
- Pipes and fittings that cause low-pressure loss should be used.

- Filters that reduce pressure less should be used.
- Old and inefficient compressors should be replaced.
- Valves that cause pressure drop should be replaced.
- A suitable design should be made and the right type and size line should be installed.

References

1. Holdsworth J (1997) Conserving energy in compressed air systems. Plant Eng 51(13):103–104
2. Compressed Air: Opportunities for Businesses, Carbon Trust, January 2012
3. Kaya D, Öztürk HH (2014) Sanayide Enerji Yönetimi ve Enerji Verimliliği, Umuttepe Yayınları, Yayın No: 114, Mühendislik Dizisi: 12, Kocaeli/Türkiye
4. Risi JD (1995) Energy savings with compressed air. Energy Eng 92(6):49–58
5. Talbott EM (1993) Compressed air systems: a guidebook on energy and cost savings, 2nd ed. The Fairmont Press, Inc., Liburn, GA 00247
6. Çerci Y, Cengel YA, Turner HT (1995) Reducing the cost of compressed air in industrial facilities. Thermodynamics and the design, analysis, and improvement of energy systems. ASME, AES 35:175–186
7. Sapmaz S, Canka Kilic F, Eyidogan M, Taylan O, Coban V, Cagman S, Kilicaslan I, Kaya D (2015) Selection of compressors for petrochemical industry in terms of reliability, energy consumption and maintenance costs examining different scenarios. Energy Explor Exploit 33(1):43–62. https://doi.org/10.1260/0144-5987.33.1.43, ISSN: 0144-5987
8. Barber A (1989) Pneumatic handbook, 7th ed, Trade and Technical Press
9. Terrell RE (1999) Improving compressed air system efficiency- know what you really need. Energy Eng 96(1):7–15
10. Kaya D, Phelan P, Chau D, Saraç Hİ (2002) Energy conservation in compressed-air systems. Int. J. Energy Res
11. Risi JD (1995) Energy savings with compressed air. Energy Eng 92(6):49–58
12. Kaya D, Saraç HI, Olgun H (2001) Energy saving in compressed air systems. In: The fourth international thermal energy congress, Çesme/Turkey, pp 69–74

Chapter 14
Energy Efficiency in Fans

In many fan systems used in the industry, the fluid flow rate needs to be adjusted due to the changes in operating conditions. On the other hand, most of the fans are used over-capacity for safety reasons. As a result, the fans generally operate below the designed flow rates (with low efficiency) and thus, lose energy. Flow control can be done in the following ways [1]:

- Inlet and outlet dampers (flaps, valves, etc.),
  - Speed control,
  - Belt-pulley mechanism,
- Frequency converter,
- Variable inlet blades, and
- Variable inclined fan blades.

When selecting the flow control system, it is important to know whether the flow rate varies over time due to operating conditions. Damper or valve-controlled systems, belt-pulley mechanisms, and frequency converter systems are commonly used for flow rate adjustment. The flow rate adjustment with the damper or valve is based on the principle of flow rate reduction due to the increase in system pressure, as can be seen in the fan characteristic curves. With this application, the fan is started to operate with lower efficiency. These methods, which control flow rate by blocking flow, result in significant energy loss due to friction.

Systems with belt-pulley mechanism and frequency converter enable energy saving by changing the fan speed. The variable inlet blades and the variable inclined fan blades are essentially based on the same principle. By changing the blades in such controls, the amount of energy that the fluid will gain is reduced. Thus, while the fan's energy consumption decreases, the flow rate decreases, accordingly. The use of these systems is quite low.

D. Kaya et al., *Energy Management and Energy Efficiency in Industry*,
Green Energy and Technology,
https://doi.org/10.1007/978-3-030-25995-2_14

14.1 Fan Laws

In fans; the relationships between flow rate (Q), pressure (p), power (P), revolutions per minute (or Speed, RPM) (n), rotor diameter (D), and density are given in the following equations [1, 2]:

$$Q_2 = Q_1(n_2/n_1) \quad \Delta p_2 = \Delta p_1(n_2/n_1)^2 \quad P_2 = P_1(n_2/n_1)^3 \quad \rho \text{ and } D = \text{Constant} \tag{14.1}$$

$$Q_2 = Q_1(D_2/D_1)^3 \quad \Delta p_2 = \Delta p_1(D_2/D_1)^2 \quad P_2 = P_1(D_2/D_1)^5 \quad n \text{ and } \rho = \text{Constant} \tag{14.2}$$

$$\Delta p_2 = \Delta p_1(\rho_2/\rho_1) \qquad P_2 = P_1(\rho_2/\rho_1) \qquad n, D \text{ and } Q = \text{Constant} \tag{14.3}$$

The results of these equations can be summarized as follows:

- Flow is proportional to RPM (speed),
- The pressure varies with the square of RPM, and
- The power varies with the cube of RPM.

Theoretically, if the revolutions per minute RPM (speed) is reduced by 80%, the flow rate is reduced by 80%, the pressure by 64% (= 0.8 × 0.8), and the power by 51.2% (= 0.8 × 0.8 × 0.8).

14.2 Flow Control Systems and Energy Economics

In accordance with the conditions required by the process, the fluid flow rate provided by the fan may vary over time. The correct selection of the flow control system, which will enable the system as a whole to operate at the highest efficiency, is particularly important in terms of energy saving. In system selection; factors such as cost, fan type, fan characteristics, and annual operating load are considered, carefully [3].

14.2.1 Damper-Controlled Systems

This control method is used to reduce the flow rate in the system. When the damper is throttled, the operating point shifts to the left, provided that it remains on the characteristic curve. In this case, while the flow rate decreases, the pressure value that the fan must meet increases unnecessarily. In other words, there is a big difference between the energy supplied by the fan and the energy required by the system. The energy required for the differential pressure is completely wasted. In Fig. 14.1, the

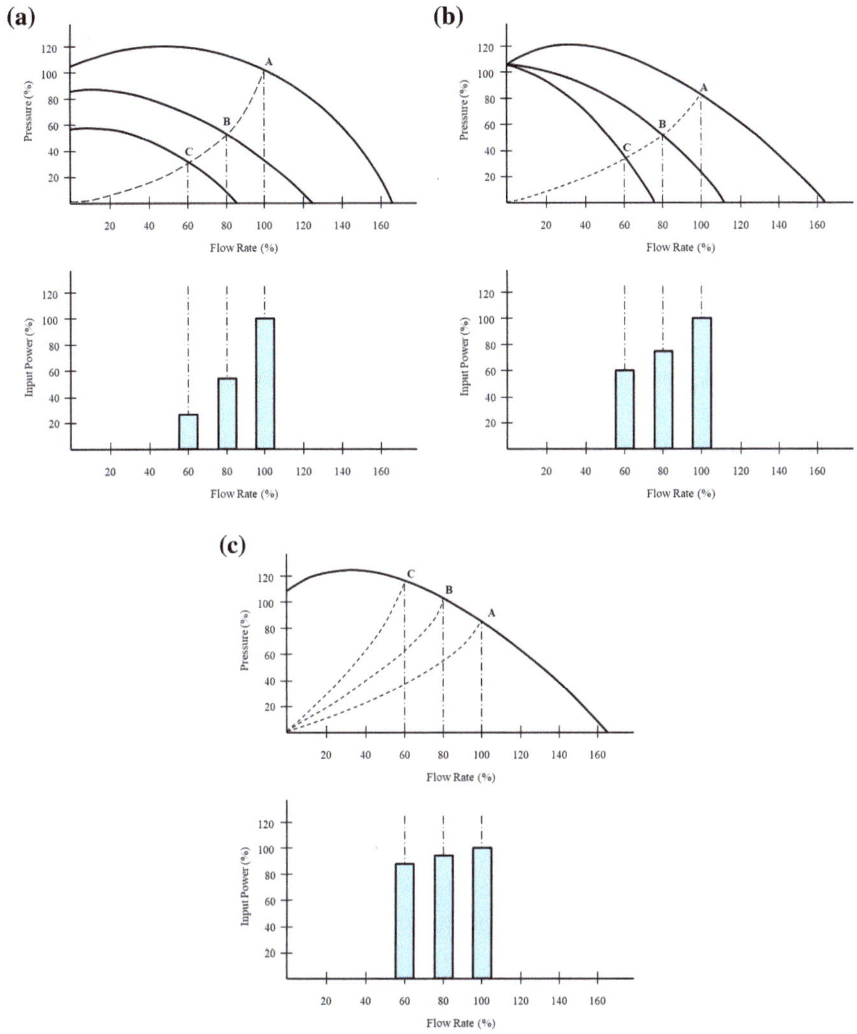

Fig. 14.1 Variation of pressure and input power with flow rate (for the application with outlet damper (**a**), inlet damper (**b**), and frequency converter (**c**))

changes in flow rate and pressure of the input and output power—flow control is made in three different systems; (a) output damper, (b) inlet damper, and (c) frequency converter are given for comparative purposes.

In the inlet damper application, the gas inlet is reduced before the fan, and therefore flow rate adjustment is made. In this type of application, the static pressure drops. The curves given in Fig. 14.1b are obtained for the different states of the valve. While the pressure difference at the working points A, B, and C where the fan speed is constant, the dynamic pressure that determines the input power remains constant. In other words, the input power will decrease more depending on the flow than the output damper.

14.2.2 Speed-Controlled Systems

The most effective way to change the efficiency of the fans is to change the speed. To change speed, it is possible to use systems with belt-pulley mechanism or variable speed controller (frequency converter or mechanical clutch).

The belt-pulley mechanism may be particularly preferred for situations where the flow is constant or there is little variation due to operating conditions. The flow rate adjustment with frequency converter (FC) is more suitable for situations where the flow changes frequently and for automation. However, due to the rapid development of electronic devices and the decrease in prices, frequency converter systems have become widely used in recent years.

When the speed control with frequency converter is examined, which is given in Fig. 14.1c, it is seen that the dynamic pressure changes according to the operating point. Since the dynamic pressure decreases considerably depending on the flow, a significant reduction in the input power and consequently, energy savings are achieved [4].

Theoretically, to emphasize the importance of speed control in terms of energy saving, it is useful to revise the fan laws once again. As it is known, the designed installation is not always expected to operate at full load. For example, as it can be seen from the equations given below, when the flow is halved, the pressure loss decreases to 1/4 of the full-load loss and the energy to be supplied to the system decreases to 1/8.

$$\frac{Q_1}{Q_2} = \frac{n_1}{n_2} \qquad \frac{\Delta p_1}{\Delta p_2} = \left(\frac{n_1}{n_2}\right)^2 \qquad P = \frac{Q_1}{2} \times \frac{\Delta p_1}{4} \tag{14.4}$$

Theoretically, it is clear that energy savings can be achieved. However, the energy to be supplied to the system is different from the energy required to operate the fan.

In a fan system, the energy savings to be achieved when the flow is controlled by the output damper and frequency converter are given in Fig. 14.2. For example, if there is a 20% reduction in flow rate, the decrease in power is 5% with the valve control, while the reduction with the frequency converter is 38.5%. It is possible to save as much as the difference, in other words, 33.5%.

Variable frequency drives (VFD) are used to achieve maximum energy savings in both direct drive and belt-pulley systems. As the building load is reduced, the VFD decreases the input power frequency, reducing engine speed, air flow rate, and power dissipated. Depending on the building usage, the fan speed can be changed and the energy consumption can be reduced. In order to examine the energy-saving economy by applying the frequency converter to the fans;

- As a first step, the load distribution of the system should be determined.
- Depending on the percentage change in the flow rate, the change in power should be defined comparatively.
- Weighted power distribution should be determined.
- The power value must be defined at full speed or under the current conditions.

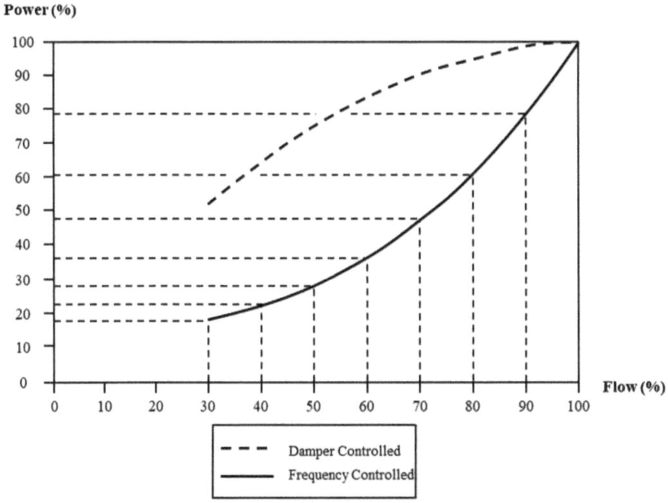

Fig. 14.2 Flow rate and power relationship in damper and frequency-controlled fan systems

- Weighted power savings should be determined.
- The monetary value of the energy to be saved should be determined depending on the annual working time.
- The total cost of the frequency converter must be determined.

14.3 Fan Selection

Fan selection is important in terms of reducing the energy consumption. Usually, impeller or axial fans are relatively better for low static pressures. Likewise, centrifugal fans are also better for high static pressure. Generally, only the initial cost is taken into account when choosing a fan. As a result, small diameter, high-speed fans are used as required. When running at high speed, small fans require more operating energy and produce more noise than a large fan at low speed. Fan static efficiency is calculated using the following equation [5]:

$$\text{Fan Static Efficiency } (\eta_{\text{static}}\%) = \frac{\left[\text{Volume in } \left(\frac{\text{m}^3}{\text{sec}}\right) \times \Delta\text{p (Static Pressure)in mmWC}\right]}{\left[102 \times \text{Power Input to the Fan Shaft in (kW)}\right]} \times 100 \qquad (14.5)$$

Where 102 is a conversion constant. Generally, there are two ways to describe fan efficiency: these are static efficiency (as mentioned above) and mechanical efficiency (it is also called the total efficiency). These efficiencies measure how well the fan converts horsepower into flow and pressure. The equation for determining mechanical efficiency is

Fan Mechanical Efficiency ($\eta_{\text{mechanical}}\%$)

$$= \frac{\left[\text{Volume in}\left(\frac{\text{m}^3}{\text{sec}}\right) \times \Delta\text{p (Total Pressure)in mmWC}\right]}{\left[102 \times \text{Power Input to the Fan Shaft in (kW)}\right]} \times 100 \qquad (14.6)$$

When the fan operates at equal flow and static pressure, it achieves the lowest energy consumption at the highest static efficiency. A small change in the system pressure results in a large change in air flow. Variable flow systems may need to be selected on the right side of the fan curve to ensure stability at low air flow.

Whether the fan is directly driven or with a belt-pulley mechanism, this affects efficiency. Although the energy loss percentage of the belt-pulley mechanism fans values are between 15% in stepper motors and 4% in big motors, they are more successful in adjusting the fan speed and balancing the system. However, fans with belt and pulley mechanisms need regular maintenance. In the air processor or fresh air preparation device, the belt pulley is worn, dust accumulates around the fan, which causes the filters to become dirty more quickly or in the absence of the filter, it causes poor air quality. In direct driven fans, there are no losses as in the belt pulley, as the impellers and propellers are directly connected to the motor shaft. In addition, the absence of belt, pulley and shaft bearing means that it will require less maintenance, and less vibration occurs due to the lack of moving parts. Recycling calculations should be made for this investment, as large fans with direct drive and large and low-speed motors are often costly.

Operating principle of fans can be seen in Fig. 14.3.

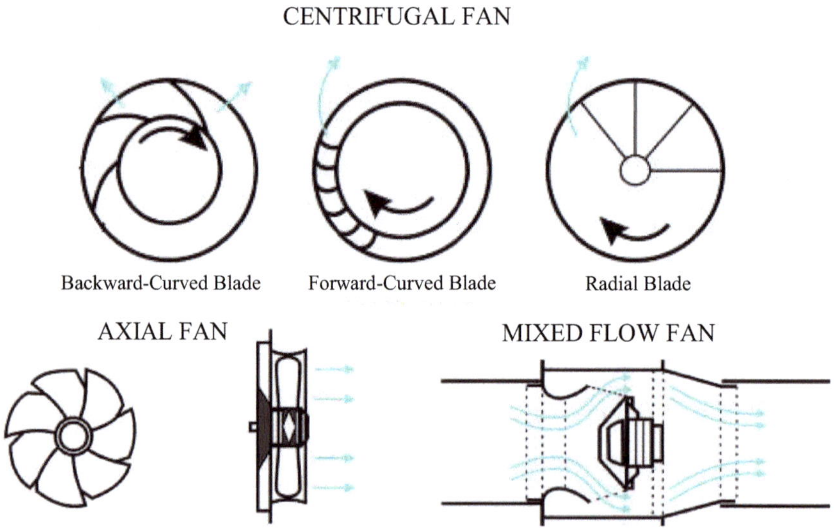

Fig. 14.3 Operating principle of fans

References

1. Kaya D, Öztürk HH (2014) Sanayide Enerji Yönetimi ve Enerji Verimliliği, Umuttepe Yayınları, Yayın No: 114, Mühendislik Dizisi: 12, Kocaeli/Türkiye
2. Hadra M, Hergül AS, Kaya D, Eyidoğan M, Çanka Kiliç F, Özdemir NC (2019) An energy audit and optimization in bar mill annealing furnace. Int J Ecosyst Ecol Sci (IJEES) 9(1):33–42. https://doi.org/10.31407/ijees https://doi.org/10.31407/ijees91
3. Uğural G, Parmaksızoğlu C (1992) Ventilators and Systems, Temas Publications 6, 1992. (In Turkish: Uğural G, Parmaksızoğlu C, Vantilatör ve Sistemleri, Temas Yayınları 6, 1992)
4. Wagenkecht J (1997) An economic overview of frequency controlled pumps, III. National plumbing engineering congress proceedings book, 2, 883, November 1997 (In Turkish: Wagenkecht J, Frekans Kontrollü Pompalara Ekonomik Bir Bakış, III. Ulusal Tesisat Mühendisliği Kongresi Bildiri Kitapçığı, Cilt 2 s. 883, Kasım 1997)
5. Akkoç H, ve Gürbüz N (1997) Factors effective in fan selection, mistakes in application and their prevention, III. National plumbing engineering congress proceedings book, 2, p 933, November 1997 (In Turkish: Akkoç H, ve Gürbüz N, Vantilatör Seçiminde Etkili olan Faktörler Uygulamada Yapılan Yanlışlıklar ve Bunların Önlenmesi, III. Ulusal Tesisat Mühendisliği Kongresi Bildiri Kitapçığı, Cilt 2 s. 933, Kasım 1997

Chapter 15
Energy Saving with Variable Speed Driver Applications

15.1 Variable Speed Drive Systems

In systems where electric motors are used as carriers, the circuits that perform motion control are defined as electric drives. At least 50% of the energy required for the whole system is consumed in electrical drive circuits. Electric drive circuits are examined in two separate groups as constant speed and variable speed. While 75% of these are used at constant speeds, the rate of use of variable speeds has increased to over 25% [1].

There are many terms used to describe speed-controlling equipment. These abbreviations are used interchangeably, but have different meanings:

Adjustable Speed Drive (ASD): It is a very general definition, which can be used for mechanical or electronic speed control methods.

Variable Speed Drive (VSD): This more general definition is used for the equipment that controls the speed of a motor or an equipment driven by a motor like fan, pump, compressor, etc. Such equipment may be electronic or mechanical.

Variable Frequency Drive (VFD): This equipment uses power electronics to change the frequency of the motor input power to control the motor speed.

Variable speed drive (VSD) systems, inverter, frequency converter, also called a variable frequency drive. These systems prevent the motor from overloading by changing the frequency of the alternating current (AC), and therefore the rotational speed of the motor. This allows the same work to be done with much less energy. VSD systems are converted to direct current (DC) from the grid or converts 1–3 phase AC mains voltage to DC first in motor drive applications. These are high technology motor speed controllers that can clean the disturbing electric fluctuations (they clean voltage fluctuations, peaks, etc., from the network through the filter circuit) and adjust the speed of the AC or servo motor from zero to the desired value and time with a high starting torque. These systems, which adjust the automatic or manual frequency

D. Kaya et al., *Energy Management and Energy Efficiency in Industry*, Green Energy and Technology,
https://doi.org/10.1007/978-3-030-25995-2_15

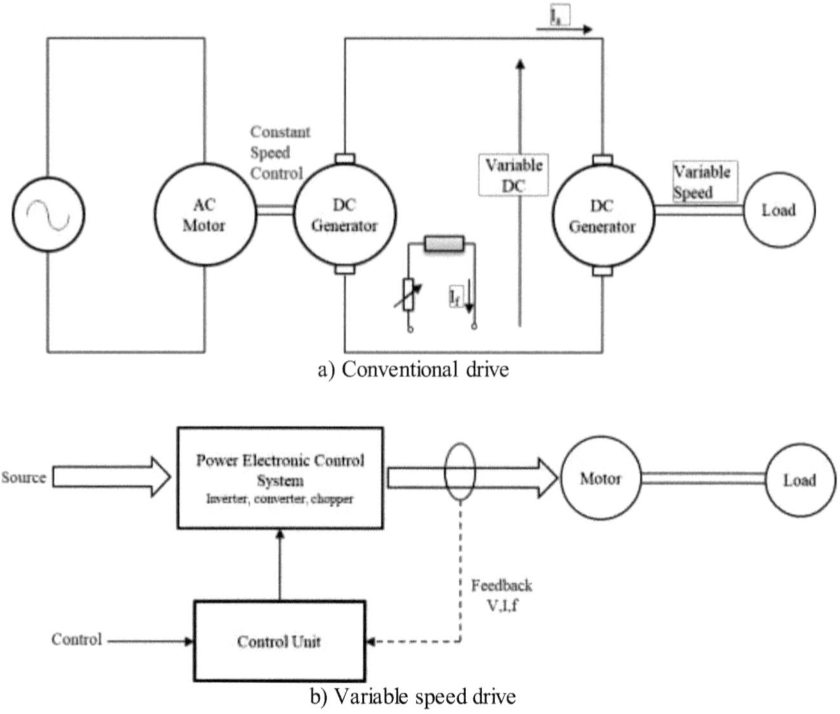

Fig. 15.1 Conventional and variable speed drive [a) and b)]

and enable the motor and mechanical order in the machines to be changed or fixed at the desired speed, are also called as variable frequency drive (VFD) systems [2].

Conventional and variable speed drive can be seen in Fig. 15.1 [a) and b)].

By adding a VSD system to the motors, energy saving up to 50% is possible. In other words, the amount of electricity the motor consumes for the same job can be reduced by half. The cost of engines equipped with VSD is definitely higher but, in correctly selected applications—such as pumps and compressors—VSD systems can pay for their expense by assuring energy savings in a very short time.

The purpose and benefits of using a variable speed drive are

- The variable frequency drive is used to adjust a current or pressure according to the actual demand. It controls the frequency of the electrical power supplied to a pump or fan. Significant power savings can be achieved when a VFD is used.
- It ensures that the repair and maintenance costs of the motor and mechanical parts are kept to a minimum level by minimizing the errors of the motor and mechanical parts due to the disturbing effects in the network frequency and thus, the motor/mechanical parts' life is extended.
- It reduces the reactive energy drawn from the grid by setting the Cos φ value in the grid to 0.98–1.

- In some applications, energy savings of 30–40% can be achieved, successfully.
- Since VSD does not require gear or pulley mechanisms for speed changes, it enables rapid and cost-free speed changes.
- Due to its soft starting feature, it ensures that mechanical maintenance and breakdown costs are reduced, maintenance times and lifetimes are extended due to the lack of pulsed start and stop features made to mechanical devices at the time of initial departure and stop.

15.1.1 Variable Frequency Drive

The variable frequency drive (VFD) is a system for controlling the rotational speed of an AC electric motor (Fig. 15.2) [3]. VFD technology is widely applied in air conditioners, pumps, coolers and tower fans and controls the frequency of the electrical power supplied to the motor. The variable frequency drive is a special type of adjustable speed drive. Variable frequency drives are also known as adjustable frequency drives (AFD), variable speed drives (VSD), or inverter drives. Variable frequency drives are widely used on pumps, machine tool drives, compressors, and ventilation systems of large buildings. The variable frequency motors on the fans adjust the amount of air driven to suit the requirements of the system, saving energy. For example, if the cooling systems are controlled by the frequency converter of the fan motors depending on the seasonal temperatures, energy savings of 60% are achieved annually. Therefore, it is recommended that the fan, pump, and compressor motors are operated under frequency converter control [3, 4].

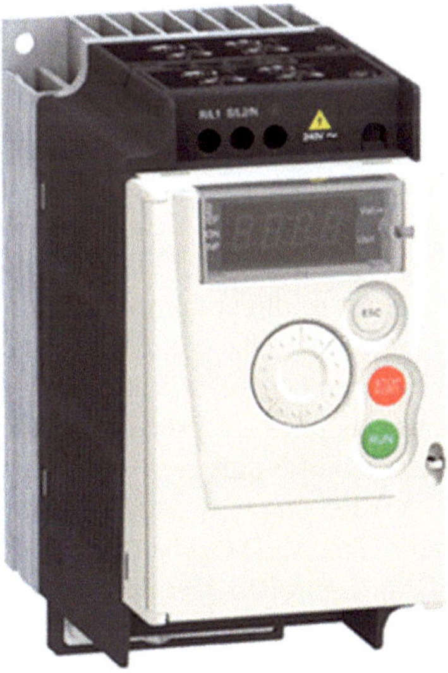

Fig. 15.2 Variable frequency drive [3]

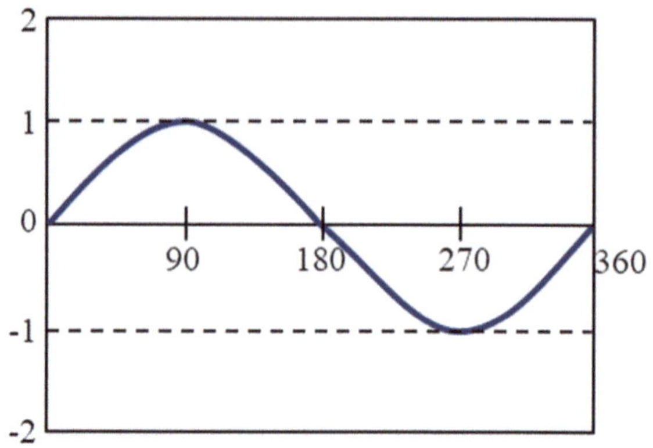

Fig. 15.3 AC sine curve

Automatic frequency control consists of a primary electric circuit that first converts alternating current into direct current and then back to an alternating current of the required frequency. The internal energy loss in automatic frequency control is defined as ~3.5%. The reasons for applying automatic frequency control can be related to both the functionality of the application and energy saving. For example, automatic frequency control is used in pump applications where the current is matched by volume or pressure. The pump regulates its speed to a certain set point by means of a regulating cycle. Adjusting current or pressure to actual demand reduces power consumption [5, 6].

In order to understand the basic principle of VFD technology, it is necessary to first recognize the three basic units of VFD. These are

(1) Rectifier (rectifier),
(2) Direct current (DC) distribution bar (bus bar), and
(3) Transducer (inverter).

The voltage of the AC power source increases and decreases as a sine curve (Fig. 15.3). When the voltage is positive, the current flows in one direction and when the voltage is negative, the current flows in the opposite direction. Such power systems ensure the efficient transfer of large amounts of energy over long distances.

The rectifier in a VFD is used to convert the alternating current (AC) power at the input to direct current (DC) power. Two rectifiers are needed for each phase of the power. One rectifier only allows power to pass through when the voltage is positive and the second rectifier only when the voltage is negative. Since most large power supplies are three-phase, a minimum of six rectifiers are used (Fig. 15.4). The definition of 6-pulse is used to identify a drive with six rectifiers.

A VFD can have multiple rectifier sections, each of which consists of 6 rectifiers. Thus, VFD can be "12-pulse", "18-pulse", or "24-pulse". Rectifiers use diode, silicon-controlled rectifier (SCR), or transistor to rectify power. Diodes are the simplest devices that allow power to flow at any time when the voltage is of appropriate polarity. The silicon-controlled rectifiers have a gate circuit that makes it possible to control a microprocessor when power flows, which makes it possible to

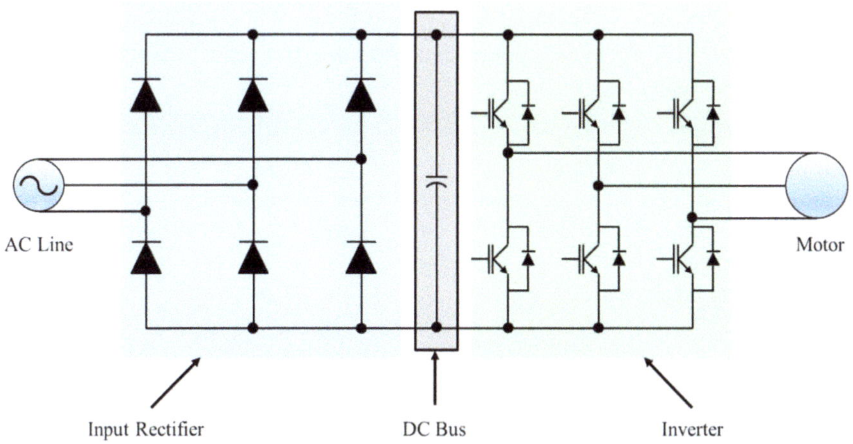

Fig. 15.4 VFD operating principle

use these rectifiers in electronic starters as well. Transistors, which include a gate circuit that allows a microprocessor to be switched on/off at any time, are the most useful of these three devices. The VFD has an active front end using a transistor in the rectifier section.

The power passing through the rectifiers is stored in the direct current (DC) distribution bar (bus bar). The bus bar has capacitors to hold, store, and then send the power from the rectifiers to the converter. The bus bar can also include similar elements that filter the input power supply to the bus bar by increasing the inductance, such as the inductor, direct current bridges, and shock coil. The last part of the VFD is a wave rectifier. The rectifier includes transistors that transmit power to the motor. Insulated-gate bipolar transistor (IGBT) is the most widely used option in modern VFDs. The IGBT can be switched on/off thousands of times per second and very precisely controls the power delivered to the motor. The IGBT uses a method called pulse width modulation (PWM) to simulate the current sine wave at the desired frequency to the motor.

The motor speed (revolutions per minute (abbreviated as; rpm, RPM, rev/min, r/min)) depends on the frequency. The motor speed depends on the change in VFD output frequency. The speed variation is determined as follows:

$$\text{Speed (rpm)} = \text{Frequency (Hz)} \times 120/\text{Number of Poles} \qquad (15.1)$$

Example: 2-pole motor at different frequencies:

$$3600\ \text{rpm} = 60\ \text{Hz} \times (120/2) = 3600\ \text{rpm}$$
$$3000\ \text{rpm} = 50\ \text{Hz} \times (120/2) = 3000\ \text{rpm}$$
$$2400\ \text{rpm} = 40\ \text{Hz} \times (120/2) = 2400\ \text{rpm}$$

Benefits of VFD: VFD technology enables speed control in fans, pumps, air conditioners, and chillers in HVAC applications. Variable frequency drive mechanisms provide the following benefits.

- Energy-saving,
- Low motor starting current,
- Reduction of thermal and mechanical stress on the motor and belts during the start,
- Easy montage,
- High power factor, and
- Low kVA.

The main features of VFDs are given below:

- VFDs provide the highest energy-efficient capacity control.
- VFDs have the lowest starting current among the starter types.
- VFDs prevent thermal and mechanical stresses in motors and belts.
- VFD setup is done by connecting the power supply to VFD and it is very simple.
- VFDs with high technology meet even the most stringent harmonic standards, while reducing the support generator capacity.
- They provide a high power factor and they do not need power factor correction capacitors to be connected externally.
- VFDs provide lower KVA and help reduce voltage fluctuations and power outages.

VFD Capacity Control Saves Energy: Most applications do not require a constant flow of fluid. The capacity of the device is determined by the peak load, which corresponds to only 1% of the operating time. Only a part of the flow is needed in the remaining time of the operation. Conventionally, devices that reduce output are used to reduce flow. However, these methods are significantly less efficient compared to the speed control.

Mechanical Capacity Control: Throttling valves, blinds, or dampers can be used for capacity control of a pump or fan with constant speed. These devices force the fan or pump to operate at a point that allows less flow over the curve by increasing the pressure (Fig. 15.5). The use of power is a result of pressure and flow. Decreasing the output increases the pressure, reduces the flow, and saves some energy.

Variable Speed Capacity Control: In centrifugal pumps, fans, and compressors, the relationships among the flow, pressure, and power consumption of the speed are determined according to the fan laws. When speed is used to reduce capacity, pressure and flow are also reduced. This ensures maximum energy savings. When the mechanical and speed control methods for capacity reduction are compared (Fig. 15.6), it can be seen that the variable speed is the most efficient capacity control.

Low Sudden Discharged Motor Start: In designs optimized for low starting current, efficiency, power factor, capacity, and cost are generally ignored. It is quite normal

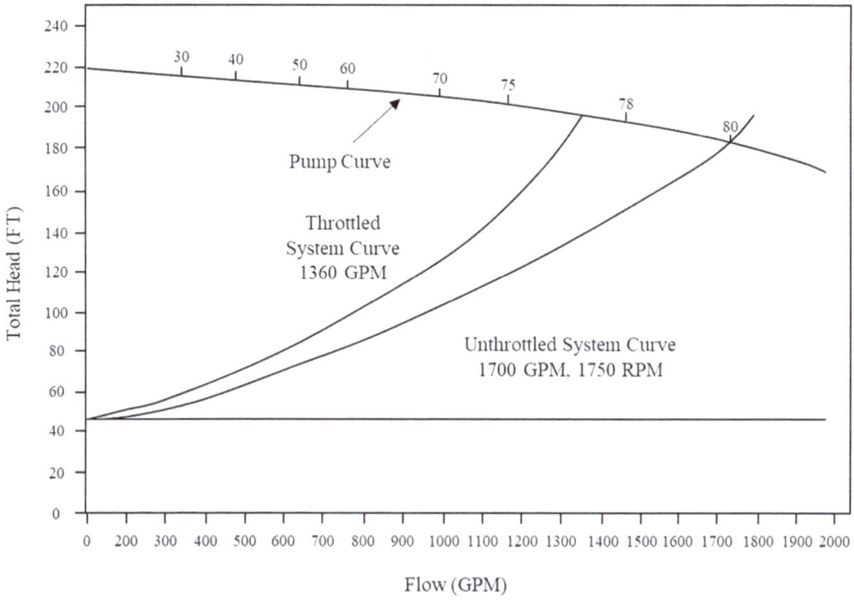

Fig. 15.5 Mechanical capacity control

Fig. 15.6 The comparison of capacity control

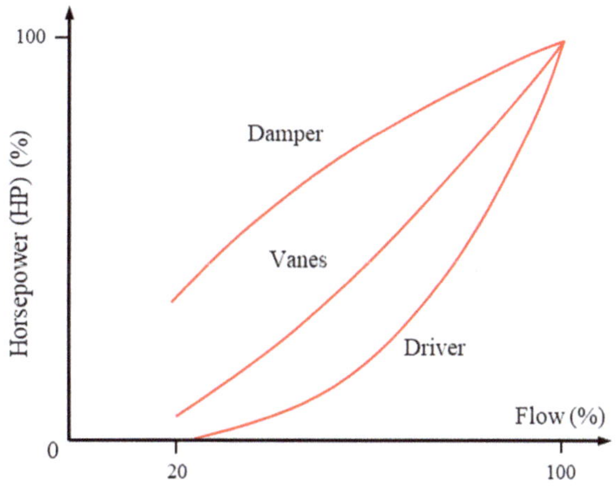

for AC induction motors to pull full-load amps 6–8 times during starting. If a large quantity of current passes through the transformer, the voltage drops. This adversely affects other devices operating on the same electrical system. Some systems with voltage sensitivity may even be de-energized. Therefore, methods for reducing the starting current of AC induction motors are developed.

Table 15.1 Comparison of starting requests

Starting method	Starting current (The percentage of full-load current)
VFD	100
Star-delta starting	200–275
Electronic soft starting	200
Starting with autotransformer	400–500
Partial winding starting	400–500
Direct starting	600–800

Soft Starters: Star-delta, partial winding, auto-transformer, and electronically controlled starters are widely used to reduce sudden discharge during motor start-up. All of these starters transmit power to the motor at a constant frequency. Therefore, the voltage applied to the motor must be controlled to keep the current within a certain limit. Star-delta, partial winding, and auto-transformer starters use special electrical connections to reduce the voltage. Electronic starters use SCRs to reduce voltage. The engine must reach a certain torque in order to accelerate and sufficient voltage is needed for this. Therefore, voltage reduction is only possible up to a certain limit. Even with the greatest possible voltage reduction, the motor pulls full-load-amperage (FLA: Full-Load-Amperage) two or four times during starting. In addition, the rapid acceleration associated with star-delta starting causes wear on the belt and other powertrain elements.

VFDs as Starters: Comparison of starting demands based on sudden discharge is given in Table 15.1. As shown in Table 15.1, VFD is the ideal soft starter as it provides the lowest instantaneous discharge compared to other types of starters. Unlike all other types of starters, the VFD can use the frequency to limit the power and the current transmitted to the motor. VFD sends the power at a low frequency to start the motor. At this low frequency, the motor does not require a high level of current. The VFD gradually increases the frequency and motor speed until it reaches the desired speed. During starting, the motor current level never exceeds the full-load-amperage. In addition to the benefits of a low starting current, the motor designs can now be optimized for high efficiency.

Easy Installation: Most of the devices are pre-programmed, made winding and equipped with VFDs at the factory. Motor connection cables, control power, and communication lines for spare parts are all manufactured in the factory. The cooling line VFDs of the coolers are also factory mounted and placed on the device. All the installer has to do is connect the power line to the VFD.

High Power Factor: Power converted to factors such as motion, heat, sound, and so on is called real power and is measured in “kilowatts (kW)”. The power that charges the capacitors and magnetic fields is called reactive power and is measured in “kilovolt ampere reactive (kVAr)” units. The vector sum of the real and reactive power (kW

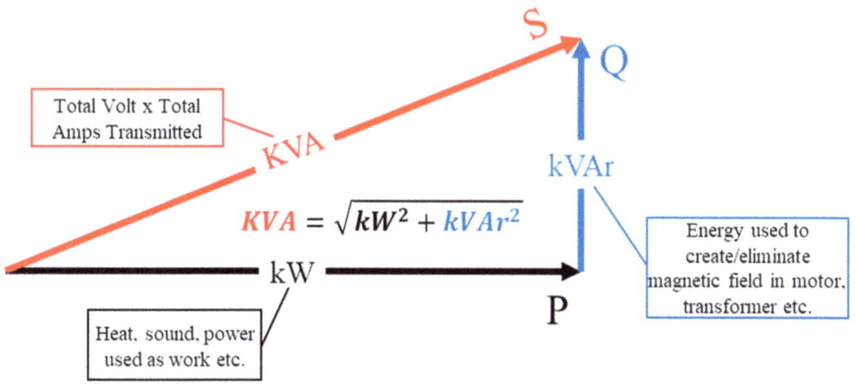

Fig. 15.7 The calculation of the electrical power

and kVAr) value is the total power (Energy) and is measured in "kilovolt amperes (kVA)" (Fig. 15.7). The ratio of actual power to total power (kW/kVA) is defined as the power factor.

The motors draw reactive current in order to strengthen the magnetic fields to ensure rotation. Excess reactive current is not preferred because it causes additional resistance losses and requires the use of larger transformers and windings. In addition, public institutions may also prevent the use of low- power-factor engines. Lowering the reactive current increases the power factor.

Typical direct current (DC) motors have a full-load power factor ranging from 0.84 to 0.88. The power factor decreases as the engine power is reduced. Power factor regulators can be added to reduce the measured reactive current at the capacitor input and increase the measured power factor. In order to avoid damaging the motor, the selection of the power factor regulating capacitors should not go beyond the recommendations of the motor manufacturer. In many applications, this provides an arrangement of 0.90–0.95.

The capacitors of the VFDs are located on the direct current (DC) line and serve the same to ensure a high-power factor in the grid part of the VFDs. This eliminates the need to add power regulating equipment or expensive condenser arrays to the motor. In addition, VFDs generally provide a higher grid power factor than fixed speed motors equipped with corrective capacitors.

Low Full Load (kVA): Total power (kVA) is often the limiting factor in the amount of energy transmitted through a power tool or system. If the kVA required by a device during the peak requirement period can be reduced, this helps to reduce voltage fluctuations, voltage drops, and power cuts. In calculating the kVA value, the unit efficiency and power factor have equal weight. Therefore, devices with equal or lower efficiency but higher power factor have a significantly lower kVA (Table 15.2).

In this example, a device with a higher power factor uses 15% less kVA when doing the same job. This reduces the cost of the electrical system in new projects, while increasing the kVA capacity in existing systems without additional costs. Redundant

Table 15.2 Power factor and energy use

Power (kW)	Power factor	Current (A)	Voltage (V)	Total power (kVA) kVA = A × V × 1.732
350.4	0.84	502	480	417
350.4	0.99	426	480	354

generators are selected to meet the load approximately. Reducing the kVA allows a smaller generator to be needed. When active front-end VFDs are used, the generator size approaches 1:1, which is the ideal kW/kVA ratio, since the power factor is close to the value of 1 and the harmony generated by the VFDs is very low. Low kVA values are also generally required by public institutions. The higher the power factor, the more power (kW) is transmitted from the same device.

15.2 Application in Air-Conditioning Rooms

Especially in fan and pump motors, energy saving is very common with VSD applications. VSD applications are recommended for fan and pump motors in air-conditioning systems in a spinning mill. As an example of this application, the control of the fan (aspirator and ventilator) motors which are still applied in the spinning air conditioner with VSD and the modernization of the water spray system with the humidification system were examined and the measurement data obtained from 1 to 9-barrel spinning air conditioner operating in the same circle and which were operated in a traditional way are compared in Table 15.3 and their potential energy savings are explained. When Table 15.3 is examined, the potential savings is clearly seen with the VSD application for the air conditioners in the spinning circle. With the VSD application, a reduction of 50% is seen in the power values drawn by the motor. If the reduction rate of the power ratio is taken as 30%, considering the changes in the year base of these measurements, the annual saving amounts for spinning air conditioners are given in Table 15.3 and for twisting and preparation air conditioners in Table 15.4.

Annual Energy Use (AEU; kWh/year), Recommended Energy Use with VSD (REU; kWh/year), and Annual Possible Savings (APS; $/year) values are calculated from the following equations.

$$\text{AEU} = \text{Engine Power} \times \text{LF} \times \text{UF} \times \text{OT} \tag{15.2}$$

$$\text{REU} = \text{Power drawn by the VSD} \times \text{UF} \times \text{OT} \tag{15.3}$$

Table 15.3 Energy saving with VSD application in spinning unit air-conditioning systems

Engine	Tag power (kW)	Power drawn (kW)	VSD	Load factor (LF)	Use factor (UF)	Operating time (OT) (h/year)	Annual energy use (AEU) (kWh/year)	Recommended energy use (REU) (kWh/year)	Annual possible savings (APS) ($/year)
Fan	30.00	16.60	Yes	0.50	1.00	7000	105,000		
Aspirator	30.00	7.60	Yes	0.30	1.00	7000	63,000		
Fan	22.00	5.10	Yes	0.30	1.00	7000	46,200		
Aspirator	30.00	12.00	Yes	0.40	1.00	7000	84,000		
Engine	1.47	1.20	No		1.00	7000	0.00		
Fan	22.00	19.00	No	0.86	1.00	7000	132,440	39,732	6953.10
Aspirator	15.00	14.20	No	0.95	1.00	7000	99,750	29,925	5236.88
Fan	22.00	22.00	No	1.00	1.00	7000	154,000	46,200	8085.00
Aspirator	18.50	18.00	No	0.98	1.00	7000	126,910	38,073	6662.78
Pump	5.50			0.80	0.50	7000			
Fan	22.00	20.70	No	0.94	0.50	7000	72,380	21,714	3799.95
Aspirator	22.00	19.50	No	0.88	0.50	7000	67,760	20,328	3557.40
Fan	30.00	27.00	No	0.90	0.50	7000	94,500	28,350	4961.25
Aspirator	30.00		No	0.00	0.50	7000	0.00	0.00	0.00
Pump	Canceled								
						Total savings ($/year)			**39,256.35**

Table 15.4 Energy saving with VSD in air-conditioning

Motor	Label power (kW)	Power drawn (kW)	VSD	Load coefficient (LC)	Using factor (UF)	Operating time (OT) (h/year)	Annual energy use (AEU) (kWh/year)	Recommended energy use (REU) (kWh/year)	Annual probable savings (APS) ($/year)
Fan	18.50	16	No	0.86	1.00	7000	111,370	33,411	5,846.93
Aspirator	30.00	24	No	0.80	1.00	7000	168,000	50,400	8,820.00
Fan	18.50	13	No	0.70	1.00	7000	90,650	27,195	4,759.13
Aspirator	30.00	22	No	0.73	1.00	7000	153,300	45,990	8,048.25
Pump	5.50		No			7000	0.00	0.00	0.00
Fan	30.00	27	No	0.90	1.00	7000	189,000	56,700	9,922.50
Aspirator	30.00	27	No	0.90	1.00	7000	189,000	56,700	9,922.50
Fan	30.00	29	No	0.90	1.00	7000	**189,000**	**56,700**	**9,922.50**[a]
Aspirator	30.00	29	No	0.98	1.00	7000	**205,800**	**61,740**	**10,804.50**[a]
Fan	30.00	29	No	0.90	1.00	7000	**189,000**	**56,700**	**9,922.50**[a]
Aspirator	30.00	30	No	1.00	1.00	7000	**210,000**	**63,000**	**11,025.00**[a]
Pump	5.50						0.00	0.00	0.00
						Total possible savings ($/year)			88,993.80
						Recommended total savings ($/year)			**41,674.5**

[a]Suggested savings potential

$$\text{APS} = (\text{AEU} - \text{REU}) \times (0.075\$/\text{kWh}) \tag{15.4}$$

where

LF = Load Factor,
UF = Use Factor, and
OT = Operating Time (h/year).

Since VSD applications can start ramping the motor in a suitable way, star-delta starting system is eliminated, motor starting currents are reduced and motor life is increased. There is also the potential for additional savings through proportional control applications. For VSD applications, measurement applications and monitoring studies should be performed. The savings potential determined in this example study is taken approximately. Actual energy savings can be determined more precisely by additional applications and trial runs.

15.2.1 Payback Period

In the calculations, the energy saving that can be done by VSD application is taken into consideration only in the ventilator and aspirator motors in the air-conditioning room. There is also an additional saving potential with the water pump motors currently in use. Since the running times of these motors are not too long, they are not taken into account in the calculations. For VSD applications, a payback period of approximately 8.23 months has been reached. Annual working hours were taken as 7000 h/year on average. The total possible savings and repayment period to be made with VSD application in ventilator and aspirator motors in air conditioner rooms that do not have VSD application are given in Table 15.5.

However, in some of the preparation and spinning rooms, due to the extreme weather problem, the application of variable speed with VSD was not considered appropriate because it would bring some problems such as product quality and fire risk. Therefore, the realistic saving potential is given in Table 15.6. Motor power and the number of repayment periods were calculated separately and on average, based on total savings and total investment amounts. The following method was used in the calculations.

Table 15.5 Energy saving and payback period with VSD application

Motor power (kW)	Number of motors (Piece)	Total savings (USD/year)	Total investment (USD)	Payback period (month)
30	9	83,348	52,000	7.48
22	4	22,394	20,000	10.71
18.5	4	22,505	16,000	8.53
Total	**17**	**128,247**	**88,000**	**8.23**

Table 15.6 Energy saving and payback period with VSD

Motor power (kW)	Number of motors (Piece)	Total saving (USD/year)	Total investment (USD)	Payback period (Month)
30	5	46,635	29	7.4
22	4	22,394	20	10
18.5	2	11,900	8	8.06
Total	**11**	**80,929**	**57**	**8.4**

$$\text{Total investment} = \text{Number of motors} \times \text{ VSD investment cost}$$
$$\text{Payback Period} = (\text{Total investment/Total savings}) \times 12 \text{ months}$$

15.2.2 Motor Driver Use in Twisting Room

In the detailed electrical measurements performed on twisting machines with 36 motors in the twisting room, when the currents, power, and power factor values drawn by the motors are examined carefully, it is determined that there are significant differences between the motor power and power factor values according to the type and quantity of the goods in machines. The power of these engines is 45 kW and the power they draw varies between 13 and 35 kW, while the power factor values range from 0.5 to 0.87. In this case, it can be said that the motor loading rates vary between 30 and 80%. In the three-phase measurements, it is seen that the motor power factor decreases significantly with the power decrease. Measurement results are given in Table 15.7.

Since speed adjustment is not required in twisting machines in the twisting room, it is not suitable to save energy with VSD application. However, it is recommended to use soft starters as well as energy-saving starters by limiting the energy saving and motor initial development current. In this context, for a 70% loaded engine, it saves about 20%, which means 45 kW × 0.7 = 31.5 kW → 31.5 kW × 20% = **6.3 kW** can be saved. Considering that the engine is used for 7000 h × 0.8 = 5600 h per year, the annual savings are 5600 h × 6.3 kW = 35,280 kWh/year. Monetary value of annual savings: 35,280 kWh/year × 0.075 USD = 2646 USD. Savings for 36 machines = 2646 USD × 36 = 95,256 USD/year.

Table 15.7 Motor power measurement results in twisting room

Machine name	Motor	Label power (kW)	Measured power (kW)	Load coefficient	Using factor	Load factor	Belt pulley
Twisting 1	BBC	45	34	0.82	0.8	0.75	1 × 5 belt
Twisting 2	BBC	45	32	0.80	0.8	0.71	1 × 5 belt
Twisting 3	BBC	45	34	0.82	0.8	0.75	1 × 5 belt
Twisting 4	BBC	45	21	0.78	0.8	0.46	1 × 5 belt
Twisting 5	BBC	45	Not working		0.8		1 × 5 belt
Twisting 6	BBC	45	32	0.82	0.8	0.71	1 × 5 belt
Twisting 7	BBC	45	16	0.75	0.8	0.35	1 × 5 belt
Twisting 8	BBC	45	30	0.80	0.8	0.75	1 × 5 belt
Twisting 9	BBC	45	34	0.82	0.8	0.75	1 × 5 belt
Twisting 10	BBC	45	22	0.74	0.8	0.48	1 × 5 belt
Twisting 11	BBC	45	31	0.80	0.8	0.68	1 × 5 belt
Twisting 12	BBC	45	32	0.80	0.8	0.71	1 × 5 belt
Twisting 13	BBC	45	15	0.60	0.8	0.33	1 × 5 belt
Twisting 14	BBC	45	Not working		0.8		1 × 5 belt
Twisting 15	BBC	45	30	0.78	0.8	0.66	1 × 5 belt
Twisting 16	BBC	45	35	0.82	0.8	0.77	1 × 5 belt
Twisting 17	BBC	45	13	0.50	0.8	0.28	1 × 5 belt
Twisting 18	BBC	45	Not working		0.8		1 × 5 belt
Twisting 19	BBC	45	32	0.75	0.8	0.71	1 × 5 belt
Twisting 20	BBC	45	Not working		0.8		1 × 5 belt
Twisting 21	BBC	45	Not working		0.8		1 × 5 belt
Twisting 22	BBC	45	19	0.74	0.8	0.42	1 × 5 belt
Twisting 23	BBC	45	Not working		0.8		1 × 5 belt

(continued)

Table 15.7 (continued)

Machine name	Motor	Label power (kW)	Measured power (kW)	Load coefficient	Using factor	Load factor	Belt pulley
Twisting 24	BBC	45	30	0.87	0.8	0.66	1 × 5 belt
Twisting 25	BBC	45	22	0.75	0.8	0.48	1 × 5 belt
Twisting 26	BBC	45	Not measured		0.8		1 × 5 belt
Twisting 27	BBC	45	Not measured		0.8		1 × 5 belt
Twisting 28	BBC	45	Not measured		0.8		1 × 5 belt
Twisting 29	BBC	45	Not measured		0.8		1 × 5 belt
Twisting 30	BBC	37	23	0.74	0.8	0.62	1 × 5 belt
Twisting 31	BBC	45	34	0.82	0.8	0.75	1 × 5 belt
Twisting 32	BBC	45	34	0.82	0.8	0.75	1 × 5 belt
Twisting 33	BBC	37	25	0.70	0.8	0.67	1 × 5 belt
Twisting 34	BBC	45	Not measured		0.8		1 × 5 belt
Twisting 35	BBC	45	Not measured		0.8		1 × 5 belt
Twisting 36	BBC	45	Not measured		0.8		1 × 5 belt

Payback period: When the engine drive system is taken as about 2500 USD for 45 kW engine, Investment cost: 36 engines × 2500 USD = 90,000 USD. Period of repayment: Investment Cost 90,000 USD/Annual Savings 95,256 USD × 12 months. Repayment period: 11.3 months

References

1. Kaya D, Öztürk HH (2014) Sanayide Enerji Yönetimi ve Enerji Verimliliği, Umuttepe Yayınları, Yayın No: 114, Mühendislik Dizisi: 12, Kocaeli/Türkiye
2. AMCA International (2007) Fans and systems. AMCA Publication 201-02. Air Movement and Control Association International. Arlington Heights, IL
3. https://www.se.com/tr/tr/
4. AMCA International (2007) Field performance measurement of fan systems. AMCA Publication 203-90. Air Movement and Control Association International, Arlington Heights, IL
5. AMCA International (2010) Standards handbook. ANSI/AMCA Standard 99-10. Air Movement and Control Association International, Arlington Heights, IL
6. https://www.acdrive.org/ac-drives-basics.html

Chapter 16
Energy Saving with Heat Insulation

16.1 The Aim of Heat (Thermal) Insulation

Heat is an energy and in places with different temperatures; it tends to move from the high-temperature side to the low-temperature side, naturally. The heat flow occurs along the path of least resistance. Heat transfer occurs in three ways: by means of conduction, convection, and radiation. During this transition, the heat meets a resistance depending on the thermal conductivity coefficients (symbolized as λ or k) and thicknesses of the materials between the spaces. Generally, thermal insulation is a resistance that reduces heat transfer.

The most important purpose of thermal insulation is to reduce heat flows from a hot source to the environment or from the environment to cold processes or mediums. The basic principle for this is low thermal conductivity or vice versa. The reason for the application of thermal insulation is that some materials need to be kept warm and some cold, depending on the environment. In buildings without thermal insulation; If the inside of the house is warm in winter, the heat flow is from the inside to the outside and in summer, if the inside of the house is cold, the heat flow is from the outside to the inside. Thermal insulation is required in the buildings; especially in windows, roofs, attics, basements (specifically between the floors), and entrances of the buildings (see Fig. 16.1). Heating can be provided successfully by insulating the buildings correctly, and that assures using 25–50% less fuel [1].

Thermal insulation applications can be of very different types and features depending on the operating conditions at the application site and their costs vary accordingly. The following factors should be taken into consideration in determining the way of applying thermal insulation:

- Operating temperature,
- Environmental (Ambient) conditions,
- Risk of damage to insulation,

D. Kaya et al., *Energy Management and Energy Efficiency in Industry*,
Green Energy and Technology,
https://doi.org/10.1007/978-3-030-25995-2_16

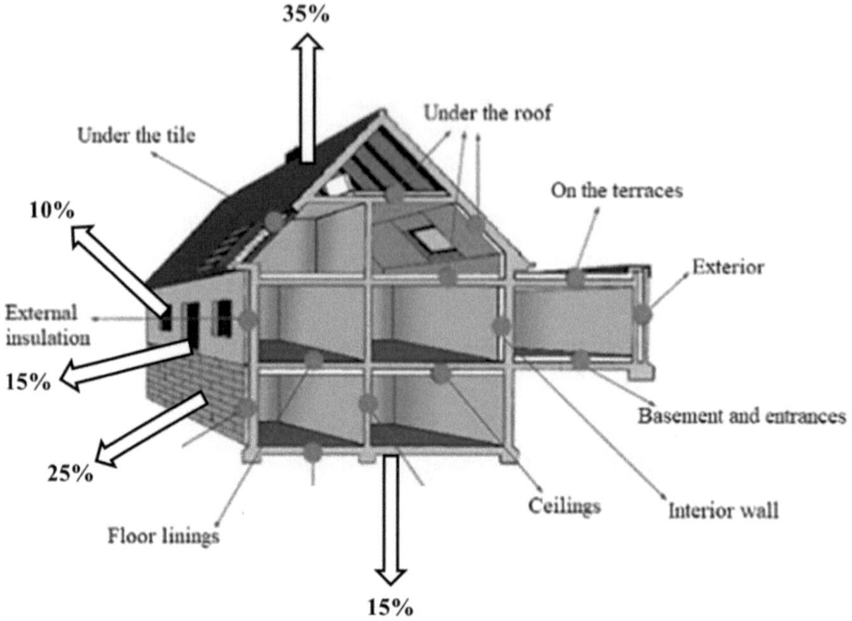

Fig. 16.1 Heat loss rates in buildings and heat insulation applications

- The type of operation (continuous or intermittent),
- Temperature difference, and
- Air movements.

16.2 Benefits of Thermal Insulation

Thermal insulation is one of the energy-saving methods that can be applied to hot and cold pipelines, facilities, and buildings that have heat loss or heat gain, not requiring a lot of investment costs, but can save a considerable amount of energy and reimburse itself in short periods by providing the great savings [2].

The insulation provided by the insulation materials used in the home and workplaces contributes to the reduction of energy consumption, the conservation of natural resources and their balance in nature, and the economy of a country. It also allows less carbon dioxide and other harmful gases to be dissipated into the atmosphere as less fuel is burned due to heat insulation. In this way, the adverse effects of global warming, which is the result of the greenhouse effect, will be reduced. Thanks to the insulation, winter fuel materials (coal, oil, natural gas) are used less. Other benefits of thermal insulation, which is mainly energy saving are

- Heat losses are reduced by 30–60% depending on the insulation thickness and the thermal conductivity of the material used.

- Molding events are prevented by reducing the risk of condensation.
- Thermal stresses are reduced on the outer walls and thus, cracks caused by the effect of temperature are prevented.
- Depending on the fuel savings, the initial investment and operating costs of the heating system are reduced.
- As less fuel is used, less air pollution occurs.
- Thermal comfort is achieved by creating an internal temperature at ideal values in the room.
- Thermal insulation has guaranteed the reduction of the cost of heating and cooling by over 40% and the insulation pays itself (depending on the conditions and materials) around 5–6 years.

16.3 Heat (Thermal) Insulating Materials

The use of insulating materials to slow the flow of heat in the materials is called insulation. Materials that cannot transmit heat well are called heat insulators. The gap between the particles forming the heat insulators is too large and the particles are irregular. Heat insulators transmit heat in a very short amount of time. Wood, plastic, concrete, and air are heat insulators. Plastic foam, glass wool, asbestos, air gap in double glazing, airless environment between the inner and outer surface of the thermos is used for thermal insulation.

A common feature of materials used for thermal insulation is the presence of air in the structure of these materials. The air is a good insulator. Because the amount of air between the particles forming the air is very large. Therefore, heat conduction between the particles forming the air is very slow. Since there is air in the spaces in the structure of the plastic foam, it is used as a thermal insulator in plastic foam and heat insulation is provided [3].

Materials used to provide thermal insulation are called thermal insulation materials. Thanks to the insulation materials, the nutrients can be kept at the desired temperatures, the drinking water temperature is ensured, and the houses are provided with thermal insulation. The purpose of each insulation material is different. Different insulation materials can be used to insulate different areas or different places within a region. Insulation materials can be used for good thermal insulation, should slow down the flow of heat, and should not lose their characteristics when exposed to very high or very low temperatures. General characteristics of thermal insulation materials are

- The thermal conductivity coefficients should be low.
- It should be lightweight.
- It should be odorless.
- It should have the feature of water and moisture absorption.
- It should not be suitable for the bacterial increase and insect nesting.
- It must be resistant to rotting.
- It should not lose its first feature.

- It should not be flammable.
- It must be long-lasting.
- It should be suitable for carriage and should not harm human health.
- It must be economical and affordable.

Thermal insulation materials are used to increase the heat transfer resistance of adjacent building materials or to reduce the amount of heat escaping to the outside. Some properties are sought in the selection of heat insulation materials. Features to consider when choosing insulating materials are

- Thermal conductivity coefficient,
- Resistance to different operating temperatures,
- Physical strength,
- Pressure resistance (Compressive strength),
- Mechanical strength,
- Water vapor resistance,
- Taking into consideration whether it is having hazardous emission status,
- Fire resistance,
- Resistance to corrosive effects,
- Density,
- Resistance to chemical effects, and
- Having the feature of water absorption.

Heat-insulating material with a thermal conductivity coefficient of less than 0.060 kcal/mh°C according to German DIN 4108 norms, is used to increase the heat transfer resistance of adjacent building materials or reduce the amount of heat escaping to the outside materials. Classification according to ISO and CEN standards:

- Thermal conductivity coefficient $\lambda > 0.065$ W/mK for building material,
- Thermal conductivity coefficient: $\lambda < 0.065$ W/mK heat-insulating material.

Heat-insulating materials can be divided into 3 main groups:

(1) *Organic materials*: Cork, flax stalks.
(2) *Inorganic materials*: Perlite, glass wool, rock wool, glass foam, aluminum silicate, calcium silicate, asbestos.
(3) *Synthetic materials*: Polystyrene, polyurethane, polyisocyanate.

The thermal insulation materials and standards used in buildings and installations are given in Table 16.1.

16.3.1 Glass Wool

Silica sand, an inorganic raw material, is melted at between 1200 and 1250 °C and made into fiber, which is a locally produced heat and sound insulation material. It is produced in the form of mattress, sheet, pipe, and casting with different coating

Table 16.1 Thermal insulation materials and standards

Insulation material used on buildings	Thermal insulation materials and standards used in plumbing
• Ceramic wool, EN 13161-EQV • Glass wool, EN 13162+A1-EQV • Stone wool, EN 13162+A1-EQV • Expanded Polystyrene (EPS), EN 13163:2012+A2:2016 • Extruded Polystyrene (XPS), EN 13164:2012+A1-EQV • Polyurethane (PUR), EN 13165:2012+A2:2016 • Phenol foam, EN 13166:2012+A2:2016 • Glass foam, EN 13167:2012+A1-EQV • Sheathing, EN 13167:2012+A1-EQV • Double glass, EN 13168:2012+A1-EQV • Wooden fiber plates, EN 13169+A1-EQV • Expanded perlite (EPB), 13170+A1-EQV • Expanded cork plates (ICB) EN 13171:2012+A1-EQV	• Glass wool, EN 14303:2015 • Stone wool, EN 14303:2015 • Elastomeric rubber (FEF) EN 14304:2015 • Glass foam (CG) EN 14305:2015 • Calcium silicate (CS) EN 14306:2015 • Extruded Polystyrene (XPS) EN 14307:2015 • Polyurethane (PUR/PIR) EN 14308:2015 • Expanded Polystyrene (EPS), EN 14309:2015 • Polyethylene foam (PEF), EN 14314:2015 • Phenol foam EN 14315-1:2013

and additive materials in different sizes and intensities according to usage place and purpose. The thermal conductivity coefficient is $\lambda = 0.040$ W/mK and the water vapor resistance factor is $\mu = 1$. The usage temperature is between −50 and + 250 °C. Unbonded (without binder) glass wool products can be used up to 500 °C. In addition, special glass wool products are produced to be used at the different usage temperature range which can be −200 and +450 °C. It does not deteriorate with time, does not rot, does not mold, does not corrode and rust, it is not destroyed by insects and microorganisms. It is an A-Class fireproof material.

16.3.2 Rock Wool

It is a heat and sound insulation material produced by melting a basalt stone, which is an inorganic raw material, at 1350–1400 °C and fiberizing it. It is produced as mattress, sheet, pipe, and casting with different coating materials in different sizes and densities according to usage place and purpose. It is used as heat isolation, sound isolation, acoustic regulation, and fire isolation purpose. The thermal conductivity coefficient is $\lambda = 0.040$ W/mK. The water vapor resistance factor is $\mu = 1$. Usage temperature is between −50/+700 and 750 °C. Even if exposed to heat and moisture, there is no change in size. It does not deteriorate with time, does not rot, does not hold mold, does not corrode and rust. It is not destroyed by insects and microorganisms. They are not hygroscopic and capillary. They are A-class non-combustible materials.

16.3.3 Expanded Polystyrene Sheet

The expanded polystyrene sheet (EPS), which is a heat insulation material (hard foam) obtained from petroleum. It is plastic material produced from solid beads of polystyrene, which is typically white colored, thermoplastic, in the form of closed pore. The granules of the pentane gas in the raw material granules of the polystyrene raw materials, which are associated with the water vapor, are produced as a block by swelling and gluing them together then heat insulation plates produced by cutting the desired thickness with resistance metal.

EPS, which is obtained by inflating and fusing polystyrene particles, is a blowing gas pentane used to inflate particles and foam in products. The pentane is displaced by air during production and very shortly after production, after allowing numerous small pores to form within the particles. Thus, stagnant air is trapped in very small closed-pore cells (3–6 billion in 1 m^3 EPS) found in the EPS plates. 98% of the material is still and dry air.

Stationary and dry air is the most economical, environmentally friendly, and excellent thermal insulation material known. The economic and superior thermal insulation properties of EPS thermal insulation sheets are provided in this way. Energy-intensive production is another important reason to be economical despite its superior technical characteristics. In addition to the effective mechanical strength, the replacement of the blowing gas with air in a very short period of time ensures that the performance of the product remains constant throughout its useful life. Thickness does not decrease, thermal conductivity does not increase, and its properties do not deteriorate.

The EPS is produced at the desired density. Since the features can be changed intensively in the desired direction, it does not cause material waste and unnecessary cost increases. For isolation purposes, EPS plates, which are generally used at 15–30 kg/m^3 density, can be produced as composite elements by covering with many other materials.

EPS products are used extensively in heat and sound insulation of buildings and in packaging industry, as sheets, pipes, or preformed elements. EPS products can also be used as wall materials for buildings, for insulation of cold-air warehouses, life preserver, and life jackets for ships, there is unlimited use in all applications where lightness, strength, easy formability, easy application, and low thermal conductivity are important.

It is produced as plate and mold in different sizes and densities, different edges and surface shapes according to use place and purpose. It is used with heat insulation board and packaging purposes. The thermal conductivity coefficient is λ

= 0.040 W/mK. The water vapor resistance factor is μ = 20–250. The usage temperature is −50/+75 °C. The capillaries have no absorption.

Flue gases like methane gas, gasoline, ether, ester, and amine groups are sensitive to chemicals. They are also sensitive to the ultraviolet rays of the sun. Hard flammable materials are classified as B1, and normal flammable materials are classified as B2.

16.3.3.1 Superiority of the Expanded Polystyrene Plates

The main reasons for preference of EPS are it has superior technical features, it can be changed according to the density depending on the density, the cost is low due to the ideal production technology, the performance can be maintained without deterioration throughout its life, and it is an environmental-friendly material. EPS products provide the required performance with no waste of materials, and therefore the most economical solution: EPS insulation plates; provides high thermal insulation (λ = 0.033–0.040 W/mK). As the density increases, the thermal conductivity decreases.

As the thermal conductivity of EPS is low, it is fixed; it does not change depending on the inflating container and time. The press is durable. As the density increases, the pressure resistance increases. It's not fragile. Thermal insulation material has a high bending strength. Because it is closed pores, it does not get wet practically, it makes the insulation continuous. The capillary has no water permeability and is not hygroscopic. Vapor impermeability can be set at desired values. As the density increases, the vapor impermeability also increases. Thickness does not diminish over time, remains constant. Very light, easy to carry, easy to apply. Economic insulation material.

Endless life: The building continues with the first day's performance to the task of insuring the insulation. EPS can be produced in a wide density range, offering application options. It provides the opportunity to choose the most suitable product for thc job and prevents wastage of resources.

16.3.3.2 Usage Areas of the Expanded Polystyrene Plates

The main uses of the expanded polystyrene sheets for insulation purposes are

- Thermal insulation of walls in buildings.
- Sloped and roofed terraces and thermal insulation of terrace gardens.
- In the heat insulation of the upholstery.
- Heat insulation of ceilings.
- Thermal insulation of outings.
- Floating floor insulation in impact sound.
- In sound insulation.
- For the creation of many covered elements.
- Heat insulation of cold-air stores.
- Heat insulation of poultry houses.

- Isolation of pipes.
- Heat insulation of Venetian blinds.
- Isolation of pipes, tanks, and storehouses.

Isolation of pipes: EPS products can be exposed to 80–85 °C for long periods. Temperatures up to 100 °C for a short time also do not damage EPS products. EPS thermal insulation products can be successfully used for conditions of use and for pipes remaining at these intervals.

In hot water and heating systems, in order to reduce heat loss from the piping, to prevent sweating on the surface in cold-water systems and to protect against frost, to reduce the loss of cooling from the cooling pipes in the cooling elements and to prevent condensation on the surface, to ensure the homogeneity of the temperature distribution in the ventilation channels, EPS thermal insulation products are used internationally to reduce heat transfer to 180 °C. For insulation in pipes, products with a density of 20–30 kg/m^3 are preferred. Insulation thicknesses are used for the purpose of use, pipe material used, pipe diameter, etc., many factors are connected.

16.3.4 Extruded Polystyrene Sheet

It is a heat insulation material brought from the polystyrene raw materials through the extrusion line and produced domestically. It is produced as sheet in different sizes and densities with different edge and surface shapes according to use place and purpose.

Thermal conductivity coefficient is $\lambda = 0.028$–0.031 W/mK. The water vapor resistance factor is $\mu = 80$–250. The usage temperature is $-50/+75$ °C. It has a homogeneous cell structure with 100% closed porosity and does not take water in itself. The capillaries have no absorption. Dimensional stability and compression strength are very high. It is Class B1, which is a very exhilarating material. The main reason to prefer extruded polystyrene foam (XPS) is can be listed down below:

- It has low thermal conductivity and preserves this feature throughout its life.
- It has a closed porous and homogeneous cell structure.
- It does not absorb water.
- No additional vapor barrier is required.
- It is a breathable material.
- Size invariance is very good, it does not change shape over time.
- Compressive strength is high.
- It does not rot and crumble.
- Thanks to the overlapped edge profile, it prevents the formation of heat bridges.
- It is easy to apply, provides the opportunity to work without waste.
- Since it is light, it can be easily carried and applied.

16.3.4.1 Specifications of Extruded Polystyrene Plate

Thermal conductivity coefficient: It is the amount of heat passing from 1 m^2 of a homogeneous material with a temperature difference between two surfaces parallel to each other, 1 K = 1 °C, in 1 h and 1 m thickness. Its symbol (λ) is its unit (W/mK). It is the most determining feature in the selection of thermal insulation materials. The thermal conductivity coefficient of extruded polystyrene foam thermal insulation boards is $\lambda = 0.028$–0.031 W/mK.

Water Vapor Diffusion Resistance Factor: The water vapor pressure tends to move from the environment where the pressure is high to the environment where it is low. Each material, as in heat transfer, resists water vapor transmission, depending on its thickness. The ratio of this resistance of the material to the vapor diffusion resistance of the air is called the water vapor diffusion resistance factor and its symbol is (μ). The water vapor diffusion resistance factor value of the extruded polystyrene foam thermal insulation boards is at the optimum value that prevents condensation but allows breathing ($\mu = 100$).

Density: The density of extruded polystyrene foam heat insulation plates is between 28 and 32 kg/m^3. Density is very important because many properties of thermal insulation materials change depending on density. These are

- Thermal conductivity coefficient,
- Compressive strength,
- Water absorption and diffusion resistance by volume,
- The percentage of closed cells, and
- Freeze–thaw resistance.

Water Absorption by Volume: Extruded polystyrene foam heat insulation boards have a very low water absorption value of 0.1% thanks to their closed porous structure.

16.3.5 Glass Foam

Glass foam boards are very hard, very resistant to pressure, easily broken, resistant to abrasion, surfaces can be easily dusted by friction. Glass foam is an insulation material that does not pass the steam ($\mu = \infty$).

Glass foam, which is closed porous, does not absorb water, only water can be filled in the recesses on its surface. It is not hygroscopic and capillary. However, if it is constantly exposed to water, it will corrode the material in a small amount. It does not rot, mold, and does not contain vermin. The pore structure of the material is around 93–94%. Sheets can be produced in the form of large panels as well as small size. It can be used by coating various materials (aluminum foil, glass, plaster-cardboard board, etc.) to the boards.

16.3.6 Calcium Silicate

Calcium silicate is a mineral-based insulating material that is used as a sheet, pipe, spray, or special-shaped parts. It is also available in powder form which hardens with the addition of water. It is generally used in high-temperature insulations since it has types up to 1100 °C. Their density is between 190 and 200 kg/m^3. Its compressive strength is very high. Calcium silicate materials are also a suitable material for fire insulation. The heat transmission coefficient is very low for such materials.

16.3.7 Melamine Foam

Melamine foam is a material with high sound absorption and excellent thermal insulation properties. They are light and easy to apply, and it is a material that is widely used in constructions today with its decoration. There is no fiber and fiber erosion. They are decorative, clean, and flexible. There are various types and sizes on the market.

16.3.8 PVC Foam

PVC foam is a thermoplastic material based on polyvinyl chloride. It can be produced as hard, semi-hard, or soft. The pore structure depends on the production method. In production with high-pressure system, it is produced with closed pores, mixed pores in low-pressure system, or low pores in unpressurized production.

Its density can be adjusted between 30 and 300 kg/m^3. Generally, 30–40 kg/m^3 is used in the construction sector. Hard boards are fragile, soft ones are elastic. The durability of PVC foam can be significantly increased with thin coatings.

Water Sensitivity: μ value is between 40 and 80, closed pores do not absorb water. Mixed or open pores absorb water. It is resistant to corrosion and decay. Does not contain vermin (pests), is resistant to certain chemical substances. It begins to soften at temperatures between 50 and 60 °C. It is hardly flammable and hard plates can be easily cut, drilled, and scrapped.

16.3.9 Polyethylene Foam

Polyethylene based materials are flexible and semi-flexible, porous, plastic-based materials made from polymers prepared from ethylene and propylene. The product made of polyethylene foam is produced in the form of pipe and plate (board) by extrusion method from the mold. The outer surface can be obtained smoothly, the pipes are produced with an inner diameter of 10–139 mm, a length of 2 m, and an insulation thickness of 5–30 mm.

The plates (boards) are produced in different sizes with thicknesses of 10, 15, and 20 mm. Closed-cell, extrusion-produced polyethylene foam product is a durable, reliable, economic, easy-to-use insulation material. It can be installed in a short time in a very easy operation. It does not contain toxic gas, it is chemically neutral and odorless. Polyethylene now has a wide range of applications both in the industry and in the subsidiary industry.

Polyethylene based materials have features like low density, elasticity, low thermal conductivity, high water vapor resistance, no water intake in the body, high impact resistance, etc. It is a sought-after product especially in the field of insulation with its superior properties and mechanical properties. It is used in many insulation areas as an auxiliary material for thermal insulation, impact sound insulation in floors, and waterproofing.

Polyethylene pipe: Polyethylene pipes are used for heat insulation of pipes and installation of pipes. They are produced in the range of 1/4"-4" nominal diameters, 6-10-5-20-30 mm wall thickness.

Filling cord: It is an auxiliary material produced in the form of a cylinder and used in water and heat insulation.

Polyethylene tape: Polyethylene tape is in roll form and is self-adhesive. It is manufactured as thin and narrow. It is a thermal insulation auxiliary material.

Polyethylene sheet: It is heat and impact sound insulation material produced in rolls. These materials are divided into two groups as polyethylene impact sounder and polyethylene sheets. Polyethylene sheets are also divided into three groups as standard sheets, self-adhesive polyethylene sheets, and polyethylene foil coated sheets.

16.3.10 Elastomeric Rubber Foam

It is available in the form of fully flexible, closed cell, expanded black synthetic pipes, and boards. It provides elasticity and flexibility to be used in different application areas, thanks to the high proportion of synthetic rubber it contains. It significantly reduces heat loss in hot pipes and heat gain in cold pipes.

Its density is between 60 and 80 kg/m^3 and it has excellent flexibility. Since it is closed pores, there is practically no water in it. Elastomeric rubber foam is generally resistant to chemicals (oil, mineral oil).

Elastomeric rubber foam is a material with high vapor impermeability among insulation materials. In thermal insulation materials, it is a material that is qualified in terms of water vapor impermeability and it is especially recommended in places with condensation problems.

16.3.11 Polyurethane Foam

Polyurethane is a plastic-based foam obtained by the combination of two chemical substances, polyol and isocyanate, foamed and hardened with the aid of air. They are usually in the form of plates, but they can also be found in the form of prefabricated pipes as well as different shaped forms. It is also applied by the on-site spraying method.

The polyurethane is yellow in color. The cells are 95% closed porous. They are produced for various purposes between the densities of 30 and 200 kg/m^3. For building insulation, a density of 30–40 kg/m^3 is used.

The thermal conductivity of the polyurethane is very low. The thermal conductivity value according to ambient temperature and density varies between 0.016 and 0.032. The operating temperature range is from -180 to 110 °C. The density of polyurethane foam can be adjusted between 30 and 200 kg/m^3. It is recommended that the plates used for insulation should not be less than 32 kg/m^3. Otherwise, it may show shape changes. The densities used in the buildings are generally 30–40 kg/m^3 and sometimes 50 kg/m^3.

Albeit with little water intake in the material itself, it is still more than EPS. While 24 h water is absorbed to 0.2–1.0% of the sample volume, in several weeks samples this ratio is about 3–5%. Because the wetting condition by steam diffusion μ (the water vapor resistance factor) is between 40 and 50, the plates must be sealed or precautions must be taken against possible stresses. For on-site sprayed foams μ is between 3 and 8. Polyurethane foam is resistant to light acids, gasoline, diesel, alkalis, and sea water.

16.3.12 Ceramic Wool

Ceramic wool is a fibrous material used at very high temperatures. It is used for temperatures of 1200–1400 °C, where the stone cannot be used. It is found in rolls, sheets, cast shapes. It is white. The density varies from 100 to 150 kg/m^3 depending on the shape of the material. It is a soft material, even the strength of the plate type is not high. The most important feature is that it can withstand high temperatures. Ceramic wool does not burn. It is not affected by acids other than hydrofluoric acid and phosphoric acid. Ceramic wool is used in other fiber materials, such as rolls, plates, ropes, etc. Ceramic wool cannot be produced as prefabricated pipes.

16.3.13 Vermiculite

Vermiculite is a natural aluminum-magnesium silicate, derived from mica mineral. The mica residues are heated and expanded. After its expansion, its density decreases from 1400–1500 kg/m^3 to 17–60 kg/m^3. Grain diameters range from 0 to 15 mm.

Vermiculite is turned into a sheet with the addition of glass water and silicofluoride by pressure. The plates are resistant to a maximum pressure of 5–6 kPa/cm^2. Bulk vermiculite lasts up to 1200–1400 °C. Plates made by adding cement are in the category of heat-insulated lightweight concrete and last up to 800 °C. The thermal conductivity value varies with density. The λ value of a lightweight concrete at a density of 300 kg/m^3 is 0.08 W/mK while the λ value of a density of 600 kg/m^3 reaches 0.15 W/mK. Vermiculite gets moisture. It is resistant to acids and alkalis and does not burn.

16.3.14 Elastomeric Rubber

The elastomeric rubber used in heating and cooling systems is a thermal insulation material produced in the form of pipes and sheets with a closed cell and a smooth cell structure. It is very flexible and long-lasting. Heat insulation value is high, it is resistant to water and moisture.

Elastomeric rubber pipe and sheet insulation material are used in heating and cooling systems for hot- and cold-water circuits. The operating temperature is usually between −60 and 85 °C. The thermal conductivity value is around 0.036 W/mK.

16.3.15 Plastic Pipe and Sheet Insulation Materials

16.3.15.1 Polyethylene Foam Insulated Pipes

Polyethylene is very thin closed-cell foam. Its density is between 30 and 40 kg/m^3 and it can be applied between −80 and 95 °C. The thermal insulation value is $\lambda =$ 0.040 W/mK and it maintains this value throughout the life of installation. The closed-cell structure carries protective properties along the outer surface of the material. The water vapor resistance factor is $\mu > 3500$. It does not accept water within its structure. Therefore, it protects the surface it covers from rotting, mildew, and corrosion.

16.3.15.2 Polyethylene Foam Insulation

The panels are used in water and liquid cooling systems, large and extra diameter pipes, tanks, valves, and flanges in indoor and outdoor insulation of air-conditioning ducts of split and central ventilation systems from −80 to 95 °C. The thermal conductivity value is $\lambda = 0.040$ W/mK. Its density is 30–40 kg/m^3.

16.3.15.3 Laminated Form-Flex Plates

It is formed by aluminum laminating to a thickness of 0.1 mm on polyethylene insulation boards. It is used for the isolation of air-conditioning plants and ventilation ducts. It is resistant to UV rays and high temperature. It is resistant to any kind of mechanical impact.

16.3.16 Fiber Insulation Materials

Fibrous insulation material is a substance made of organic and inorganic fibers, used to reduce heat transmission. These materials are made of organic or inorganic fibers bonded with bare or mineral oils or chemical binders or sewn on cardboard or paper or a carrier material. Generally, they are heat-insulating materials put on the market in the form of plates, felt, or mattresses in bulk.

16.4 Energy Saving by Insulating Hot Surfaces

Thermal insulation is applied to hot pipelines, facilities, and buildings that may cause heat losses. Although it does not require much investment costs, it provides a considerable amount of energy efficiency. It is one of the successful energy-saving methods that can repay itself in a short time [4]. In order to determine the amount of energy savings to be provided with this insulation, it is necessary to calculate the heat losses from the hot surfaces for the situation before and after the insulation. The annual energy savings can then be calculated from the formula which is given below.

$$\text{Energy saving} = [(Q_{un} - Q_{ins}) \times AOT \times 3600]/\eta \tag{16.1}$$

Here Q_{un} is the heat loss from uninsulated surfaces in the current state (W), Q_{ins} is the heat loss from insulated surfaces (It is the heat flowrate through the pipe and insulation) (W), AOT is the annual operating time (h/year), and η is the boiler efficiency. Heat loss from uninsulated and insulated steam pipes can be calculated from the following formulas:

$$Q_{un} = (h_c + h_r) \times \pi \times d_1 \times L \times (T_s - T_a) \tag{16.2}$$

$$Q_{ins} = \pi \times L \times (T_f - T_a)/[[\ln(d_2/d_1)/(2 \times k)] + [1/(h_{ai} \times d_2)]] \tag{16.3}$$

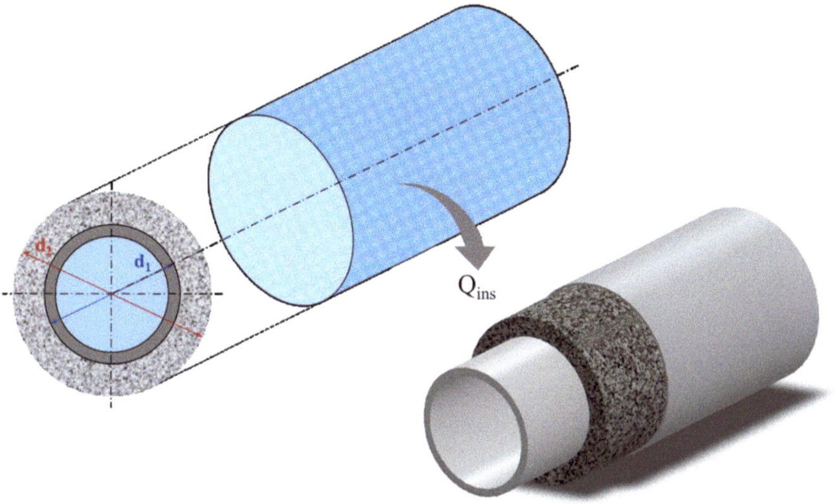

Fig. 16.2 Schematical demonstration for the heat loss from insulated surfaces

where T_s: The pipe surface temperature (at the surface of the insulation) (°C or K), T_f : The fluid temperature inside of the pipe (°C or K), T_a: The ambient temperature (°C or K), T_{surr}: The average temperature of the surrounding surfaces (K), d_1: The outer diameter of the pipe (m), d_2: The outer diameter of the pipe, after being insulated (m), h_c: The convective heat transfer coefficient (natural convection from a horizontal cylinder (approximately)) (W/m^2 K), ε: The emissivity, h_r: The radiation heat transfer coefficient (W/m^2 K), h_{ai}: The heat transfer coefficient of the surface after the insulation (the insulation-to-air heat surface heat transfer coefficient) (W/m^2 K), L: The pipe length (m), k: The thermal conductivity of the insulating material (W/m K), and $l_{insulation}$: The insulation thickness (m) (Fig. 16.2).

$$d_2 = d_1 + 2 \times l_{insulation}$$

$$h_r = 5.67 \times 10^{-8} \times \varepsilon \times (T_s^2 + T_{surr}^2) \times (T_s + T_{surr})$$

$$h_c = 1.32 \times [(T_s - T_a)/d_1]^{0.25}$$

$$h_c = [\text{Natural convection from a horizontal cylinder (approximately)}]$$

Thermal camera measurements made on the plant have been determined that some faults, vans, condensate tanks, and collectors are not insulated. The variables used in the insulation and saving calculations are given in Table 16.2 and uninsulated equipment and dimensional properties are given in Table 16.3.

Table 16.2 Values used for calculations

Parameters	
Ambient temperature T_a, °C	18
Emissivity of uninsulated surface, ε	0.90
Heat transfer coefficient after insulation, W/m^2 K	10
Thermal conductivity of insulating material, W/mK	0.04
Efficiency of boiler	0.92
Lower heating value of natural gas, kcal/Sm^3	8250
Unit price of natural gas, \$/Sm^3	0.25
Annual working hour, h	7000

Table 16.3 Uninsulated equipment and dimensional properties

Room type	Part/equipment	Quantity	Diameter (mm)	Size (m)	Surface temperature (°C)
Boiler room	No. 4 boiler outlet distribution collector degasser side/pipe	1	250	5	100
Boiler room	Degasser connection/pipe	1	200	5	100
Boiler room	Pump connection	1	50	20	100
Dyeing room	Special line machines	9	500	0.5	115
Zipper room	Zipper roller drying (*Related Brand Name)/dryer	1	1200	2	125
Zipper room	Zipper roller drying (*)/steam drying machine	2	500	3	125
Zipper room	Zipper roller drying (*)/Connecting pipes	1	50	8	125
Dye house	Hot water tank behind heat exchanger inlet/Steam line pipe	1	50	4	115
Dyeing room	(*) 3 machine behind/steam pipeline	1	100	2.5	150
Dyeing room	Steam line/pipe	1	100	2	125
Dyeing room	(*) machine/steam collector	1	80	12	114
Boiler room	Boiler room/valve	5	50		100
Boiler room	Condensate/valves	3	85		100

(continued)

Table 16.3 (continued)

Room type	Part/equipment	Quantity	Diameter (mm)	Size (m)	Surface temperature (°C)
Boiler room	Degasser/valves	3	50		100
Zipper room	Back of zipper dyeing machines/valve	2	85		140
Zipper room	Zipper dyeing/valve (1–2)	3	32		140
Zipper room	Inter dyeing zipper—Main steam distribution line/valve	1	100		150
Zipper room	Inter dyeing zipper—Main steam distribution line on the top of the door/valve	4	50		150
Spinning room	Air-conditioning room/Valve	15	32		150
Spinning room	Boiler Air-conditioning	7	32		150
Dye house	Heat exchanger behind the hot water tank/Steam line valve	10	50		115
Dye house	Heat exchanger behind the hot water tank/Steam line valve	4	28		115
Dyeing room	Hank (Skein) drying top/valve	4	50		155
Dyeing room	Hank (Skein) drying top/valve	2	50		130
Dye house corridor	Valve	1	80		125
Dye house corridor	Valve	2	50		125
Dyeing room	Valve	4	32		130
Dyeing room	Entrance (*) machine top/valve	1	100		125
Dyeing room	Hank (Skein) drying top/valve	3	100		135
Dyeing room	Hank (Skein) drying top/valve	3	50		130

By replacing the data in Tables 16.2 and 16.3 with Eqs. (16.1), (16.2), and (16.3), the amount of pre-insulation heat losses, post-insulation heat losses, and post-insulation fuel savings were calculated for each equipment, and the results are given and presented in Table 16.4.

The amount of saved fuel and the cost of saved fuel:

The amount of saved fuel = [(Recovered heat/Lower Heating Value of natural gas)/Boiler efficiency] × [(Annual working hours)/(Conversion coefficient)]

$= [(60{,}587\ \text{W}/8250(\text{kcal}/\text{Sm}^3))/0.92] \times [(7000\ \text{h/year}) \times (3600\ \text{s/h})/(4184\ \text{J/kcal})]$

$= 48{,}078.01\ \text{Sm}^3/\text{year}$

The price of saved fuel = The volume of recovered fuel × The unit price of the fuel

$= 48{,}078.01\ \text{Sm}^3/\text{year}\ 0.25\ \$/\text{m}^3$

$= \sim 12{,}019.5\$/\text{year}$

Cost of Realization: In the market research conducted for the insulation materials to be used in the application and the workmanship, it was seen that an investment of 5.000 USD would be sufficient.

Payback Period: It can be found by dividing the realization cost by the calculated annual saving amount.

$$\begin{aligned}\text{Repayment Period} &= (\$5000)/(12,019\$/\text{year}) \\ &= 0.41\ \text{years}\quad (5\ \text{months})\end{aligned}$$

Table 16.4 Heat loss before/after insulation and saving amount after insulation

Room type	Part/equipment	Quantity	Q_m (W)	Q_y (W)	Savings (W)
Boiler room	No. 4 boiler outlet distribution collector degasser side/pipe	1	4254	428	3826
Boiler room	Degasser connection/pipe	1	3486	352	3135
Boiler room	Pump connection	1	4120	479	3642
Dyeing room	Special line machines	9	8973	860	8114
Zipper room	Zipper (*) roller drying (*)/dryer machine	1	10,177	975	9201
Zipper room	Zipper (*) roller drying (*)/steam drying machine	2	13,721	1264	12,457
Zipper room	Zipper (*) roller drying (*)/connecting pipes	1	2358	250	2108
Dye house	Hot water tank behind heat exchanger inlet/Steam line pipe	1	1031	113	918

(continued)

Table 16.4 (continued)

Room type	Part/equipment	Quantity	Q_m (W)	Q_y (W)	Savings (W)
Dyeing room	(*) 3 machine behind/steam pipeline	1	1822	159	1662
Dyeing room	Steam line/pipe	1	755	79	676
Dyeing room	(*) machine/steam collector	1	4601	469	4132
Boiler room	Boiler room/valve	5	518	53	465
Boiler room	Condensate valves	3	532	54	478
Boiler room	Degasser valves	3	311	32	279
Zipper room	Back of zipper (*) dyeing machines/valve	2	613	54	559
Zipper room	Zipper (*) drying/valve (1–2)	3	337	31	307
Zipper room	Between dyeing zipper (*)—Main steam distribution line/valve	1	431	36	395
Zipper room	Between dyeing zipper (*)—Main steam distribution line top of the door/valve	4	801	69	732
Spinning room	Air-conditioning/valve	15	1889	165	1724
Spinning room	Boiler—Air-conditioning side/valve	7	881	77	804
Dye house	Heat exchanger behind the hot water tank/Steam line valve	10	1299	126	1173
Dye house	Heat exchanger behind the hot water tank/Steam line valve	4	244	25	219
Dyeing room	Hank (Skein) drying top/valve	4	846	71	774
Dyeing room	Hank (Skein) drying top/valve	2	260	25	235
Dye house corridor	Valve	1	255	23	231
Dye house corridor	Valve	2	297	28	269
Dyeing room	Valve	4	399	37	361
Dyeing room	Inlet (*) machines top/valve	1	319	29	290
Dyeing room	Hank (Skein) drying top/valve	3	1086	96	990
Dyeing room	Hank (Skein) drying top/valve	3	475	44	432
Total					60,587

References

1. Chen WH, Chung YC, Liu JL (2005) Analysis on energy consumption and performance of reheating furnaces in a hot strip mill. Int Commun Heat Mass Transfer 32:695–706
2. Kaya D, Güngör C (2002) Energy saving potential in Industry-I. Eng Mach 514:20–30. (In Turkish: Kaya D, Güngör C, Sanayide Enerji Tasarruf Potansiyeli-I, Mühendis ve Makina, 514, 20–30)
3. Kaya D, Güngör C (2002) Energy saving potential in Industry-II. Eng Mach 515:36–44. (In Turkish: Kaya D, Güngör C, Sanayide Enerji Tasarruf Potansiyeli-II, Mühendis ve Makina, 515, 36–44, 2002)
4. Kaya D, Öztürk HH (2014) Sanayide Enerji Yönetimi ve Enerji Verimliliği, Umuttepe Yayınları, Yayın No: 114, Mühendislik Dizisi: 12, Kocaeli/Türkiye

Chapter 17
Waste Heat Recovery

Today, in many countries, about 26% of industrial energy consumption is not effectively utilized because it is disposed of away in the form of hot gases and liquids. Energy loss can be significantly reduced by applying various methods for waste heat recovery [1, 2].

17.1 Heat Exchangers

A heat exchanger is a device that provides heat energy exchange between two or more fluids at different temperatures (Fig. 17.1). Heat exchangers are classified as transfer process, surface compactness, geometry design, flow regulation, number of fluids, heat transfer mechanisms, and application areas. The heat exchangers are divided into 4 groups according to the design geometry:

(1) Tubular heat exchanger,
(2) Plate heat exchanger,
(3) Expanded surface heat exchanger, and
(4) Regenerative heat exchanger.

Factors to consider when choosing a heat exchanger can be summarized as follows:

- Design materials,
- Pressure and temperature,
- Activity indicators (temperature, flow rate, pressure drop),
- Pollution trends,
- Control, cleaning, repair, and maintenance,
- Fluid types and phases,
- Heat exchanger size,
- Availability, and
- Economic factors.

D. Kaya et al., *Energy Management and Energy Efficiency in Industry*,
Green Energy and Technology,
https://doi.org/10.1007/978-3-030-25995-2_17

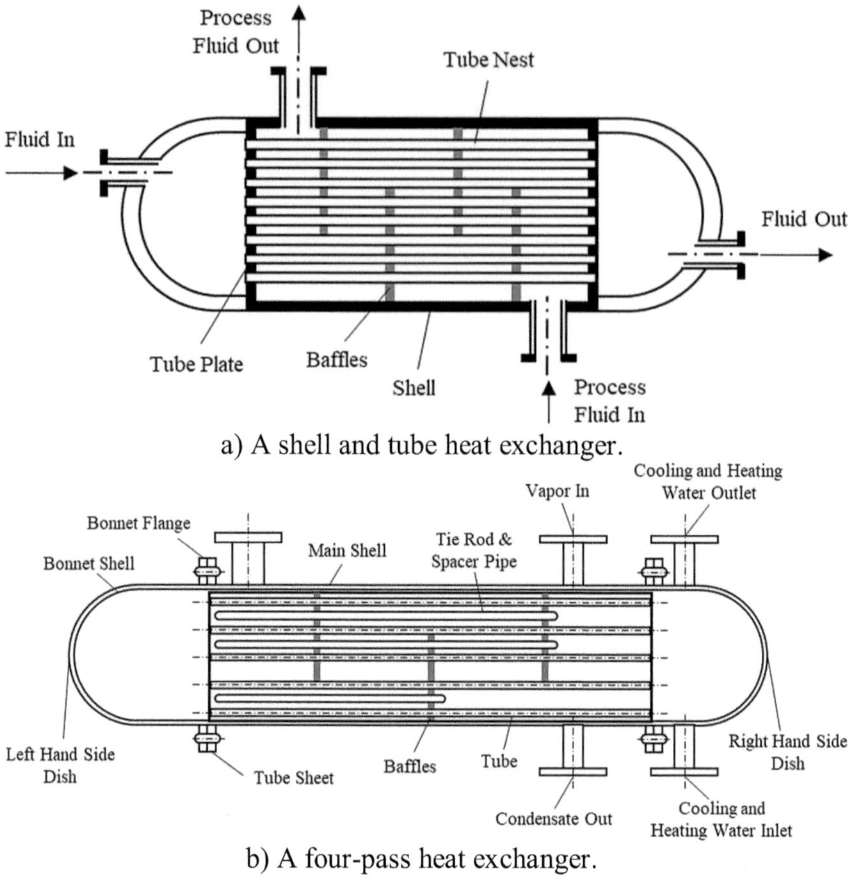

Fig. 17.1 Different types of heat exchangers [a) and b)]

17.1.1 Tubular Heat Exchanger

The tubular heat exchangers are mainly made of pipes. While one fluid flows through the pipe, the other fluid flows out of the pipe. Pipe diameter, number of pipes, pipe length, pipe pitch, and pipe arrangement can be changed. The tubular heat exchangers are grouped according to design specifications as follows:

(1) Flat tube heat exchanger,
(2) Spiral tube heat exchanger,
(3) Body heat exchanger, and
(4) Double tube heat exchanger.

Different types of heat exchangers can be seen in Fig. 17.1a and b.

17.1.1.1 Finned Pipe Heat Exchanger

The heat exchangers in the form of finned tubes are made of steel or copper pipes with steel or aluminum fins attached to the outside. The fins can be circular, square, or rectangular (Fig. 17.2). On one side is the gas, on the other side is the liquid flowing in the heat exchanger, the heat transport coefficient on the liquid side is high. For this reason, the liquid fluid side usually does not require fin. High-pressure fluid usually flows through the pipe. In practice, finned surfaces outside circular or oval cross-sectional pipes are more commonly used (Table 17.1). The fins can be designed together with the pipe, and subsequently can be placed on the pipe by casting, welding, soldering, or tight-fitting technique [1].

Finned pipe heat exchangers with different fin profiles can be seen in Fig. 17.2.

Fig. 17.2 Finned pipe heat exchangers with different fin profiles

Table 17.1 Features of finned tubular heat exchangers

<table>
<tr><th>Features</th><td rowspan="4">
A finned tube heat exchanger</td></tr>
<tr><td>• Construction is simple. The compactness can reach up to 3300 m^2/m^3.
• Usage areas; power stations, propeller cooling groups, vehicles, air conditioning and cooling plants.
• The useable temperature of the heat exchanger depends on the way the fins are connected to the pipe.</td></tr>
<tr><th>Negativity</th></tr>
<tr><td>➢ Additional pressure losses created by the fins must be taken into consideration.</td></tr>
</table>

17.1.2 Plate Heat Exchanger

In such heat exchangers, the related fluid and the secondary water pass through the channels between the plates attached to a particular roof in succession (Fig. 17.3). Settlement of plates determines the model and the size of the flow. Their cleaning and disassembly are easy. Also, crusting that will occur in every form can be eliminated. The surface area that provides heat exchange can be increased by adding a large number of plates.

Fluid circulation in a plate heat exchanger is given in Fig. 17.3.

Plate heat exchangers have high thermal efficiency. Maintenance can be done at an acceptable cost and easily. Today, the features that make plate heat exchangers attractive are

- Their thermal activities are high,
- They are designed from corrosion-resistant alloys,
- Their maintenance is easy,
- Their dimensions can be changed,
- Their designs are simple.

The plate heat exchangers consist essentially of a combination of a number of individual plates (Fig. 17.4). Plates are connected to each other by fasteners. The individual plates are hung on the upper carrier bar and guided by the lower carrier bar. For one-pass fluid circulation, the fluid connections on the hot and cold sides are placed on the fixed end cover. For multi-pass circulation, the fluid connections are placed in fixed and moving end covers.

The fluid circulation in the plate heat exchanger is shown in Fig. 17.4. The primary and secondary fluid flow is in opposite directions on both sides of the plates. Fluid flow and circulation are controlled by placing seals between the plates. The seals are positioned so as not to cause mixing of fluids. The outer periphery of all gaskets is directed to the atmosphere. Plate heat exchangers are partly low-pressure and low-temperature devices. In the plate heat exchanger, material selection is especially important for plates and gaskets.

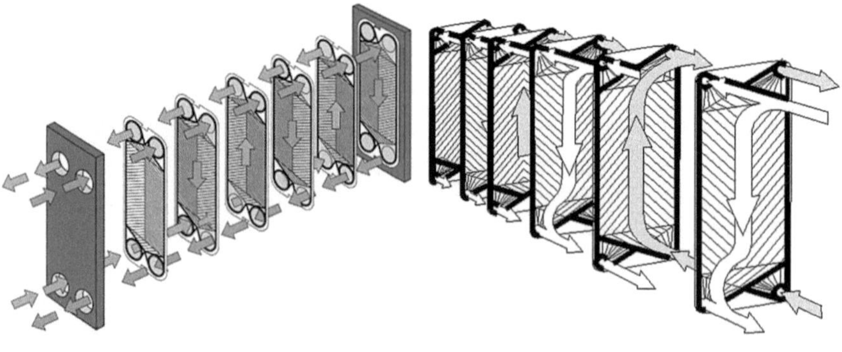

Fig. 17.3 Fluid circulation in a plate heat exchanger

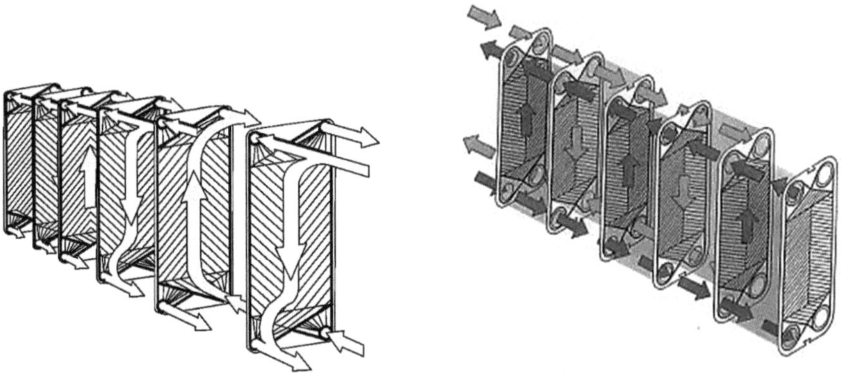

Fig. 17.4 Arrangement of plates in plate heat exchangers

The gasket plate heat exchangers (Table 17.2) are designed by packaging the thin metal plates in a frame. On each side of each metal plate, there are holes for fluid to pass through. When the plates are assembled and packed, using suitable seals prevents the fluids from intermixing and leaking out. From the spaces between the plates, hot and cold fluids flow without interfering with each other. The plates are designed wavy to provide rigidity, to stabilize the distance between the plates, and to improve heat transfer.

17.1.3 Heat Pipe Heat Exchanger

Heat pipe is a device that efficiently transfers heat between the two regions by evaporation and condensation of the working fluid contained therein. Heat pipe is a closed-loop heat transfer device with no moving parts and capable of efficiently transmitting heat. The heat pipe is a simple device that can quickly transfer heat from one point to another (Table 17.3). They are superconducting units that can transfer heat without any loss of heat. The heat pipe principle was originally proposed by *R.S. Gaugler* in 1942. However, the heat pipe was discovered by *G.M. Grover* in 1962.

Heat pipe is a closed pipe or a different type of heat exchanger with vacuum treatment applied, inner surfaces covered with porous capillary wick (Table 17.3). The main elements of the heat pipe are a sealed container, working fluid, and porous capillary wick. One end of the heat pipe acts as an evaporator and the other as a condenser. The working fluid evaporates through the heat taken from the external heat source in the evaporator and flows into the condenser. In the condenser, the working fluid condenses by giving its latent heat to the cold environment. The condensate fluid returns to the evaporator along the wick with the help of capillary action. Thus, the heat pipe is a heat exchanger that continuously transfers the evaporation latent heat from the evaporator part to the condenser part.

Table 17.2 Features of gasket plate heat exchangers

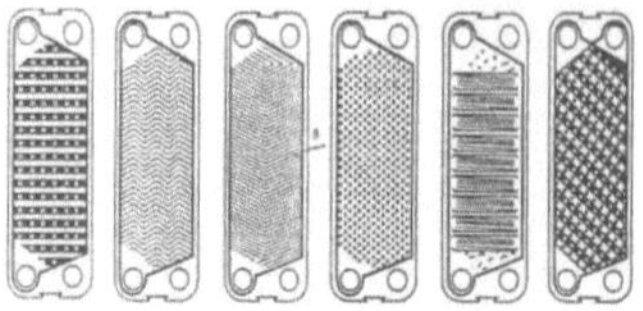

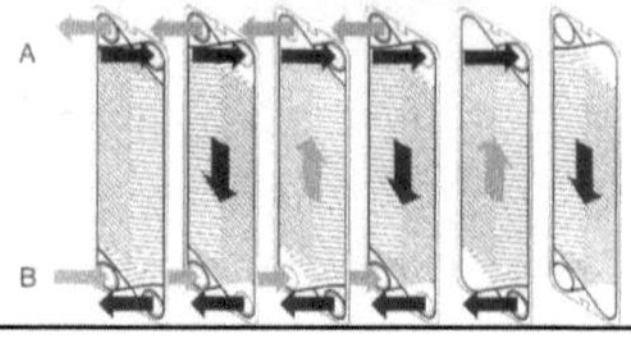

Properties

- The highest pressure is normally 25 bars, with special designs this value can be increased up to 40 bars.
- Temperature values range from (-25 °C) to (+175 °C). The temperature values vary with special materials from (-40 °C) to (+200 °C).
- Typical surface area for a single unit ranges from 1 m^2 to 1200 m^2.
- The highest efficiency is 0.95.
- The lowest temperature difference is 1 K.
- Plates are made from stainless steel or pressed from materials such as titanium.
- Gaskets are made from nitrile rubber, Hypalon, fluorinated rubber, or neoprene.
- Fluid restriction is carried out through the gasket.
- It is used in food, synthetic rubber, chemical, pulp (cellulose) and paper industries, and combi devices.

Advantages	Disadvantages
✓ Heat transfer coefficient is high. ✓ Thermal efficiency is high. Their efficiencies can range from 0.90 to 0.95. ✓ Temperature difference is low. ✓ Structure is simpler than tube heat exchanger. ✓ Plates are thin and low cost ✓ Control and cleaning procedures are easy. ✓ Extra plates can be added. ✓ Suitable for hot or expensive fluids. ✓ The possibility of contamination is low. ✓ They have a modular design. ✓ It is easy to maintain and clean. ✓ There is little chance of intermixing between fluids. The leaks are right on the outside. ✓ There is no vibration. ✓ Hot and cold zones are not formed in the heat exchanger.	- Due to the planarity (flatness) of the plates, the design of the sealing materials and the frame, high pressures cannot be reached. - Capacity is limited by the size of the ports. - If special wide cavity plates are not used, they can be easily blocked by the solids in suspension. - Plates are suitable for organic solvents. Seals may not be suitable. - There is a possibility of leakage. Seals continuously increase the possibility of leakage. - Gasket materials may limit the use of certain fluids. - Gasket materials limit the temperature. - Gasket loses its characteristic over time. For this reason, it may be necessary to change these gaskets at certain times. - Gas is not suitable for two-phase flow.

Table 17.3 Features of heat pipe heat exchangers

Features	Heat Pipe Heat Exchanger
• The highest pressure is 40 bar. • The temperature ranges from (-100 °C) to (+400 °C). Higher temperatures can be achieved with special working fluids. • The surface area for a single unit varies between 100 and 1000 m^2. • They are suitable for heat exchange between large volume gases at low pressure. • Applied in heat-rejection from electrical-electronic devices, de-icing units for bridges, roads, etc., solar collectors, air-conditioning applications, drying and heat recovery systems. • They are used in the applications like removing heat from electrical-electronic devices, for de-icing units in the bridges, roads, etc., in solar collectors, in air-conditioning systems, in drying and heat recovery systems.	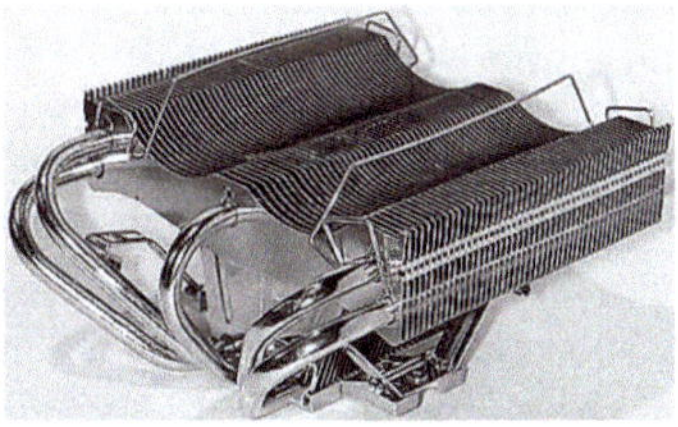

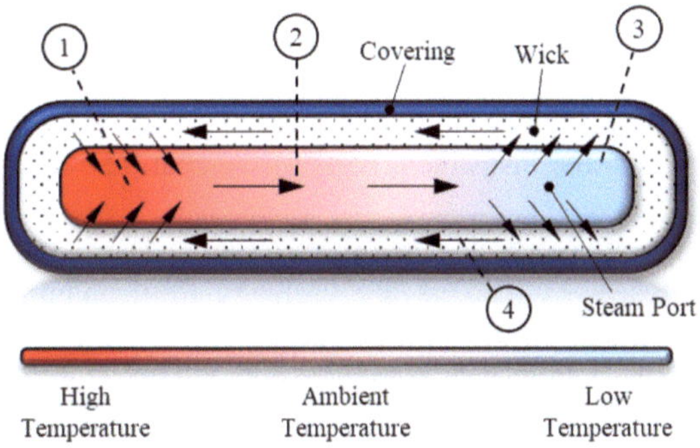

Working Principle of Heat Pipes

1. The working fluid evaporates by absorbing heat energy.
2. The steam is transported toward the space in the low-temperature end.
3. The steam condenses back into a liquid state and heat energy is released.
4. The working fluid flows back to the high-temperature end.

Advantages
• They have no moving parts, they work quietly. • Extended surfaces can be used on both gas sides. • Their structure is simple. • They have extremely good flexibility. • With very little temperature drop, it has a high amount of heat transfer ability at significant distances. • Their controllability is good. • Does not require external pump power.

17.2 Energy Saving in Air-Conditioning Systems

Pre-cooling or preheating of the outside air can be carried out with an exhaust air from the air-conditioning system or with an energy recovery unit to be added to the system in summer or winter air-conditioning situations. In this way, the heating unit can be combined with the cooling unit to reduce the capacities of the other heating and cooling group, thus reducing both the investment costs and operating costs of these devices. The following four types of energy recovery systems are used in air-conditioning plants:

(1) Circulation of water with the help of a pump between the finned tube serpentines placed in the hot and cold (flow and return) air ducts,
(2) Using plate-type heat exchangers,
(3) Using rotary-type heat exchangers, and
(4) Using a heat pipe heat recovery device.

Primarily, it should be remembered that an energy recovery device to be added to the system will cause an additional pressure loss load. However, each of the above-described energy recovery devices has advantages and disadvantages. For example, the energy recovery cannot be too much (about 40%) in the case of recirculation of water between two lines (battery-operated). Because of the continuous rotation of the heat wheel, the additional electrical energy and the presence of rotating elements, maintenance and control must be done frequently in the rotary-type energy recovery device. In this type of system, there is a risk of contamination of fresh air into the dirty air. Along with this, however, it is possible to achieve both sensory and latent heat. It can be said that the investment costs of the heat pipe recovery systems are high [3].

Air-conditioning with renewable energy systems can be seen in Fig. 17.5.

Generally, plate heat exchangers are designed from aluminum special plate profiles. If latent heat recovery is desired, cellulose-based material (allowing moisture transfer) can be used. Plate heat exchanger does not need lubrication, adjustment and part replacement, heat transfer gas, special coating, etc., as in heat pipe or rotary-type systems. Installation is easy and there is no additional operating cost beyond pressure loss. In the case of plate-type energy recovery devices, the recovered energy ratio varies between 65 and 75%.

Heat exchanger in boilers can be seen in Fig. 17.6.

17.3 Heating of Combustion Air

An important application area where the waste heat can be evaluated in the plants is the heating of the combustion air used in the boiler or furnace. The combustion efficiency of the combustion air increases by approximately 1% with every 28 °C temperature increase. In a study, it was stated that if a production excess of

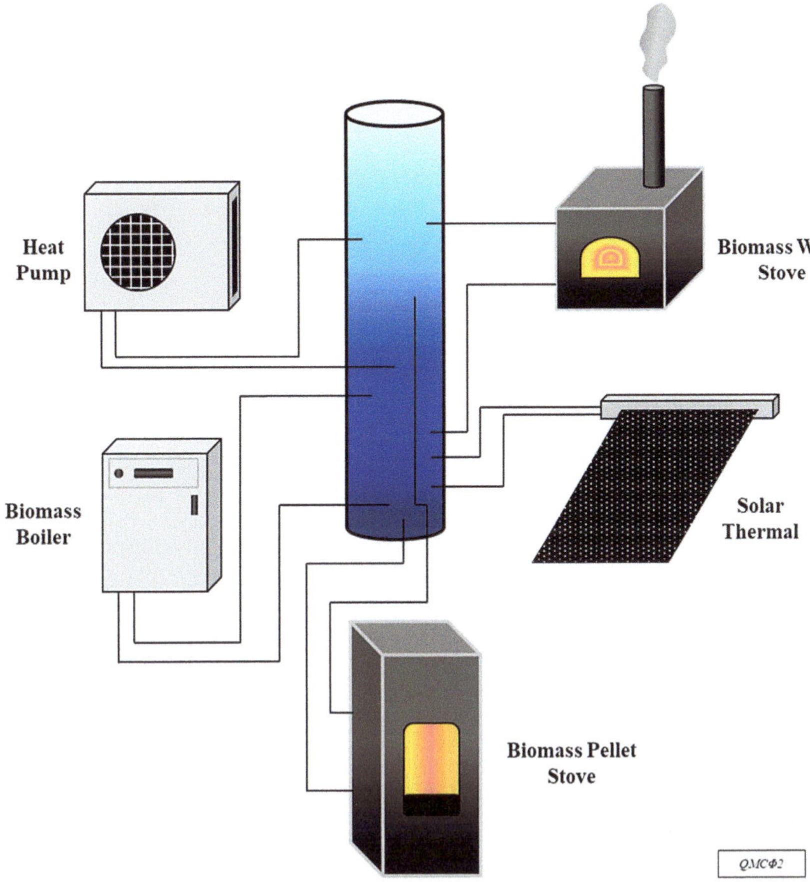

Fig. 17.5 Air–conditioning with renewable energy systems

14,911 kg/h of steam (at the pressure of 350 kPa and at the temperature of 428.16 K) is to be used for the preheating of the combustion air of the furnaces, it would save $1,093,570 annually. The payback period for this investment is calculated as only one month.

The amount of energy savings to be achieved as a result of boiler combustion air preheating can be calculated using the following equations down below:

$$\text{Annual Fuel Saving} = \text{Annual Fuel Consumption} \times [1 - (\eta_c \times \eta_t)] \qquad (17.1)$$

$$\text{Targeted Boiler Efficiency } (\eta_t) = \eta_c + (\text{EIC} \times \Delta \text{T}) \qquad (17.2)$$

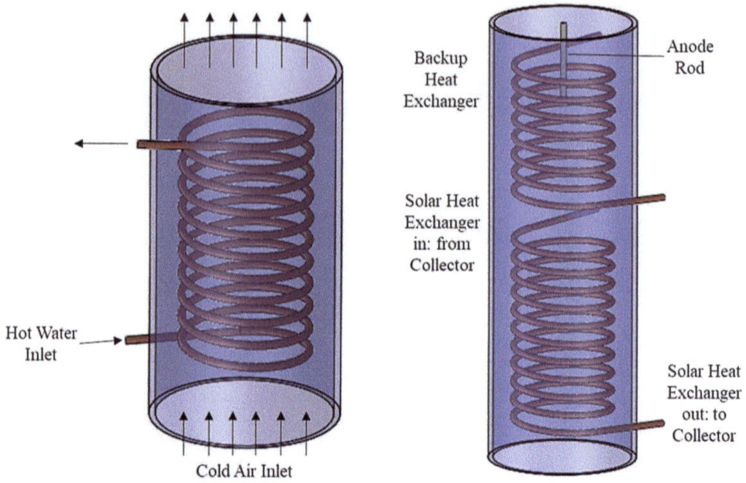

Fig. 17.6 Heat exchanger in boilers

where η_c—Current (existing) boiler efficiency, η_t—Targeted boiler efficiency, EIC—Efficiency increase coefficient and ΔT—The difference between the present and the targeted air temperatures.

17.4 Heat Recovery from Contaminated Fluid

The most energy-saving potential from the contaminated fluid is from the textile sector. In this sector, the most energy-intensive part is the wet processing part. Wet process consists of dyeing, bleaching, chemical application, and washing processes. Heat recovery from a fluid with a temperature of 70 °C is a known method in energy saving [4–7]. Water-to-water heat exchangers are effective in heat recovery from hot fluids with temperatures between 40 and 100 °C. Most of the water-to-water heat recovery systems use recycled heat to preheat the incoming cold water. The required process temperature is provided by additional steam or gas heating. Thus, the energy savings provided directly corresponds to the reduction in steam or gas consumption. Relevant to the establishment of such a system is given below:

$$\text{Waste Water Flow Rate} = \text{Amount of Water Used} \times \text{Total Time} \tag{17.3}$$

$$\text{Recovered Waste Heat} = \text{Waste Water Flow rate} \times (\text{Outlet Temperature} - \text{Inlet Temperature}) \tag{17.4}$$

$$\begin{aligned}\text{Amount of Recovered Fuel} = {} & \text{Recovered Heat} \times \text{Fuel Specific Heat} \\ & \times \text{Boiler Efficiency} \times \text{Annual Operating Time} \\ & \times \text{Coefficient of Use}\end{aligned} \tag{17.5}$$

$$\text{Saving Fuel Cost} = \text{Amount of Recovered Fuel} \times \text{Unit Price of the Fuel} \tag{17.6}$$

17.5 Waste Heat Recovery Application

It is determined that the hot air outlet has a high thermal capacity in the course of making energy-saving activities in the factory, from the 25 pieces of operating machines in the operation (Fig. 17.7).

In order to recover heat in hot gases, devices such as plate and serpentine heat exchangers, heat pipe, heat pump, gas-liquid heat exchangers have been designed and applied. In industrial enterprises, while the process requires hot and liquid residues to escape, considerable amounts of energy are also thrown away. The portion of this energy that can be recovered economically can be re-used with the above-mentioned devices and facilities [8–10].

In the textile sector, the most important problem for waste heat recovery systems is fiber, cotton thread, and manufacturing waste in the fluid to be used in the recovery of the waste heat. The fibers, cotton, and manufacturing waste that are involved in the flow somewhat during production must be kept in the filters. Filtration is the most important problem for heat recovery. In every plant, the flow rate and the type of solids are different. For this reason, it should be careful when selecting the filter. Automatic clean filter types should be preferred. Despite the high costs of automatic-type filters, they may not perform the required filtration. If user errors take into consideration, operational problems can occur. In the face of this problem, companies that manufacture plate heat exchangers have had to take some precautions.

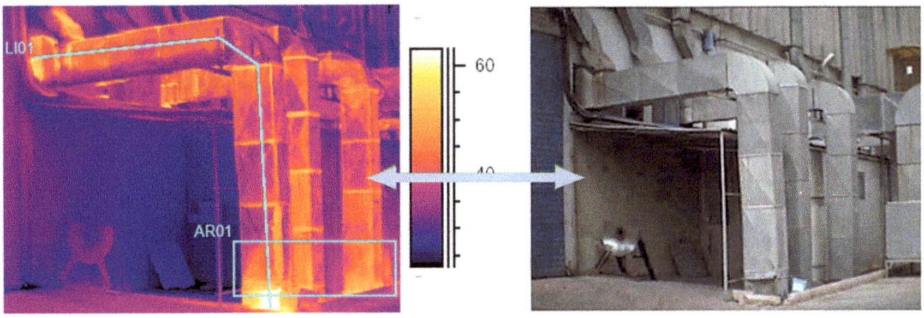

Fig. 17.7 Temperature values of hot air

The distance between the two plates through which the waste fluid passes is increased and the design of plate surfaces was changed. Thus, the spacing between the plates is large enough that the fibers and waste will not clog. The other side where the clean fluid is circulated is designed in the appropriate range for heat transfer. In front of the waste fluid, a mesh filter with a 4 mm grid spacing is also put in, and coarse wastes are also kept.

In order to make it economical to add a standard heat recovery system, it is necessary that the amount of waste heat that can be recovered is at a sufficient level and this waste heat must be useful [11–13]. The following calculations have been made to realize this aim. For the evaluation of the hot air from the dryer to out, it is considered to install a waste heat exchanger in the area where the air outlet is located.

17.5.1 Waste Heat Saving Potential

The total energy capacity that can be obtained from the hot air from the drying machine at different temperatures can be calculated taking into account the following assumptions and formulas.

$$\text{Hot air volumetric flow rate}: \dot{V} = V \times A \tag{17.7}$$

where *V*—Average velocity of the hot air in the channel at the outlet from the dryer (m/s), *A*—The cross-sectional area of the duct at the outlet of the hot air dryer (m^2), and $\dot{V}$—Volumetric flow rate of hot air (m^3/s).

$$\text{Hot air mass flow rate}: \dot{m} = 1.2 \times \dot{V} \tag{17.8}$$

where $\dot{m}$—mass flow rate (kg/s) and 1.2-density of air (kg/m^3).

$$\text{Potential energy saving}: \dot{Q} = \dot{m} \times C_p \times \Delta T \tag{17.9}$$

where $\dot{Q}$—Heat capacity of air (kW), C_p–Specific heat of the air (1.0035 kJ/kg ºC), and ΔT—Average temperature difference (ºC). In calculations, the outlet temperature was taken as 15 °C higher than ambient temperature.

As seen in Table 17.4, the power of the heat recovery system to be installed is 2164 kW (1,860,887 kcal/h). with the acceptance of Drying machine working time: 6,700 h/year and heat recovery system efficiency: 80%, yearly energy saving is 9,974,351,640.00 kcal.

When the lower heat value of the natural gas is taken as 8169 kcal/Nm^3, the natural gas amount to be saved is determined as follows:

$$9,974,351,640(\text{kcal/year})/8169\left(\text{kcal/Nm}^3\right) = 1,221,000\left(\text{Nm}^3/\text{year}\right)$$

Table 17.4 Thermal capacity of waste hot air

Dryer number	Average speed of hot air (m/s)	Temperature difference (ΔT °C)	The Duct cross-sectional area where hot air passes (m^2)	Volumetric flow rate of hot air (m^3/s)	Mass flow rate of hot air (kg/s)	Thermal capacity (kW)
No: 1	18.50909091	35.5	0.06	1.110545455	1.332654545	47.47481869
No: 2	16.52857143	46.5	0.1152	1.904091429	2.284909714	106.6201708
No: 3	15.89	43.20	0.13	1.99	2.38	103.33
No: 4	37.47142857	46.7	0.06	2.248285714	2.697942857	126.4349102
No: 5	30.48888889	40.5	0.06	1.829333333	2.1952	89.2167696
No: 6	35.46666667	45	0.06	2.128	2.5536	115.314192
No: 7	35.17777778	47.7	0.06	2.110666667	2.5328	121.237411
No: 8	10.125	39	0.096	0.972	1.1664	45.6488136
No: 9	9.176	4	0.204	1.871904	2.2462848	9.016587187
No: 10	21.41111111	30	0.1152	2.46656	2.959872	89.10694656
No: 11	20.06	33	0.1152	2.310912	2.7730944	91.8324076
No: 12	21.05	35	0.1152	2.42496	2.909952	102.2047891
No: 13	20.88333333	40	0.153	3.19515	3.83418	153.9039852
No: 14	27.86666667	42.5	0.064	1.783466667	2.14016	91.2751488
No: 15	29.12222222	31.5	0.064	1.863822222	2.236586667	70.69906368
No: 16	15.36	48	0.1152	1.769472	2.1233664	102.2783128
No: 18	13.11428571	41	0.1152	1.510765714	1.812918857	74.589827
No: 19	11.64285714	43	0.1152	1.341257143	1.609508571	69.45109961
No: 20	11.15	46	0.1152	1.28448	1.541376	71.15145754

(continued)

Table 17.4 (continued)

Dryer number	Average speed of hot air (m/s)	Temperature difference (ΔT °C)	The Duct cross-sectional area where hot air passes (m^2)	Volumetric flow rate of hot air (m^3/s)	Mass flow rate of hot air (kg/s)	Thermal capacity (kW)
No: 21	9.1625	55	0.1152	1.05552	1.266624	69.90814512
No: 22	9.616667	67.5	0.1152	1.10784	1.329408	90.04911264
No: 23	12.4	24.4	0.1152	1.42848	1.714176	41.97228503
No: 24	13	50	0.1152	1.4976	1.79712	90.170496
No: 25	13.5625	56.8	0.1152	1.5624	1.87488	106.8659101
No: 26	14.06	43.3	0.1152	1.619712	1.9436544	84.45479634
				Total Thermal Capacity		2164.210953

When the natural gas price is taken as 0.35 \$/Nm^3, the total annual saving potential is calculated as 427,350 \$/year. This value is only the cost of the energy that is thrown into the environment with hot air.

Using waste heat in ambient air heating: Radiation heaters are used in the facility for ambient heating purposes. The total installed power of natural gas radiation heaters used in the enterprise is 4835 kW. These heaters operate for about 5 months a year depending on the seasonal conditions of the region. During this period, they work on average 8 h a day.

$$\text{Daily consumption: } 4835\ \text{kW} \times 8\ \text{h} = 38{,}680\ \text{kWh}$$

$$\text{Daily waste heat that can be earned: } 2164\ \text{kW} \times 24\ \text{h} = 51{,}936\ \text{kWh}$$

In other words, a waste heat system to be installed will be able to meet the entire environment heating needs of the enterprise.

$$\text{Annual savings to be achieved} : 38,680\ \text{kWh/day} \times 150\ \text{day/year} = 5,802,000\ \text{kWh}$$

$$\begin{aligned}\text{Monetary savings} &= [5,802,000\ \text{kWh} \times 860\ (\text{kcal / kWh})]\ /8,169\ \text{kcal/Nm}^3 \\ &\times 0.35\ \text{USD/Nm}^3 = 213,784\ \text{USD}\end{aligned}$$

The price offered by the companies for the installation of the system proposed above is 270,000 USD. In this case, the repayment period of the investment is 270,000/213,784 = 1.26 year (~16 months).

References

1. Kaya D (1996) Tüpraş İzmit Refinery research of process waste steam heat energy recovery, Master Thesis, Kocaeli University, Kocaeli, Turkey. (In Turkish: Kaya, D., Tüpraş İzmit Rafinerisi Proses Atık Buharı Isı Enerjisinin Geri Kazanılmasının Araştırılması, Yüksek Lisans Tezi, Kocaeli Üniversitesi, Kocaeli/Türkiye, 1996)
2. Saraç Hİ, Kaya D, Sözbir N, Çallı İ (1997) Tüpraş İzmit refinery research of process waste steam heat energy recovery. In: Fifth symposium on combustion, Kirazliyayla, Bursa, Turkey. (In Turkish: Saraç, H. İ., Kaya, D., Sözbir, N., Çallı, İ., Tüpraş İzmit Rafinerisi Proses Atık Buharı Isı Enerjisinin Geri Kazanılmasının Araştırılması, Beşinci Yanma Sempozyumu, Kirazlıyayla, Bursa/Türkiye, 1997)
3. Yıldırım İM, Kaya D, Eyidoğan M, Çanka Kılıç F, Özdemir NC (2017) Waste heat recovery through an absorption cooler in a chemical production plant. In: First international conference on energy systems engineering (ICESE 17), 2–4 Nov 2017, KBU, Karabuk, Turkey
4. Bilgin A (2001) Evaluation of chimney gas analysis in boilers, investigation of internal cooling losses, V. National Plumbing Engineering Congress and Exhibition, 671–622. (In Turkish: Bilgin, A., Kazanlarda Baca Gazı Analizlerinin Değerlendirilmesi, İç Soğuma Kayıplarının İrdelenmesi, V. Ulusal Tesisat Mühendisliği Kongresi ve Sergisi, 671–622, 2001)

5. Si M, Thompson S, Calder K (2011) Energy efficiency assessment by process heating assessment and survey tool (PHAST) and feasibility analysis of waste heat recovery in the reheat furnace at a steel company. Renew Sustain Energy Rev 15:2904–2908
6. Terzi ÜK, Baykal R (2011) Efficient and effective use of energy: a case study of TOFAS. Environ Res Eng Manag 1(55):29–33
7. Ünlü C (2001) Energy saving and recovery in industry. Tesisat J. (In Turkish: Ünlü, C., Sanayide Enerji Tasarrufu ve Geri Kazanımı, Tesisat Dergisi, 2001)
8. Bilgin H, Çanka Kiliç F, Eyidoğan M, Kaya D (2017) Electrical energy efficiency in fiberboard production in Turkey. (In Turkish: Türkiye'de Lif Levha (MDF) Üretiminde Elektrik Enerjisi Verimliliği). In: First international conference on energy systems engineering (ICESE 17), 2–4 Nov 2017, KBU, Karabuk, Turkey
9. Bilgin H, Eyidoğan M, Canka Kılıç F, Kaya D. Electricity energy efficiency in medium-density fiberboard (MDF) production in Turkey. In: First international conference on energy systems engineering (ICESE 17), 2–4 Nov 2017, KBU Karabuk, Turkey
10. Çanka F (2000) Investigation of total pressure drop in loop heat pipes. PhD Thesis. Clemson University, Department of Mechanical Engineering, Clemson, SC USA
11. Canka Kilic F, Eyidogan M, Yuce M, Uzun E (2017) Energy efficiency studies in ceramic sanitary ware industry in Turkey. In: UEMK 2017 2nd international energy and engineering conference 12–13 Oct 2017 Gaziantep University, Turkey, Paper ID: A280917-1, p 512
12. Kaya D, Çanka Kiliç F, ve Eyidoğan M (2017) Evaluation of energy efficiency in ceramic tile production in Turkey, Toprak İşveren J 113:14–21. (In Turkish: Kaya D., Çanka Kiliç F. ve Eyidoğan M., Türkiye'de Seramik Kaplama Malzemeleri Üretiminde Enerji Verimliliğinin Değerlendirilmesi, Toprak İşveren Dergisi, Sayı:113, pp. 14–21, Nisan 2017)
13. Sert MÖ, Eyidoğan M, Çanka Kiliç F, Kaya D (2017) An energy efficiency study in an integrated textile production company. In: First international conference on energy systems engineering (ICESE 17), 2–4 Nov 2017, KBU, Karabuk, Turkey

Chapter 18
Energy Efficiency in Water Heating-Distribution-Pressurizing Systems

18.1 Energy Efficiency in Water Heating Systems

A significant amount of energy is consumed to heat water because energy is consumed all year round for hot water needs. The amount of heat energy required for hot water heating in residences varies between 10 and 20% of the annual heating requirement. Depending on the technological developments, effective thermal insulation applications, precautions for heat losses and automation, etc., in buildings, the amount of energy required for heating is reduced. In large commercial buildings, the amount of energy consumed to heat water for use can reach 4% of annual energy consumption. On the other hand, this value is between 20 and 35% of the annual heating requirement, depending on the heat insulation and the systems on zones [1]. By reducing the consumption of hot water, water heating energy is saved. However, inefficiencies in water heating systems can result from the following factors:

- Features of the boiler,
- Distribution and circulation pipe installation,
- Selection of water temperature at high values in the boiler,
- Selection of hydrophore system pressure at high values, and
- Faucet and battery features.

In Fig. 18.1 single- and double-coil boilers and in Fig. 18.2 the main components of single- and double-coil boilers can be seen, respectively.

18.1.1 Potable Water Temperature

The following factors should be taken into account for the potable water temperature:

D. Kaya et al., *Energy Management and Energy Efficiency in Industry*,
Green Energy and Technology,
https://doi.org/10.1007/978-3-030-25995-2_18

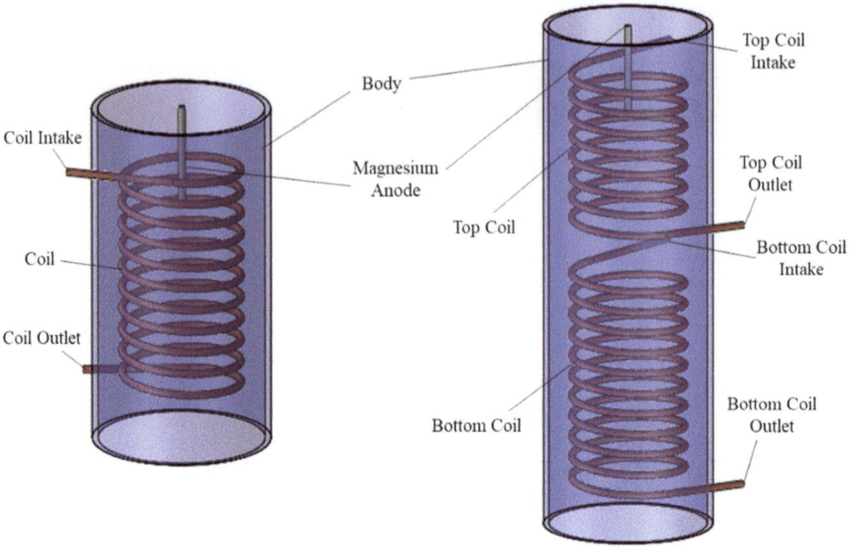

Fig. 18.1 Single- and double-coil boilers

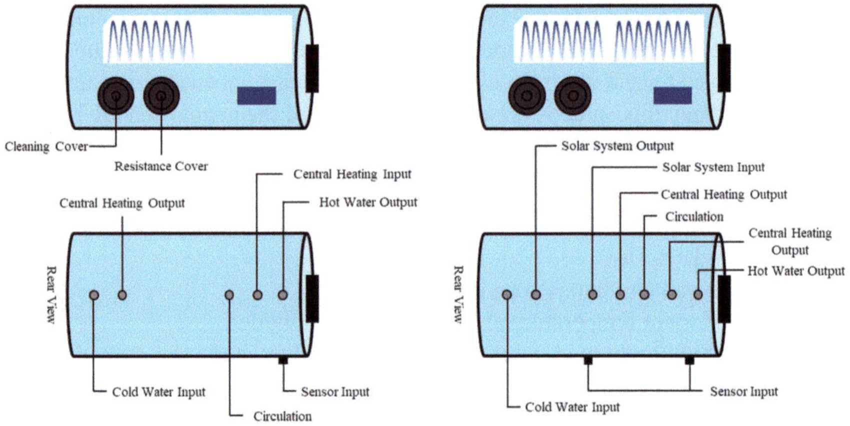

Fig. 18.2 Main components of single- and double-coil boilers

- The water temperature of the boiler for residences must be 45 °C due to economic reasons.
- The water temperature of the boiler should not be lower than 42 °C at the end points (at the tap entrance).
- Equal resistance should be established in the usage water and circulation pipes to ensure full circulation.
- There should be no blind spots without water circulation.
- Pipe insulation should be done efficiently.

If the water temperature of the boiler is high, the following drawbacks are encountered:

- The amount of energy carried by the unit water body increases.
- If water temperature is not controlled automatically during use, more energy is consumed with high-temperature water.
- More water and energy are consumed until you adjust the water temperature during use.
- More energy is lost in water distribution and circulation lines.
- More energy is lost from the boiler surface.
- The boiler must work at a higher temperature. The boiler operates at a higher temperature, with lower efficiency. In this case, more fuel is consumed due to reduced boiler efficiency.

18.1.2 Energy-Saving Measures

In the production of hot water for use, waste heat energy or renewable energy sources should be used instead of directly consuming fuel. For this purpose, the following applications can be used for the evaluation of waste heat:

1. Condensers of water-cooling groups can be used in commercial buildings.
2. Waste heat released from the condensate return heat exchangers (front boilers) of the steam system in laundries can be used to heat water for use.

In order to meet the need for hot water for laundry, the following factors should be considered:

- A separate boiler should be selected for the laundry.
- The boiler temperature should be adjusted to 55–60 °C.
- Condensate returned from laundry devices must be preheated by passing through a separate front boiler. Thus, both heat recovery is ensured and the heat and water loss, calcification risk, etc., which will occur as a result of evaporation of the condensate entering the condensation tank problems are avoided.
- The front boilers must be connected in series to the laundry boiler (or boilers).

Example: Recovery of Steam Discharged into the Atmosphere in the Condensate System of Steam Generators.

If the condensate at 150 °C and under pressure is cooled and turned into water at 90 °C, the amount of heat gained from 1 L steam condensate will be 60 kcal/L steam. From a steam generator condenser with a capacity of 600 L/h in a day;

$\dot{Q} = 600$ L/h × 60 kcal/L × 8 h/day = 288,000 kcal/day heat energy will be recovered. In the case of using this heat energy in the utilization of water for heating, the amount of water that can be heated on a daily basis is determined as follows:

Mains water temperature: $T_1 = 15\ °C$.
Potable water temperature: $T_2 = 45\ °C$.
Temperature difference: $\Delta T = 45\ °C - 15\ °C = 30\ °C$.
$\dot{Q}$ = The amount of water that the 288,000 kcal/day energy can heat from.
$T_1 = 15\ °C$ to $T_2 = 45\ °C$;

$$\dot{Q} = \dot{m} \times C_p \times \Delta T$$

For 1 L (L) Water;

$$= 0.9979 \frac{\text{kcal}}{\text{L} \times^\circ \text{C}} \times (45 - 15)^\circ\text{C}$$

$$=\sim 30 \frac{\text{kcal}}{\text{L}}$$

$\dot{V}$ = (288,000 kcal/day) / (30 kcal/L) = 9600 L/day.

If 80 L of bathing water is required at 40 °C per person, 9600/80 = 120 people's daily hot water need for bathing will be covered.

18.1.3 Selection of the Boiler

The following factors must be taken into account when choosing a boiler to heat water for use:

- The boiler with low volume of water should be selected due to its advantages entering in a short time regime, less waste heat generation, and less loss of stopping (halt) losses.
- For the hot water to be used, a separate boiler must be used from the main boiler. If the boiler is separate, the fuel consumption is reduced. In the intermediate seasons, the building water heating equipment for the boiler is operated at high temperatures and it is prevented to consume a large amount of fuel.
- A high-efficient self-condensing boiler (Fig. 18.3) should be used for heating.
- Energy consumption can be reduced by using a cascade system in self-condensing boilers.

In quality water heating equipment, since the heating coil is at the bottom of the water heating equipment, the heating return water temperature decreases, and the boiler efficiency increases.

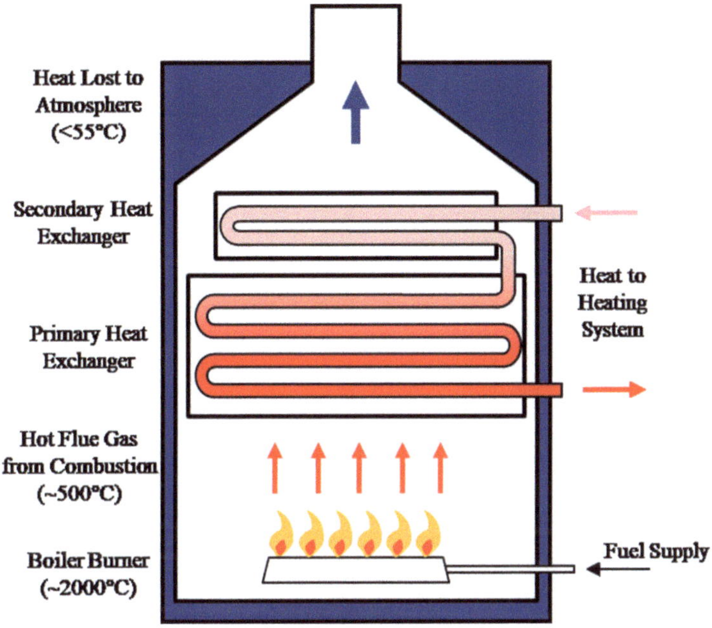

Fig. 18.3 Main components of a self-condensing boiler

18.2 Water Distribution Systems

The main function of the water distribution systems (Figs. 18.4 and 18.5) is to deliver hot and cold water to the usage appliances of the building at the appropriate pressure and temperature. Water distribution systems must have the following basic features:

- Water dispensing systems should deliver the water farthest, with a volumetric flow of suitable value, minimum pressure loss, and highest flow conditions.
- Provide water at the highest and lowest pressure conditions, the farthest and closest element, in the pressure range sufficient to meet the requirements.
- The system must be protected from excessive pressures.
- Pressure losses, which are very important in high structures, should be designed and implemented in a way to minimize them.

18.3 Water Pressurization Systems

In water-pressurizing systems (see Fig. 18.6) in high-rise structures, it is necessary to separate the system into pressure stages and to use special hydrophores for each stage. Approximately every 35 m static height must be a pressure zone. In the case of a superfluous pressure increase:

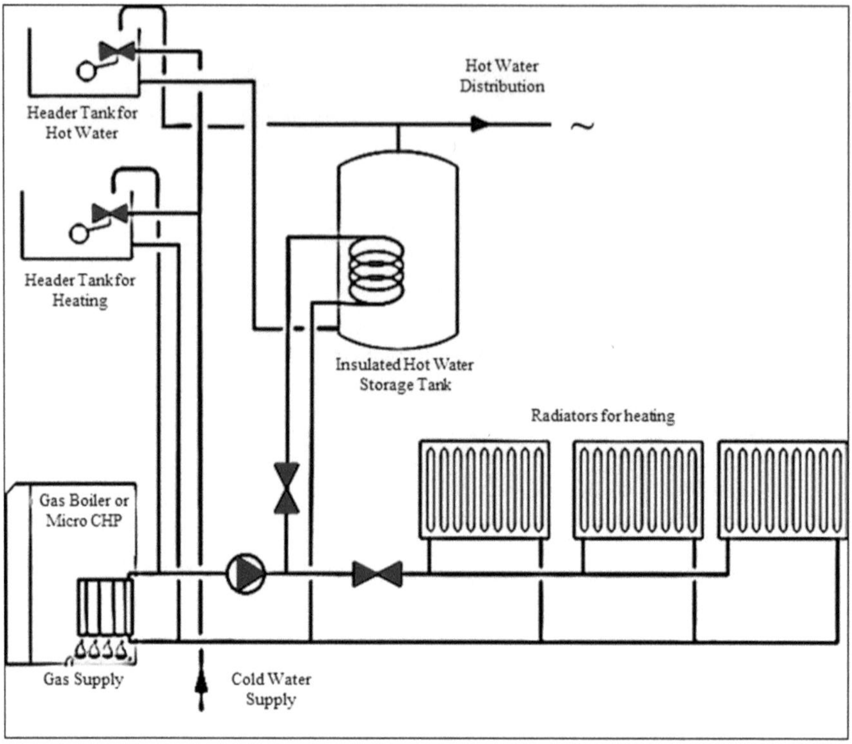

Fig. 18.4 Natural flow hot water distribution system

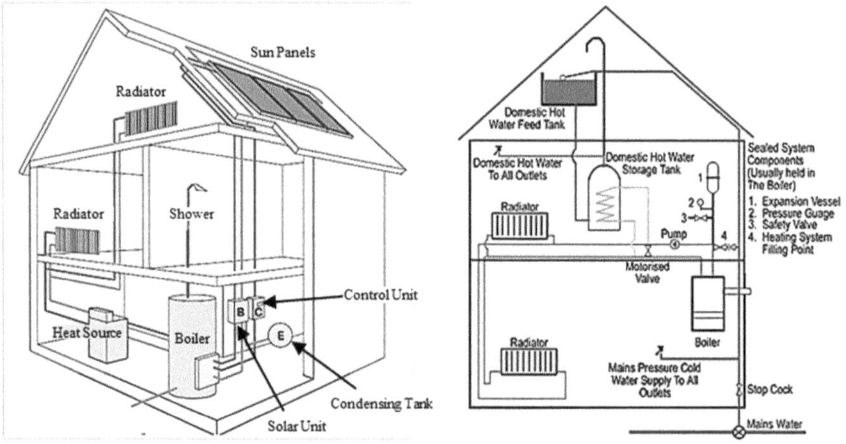

Fig. 18.5 Boiler usage in residences

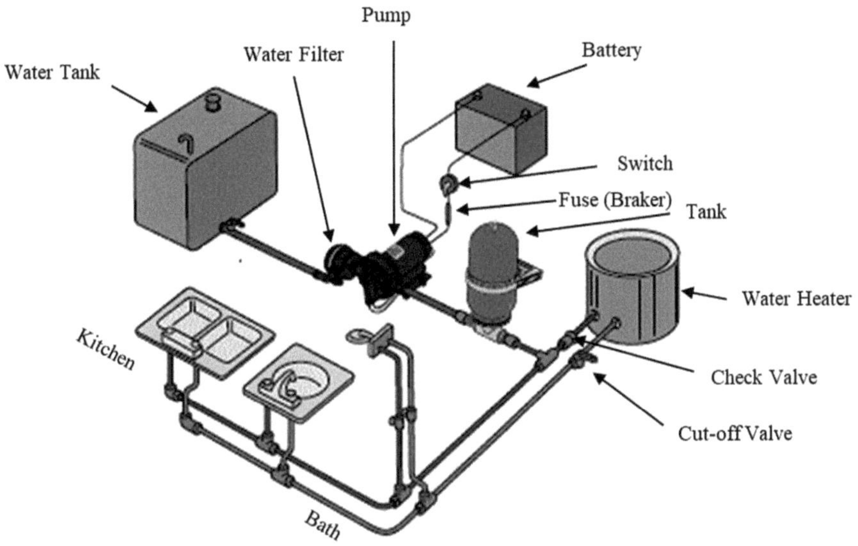

Fig. 18.6 Main parts for water-pressurizing systems

- The pump efficiency is reduced,
- The electricity consumption of the water booster (hydrophore) pumps increases,
- The amount of water flowing from the tap increases. In other words, the pressure increase causes unnecessary water consumption,
- Excessive pressure creates water splashes, sound, and strong flow. As a result of these effects, comfort is degraded,
- The water velocity in the pipes should be less than 2 m/s. If the water velocity is higher than 3 m/s, excessive flow occurs, and noise is generated, and
- Leaks may occur at the weak points of the installation, such as the connection points of the pipe installation and tap pipes.

Three different hydrophores were considered in Table 18.1 to assess the energy loss caused by the unnecessary increase of the hydrophore pressure. The nominal pressure booster (hydrophore) explained in Table 18.1 essentially represents the nominal pressure hydrophore selected in accordance with the requirement shows the values in the same hydrophore when the pressure is increased by 1 bar. Here, the annual electricity consumption was calculated by assuming that the flow rate of the taps was adjusted and the pressure was increased, the water consumption did not change. By increasing the pressure of the hydrophore by 1 bar, depending on the operating pressures and the feature of the hydrophore pump, electricity consumption can increase by 20–52%. As the nominal pressure value decreases, electricity consumption increase corresponding to 1 bar pressure increase is higher [2, 3].

In pressurization systems (Fig. 18.7), the following measures must be taken into account in order to provide energy management in an effective way to reduce energy consumption:

Table 18.1 The effect of Increasing the Pressure of Hydrophore 1 Bar on Annual Energy Consumption

Variables	Example 1		Example 2		Example 3	
	Nominal pressure	High pressure	Nominal pressure	High pressure	Nominal pressure	High pressure
Pressure (mSS)	60	70	50	60	40	50
Flow rate (m^3/h)	12.6	10.0	23.0	16.0	5.60	3.00
Efficiency (%)	55	54	57	54	55	45
Power (kW)	3.72	3.55	5.50	4.90	1.10	0.90
Consumption (kWh)	1313		2620		982	
Increase in electricity consumption (%)	20		28		52	

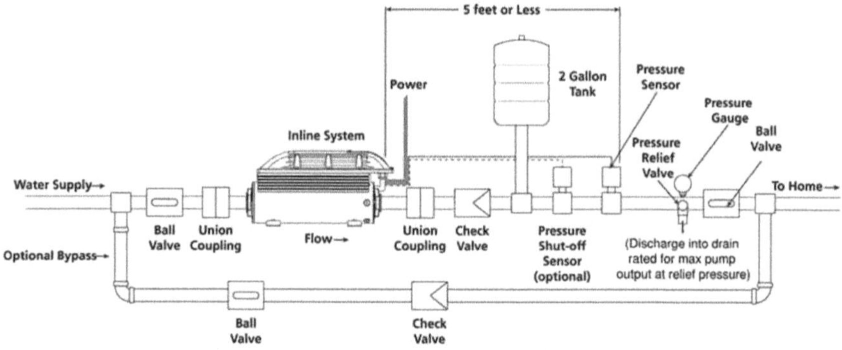

Fig. 18.7 Constant pressure flow system for residences

- To prevent pressure changes and to reduce water consumption by sending water at a constant pressure to the facility, a pressure regulator should be used at the outlet of the booster.
- Multi-pumped (stepped) hydrophore should be used for boiler water with a flow rate higher than approximately 5 m^3/h.
- A multi-pump (stepwise) booster (hydrophore) should be used for the flow rates higher than approximately 5 m^3/h for tap water,
- The pressure in the city network should be benefited from as much as possible.
- In the large plants, variable speed pump hydrophore should be used.
- Pipe diameters must be chosen correctly and sufficiently large enough to prevent pressure changes in the water piping.
- Thermal insulation should be applied to the cold water, hot water, and circulation pipes installed in the wall against sweating and heat loss.

- The clean water pipe must be passed through inside the protection pipe (sleeve) over the entire wall thickness in the outer wall passages.
- Pipes should be insulated against freezing in the areas where it is exposed to open air or in the danger of freezing.
- Lines that are underused or at risk of freezing should have a zoning valve and a discharge valve.
- In order to avoid corrosion problems, it should be ensured that steel pipes do not come into contact with materials that having humidity.

In water pressurization (hydrophore) applications, the use of a variable speed single pump is usually not feasible since the required pressure is usually constant [3, 4]. For this reason, it is necessary to use multiple pumps and one of the pumps must have variable speed. Instead of a single pump system, an optimized multi-pump pressurization system should be preferred. In the pressurization systems, if the entire building is fed by a single pump, the following drawbacks are encountered:

- Due to the variable flow rate, the pump does not operate at the highest efficiency point.
- Due to the pressure stages in the system, water is sent with high pressure to places where low pressure is required.
- Pressure energy is wasted in pressure reducers or faucets.
- Motor efficiency is also significantly reduced at partial loads.

References

1. Kaya D, Öztürk HH (2014) Sanayide Enerji Yönetimi ve Enerji Verimliliği, Umuttepe Yayınları, Yayın No: 114, Mühendislik Dizisi: 12, Kocaeli/Türkiye.
2. Enerkon, Energy Efficiency in Steam Distribution System (1987)
3. Kucukcali R (2006) Economy in water distribution and pressurization (Booster) systems, Gelişim Teknik, Su Teknolojisi Journal, January 2006. (In Turkish: Kucukcali, R. Su Dağıtım ve Basınçlandırma (Hidrofor) Sistemlerinde Ekonomi, Gelişim Teknik, Su Teknolojisi Dergisi. Ocak, 2006)
4. TMMOB (2005) The main lines of the effective and efficient use of energy, p 21–22. (In Turkish: TMMOB, Enerjinin Etkin ve Verimli Kullanılmasının Ana Hatları, sf. 21–22, 2005)

Chapter 19
Energy Efficiency in Illumination (Lighting)

It is expected from a good quality lighting system to send enough light to the areas intended for illumination. Illumination of unused areas or excessive illumination in the areas that are being used for different purposes will also cause energy losses. Inadequate lighting is inconvenient for safety and comfort. Likewise, excessive illumination can completely disrupt the vision due to the problem of glare.

Approximately 20% of the monthly electricity bills in the houses belong to lighting use. Appropriate and efficient lighting will provide relief both in bills and in terms of human health and comfort. Thus, satisfying results will be obtained with lower invoices and more appropriate lighting. It is an important advantage to achieve energy saving in lighting with simple measures. The important thing here is to pay due attention to the subject [1].

Energy saving in lighting should be done by fulfilling the requirements of a good lighting without degrading the quality of the lighting. As good lighting is provided with more efficient lighting elements, the same lighting levels are achieved with less energy consumption.

19.1 Energy Saving in Lighting

Lighting has an important place in electricity consumption. Lighting systems are the biggest source of energy consumption after heating–cooling systems. Electricity: 20% is consumed in industrial enterprises, 30% in stores (shops), and about 40% in offices for lighting purposes. These figures clearly demonstrate the necessity of economic solutions in lighting systems. Significant energy savings can be achieved in lighting by using efficient light sources and efficient fixtures [2, 3]. However, a lighting system that does not have the ability to make as much lighting as needed, when required, in other words, is not controlled, is not suitable for today's economic and technological conditions.

D. Kaya et al., *Energy Management and Energy Efficiency in Industry*,
Green Energy and Technology,
https://doi.org/10.1007/978-3-030-25995-2_19

19.1.1 Lamp Types

19.1.1.1 Incandescent Lamps

The way of obtaining the light is the incandescent lamp, which is the thermal radiation, bringing the electric current passing through the tungsten conductor into the incandescent state by heating [1]. As the wire begins to warm up, it becomes an electrical radiation energy. Most of the radiation emitted by these lamps is heat, and a small part is visible radiation. For this reason, the efficiency is very low between 10 and 20 lm/W.

19.1.1.2 Incandescent Halogen Lamps

The incandescent halogen lamp is a thermal light source created by the exchange of the gas mixture in the atmosphere of the incandescent lamp (adding halogen). The halogen, which is used in the atmosphere of such lamps, can increase the wire temperature due to the tungsten-metal renewal of the molecules. As a result, both the light emission and the color temperature can be increased slightly, according to the same powerful incandescent lamp.

19.1.1.3 Fluorescent Lamps

In the case of fluorescent lamps, the mode of radiation generation is thermal radiation, light production occurs in two stages. The first step is the generation of radiation by the discharge of electric current through the application of fluorescent material to the inner surface of the lamp in the low-pressure mercury vapor environment. The emission of fluorescent lamps is basically increased as the lamp power increases. However, when the same power lamps are considered, the change in the efficiency depends directly on the structure of the fluorescent powder [1].

As a result of the studies carried out for the energy-efficient production of light sources, great improvements have also been made in tubular fluorescent lamps. Instead of lamps with a power of 20, 40, and 65 W with a diameter of 38 mm, fluorescent lamps with a diameter of 26 mm have a power of 18, 36, and 58 W are available for use, respectively. The diameters of the lamps have been reduced, the luminous fluxes have been increased, and lamps with many different color temperatures and color separation features have started to be produced. Small diameter lamps are more economical.

19.1.1.4 Compact Fluorescent Lamps

Compact fluorescent lamps are suitable for residential and office use, and it is easy to convert incandescent lamps into compact fluorescent lighting. Compact fluorescent lamps can be used almost everywhere using incandescent lamps. For example, using a 15 W compact fluorescent lamp instead of a 75 W incandescent lamp, the same illumination is achieved by consuming 80% less energy.

In many countries, these lamps have not become widely used compared to developed countries due to their high cost. There is a considerable price difference for the efficient lamps. However, considering their total costs, it is seen that the cost of compact fluorescent lamps is lower during their lifetime. Two factors confirm this, the first is that the lifetime is 8 times longer than the incandescent lamp, and the second is that they use 20% of the energy of the incandescent lamp.

Magnetic and Electronic Ballasts for Fluorescent Lamps

The ballasts command and operate the fluorescent tubes and compact fluorescent lamps for the first time. Fluorescent lamps are filled with mercury and argon gas. Electrodes at the end of the lamp powered by the ballast generate an electric discharge to ionize the gas. Mercury atoms return photovoltaic energy to emit ultraviolet photons. The phosphor coating of the lamp absorbs photons, fluorescents and produces visible light.

Magnetic Ballasts: These ballasts are called as magnetic or electromagnetic, they have aluminum or copper wire around an iron ore. Copper wire ballasts are 10% more efficient than aluminum ones. Magnetic ballasts operate at the AC alternating current standard frequency, 50 Hz. Electricity consumers are recommended to use ballasts with very low loss.

Electronic Ballasts: Electronic ballasts operate at a higher frequency than magnetic ballasts, which reduces the flickering and noise of the lamp. Electronic ballasts use 25% less electricity than magnetic ballasts. More energy savings are possible with many electronic ballasts allowing the lamp to dim. More efficient lighting systems produce less heat. It should be ensured that electronic ballasts comply with IEC 928 and 929 international standards and have a high power factor.

Electronic and magnetic ballasts may have a low power factor due to deformation or phase jumps leading to higher line losses. To correct the phase power factor, capacitors can be installed in ballasts or produced in conjunction with the ballast during production.

19.1.2 Accurate and Efficient Lighting

One or more of the following is used to benefit from the sunlight at a maximum level in the buildings, etc., and avoid unnecessary artificial lighting:

- Easy to access manually controlled switches,
- In places where it is possible to benefit from daylight, photoelectric switches connected with daylight and telephone, infrared, sonic, and ultrasonic controlled remote-control switches,
- Automatic switches and systems that can detect that the environment (room, office, etc.) whether is empty or not and then turn off artificial lighting if it is necessary,
- Time adjusted switches.

Usually, photoelectric switches with timer or daylight connected are used in the buildings where continuous lighting is required during the working hours and general lighting fixtures (lamps) are used in the buildings with the specifications determined by the regulation. Except for buildings used for residential purposes, whenever there are people in the buildings, all kinds of places that are under the authority of the administrative personnel must have a device that allows the lighting to be turned on and off. If this device is not located in the place in question, it should allow the lighting status in the place to be seen from the control point.

In buildings where there are different lighting levels such as sportive and multi-purpose halls, and where there are at least two or more different usage areas, measures should be taken that only authorized personnel can increase the basic lighting level. In the same space, when the total installed power in each of the artificially illuminated points closer than 5 m to a window opening exceeds 200 W, these points should be controlled independently from other lighting points.

When the natural lighting is sufficient in the buildings, artificial lighting should not be activated automatically with the device that detects time, adjusted, or human presence. Unless there are special circumstances, incandescent wire lamps are not used, and in cases where the color temperature is not important, A- and B-class electronic ballast tubular fluorescent, compact-type fluorescent, or sodium-vapor lamps should be preferred.

Decorative lighting devices with high energy consumption cannot be used for general lighting purposes. Armature selection and distribution should be made in the working areas that will provide adequate brightness. It is necessary to use motion, heat, or light-sensitive equipment in places where feasibility is appropriate, especially in places such as toilets, sinks, corridors located in stairwells and working environments, and unnecessary uses should be avoided. Light-colored furniture and wall colors should be preferred to achieve the desired brightness levels with fewer fixtures and consequently less electricity consumption. Lighting devices should be cleaned periodically to increase the efficiency of the luminaires and the brightness level in the rooms.

In determining the lighting energy need of non-residential buildings, in addition to the building's internal lighting load, the building's external lighting load should be taken into account, with the exception of security lighting in buildings other than buildings used for residential purposes where there are different lighting levels, systems should be installed that will allow only authorized personnel to increase the minimum lighting level.

In the design of the electrical installation of the buildings, it is necessary to comply with the related Laws and Regulations and relevant legislative provisions. In the electrical systems of non-residential buildings, power compensation should be made at the central and/or local level in accordance with the relevant legislation.

Lighting should be done according to the purpose of the use. The architectural and functional features of the buildings should be examined, the level of brightness needs of the environment should be defined and the luminaires should be determined according to these needs. The aim of illumination is to obtain a certain level of illumination and provide good vision conditions. The use of LED (Light Emitting Diode) and fluorescent lamps should be made widespread in both commercial and households to improve lighting quality and to save energy. The first and biggest step in energy saving in indoor lighting is to use LED and fluorescent lamps in suitable places. The second step is to turn toward energy saving in these systems. Prices of high-efficiency lamps are higher than other lamps. However, the lifetime of compact fluorescent lamps is 8 times longer than incandescent lamps, and the lifespan of LEDs is 15–20 times longer than incandescent lamps. Various properties of incandescent lamps, fluorescent lamps, and LEDs are compared in Table 19.1.

Table 19.1 Specifications of incandescent lamps, fluorescent lamps, and LEDs

Type	Incandescent filament		Fluorescent		LED (light emitting diode)
	Normal	Halogen	Tube	Compact	Normal
Power (W)	15–1000	20–2000	6–65	9–25	1–500
Efficiency (lumens/W)	10–20	20–25	50–95	45–80	70–90
Lifetime (hours)	1000	2000–3000	4000–7000	8000–10,000	15,000–50,000
Color of light	Hot	Hot	Various colors	Hot	Various colors
Color Rendering	Good	Very good	Between mid to good	Very good	Very good
Cost	Low	Medium	Medium	Medium	Medium
Usage suggestions	In short-term work and general-purpose places	In high-intensity lighting and where good color rendering is required	In continuous or intermittent lighting, for general purposes, where good color rendering is required	In continuous or intermittent lighting, for general purposes, where good color rendering is required	In continuous or intermittent lighting, for general purposes, where good color rendering is required

Table 19.2 Cost Analysis in Incandescent Bulb and Compact Fluorescent Lamps

Lamp type	Incandescent bulb (100 W)	Compact fluorescent lamp (23 W)
Purchase price ($)	0.75	11.00
Lamp life (h)	750	10,000
Daily usage time (h)	4	4
Number of lamps needed (pieces)	6 pieces in 3 years	1 pieces in 6.8 years
Total lamp cost ($)	4.50	11.00
Lighting level (Luminous flux (Lumen)	1690	1500
Total electric cost (8 cent/kWh) ($)	35.04	8.06
Total cost (3 years) ($)	39.54	19.06

Example: Comparison of Electricity Consumption Expenditures of Incandescent and Compact Fluorescent Lamps.

The electricity consumption expenditures of the lamps are calculated as follows.

$$\text{Electricity Cost}(\$) = [\text{Electricity cost}(\$/\text{kWh}) \times \text{Lamp power}(\text{W}) \times \text{Life}(\text{h})]/1000 \tag{19.1}$$

A simple comparison of the two lamp types for three years (4380 h), 4 h per day and the same amount of illumination is given in Table 19.2. During this period, while using 6 incandescent lamps, the compact fluorescent a lamp will continue to last for another 3.8 years more.

19.1.3 Selection of Lamps

The right choice of the lamp depends on what purpose and where it will be used. Along with factors such as lighting level, operating time, and ease of replacement, the following factors should also be considered when choosing a lamp:

Hourly cost efficiency: As the piece of lamps increases, investing in longer lasting and efficient ones is the right choice. It allows the investment to be repaid in a shorter time.

Starting Features: The initial operating characteristics of various types of lamps are different. For example, during the first run, fluorescent lamps with magnetic ballasts delayed illumination, while fluorescent lamps with electronic ballasts illuminate instantly. Both magnetic and electronic Ballast fluorescents reach full brightness after 1–2 min.

Color: Standard incandescent string lamps give a warm yellow-white color. Halogen lamps are whiter. Fluorescent lamps illuminate in different colors, from warm-yellow to cool-white.

Mass: Lightweight and fragile fixtures and fittings can be an important factor in decision-making. The mass of compact fluorescent lamps with magnetic ballast is more than that of compact fluorescent lamps with electronic ballast.

Installation (Placement): Lamps designed for indoor areas should not be used in outdoor lighting. For example, incandescent string lamps are not recommended for outdoor lighting, as the thin-glass lamp may break in contact with cold water.

Orientation (Guidance): When choosing a lamp, it is important to consider how much useful lighting can be directed to where it is needed. In special applications, low-voltage halogen lamps are suitable for lighting partially spotlight locations.

Adjustment: The adjustment switch allows energy saving in incandescent string lamps through mode adjustment (dimming). However, not all lamp types are suitable for use with a setting switch. Instead of the adjustment (setting) switch, optical or ultrasonic sensors can be used that turn on or off a watch or lamps that automatically turn off the lamps at a specific time.

Price: Choosing the cheapest lamp does not save money in the long run. Because, during the lifetime of the lamp, the lighting energy cost will be ten times higher than the purchase cost. Therefore, while energy-efficient lamps are costly at the beginning, the payback period of the investment is much shorter with low invoices.

Efficiency Factor: Lamps with high efficiency factor (lumen/Watt), long life, and low light flux over time should be preferred in the selection of the lamps. The lumen/Watt ratio of a typical incandescent lamp is 15/1, 60/1 for fluorescent lamp and 75/1 for LED.

19.1.4 Lighting Control Systems

The increase in the number and type of luminaires used in lighting systems has made the control of illumination quite complicated. Lighting control systems are used in order to simplify the control of lighting units and to use lighting in the most efficient way [4, 5]. The intended use of lighting control systems can be summarized in four main categories:

(1) Efficiency,
(2) Energy saving,
(3) Aesthetic, and
(4) Flexibility.

The most important benefit of lighting control is energy saving. Lighting control can save 30% in energy consumption and 10% in operating costs. A comfortable working environment is created with a correctly selected and applied lighting control system. In addition, an efficient lighting control will increase the productivity of the employees.

In areas such as meeting rooms, design offices, textile workshops, factories, lighting control is very important for the highest level of working efficiency. With a well-programmed lighting automation system, it is possible to provide the highest level of working efficiency with the most suitable lighting level according to the working hours, the location of the daylight, and the nature of the work performed.

Lighting automation systems eliminate the waste of time due to lighting adjustments by realizing sudden changes according to the workings of the lighting programs in a very short time. Thanks to the dimmer units used in lighting automation systems, it is possible to save energy and prolong the life of the light sources in the presence of illumination.

Light sensors are used for maximum use of daylight, *motion detectors are used* to prevent energy consumption in areas where nobody is available, *astrological time clocks* with the lighting automation system are used in order to economically program environmental lighting and to regulate lighting control according to working hours energy can be saved successfully. In addition, energy savings can be achieved by automatically activating different lighting programs for times when electricity is expensive or cheap.

Significant energy savings are achieved by controlling the lighting with motion sensors in places that do not have regular working hours and where the lamps are left on position when not in use. With this method; Energy savings of 20% in open offices, 60% in personal offices, 70% in toilets, 40% in warehouses, 50% in classrooms, and 65% in toilets in hotel rooms can be achieved successfully.

A well designed and daylight-saving lighting system saves 30% of the lighting energy in daylight hours of the day. However, this rate can reach a level of 70% by applying the saving methods mentioned below:

- Using lighting only when necessary, thanks to the application of the time clock in the system and the use of access control systems.
- Use of motion detectors in the areas where the number of employees is low.
- System dimmers are compatible with daylight thanks to the use of light-level sensors.
- When energy usage reaches the peak values, the light control system turns off or dims the lighting of unimportant areas.

19.1.5 LED Lighting

LEDs seem to be perhaps the most important innovation ever compared to the traditional lighting techniques. This innovation shows itself not only in appearance but also in effectiveness.

Theoretical calculations and experiments show that the lifetimes of LEDs are more than 100,000 h. When considering the factors like electrical and environmental effects, environmental elements that are used, material structure of the cover, etc., it can be assumed to have a service life of around 50,000 h. Their low-voltage working capabilities, no warming up problems, and being in a completely enclosed structure make LEDs an excellent alternative to outdoor lighting with simple luminaires. Also, LEDs have a high potential in terms of market opportunities.

LEDs are completely different light sources, and it is unnecessary to recreate structures similar to armatures created with other systems with LEDs. The traditional approach at this point is to create the luminaire around a lamp. Therefore, the image of a luminaire is largely determined by the lamp. However, with their compact dimensions, precise light beam controls, low warming abilities, and long lifetimes, LEDs have overridden all of these restrictions.

19.1.6 Energy-Saving Opportunities in Lighting

Particular attention should be given to the energy savings that will be provided by the smart combination of using daylight and artificial light in an intelligent way. Energy saving in lighting can be achieved by using high-efficiency light sources instead of low-efficiency light sources and by taking some simple precautions [6, 7]. Some of these measures can be listed as follows:

- While buying a lamp, highly efficient ones should be preferred. Lamp selections should be made based on the highest lumen/Watt ratio (efficiency factor).
- Unused areas should not be illuminated.
- Daylight should be used as much as possible.
- Periodic maintenance of lighting fixtures should be made. Dirty and dusty fixtures absorb some of the light, causing inefficient lighting.
- The lamp light output should be used efficiently. One of the most important factors affecting the efficiency of the lighting system is the fact that the light output of the lamp is not reached at the highest level on the surfaces to be illuminated.
- It is important for energy saving to control illumination by means of timers, photosensors, or proximity sensors.
- Wall, ceiling, and decoration materials should be selected as light as possible.
- In areas where more lighting is needed, the use of a single high-power lamp rather than a large number of low-power lamps provides more efficient lighting.
- Care should be taken to use small powerful bulbs in stair lighting.
- Decorative lamps send light in undesirable directions. Light-colored, transparent-shade lampshades transmit light better.
- The lamps should be turned off when leaving the room.
- The working table lamp should be used.
- To prevent energy loss, fluorescent bulbs and LEDs should be used instead of halogen and normal bulbs. Thus, 40% of energy can be saved.

- Energy savings of up to 80–85% is achieved by using compact fluorescent lamps and LEDs instead of incandescent lamps.
- If high-pressure sodium-vapor lamps are used in road lighting instead of high-pressure mercury vapor lamps, approximately 60% saving is achieved at the same level of illumination.
- In the case of garden and surroundings lighting, instead of high-pressure mercury vapor lamps, low-pressure sodium-vapor lamps are preferred, which can save about 70% energy at the same luminous intensity.
- With LED, developments in lighting technology should be closely monitored.

References

1. Kaya D (2003) Energy Conservation Opportunities in Lighting Systems. Energy Eng J Assoc Energy Eng 100(4):37–57
2. Uyanık M, Sarıbaş N (2005) Energy efficiency in lighting-ballast relationship, 3e Electrotech J 129. (In Turkish: Uyanık, M., Sarıbaş, N., Aydınlatmada Enerji Verimliliği- Balast İlişkisi, 3e Electrotech Dergisi, Sayı: 129, 2005)
3. Yorulmaz İ (2006) Efficiency in lighting, Yenilenebilir Enerji & Enerji Verimliliği Journal, 200:76–79. (In Turkish: Yorulmaz, İ., Aydınlatmada Verimlilik, Yenilenebilir Enerji & Enerji Verimliliği Dergisi, 200:76–79, 2006)
4. Çolak N (2003) Lighting control and benefits. Best J 19. (In Turkish: Çolak, N., Aydınlatma Kontrolü ve Faydaları, Best Dergisi, Sayı 19, 2003)
5. Çolak N (2003) Control of lighting with motion sensors, 3e Electrotech J 105. (In Turkish: Çolak, N., Hareket Sensörleri ile Aydınlatmanın Kontrolü, 3e Electrotech Dergisi, Sayı 105, 2003)
6. Erciyes S (2004) Lighting design should be made according to purpose, İnşaat Dünyası J 249. (In Turkish: Erciyes, S., Aydınlatma Tasarımı Amaca Göre Yapılmalı, İnşaat Dünyası Dergisi, Sayı 249, 2004)
7. Onaygil S (2001) Urban lighting, Kaynak Elektirik Dergisi, Haziran. (In Turkish: Onaygil, S., Kent İçi Aydınlatma, Kaynak Elektrik Dergisi, Haziran, 2001)

Chapter 20
Energy Saving in Residences

When purchasing new lamps, lighting equipment, appliances, tools, devices, etc., for a new house or to replace the existing ones in the residencies, the most efficient ones that can afford economic power should be bought. There are a few important things to consider when making a decision. Knowing the efficiency ratios of the tools, where they can be found and how the energy cost will affect the user during the purchasing and usage are important points. At the same time, it is important to know which appliances are consuming the most energy in homes and how to estimate the energy costs of certain appliances [1].

Sometimes energy-efficient appliances are required to be paid more, sometimes less when purchasing. But in both cases, energy efficiency saves money for long-term use. The user may initially prefer a cheap, but a low-efficient tool to consider saving, but this will result in a higher electricity bill every month. As long as one has a appliances or a device, it is being entered into the obligation to pay the usage cost. These usage costs can increase exponentially. For example, using a refrigerator for 13–15 years usually costs 1.5–2 times the purchase cost. The sum of costs during the purchase and use of an efficient device is less than the total cost of a low-efficient device.

The electrical appliances we use in our homes can be used with less energy without affecting the desired level of service and comfort. Electricity bills can be reduced by using efficient lighting systems and home appliances. Prices of energy-efficient devices can be expensive than their similar common models. However, the difference in price paid during the purchase of efficient appliances is then repaid back to the user with a reduction in electricity bills [2].

D. Kaya et al., *Energy Management and Energy Efficiency in Industry*,
Green Energy and Technology,
https://doi.org/10.1007/978-3-030-25995-2_20

20.1 Energy Savings in the Use of Electrical Equipment

The group that needs to be emphasized most is the family in terms of energy needs and usage. Because, in the society, families are one of the most important consumer groups that play a role in the use and consumption of energy sources as well as other sources. In residences, a considerable amount of energy is consumed by devices like home appliances and other electrical devices, etc. Families use energy resources to carry out heating, lighting, cleaning, personal care, entertainment, recreation, and similar activities.

Nowadays, while energy resources are becoming scarce and prices are increasing, they affect families economically in carrying out activities related to home, while limiting the sustainability of social resources. Therefore, it is imperative that families pay attention to some savings points when purchasing and using these devices. For example, electricity consumption for heating, cooling, cooking and storing meal, washing dishes and laundry, lighting, using entertainment devices for a family of four living in a 120 m^2 house is approximately 3500 kWh per year based on the world average. However, it is possible to save at least 30% of energy in our homes by taking simple precautions.

On average, 80% of the electricity we use at home is consumed by electrical appliances. There are 10 different energy classes in electrical appliances known as G, F, E, D, C, B, A, A+, A++, and A+++. While A+++ represents the highest energy-saving product groups, G-class electrical appliances are expressed as products that do not save energy but rather consume a lot of energy. Today, it is not possible to see the lower class products in the market than the B-class products with the lowest energy savings. Therefore, in terms of energy saving, comparisons are made mostly between A+++, A++, A+, A, and B classes. Therefore, those who consume the least amount of energy should be preferred when purchasing electrical appliances.

In addition to protecting the environment, using natural resources efficiently, it is important to use energy efficiently to produce energy without harming the environment. In order to provide energy efficiency in home electronics, it is necessary to create social awareness about energy classes and to use the energy class products A and above widely. According to researches, an electronic item with an A++ energy class saves 40% energy compared to the A energy class product. Conversely, an energy Class A product consumes 66% more energy than an A++ energy class product. Therefore, each increasing "+" is an indicator of the developing technology and the efficiency to be obtained in energy saving. Mostly, A+ and A++ energy class products are preferred, thus saving a great amount of energy. It is possible to encounter energy class technology in refrigerators, washing and dishwashers, televisions, air conditioners, and built-in ovens (stoves, cookers).

Class A products are the best products for providing energy efficiency. The electricity used by different electrical appliances of the same energy class varies. A refrigerator in A-class consumes approximately 1.23 kWh per day, a dishwasher in the same class 1.05 kWh per day, and a washing machine in the same class consumes 0.95 kWh per day. There is a refrigerator compared with A++ and B classes in terms of daily electricity consumption as you can see below:

- A refrigerator in A++ energy class consumes approximately 0.5 kWh of electricity per day.
- A refrigerator in A+ energy class consumes approximately 1.07 kWh per day.
- A refrigerator of energy Class A consumes approximately 1.23 kWh of electricity per day.
- A Class B refrigerator consumes approximately 1.7 kWh of electricity per day.

20.1.1 Refrigerator

Refrigerators, which are important energy users, operate 24 h a day, 365 days a year, as opposed to most devices operating periodically (Fig. 20.1). 15% of the energy consumed at home is consumed for the operation of the refrigerator. Factors such as the thickness and quality of the insulating material used in a cooler or freezer, the ice melting system, and the door-cabinet design affect the electricity consumption spent to maintain the desired temperature level.

Instead of waiting for a refrigerator to break down or reach the end of its working life, it is possible to replace the old one with the new one to get rid of many troubles and overpayments. Because a new refrigerator will consume far less energy than the one that is 8–10 years old. When a new refrigerator is to be bought, the model that has the desired characteristics, the appropriate measures and the model that can best comply with the kitchen must be searched and the refrigerator that consumes the least electricity and other desired properties should be selected.

Fig. 20.1 Refrigerator

Refrigerators similarly have different efficiencies according to properties such as defrosting characteristics (manual, partial automatic, automatic), door style, and size. Since large refrigerators will consume more energy than smaller ones, it will be useful to choose a refrigerator as large as it is needed. Those consuming less energy should be preferred. Multiple-door models often cause more energy consumption, as being large refrigerators, they need longer cooling tubes and larger compressors this means more energy consumption. The top freezer models precede the side by side models and manual or partial automatic defrost models are the most efficient [3].

20.1.1.1 Energy Saving in the Use of Refrigerator

In addition to the consumption value that depending on the design of the refrigerator, the way the refrigerator is used also affects electricity consumption. For this reason, the following points should be considered by the user:

- Regular maintenance is essential for the efficient use of refrigerators. A regular maintenance program will increase the efficiency of the system and prolong the life of the device. For example, the serpentines should be cleaned monthly with a non-metallic brush, the automatic defrost should be adjusted once a year by the service personnel and the level should be adjusted to automatically close the doors from the open position.
- The surroundings of the refrigerator should be kept away from dust and other substances that may affect air circulation. The more easily the air circulates, the better the serpentines will reject the heat. The refrigerator should be placed in such a way as to have airflow. The more comfortable the weather is, the greater the efficiency of the coolers. For this, the distance between the wall behind the coolers must be at least 10 cm.
- The hot air from the oven and other heat sources will cause the refrigerator to work harder to stay at the set temperature. Refrigerators should not be exposed to direct or indirect sunlight, should not be placed near a stove or radiator. If the refrigerator is placed in a warm place, the total energy consumption increases.
- Excessive packages of foods must be removed before putting them in the refrigerator. As the thickness of the protectors' increases, the refrigerator will work harder to keep the contents cool. It will be better to use thin plastic films, as it will save energy as much space.
- Taking the material from the freezer section to the inside of the refrigerator one day before and leaving it there to thaw will help the refrigerator to consume less energy as it will cool the refrigerator.
- Make sure that the ice thickness in the refrigerator does not exceed 5 mm.
- Significant cold losses occur during the opening and closing of the refrigerator's freezer and body door. Don't open the refrigerator door often as much as possible, also should not keep it open position for a long time.
- The most commonly used materials should be placed in the refrigerator cabinet carefully and can be taken easily. The bottom and back of the cabinet should

be cleaned twice a year with a vacuum cleaner. This will result in less energy consumption.

- Do not put hot material in the refrigerator. Otherwise, your device will consume much more energy by working longer to remove the heat.
- Liquid food should be covered. Otherwise, it increases the humidity in the cabinet and causes the compressor to work more and that means it will consume more energy.
- The door seal that provides a sealing between the refrigerator body and its door should be kept clean and replaced if it is broken.
- The recommended temperature in the refrigerator is 4–5 °C, and the freezing temperature is –20 °C. In this regard, the manufacturer's instructions must be followed.
- Food should be placed in the freezers so that they do not block the air flow in the cupboard. Also, the cabinet of the refrigerator should fill as it is needed, but it should be paid attention not to be obstructed the air circulation in any way.

20.1.2 Energy Saving in an Air-Conditioning Unit Usage

Some of the countries are extremely hot in summer, which affects working and living conditions negatively. For this reason, coolers and air conditioners are used to reduce temperature and humidity to provide a more comfortable living environment. In the use of these units, in order to save electricity, the right choice must be done. Using a much larger air conditioner than needed means a waste of money and energy. In proper sizes, an efficient air conditioner consumes less energy [4, 5].

Recommended air-conditioning sizes depending on the area can be seen in Table 20.1.

Electronic thermostat control can increase efficiency and save up to 30% in energy usage. A rotary compressor may be more efficient than a piston compressor. But the most important factor for the right decision is to choose the appropriate size.

The least energy-consuming models with the lowest ***Energy Efficiency Ratio (EER)***, in other words in the range of Btu/hour, with the lowest wattage value, should be selected. Watt value and cooling capacity ratio (Btu/h) of compressor motor can be found in sales catalogs and air conditioner labels.

Table 20.1 Recommended air-conditioning sizes depending on the area

Area (m^2)	Air conditioner (BTU/h)
13–15	7000–9000
16–17	9000–11,000
18–22	11,000–13,000
23–24	13,000–16,000
30	18,000–20,000
40	24,000

The easiest way to estimate energy efficiency in the air conditioner is to calculate the EER value. EER value may not be included on the label of the air conditioner. This ratio can be found by dividing the Btu/hour rate by the Watt value.

$$\text{EER} = \text{Cooling Capacity (BTU/hour)/Watt Value} \quad (20.1)$$

For example, in an air conditioner with 1200 W and 12000 Btu/hour, the EER value is calculated as 10. Higher EER is better. When choosing, models with higher ratios or close to EER values of 10 should be preferred. If a split system is selected, in the above equation, the Watt value of the outdoor unit is added to the Watt value of the indoor unit to obtain the total Watt value.

For the highest efficiency, it should be ensured whether the efficiency ratios of the internal and external coils are compatible or not. Moving the indoor unit air inlet panels should be mounted so that the service personnel can easily clean the cooling coils.

A high-efficiency air-conditioning unit can be purchased cheaper than a lower efficient unit of the same size. For this, careful price research should be done.

The outdoor unit should be conveniently placed in a cool, shady place at a distance of about 50–60 cm from the home or office. Avoid direct sunlight, which will cause the air-conditioning unit to work harder. Also, the air-conditioning unit should not be prevented from getting air, for example, it should not be close to the bush or secluded places.

Energy Saving Possibilities in an Air Conditioner Use

Preventive Maintenance: In the use of an air-conditioning unit, it is necessary to regularly change the filters, check the entire system, and adjust the fan belt. These processes save the energy, increase the efficiency, and extend the life of the air conditioner. In order to obtain the best energy-saving results, the maintenance leaflet should be applied properly.

Shading: Processes such as shading the sun-exposed parts of the house with trees, having window films, window, and wall awnings, and the application of passive cooling methods will significantly reduce the cooling load.

20.1.3 Energy Saving in the Oven and Burner Usage

In food preparation and cooking processes, energy saving can be achieved by the use of energy-efficient cooking tools. Whether cooking with electricity or gas, cooking in stoves is more economical than cooking in the oven. Factors to be considered for energy saving in the use of stoves and ovens are

- Containers with flat bottom should be preferred in the cooking process. The bottom of the container should be clean, and also placed directly above the fire.

- Cooking in closed pots and pans speeds up the cooking process and reduces energy consumption. Thus, 60% of energy saving will be achieved by means of steam energy to be provided.
- Pressure cooker and steam cooker during cooking not only saves energy, but also preserves the nutritional values of the food.
- Switch off the electric stove and ovens a few minutes before the end of the cooking time and use the temperature of the container and the unit itself.
- The oven should not be preheated unless necessary, if it is necessary this time should not exceed 10 min.
- 20% heat loss occurs due to the oven door is opened each time, and therefore the door should not be opened, if it is possible this should be until the end of the cooking time.
- If there is more than a unit of the oven in the house, the smaller one should be preferred. Instead of using a large oven at half capacity, using a small oven at full capacity saves a great deal of money.
- Cooking a frozen meat takes 20 min longer than cooking a thawed meat. For this reason, frozen meat must be thawed before cooking.
- Microwave ovens use energy more efficiently than conventional ovens. In microwave ovens, cooking takes 2–10 min and heating takes 10–30 s. Therefore, they consume 50% less electricity than conventional ovens. However, when cooking large amounts of food, the situation is reversed, and they consume more electricity. It should be preferred for cooking small amounts of food.
- Use microwave heating as much as possible in food heating, saves 60–65% of energy.
- The inside of the oven, electric stovetop, and gas stove burners must always be kept clean to ensure efficient usage and combustion.
- When buying an oven, air and heat-tight insulation should be preferred and thus, low energy consumption can be assured.
- In modern ovens, more than one type of food can be cooked at the same time, this process reduces electricity bills.

20.1.4 Energy Saving in the Washing Machine Usage

Most of the electricity consumed in the washing machines is used to heat the water. The following factors must be considered in order to save energy in the use of washing machines and dryers:

- Instead of washing at high temperatures, it should be washed with warm or cold water, while rinsing should be done with cold water.
- The cost of the laundry washed in the washing machine largely depends on the amount of hot water used. Laundry should be washed with warm (30–40 °C) or cold water instead of washing at high temperatures unless necessary. 90% of electricity is spent on heating water.
- Washing programs should be run at full capacity.

- Washing machines should not be operated before they are filled in accordance with the washing programs.
- If it is possible, laundry should be dried outdoors in the sun and wind. Drying machines should not be used unless it is compulsory.
- If a washer-dryer is to be purchased or is to be used, those with a humidity sensor should be preferred, which turns off the unit when they are dry.
- Front-loading machines consume less energy than top-loading machines.
- The machine that consumes less water, energy and detergent should be preferred for purchasing.
- It should be noted that there is at least 5 cm or more of space around the washing machine.
- Do not use more detergent than recommended. Low-temperature cleaning detergents should be preferred. Excessive foaming or the use of less or more detergent means inefficient cleaning.
- Pre-soaking heavily soiled and stained laundry or canceling the second wash using the soaking function in the automatic washing machine, if any, saves considerable energy.
- The correct washing programs should be chosen by separating the laundry according to their type and dirtiness.
- The filter of the washing machine should be cleaned occasionally.
- Do not over dry the laundry. This causes extra energy consumption, also, excessive drying will lint and wear the laundry.
- The strainer should be cleaned after each drying process in the dryer. Otherwise, it fills the air ducts with laundry fibers, strands, etc., and prevents the blown dry air from passing. This results in prolonged drying and higher energy consumption.

20.1.5 Energy Saving in the Dishwasher Usage

A dishwasher can be defined as one of the most important appliances in houses and in other usage areas. It is generally used three to seven times a week in an average household. When buying a dishwasher there are several important things that should be considered carefully such as energy consumption values along with factors as the capacity and water usage. For uses that cannot always be filled at full capacity, those with a single basket washing program can be preferred.

Dishwashers are more efficient today, but their operating cycles have become much longer. There are dishwashers that have low water usage features or also, quick wash options which are practical, if you are on metered water use. Some dishwashers have energy saver features, for example, a delay start option if your electricity usage is calculated on time-of-use rates, a user can set the dishwasher to wash later when energy is cheaper. Dishwashers generally have features like:

1. Energy class: It ranges from A to A+++. As the "+" increases, it is understood that it is a more economical model. It is necessary to check the energy guide stickers to know more about the water and power consumption of the dishwasher model.

A regular dishwasher can consume three and a half gallons (approximately 13.25 L) to twelve gallons (approximately 45.43 L) of water per load. Energy costs can reach up to $65 a year. So always select a water and energy, efficient model.
2. Annual energy consumption: It is given in kWh (the lower the value, the more economical the product).
3. Annual water consumption: It is given in liters (the lower the value, the more economical the product).
4. Drying class: There are values between C and A. A wash indicates the best drying level.
5. Capacity: It is the information that shows for how many people the dishwasher can wash dishes at a time.
6. Sound level: This shows the noise level in dB. Lower values indicate a quieter product.

In order to save energy in the use of the dishwasher, the following factors should be considered:

- It is unnecessary to pre-rinse the dishes. Where necessary, cold water should be used instead of hot water.
- For sanitation of dishes, a water temperature of 55 °C is sufficient, unless a high-temperature wash is desired.
- The amount of water consumed by hand washing dishes varies from 35 to 200 L. However, when the same amount of dish can be washed in the machine by consuming 15 L of water.
- After the machine has completed the final rinse, opening the lid and drying the dishes by means of air and this will save about 10% energy.
- The machine should never be operated half or excessively loaded.
- For lightly dirty dishes; short or economic cycle, a low-temperature program should be chosen.
- If the correct detergent and dosage are not used for dishes, energy consumption increases.
- It is necessary to pay attention to the filters and connections of the machine, they should be kept clean, constantly. A clogged filter degrades the dishwasher's performance and increases energy consumption.

20.1.6 Energy Saving in the Vacuum Cleaner Usage

In order to save energy in the use of vacuum cleaners, the following factors should be considered:

- One of the important factors is how to maintain the vacuum cleaner, for example, having a full bag and/or clogged filters increase power use and reduce efficiency. Therefore, the bag of the vacuum cleaner should be discarded and renewed as often as required (It should be realized as suggested by the manufacturer). This will provide more efficient and quicker cleaning as it will increase the suction power of the cleaner.

- If a removable container and reusable filter is to be used to trap debris and dirt, the container should be emptied and the filter cleaned for re-use. In some countries, the majority of residential vacuums are realized with bagless designs.
- Old brushes should be renewed when it is necessary.
- The engine compartment should be opened at least once a year and the dust, dirt, pollens, and other debris should be cleaned here.

20.1.7 Energy Saving in Radio/Television and Other Home Appliances Usage

Advances in electronic technologies significantly affect the consumption needs of society. Accordingly, as more and more people start working at home, there is a significant increase in the number of devices that require extra power for an average home. Nowadays, in almost every house, many devices such as computers, printers, faxes, video and CD players, etc., and the home appliances are used in a widespread manner.

Computers are the fastest electrical charge constituent devices in the world. This amount will increase gradually in the coming years. However, most of the energy used by computers is wasted because it is turned on even when the computer is not being used.

TVs, videos, CD players, cordless phones, robbery alarms, microwave ovens continue to consume energy in standby mode when not in use. This energy is usually spent on the clock display, memory chip, and remote-control function. This type of energy consumption is called leakage and is about 5% of the device's own energy. In standby mode, work is carried out to reduce energy leakage. For example, new technology TVs and videos labeled Energy Star reduce lost energy by up to 75%. Home office devices such as Energy Star labeled PCs, faxes, printers, and scanners automatically switch to sleep mode when not in use, resulting in reduced power consumption and extended device life.

Hair drying should be done with a towel as much as possible without using a hair drying machine. Operating the hairdryer for 10 min consumes electricity equivalent to a 60 W lamp being on for 3 h.

When the radios are used for 100 h and color televisions are used for 5–7 h, they consume 1 kWh of electricity. Small-screen televisions consume less electricity than large-screen televisions. Keeping the volume low also reduces electricity consumption. In order to save energy in the use of television, DVD/VCD, computer and radio, the following factors should be considered:

- When unused computers are left in standby mode and unplugged, they continue to consume 10% of electricity. Therefore, unplug devices when not in use.

- The sound level of the devices like TV, radio, and so on should be at an audible level. Keeping the volume at low levels reduces electricity consumption as it is mentioned above.
- If it is possible, a laptop should be used instead of a desktop. Desktop computers consume around 3–5 times more electricity.

20.1.8 Energy Saving in Ironing

Considering the energy saving in the use of an iron, the most important factors are ironing of the laundry as in moist, choice of irons with steam, and thermostat. Ironing should be done at one sitting, so that it is not necessary to re-heat the iron. The following factors should be considered in order to save energy when using the iron:

- The iron can be unplugged five minutes before the end of the ironing process and the remaining work can be completed with the heat inside the iron.
- Care should be taken to ensure that the laundry is still damp while ironing.
- Another important point to be considered when buying an iron is to choose it with low drying power and high steam capacity, which provides less energy consumption.

20.1.9 Energy Saving in Lighting

In order to save energy in lighting applications, the following factors must be considered:

- The lights should be turned off when leaving the room, even if it is for a short time.
- It should make use of daylight as much as possible, and lighting tools should not be used unless necessary.
- Light-colored walls reflect furniture, curtains, and carpet light, reducing the need for lighting in the room.
- Low power lamps should be used at the entrance and hall of the building. Exterior door lamps can be replaced with halogen lamps.
- Instead of using a large number of low-power lamps in places where more light is needed, using a single high-power lamp provides more efficient illumination.
- Instead of lamps that illuminate the whole room while sewing, reading, and for other similar activities, the use of working lamps saves money.
- Controlled lighting with timers, photocells, or proximity sensors also saves energy.
- The level of light should be increased/decreased in desired situations by using light adjusters in the lamps.
- Economical bulbs make a great contribution to the economy and the environment. They consume 5 times less energy than ordinary lamps and last 10 times longer.

By replacing only two 100-W old type bulbs with efficient bulbs in each home, approximately $120 energy savings can be achieved annually. In addition, it will be possible to prevent 1.1 million tons of carbon dioxide emission annually, which can only be cleaned by 52 million trees. A 15-W energy-saving bulb consumes 80% less electricity than a 75-W common bulb. While 12 old bulbs are needed for 6 years, only one energy saving bulb will be sufficient in the same period.

20.1.10 Energy Saving in the Residence Heating

In order to save energy in the house heating, the following factors should be considered:

- Wearing thick clothing during the winter makes a significant contribution to energy savings in heating the house.
- Radiators should be kept clean. This prevents dust from absorbing heat.
- Any covers on the radiators prevent air circulation and collect heat on themselves. Therefore, radiators should not be covered.
- Deflate (take the air out from) the radiators at least once a year to save energy.
- Furniture and curtains should be arranged so as not to hinder the flow of heat around the radiator.
- Unused rooms, bathrooms (when it is not in use), entree, and stairwell radiators should be turned down.
- To prevent heat loss from doors and windows, insulation must be done strictly. Because windows and doors cause the loss of a quarter of the heat in the houses, approximately. For example, 20% of the heat loss in an uninsulated house arises from single glass windows. Therefore, this loss is cut in half by using double glass windows.
- Curtains should be kept closed during the winter months unless necessary and especially if the sunlight does not exist.
- Heat loss from windows can be reduced by choosing curtains made of thick materials.
- The windows should not be left open for a long time to ventilate the rooms.
- Cold is felt more in dry weather. Therefore, rooms should be moistened in the winter.
- Radiator thermostats should be used to adjust the room temperature to the desired level in order to save money in buildings heated by the central system.
- The walls should be insulated by referring to the knowledge of a specialist. The expenses to be made for this process can amortize between 2–5 years.
- It is necessary to check the heaters periodically by authorized services and to check their settings.
- Roofs should be insulated with a suitable insulating material.
- The fact that keeping the room temperature in the house is 20 °C instead of 22 °C reduces the fuel cost by 12%.

- The combi boiler should not be switched on or off at all times but should be kept in continuous operation at low setting values.
- The filters of the natural gas stove should be cleaned or replaced once a month.
- If possible, the surroundings of the house should be afforested.
- Unused flue holes should be covered.
- Chimneys should be checked and cleaned before entering the winter months.
- Vented (funneled) natural gas devices should not be placed in the cabinet. This prevents the devices from getting air from the environment and reduces their efficiency.
- Natural gas installations should be protected and controlled regularly. Care must be taken against leaks.
- The apartment doors must be kept closed. With rational consumption and energy-saving applications, natural gas can be used much cheaper and economically.
- Measures to be taken such as thermal insulation, device selection, ensuring complete combustion and controls at certain intervals provide a decrease of up to 20–30% in natural gas bills.

References

1. Yorulmaz İ (2005) Efficient use of electrical energy in buildings, energy bulletin, general directorate of electrical works survey administration 199:70–74. (In Turkish: Yorulmaz, İ., Binalarda Elektrik Enerjisinin Verimli Kullanımı, Enerji Bülteni, Elektrik İşleri Etüt İdaresi Genel Müdürlüğü, 199:70–74, 2005)
2. Kaya D, Öztürk HH (2014) Sanayide Enerji Yönetimi ve Enerji Verimliliği, Umuttepe Yayınları, Yayın No: 114, Mühendislik Dizisi: 12, Kocaeli/Türkiye
3. Kaya D (2003) Energy conservation opportunities in lighting systems. Energy Eng J Assoc Energy Eng 100(4):37–57
4. https://www.enerjigazetesi.ist/enerji-tasarruf-noktalari-ile-enerji-verimliligi/
5. ASHRAE (2007) Handbook-HVAC applications. American Society of Heating, Refrigerating and Air-Conditioning Engineers, Atlanta

MIX
Papier aus verantwortungsvollen Quellen
Paper from responsible sources
FSC® C105338

If you have any concerns about our products,
you can contact us on
ProductSafety@springernature.com

In case Publisher is established outside the EU,
the EU authorized representative is:
Springer Nature Customer Service Center GmbH
Europaplatz 3, 69115 Heidelberg, Germany

Printed by Libri Plureos GmbH
in Hamburg, Germany